THE PULSE OF THE EARTH

The Pulse of the Earth

BY

J. H. F. UMBGROVE
D. Sc. (Leyden)
Hon. F. R. S. E.; Hon. Memb. N. Y. Ac. Sc.
Memb. R. Netherl. Ac. Sc.
PROFESSOR OF GEOLOGY AT DELFT
Holland

SECOND EDITION

With 8 partly coloured plates, 204 textfigures and 12 tables

THE HAGUE
MARTINUS NIJHOFF
Photomechanical reprint 1971

DOI 10.1007/978-94-010-3017-5
www.springer.com/mycopy

In memory of
A. C. Umbgrove-Gordon
my Mother
with sincere feelings of gratitude

CONTENTS

LIST OF ILLUSTRATIONS

If no source is mentioned the illustrations are new or from recent papers by the author.

TEXTFIGURES

Figures

Figures

Figures

FIGURES

Figures

PLATES AND TABLES AT THE END OF THE BOOK

PREFACE

Problems of current interest relating to the earth's physical history will be discussed in this volume.

Each chapter constitutes a subject in itself, but the sequence I have chosen will, I hope, show and explain the deeper correlation of several terrestrial processes which, at first sight, appear to be heterogeneous.

The geologist follows the changing face of the earth, the oscillations of the sea-level, the pulsation of folding and mountain-building, the periodicity of the ice-ages, the rhythmical cadence of Life. Just as the physician will draw his conclusions from outward symptoms when examining his patient, so the geologist tries to discover the deeper significance of the sequence of observed phenomena by feeling the pulse of the earth.

The many additions and revisions which have had to be made in this second edition include three new chapters, several new sections in other chapters, 109 new textfigures, 12 tables and 2 plates.

A few fundamental geological terms have been explained in Chapter I. Geologists and older students will, I trust, forgive me for handling facts that no longer represent terra incognita *to them, realizing that younger undergraduates, biologists and other readers will welcome a concise synopsis concerning matters which, in the following chapters, are presupposed to be generally known. On the other hand the Appendix, consisting of a somewhat dry and monotonous compilation and discussion of data, is only intended for geologists who desire to examine further the fundamental points laid down in plates 1–8 and Chapters II and III.*

Most of the topics are adapted from lectures delivered in Cambridge, Delft, The Hague and Leyden in the course of the years 1939–1945.

It would necessitate too long a list if I were to mention all the names of those who, in one way or another, have generously given assistance or contributed information, to all of whom I hereby wish to express my gratitude. I owe special thanks to my colleagues and friends Prof. Arthur Holmes of Edinburgh, Dr. E. C. Bullard, and Dr. Harold Jeffreys of Cambridge, Prof. B. G. Escher, Prof. H. J. Lam and Prof. J. H. Oort of Leyden, Prof. Ph. H. Kuenen of Groningen, Prof. F. A. Vening Meinesz and the late Prof. J. A. A. Mekel of Delft for their illuminating discussions or critical reading of parts of the manuscript.

Mr. J. A. van Houten, B. A., translated the Dutch manuscript of the first edition into English with great devotion and patient perseverance. I greatly regret that he could not undertake the translation of the new chapters. I had to write them in English myself, but I am very glad to say that some of my friends in England made linguistic corrections in the manuscript, while Professor Holmes even kindly undertook the onerous task of reading the page proofs. Of course, I remain personally responsible for any mistakes or misprintings that may finally have been overlooked.

I may also mention my indebtedness to Mr. C. van Werkhoven, who executed plates 1–8 with such admirable skill and overcame numerous difficulties with great ability. It was he too who put the finishing touches to several figures in the text. Most of the latter, however, I owe to Dr. R. de Wit and Mr. G. A. de Neve, for whose valuable assistance I express my sincere appreciation.

And last but not least, I wish to extend my thanks to Mr. W. Nijhoff, who has published this book with the meticulous care which so unfailingly characterizes his work.

Wassenaar, May 12th 1946.

FOREWORD TO THE PHOTOMECHANICAL REPRINT

At the time of his death the author was preparing a new edition of this book, as in his opinion it needed a revision.

Several outstanding colleagues, some of them close friends of the author, were asked to try and complete the manuscript for this third edition. But out of respect for the personal touch which gives "The Pulse of the Earth" such a special character, all concerned came to the conclusion that it would prove to be impossible to find a suitable editor.

Because there is a continuing demand for copies of the second edition Mrs. Umbgrove consented to a completely identical photomechanical reprint.

The Publisher

Chapter I

SPACE AND TIME

"Our conception of the structure of the Universe bears all the marks of a transitory structure. Our theories are decidedly in a state of continuous and just now very rapid evolution".
(W. de Sitter)

Introduction

Geology, the science of the history of the Earth and Life, reaches back into the infinitely remote ages and depths of the Universe and extends its speculations to the origin and meaning of all organisms and inorganic matter.

One generation after another has attempted to unravel the problems of the continents and oceans, or to decipher the origin and evolution of Life, or even the mystery of Man himself, who never rests in his unceasing quest for knowledge.

The historical succession of phenomena, their correlation and meaning, form the most attractive and interesting feature of geology. This applies not only to the major outlines of the development of our globe and the evolution of life, but also to any other minor geological problem. Thus the geologist who is mapping a region keeps careful note of every detail of the rocks and strata. Nevertheless, his object is not merely to determine whether granite, limestone, schists or other rocks occur within the area, nor will the presence of folds, overthrusts or other tectonic phenomena satisfy his curiosity. What he wants to know is the sequence of events through space and time, i.e. the geological history of that particular region up to the present day. The only reward for his painstaking efforts will perhaps be what Termier so enthusiastically described as "la joie de connaître".

The geologist might be compared to the historian. The historian will make a careful study of any parchment that happens to fall into his hands and will reconstruct the past with the aid of its data if he finds them to be complete. On the other hand, he will be sure to point to the gaps in his knowledge if he finds them to be incomplete and realizes that he cannot arrive at the truth with absolute objectivity. In such circumstances he may possibly revert to other methods in an attempt to bridge over the missing parts, and will make use of temporary hypothetical constructions for lack of solid facts. It often happens that the historian, who supplies from his own imagination the missing lines of his scientific prose, allows himself to be carried away by an unbridled poetical inspiration and soars to giddy heights. This is inevitable, yet he should never forget that hypotheses constitute a necessary evil and should be discarded as soon as contradictory facts come to light.

We may expect to find a similar "geopoetical" aspect in many a geological treatise, in addition to the normal geological prose. However, authors should always keep their theories strictly separated from descriptions and conclusions of a more rigorously documented kind.

We repeat it: geology is a historical science. The history of the earth is a most absorbing one and its unknown elements — many of which will perhaps elude us forever — challenge us. The beginning of this history brings us into contact with Astronomy, particularly with Cosmogony, the science of the origin of the universe. Immeasurable space and time encompass us. "As Rama looks out upon the Ocean, its limits mingling and uniting with heaven on the horizon, and as he ponders whether a path might not be built into the Immeasurable, so we look over the Ocean of time, but nowhere do we see signs of a shore". These lines appear towards the end of the second part of Suess' masterly work *The Face of the Earth*.

Science progresses with steady strides. New facts come to light and new ideas are born every day, and our conception of the structure of the universe changes accordingly. Our views have constantly to be revised and readjusted, but occasionally new aspects of far-reaching consequence shed such an unexpectedly different light on existing problems that its effects might be compared to those of a revolution. All that had hitherto been sacrosanct crumbles to the ground; hardly anything is left untouched.

The late American geologist Barrell, whose death, alas, came so prematurely, wrote in one of his brilliant articles: "The scheme of the Universe is more profound and the unknown is a little nearer than it was recently thought to be. But such has been the progress of knowledge since man, in the days before the advent of science, naively regarded the earth, his home, as the center of the universe and the heavenly bodies as lights in a nearby firmament, created a few thousand years previously especially for his benefit".

In the light of the above, it would be advisable to begin with a brief outline of some of the features of the modern conception of space and time.

The universe, the solar-system and the earth

We will begin with a short summary of cosmic dimensions. These may help us to obtain a better idea of the earth's humble place in the universe. We will then immediately pass on to a discussion of the origin of the earth.

The earth's volume is more than a thousand times less that of Jupiter, and the volume of the latter is in turn a thousand times less than that of the sun (fig. 1). The sun looks like a small star when compared to a giant of the type of Antares. Sixty million suns could fill Antares' space, but this giant is surpassed several times by the super-giant Epsilon Aurigae, which has a diameter 3,000 times that of the sun. The sun represents only a small element of a spiral nebula, the so-called galaxy, composed of about ten to a hundred thousand milliard stars. A great many examples of this type of "nebulae" are known and their diameters vary between 1,000 and 100,000 light-years [1]).

The planets of our solar-system lie extremely far apart, but the galaxy appears to be even more thinly populated with stars than the solar-system with planets. A few comparisons from Jeans may illustrate some of these cosmic dimensions. Five apples, placed on our five continents — one on Europe, another on Asia, etc., would provide us with a scale model of the dimensions of the stars and the intervening distances. Supposing that — à la Jules Verne — we were to let ourselves be fired from the earth in a rocket, travelling at a rate of 5,000 miles an hour, it would take us two days to reach the moon. If we were to travel through the sun at the same speed, our journey would take us a week, and nine years would be required to pass through Antares. Finally it would take us no less than five thousand million years to travel in this same rocket through a spiral galaxy such as the Andromeda nebula (fig. 2, B).

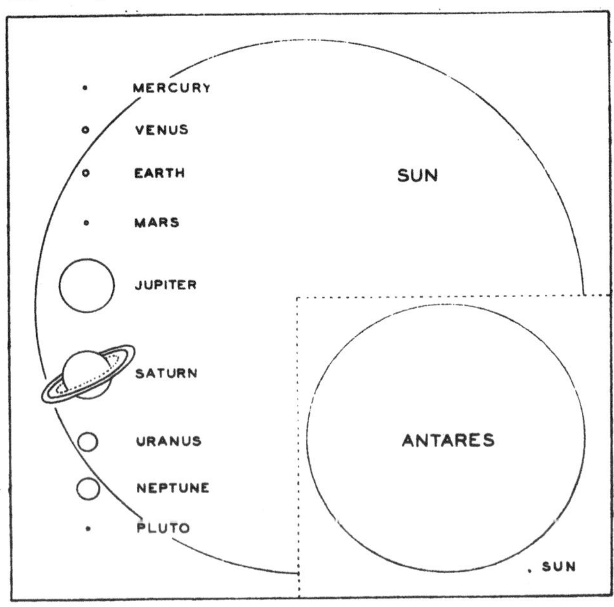

Fig. 1. The dimensions of the planets as compared to the disc of the Sun, and the Sun as compared to Antares.

Spiral galaxies are distributed less sparsely through space than the stars through a galaxy. If Amsterdam — to quote an example from de Sitter — were to represent the extent of our galaxy its next-door neighbors would be The Hague and Utrecht. Some 800,000 light-years separate us from the nearest spiral nebula. In other words, if viewed from this region today, the earth would present a somewhat unwonted aspect for Man would just be beginning to appear. The spiral nebulae are distributed fairly evenly through space. The most distant ones — representing the very extremities of that

[1]) It may be usefull to mention a few data on dimensions and distances. Earth's radius at the equator 6378 km; circumference (equator) 40,000 km; flattening at the poles $\frac{1}{297}$ or 21.5 km; surface 510 × 10^6 km²; eccentricity of the earth's orbit 0.01674; average velocity 29.76 km/sec; average distance Earth-Moon 384,403 km (60 earth's radii); light year 60 × 60 × 24 × 365 × 300,000 km (approximately 10 billion or 10^{13} km); 1 mega-parsec 3.26 million light-years; distance Sun-Earth 149.5 × 10^6 km; distance Sun-Pluto 5908 × 10^6 km; diameter of Galaxy 100,000 light-years; distance solar system to the remotest spiral galaxy known at present 150 × 10^6 light years, receding from us with a velocity of 15,000 miles per second. The mass of the universe is about 10^{55} grams.

part of the universe which is known to us at present — are separated from us by a thousand million light-years. This means that if it were at present possible to scan the earth through a super-telescope from one of these extremely distant parts, we should find that we were looking at the first and most primitive terrestrial organisms in our history, swimming around in Pre-Cambrian seas, and we should have to wait 700 million years to detect the first signs of life on the continents — that is, assuming that the distance remained unchanged during this lapse of time. These cart-wheel shaped galaxies have a central hub (fig. 2, B). Thirty thousand light-years separate the sun from the center of its sidereal system (fig. 2, A) and this galaxy rotates in approximately 200 million years according to observations by Oort, Lindblad and Plaskett.

One of the most sensational astronomical discoveries of the twentieth century was that all the star-systems appear to recede from our galactic system, as well as from one another. Everyone knows that the pitch of the whistle of a locomotive will drop as the engine rushes past and vanishes into the background. The same phenomenon is observed when a source of light recedes from us at high speed. It manifests itself to the astronomer as a red-shift of the spectral lines of a star. This enables him to calculate the velocity of the movement of the star relative to the sun from the amount of the red-shift. Observations have disclosed the remarkable fact that the velocities of the recession of the spiral systems from our own galaxy and from one another, increase proportionately with the distance between the receding systems. By reversing the picture and imagining the galaxies to travel towards instead of away from one another, we are able to conclude that a tremendous quantity of matter was packed into a considerably smaller volume of space some 2,000 to 3,000 million years ago. The fact that all the galaxies move away from us, does not mean that we are remaining stationary. We should think in this respect of bits of straw which, while floating in a swiftly-flowing river, move away from one another as the river broadens. The statement that the universe

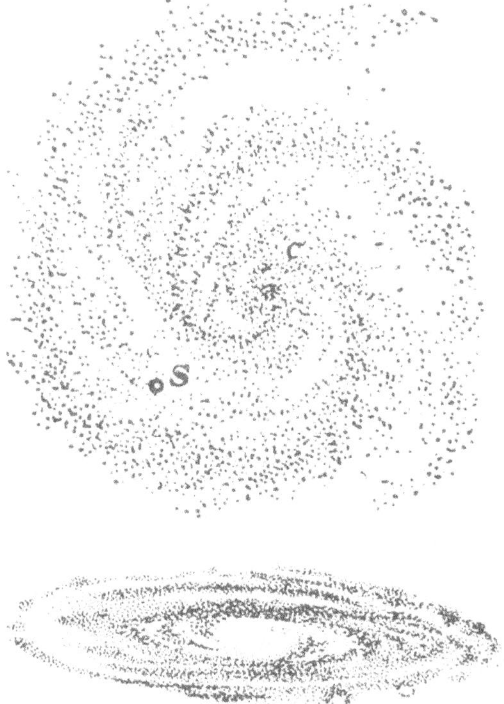

Fig. 2. Above, our galactic system (after Easton). S indicates the solar-system. Below, the Andromeda nebula, a spiral galaxy seen from the side.

is expanding (which was first alluded to by W. de Sitter in 1917 on theoretical grounds) is one of several possible interpretations of a mathematical equation. One possible solution is that the universe had once shrunk considerably in the past and that since then it has been continually expanding. Another is that the universe is subjected to alternating contraction and expansion, a pulsating movement. The last expansion must have begun at any rate some 3,000 million years ago, and this moment must have been one of major importance in the history of all cosmic matter. This question will be dealt with again presently, but it is obvious that our conceptions as to what exactly must have happened at that moment are, to say the least, of a very uncertain nature. The following reflections appear in a publication by de Sitter [1]: "It cannot be said whether the galaxies were already in existence before the catastrophe, or whether they originated from this turbulence, as it may be that the stars were distributed more evenly through space previously. We know for certain, however, that all eventual deviations from absolute homogeneity existing at that time must have been strongly augmented by the terrific commotion". It would take us too far to mention the many interesting theories on the ultimate limit of the age of matter, which astronomers estimate to amount to between 5 and 10 billion years, or to deal with the absorbing results obtained as regards curved and finite space, and the correlation of matter and radiation, space and time.

The above quotation from de Sitter shows that the problem of the origin of the galaxies is far from being solved. This is equally true of the solar-system. There are still a great many conflicting views on this subject! Russell has reviewed the *embarras de choix* in a cleverly written book. One group of hypotheses assumes that there has been a close encounter between the sun and another star during some period in the past, and that both either approached one another very closely or collided. The chances are that such an event did actually take place 3,000 million years ago. A much debated hypothesis of Jeans and Jeffreys surmises that the sun thus entered the danger zone of gravitational pull of a bigger star, and that the sun's surface rose toward it in the shape of a conical surface, from which a narrow filament would be produced (fig. 3). The ejected material would then condense into separate cooling masses revolving around the sun, and

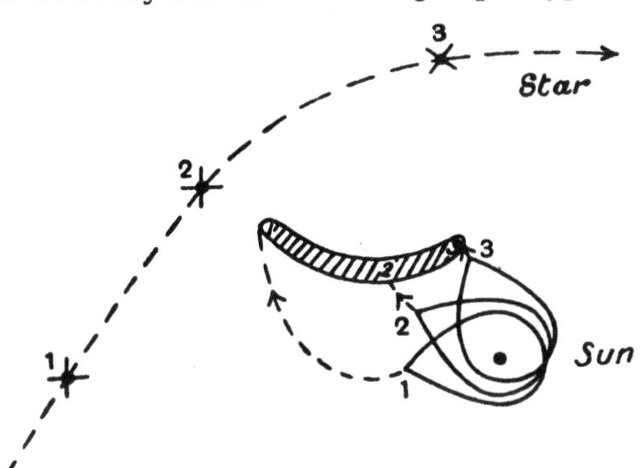

Fig. 3. Diagram of the changes in the form of the Sun and the formation of a boomerang-shaped filament of ejected matter during the passage of a star. The filament is shown in a position corresponding with position 3 of the star (After H. Jeffreys).

[1]) De Sitter, 1934, p. 377 (translated).

would thus lead to the formation of the planets. Russell and Lyttleton are of the opinion that the sun might have been a binary star at that time, and that its smaller companion broke into fragments as the result of a collision with — or the near approach to — a passing star. These fragments would then have developed into the planets.

The birth of the moon

The genesis of the solar-system presents many problems and is the subject of much conjecture. The same is true of the origin of the moon.

The ratio of the volumes of the moon and the earth is 1 : 82. This is exceptionally high compared to that of Titan and Saturn (1 : 4700, the highest ratio found among the remaining satellites and planets). Besides, the moon's orbital momentum — which, in the case of other satellites, amounts to a mere fraction of the angular momentum of the accompanying planets — is five times that of the earth. This means that the system of the moon and the earth is more like a binary star than a miniature planetary system. From this it might be assumed that the moon originated as a separate body during the catastrophe which resulted in the formation of the planets. However, this hypothesis — if true — contains a few incomprehensible elements. It is difficult to see why two bodies, so unequal in size, should have originated during the catastrophe, and why these two bodies should have formed so close together. One thing is clear, however — the moon has slowly been retreating from the earth ever since its formation. The friction of the oceanic tides has gradually slowed down the earth's rotation and diminished its angular momentum. The result is that the moon's recession from the earth amounts to about five feet per century. Higher tides are observed as we reach back into the past; the days are found to be shorter, and the moon is seen to revolve nearer the earth.

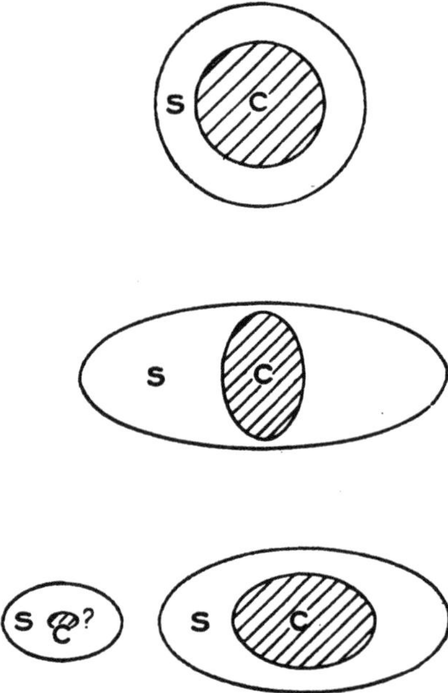

Fig. 4. Changes in the form of the earth during resonance, and the origin of the moon; c, core; s, intermediate and outer shells (After H. Jeffreys).

Of course, we sometimes wonder whether an entirely different hypothesis should not be preferred to the above, and whether the moon and the earth were not at one time united as a single body. This was the opinion of Sir George Darwin, who regarded the moon's disruption as a resonance effect, i.e. the result of a concurrence of the solar tide with the natural, free period of vibration of the earth's cooling fluid mass. The initial small amplitude of the tidal movement would continue to increase steadily according to this theory, and the globe would ultimately grow unstable and disintegrate when a certain limit was exceeded (fig. 4).

This hypothesis is based on the assumption that the moon was severed from the earth during the earliest part of their joint evolution. The liquid earth-materials were probably already more or less differentiated by that time, the heavier minerals sinking, and the lighter ones rising and accumulating in outer layers of the molten planet.

According to the resonance-hypothesis of the moon's origin a great part of the earth's outer shells clearly entered the formation of the moon. And this would not only explain why the moon's density is less than of the earth, but also why its heavy core is comparatively smaller or non-existent, and why its external silicate shell is so much thicker [1]).

Mohorovičić computed the thickness of the earth's outer silicate mantle at 60 kilometer, whereas at the moon it would amount about 400 km (see fig. 152). It should be mentioned, however, at once that — according to Jeffreys — the moon's density could be accounted for as well under the hypothesis of its formation as a separate body near the earth. And we will see later on (p. 243) that the premisses of Mohorovičić's calculation are of questionable nature.

The publication of Mohorovičić dates from the year 1925. Fourteen years later Escher, not knowing of Mohorovičić's results, once more calculated the thickness of the lunar sial shells. He started from the same premisses and came to the same result. It was on this foundation that Escher built up

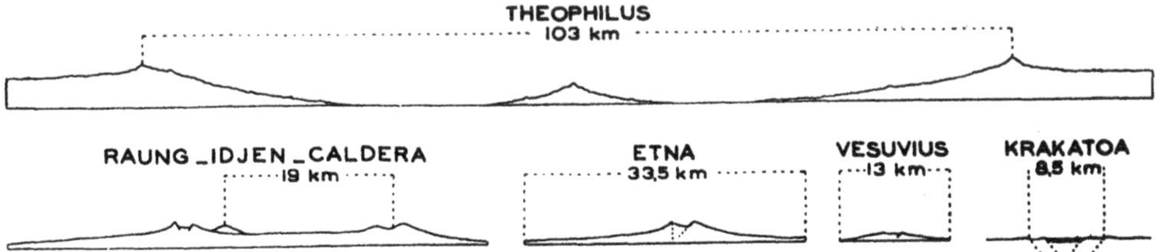

Fig. 5. Cross-section through Theophilus — a lunar crater with a central cone — as compared to the dimensions of some well known terrestrial volcanoes (After B. G. Escher).

his interesting interpretation of the moon's morphology. A granitic magma abounding in gases is known to bring about a much more explosive kind of volcanism on earth than a basic magma, and it follows that the discharge of gases from the moon's very thick sialic shell caused a huge amount of volcanism, resulting in types of volcanic explosive vents (fig. 5) such as have

---

[1]) Escher pointed out that the specific gravity of the moon's core would, on the earth's surface, amount to 3.89. This is more than the specific gravity of any rock on earth under a pressure of 1 atmosphere. "From this it would appear that some nickel-iron had passed from the earth to the moon. This must be understood in such a way, that no material of the core of the earth disappeared to the moon, but material from a ferro-sporadic shell" (Escher 1939, p. 131).

never been equalled, either in abundance or magnitude, by similar terrestrial volcanoes [1]).

On the other hand several terrestrial phenomena may perhaps find an explanation in lunar influences.

We will therefore return to the primordial history of the moon and the earth in Chapters VIII and XI, and especially in Chapter VIII, when discussing the origin of the continents and oceans.

The Earth's Interior

The supposed constitution of the earth's interior to which allusion was made in the foregoing pages, is largely based on the interpretation of seismic data.

When differences of stress in the rigid and elastic crust of the earth exceed the strength of the crust, the stress is relieved by a sudden slipping or rupture along a new or still existing fracture-plane. The spasmodic discharge of elastic stress which overcomes frictional resistance causes what is called a tectonic earthquake. Two waves with unequal velocities, even in the same medium, are propagated through the earth's body. In the swifter waves the movement corresponds with the direction of propagation; they are longitudinal waves and the first to be recorded by the seismographs at some appreciable distance from the focus; they are called primary or P-waves. In the slower waves the direction of the wave-movement is transverse to the direction of propagation. They are marked in the seismograms as secondary S-waves. The velocities of P and S vary with the density and elasticity of the rocks through which they pass. Their path is curved owing to the gradual change of density and elasticity of the rock masses through which they are propagated. Moreover, P and S are subjected to refraction and reflection at the boundary-surfaces of different kinds of rock-masses. Their arrival at the seismograph is marked by such notations as Pp, Ss (fig. 6). Finally a third type of waves, travelling along the earth's surface is recorded by the seismograph; their arrival is indicated by L (Long waves) in fig. 6 which shows some types of seismograms varying according to their distance from the focus or hypocentre and the spot situated immediately above it at the surface called the epicentre. Entering into the secrets of seismology would lead us too far from the scope of the present chapter. This short introduction will serve only to furnish an idea of a few of its fundamental principles. For seismological science has led to far reaching conclusions regarding our conceptions of the earth's internal constitution. The time of arrival of the waves is automatically recorded by many stations. Conclusions can be drawn as to the site of the epicentre and the probable depth of the focus, by studying the combined results of horizontal and vertical pendulums at these stations.

---

[1]) I endorse Escher's statements as regards the moon's volcanic morphology up to this point, but I believe that there are not enough facts with which to calculate the thickness of the lunar sial. We will return to this interesting point in Chapter VIII, when discussing the problem of the origin of continents and ocean basins.

In addition it is by a study of seismograms that conclusions are reached regarding the situation of the major discontinuities in the earth's crust and

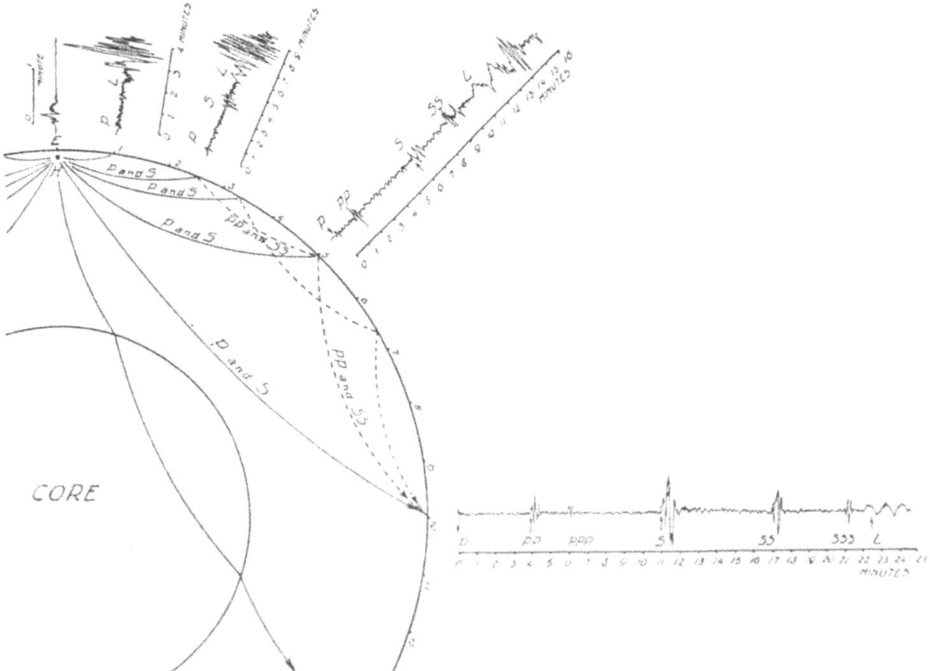

Fig. 6. Propagation of seismic waves through the earth.

its deeper interior. As shown by fig. 7 they are estimated to occur at approximately 60, 1000 and 3000 km. The first discontinuity is supposed to form the boundary between the rigid crystalline crust and the outer mantle. According to the opinion of Goldschmidt and other geochemists gravitational differentiation caused the liquid rocks of the earth to form concentric shells around a central core. In their opinion the core would consist of nickel-iron while a sulphide-oxide melt separated around it and an outer mantle was formed by the relatively light material of a silicate melt. Then the discontinuity of 2900 km was supposed to be identical with the supposed boundary between the core and the intermediate shell of sulphide-oxide. In a similar way the discontinuity of 1000 km was supposed to coincide with the boundary between the intermediate shell and the silicate melt of the outer shell.

Fig. 7 gives a diagrammatic and tentative illustration of the earth's interior based on the geochemical interpretation of the seismic data.

It is, however, doubtful whether the seismic discontinuities are due to sudden changes of the chemical composition of the earth's interior. According to Kühn it is much more probable that the major discontinuities represent changes of a purely physical nature. It is difficult to understand how a complete differentiation of the molten planet could have been accomplished in a comparatively short period of its early infancy. And, in addition gravitational differentiation as far as the earth's centre seems improbable

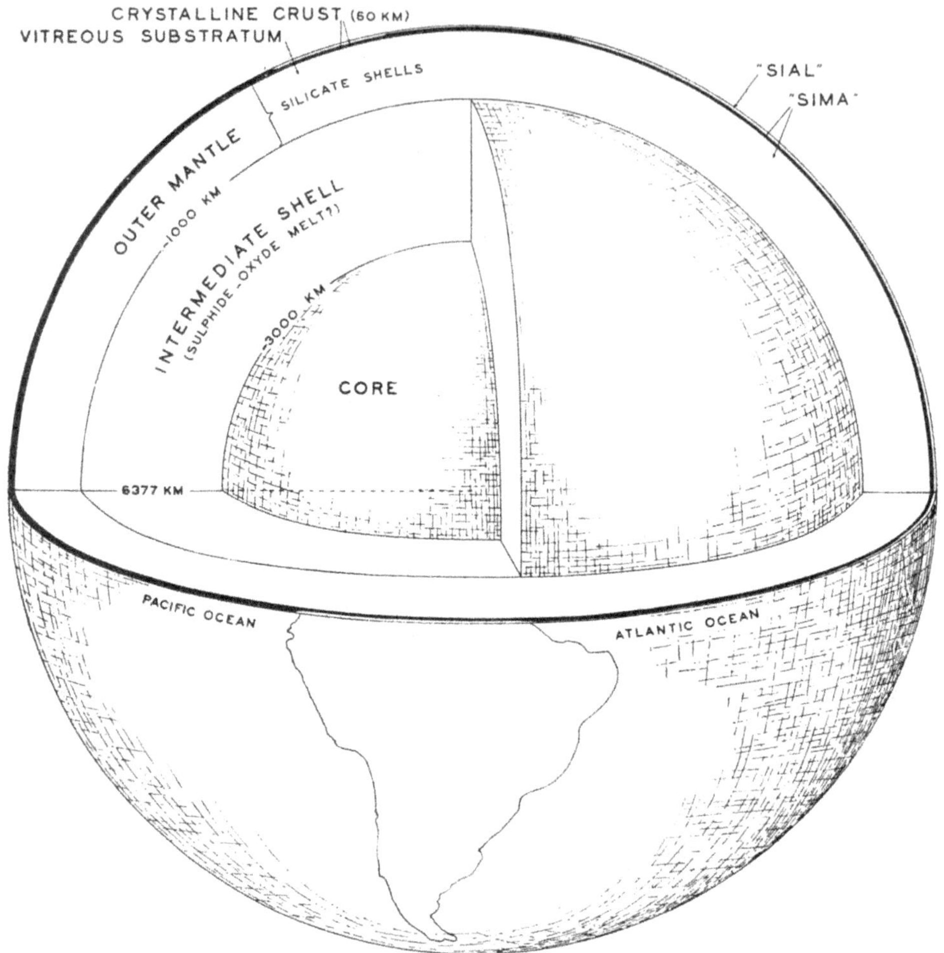

Fig. 7. Schematic and tentative illustration of the internal structure of the earth.

in as much as the value of gravity diminishes towards the interior becoming zero at the centre.

In a new hypothesis Kühn and Rittmann consider the interior of the earth to consist of undifferentiated solar material. Their opinion, in contrast to the older theory of a nickel-iron core, is clearly exhibited in fig. 8.

Perhaps the truth lies somewhere between the theory of Kühn and the older conception. Certain arguments seem to favour the new theory of a core consisting of undifferentiated solar material. Possibly, however, the discontinuity at 2900 km separates the core from the outer layers which consist of more or less differentiated material. It will be seen in Chapter VIII that convection currents possibly reached down to that level in a primordial stage of the earth. Their action would explain some striking features in the shape and distribution of the continents and ocean floors according to a recent speculation of Vening Meinesz. Convection was accompanied by a complicated process of differentiation of the outer layers. Loss of gases, especially of hydrogen at the surface, accumulation of sialic material in a comparatively thin outer layer and of heavier materials in the deeper shells are considered as the principal results of such a process of convection and differentiation.

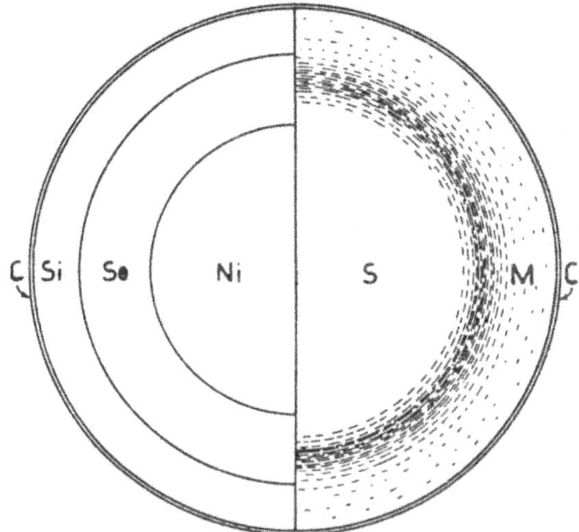

Fig. 8. Two different theories on the constitution of the earth's interior. Left side: theory of a nickel-iron core, right side: theory of a core consisting of undifferentiated solar material. C, earth's crust; Si, silicate mantle; SO, intermediate shells; Ni, nickel-iron core; M, magma zone consisting of a silicate melt with increasing content of iron and magnesium towards the depth; S, undifferentiated solar material containing much hydrogen. (After Kuhn and Rittmann).

The outer, dark-coloured zone in fig. 7 represents the earth's crust, composed of some 40 to 80 km of crystalline rocks. This solid, elastic crust presumably rests on amorphous formations, i.e. rocks which, as a result of the high temperature at this depth, must be above their melting point. They are regarded as a fluid possessing a high viscosity. Daly speaks of a "vitreous substratum". The schematic cross-sections of the continents have been drawn as white lenses in the dark zone. Petrographical, volcanological and seismic evidence has lead to the assumption that the continents are built up of light siliceous material (55-70% SiO_2), viz. granitic "acid" rocks and sediments. This constitutes the so-called sial, a term introduced by Suess (the word sial is composed of the first syllables of silicon and aluminium, the most abundant elements of the continents). The sial rests upon less acid rocks (35-55% SiO_2), ranging from basalt to peridotite. These last materials build up the so-called "sima" — another petrographic term created by Suess and mnemonic of its most abundant elements, silicon and magnesium. It is believed to extend beneath the sialic continents. One of the ocean-floors — that of the Pacific — may probably be assumed to be composed of this simatic material in a crystalline state. As regards the Atlantic and the Indian Ocean, their bottoms are thought to consist of thin layers of sial, covering — and resting upon — the crystalline sima. (More detailed cross-sections of the earth's crust will be found in fig. 41 and 42, and will be explained in Chapter IV).

The age of the earth and the universe

Similar results as regards the age of the earth are obtained by investigators in different branches of science, and show that 2,000 to 3,000 million years have elapsed since its creation. These results are arrived at by petrographers and physicists in collaboration with geologists and also by astronomers. These figures are specially impressive in as much as many of these scientists arrived at them independently and their importance is augmented by the fact that — apart from the earth — many other bodies, such as the moon, the planets, the entire solar-system and the spiral galaxies, received the impulse to perform their present movements at the same moment. This moment seems to have formed the last critical date in the universe which since then has developed gradually into its present condition.

By means of no less than ten independent methods have scientists arrived at an estimate of the age of the universe since the last great catastrophe.

Chemists have shown that uranium disintegrates spontaneously into radium and helium at known systematic rates, and that radium will in turn disintegrate into lead and helium (fig. 9). Uranium is a mixture of two isotopes UI and AcU. The first, 139 times as abundant as AcU, disintegrates through the "uranium series" to lead which has a different atomic weight (Pb 206) from lead (Pb 207) that originates via the "actinium series", while lead deriving from thorium has again a different mass number (Pb 208). Common, that is non-radiogenic lead, has again a different atomic weight (Pb 207.21). An analysis tells us the contents of radiogenic lead (we will not examine the problems affecting the determination and analysis of the various isotopes of lead), and as the amount of radiogenic lead accumulating from a given quantity of uranium or thorium per unit of time is known, the age of any mineral containing radiogenic lead can be ascertained.

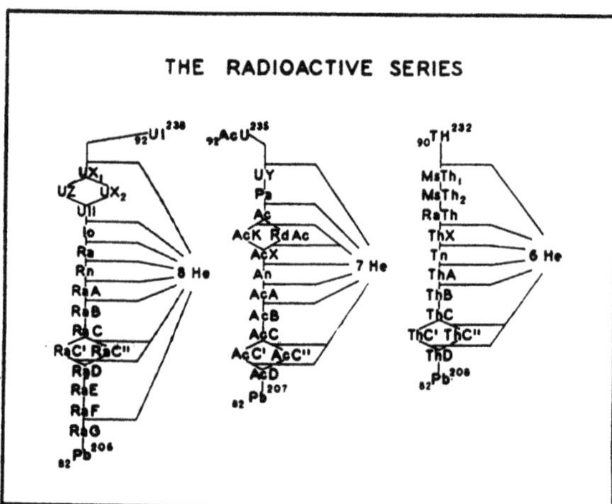

Fig. 9. The radioactive series. (From Clark Goodman).

A second method is that based on the constant ratio observed between the amount of expelled helium atoms and the initial uranium. Helium is a gas, however, and can escape as such from a rock. The values arrived at by this method will therefore probably be too low. In other words, the ages of rocks obtained by the helium method will also be too low. Many ingenious improvements have been recently introduced to remedy this deficiency, in an attempt to obtain useful results even from this method. The ages of a great many minerals have been determined so far, and their analyses show that the ages of those rocks which could indeed be said to be older because of the geological position in which they were found, were invariably higher than others. The oldest minerals ever analysed come from Manitoba, and are 1,750 million years old (fig. 10). The area in which they originated is known as one of the earth's most ancient structures, but many events preceded the formation even of this structure. A comparison between this region and other more recent and better-known areas makes it possible to calculate that the earth's age amounts to at least 1,750 plus approximately 300 million years, or some 2,000 million years.

An upper limit of 3,500 million years was found from an estimate of the entire lead content of the earth's crust, allowing for the questionable assumption that all the lead has been derived from radioactive disintegration. Hence, the true age of the earth probably is between 2 and 3,5 billion years. Recently Holmes found 3,000 m. y. as the probable age of the earth.

This figure has not only been arrived at by geologists. It was mentioned above that a similar result was deduced from the recession of the spiral galaxies.

The spiral structure, too, may serve as a basis for further conclusions. Eddington showed that a galaxy could not possibly be preserved in a steady state for more than 10^9 years without considerable collapse or dispersal, and to this Bok added that "many spiral galaxies exhibit two well developed spiral arms and it is probable that such a structure cannot persist for much longer than ten to fifteen revolutions of the nucleus". As the galactic system revolves once in every 200 million years, its age might likewise be concluded not to exceed 2,000 to 3,000 million years.

These coincidences between astronomy and geology are truly remarkable, but we

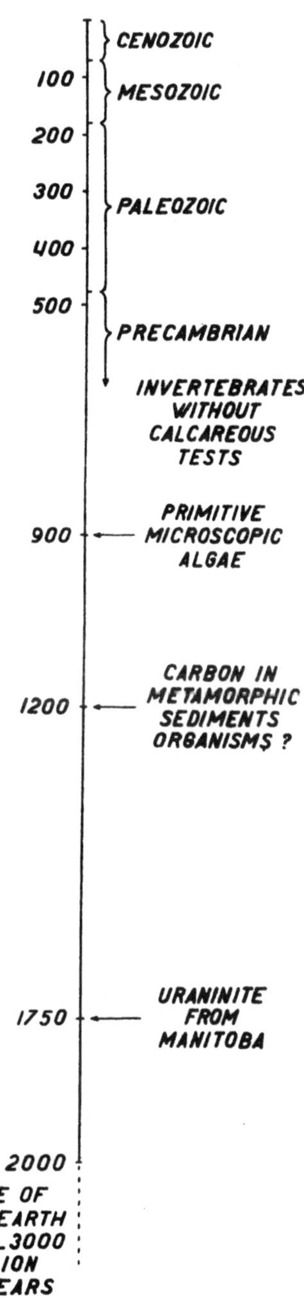

Fig. 10. Graphic representation of the age of the earth.

may add at the same time one more important argument. We know for certain that some of the nickel-iron meteorites which have dropped on to our planet can be said to have originated beyond the solar-system. A minute amount of radioactive elements has been detected in some of these bodies, and their helium ratios have been determined. The eagerly awaited results of the investigation of these fragments of cosmic matter produced various figures, none of which exceeded 2,000 to 3,000 million years, however.

Another method is based on the movement of the moon. It was seen above that this body is steadily receding from the earth. If the process were reversed, the moon would be seen to be gradually approaching the earth. Though admitting that it is impossible to tell at what moment the moon was severed from the earth, in as much as this process is itself purely hypothetical, Jeffreys estimated that the age of the moon was probably less than 4,000 million years. This figure is again remarkably near that of the 2,000 to 3,000 million years estimate mentioned above several times.

Chandrasekhar added two fresh evidences by his study of the statistics of binary stars and the dynamics of galactic star clusters. The Pleiades star cluster includes some 200 stars in a spherical volume of radius about 10 light years which is a star density about twenty times that of the background "field" stars. As a result of his calculations on the stability and dynamics of such aggregations of matter Chandrasekhar came to a rational means for an estimate of the probable rate at which a star cluster tends to disintegrate. His conclusion is that the average life of a cluster is about 3×10^9 years!

The same author made another new contribution to the question of the time-scale of the Universe by his consideration of the binaries. Binary stars are unstable because of their tendency towards disruption in as much as the distances of the nearby stars from the two components will be different, and consequently also their fluctuating tidal effects. According to a mathematical formula "binaries with separations between 1,000 and 10,000 astronomical units will be dissociated in times ranging from 7×10^{10} to 2×10^9 years". This means that during an interval of time of about 10^{10} years a tendency must be expected of ever larger separations towards a final condition of statistical equilibrium. However, it has been found that "the larger separations occur with far less frequency than should be expected under conditions approximating those of equilibrium. Accordingly, we should conclude that sufficient time has not elapsed for the tidal forces of the neighboring stars to appreciably modify the elements of binary orbits with separations in the range stated. This implies that 10^{10} years represents a true upper limit to the time scale, and would suggest a time scale of the order of say 5×10^9 years".

"The discussion of the mean lives of galactic clusters and the statistics of binary stars agree therefore in pointing to a time scale of the order of a few billion years".

Lastly, Jeffreys calculated from the eccentricity of the planet Mercury that the age of the solar-system was probably nearer 1,000 than 10,000 million years.

This concludes our summary of the various methods which make it possible to calculate a minimum value of the age of the earth and the universe. It is obvious that these methods provide converging evidence of great importance.

A few methods are dealt with more extensively by Holmes in *The Age of the Earth*. Data on the absolute age of terrestrial rocks are extremely important, for they give us an idea of the duration of the different geological eras and periods. Therefore the question of absolute age determinations will be treated at greater length in the Appendix (p. 350).

The history of the earth's crust

We usually divide the history of the continents into two distinct parts. The first deals with the remotest ages and covers the extremely long Pre-Cambrian era, of which very little is known. The second deals with the succeeding ages, from the Cambrian up to the present day. Much more is known of the earth's history since the Cambrian than of the preceding periods, for, generally speaking, its historical records have remained in a far better state of preservation than those of the earlier times. Historical geology has often been compared to a book, a torn and tattered old manuscript the pages of which lie scattered far and wide. The task of the geologist is to hunt up the missing pages and replace them in their original order. The sediment layers can be said to represent the pages of this geological history and the fossils the writing. It should be noted that fossils occur very rarely in the Pre-Cambrian and are almost entirely absent in the older parts of this era. This explains why so little is known of Pre-Cambrian history. Its pages can only be replaced in the correct sequence with the greatest difficulty (if at all), and the writing upon most of them has faded like that of an old palimpsest.

The principal events in the history of Life are listed in Table II, next to the time-scale. The Pleistocene and Holocene, encompassing the whole history of mankind from the earliest days up to our own, represent less than the thickness of the top line in fig. 10. Should our knowledge of the earth's history be communicated to a student in correct chronological proportion in the course of 50 to 60 lecture-hours, a bare two minutes would have to suffice for a discussion of the entire history of the Pleistocene-ages. A comparison between the earth and various cosmic dimensions made us realize our planet's modest size. The preceding statement, hovewer, shows that the history of the earth is at least 3,000 times as long as the whole history of *Homo sapiens*.

One of the most remarkable features of that history is the ceaseless flow of alterations. Mountains, seas, glaciers and deserts all seem immutable compared to the ephemeral span of human life. However, nothing is constant in geological time. Most visitors to the small though interesting town of Le Puy, in the south of France (fig. 11), go solely in order to admire its beautiful old cathedral and to gaze upon the enormous bronze statue of the Virgin. They stare in silent wonder at the steep rocks, each of which is crowned with a church or statue, but few realize that this strange scenery is merely due to the fact that the hard filling of a volcanic vent and some remnants of former lava flows are resisting erosion a little longer than the largely eroded tuffs and strata.

All those external forces, such as frost, solution, running water, the weather and wind, etc., which act unceasingly upon the terrestrial crust, are wearing down the mountains slowly but surely (fig. 12) and levelling them with the sea. The peneplain constitutes the final stage of the largest mountains. In other areas the sea helps to abrade the coasts (fig. 13), and this process, too, has a levelling effect. Neptune invades the land to an ever-increasing extent,

Fig. 11. The town of Le Puy in southern France, which is also interesting from a geomorphological point of view.

and the débris of the organisms which remain behind enable us to determine the relative ages of the various deposits. An illustration of one of these marine strata, which was deposited during the Cambrian in a sea invading the land over Torridon sandstone of late Pre-Cambrian origin, will be found in fig. 14.

This erosive action, or denudation, is represented schematically in blocks 1 and 2 of fig. 15. Only a few of the remnants of this erosive action, or "monadnocks", as they are called, protrude from the peneplain in block 2 as evidence of the former relief. These two blocks show more, however. Rivers transport detritus from the mountains to the sea, where it settles in large quantities. The land represents an area of erosion, but the sea is the most suitable place for the accumulation of great quantities of erosion products. One layer slowly accumulates on top of another, and this leads to a thick sequence of sedimentary strata. Block 1 shows that sedimentation takes place in a trough-shaped depression of the earth's surface. The coarsest products remain near the coast, the finer material is carried further away and settles in thin layers at a considerable distance from the shore. The coastline extends slowly but surely into the sea, and the sequence of sediments steadily increases. Block 2 shows that the floor of the basin of sedimentation also subsides. Without such subsidence the thickness of the strata would never be able to exceed the original depth of the sea at this point. The accumulation of sediments, however, increases proportionately with the subsidence of the floor. A thickness of many thousands of meters (sometimes as much as 15,000 meters or more) is not an unusual occurrence.

Everyone probably knows that not only are the sediments which settle

Fig. 12. The active Gedeh-Panggerango volcano (above) and the much eroded volcanic ruin near Plered (below), as seen from mount Bengkung, west of Bandung, Java.

near the coast unlike the strata deposited in the deep-sea, but that these separate environments also contain different kinds of organisms. Various factors such as light, food, pressure, temperature, salinity, the movement of the water and a host of other influences all effect the zonal distribution of organisms in the sea. An idea of the conditions under which these strata were originally formed can be obtained partly from their special lithological character but principally from the remains of former organisms. This total aspect of the sediments is known as the facies in geological terminology (fig. 16). The facies of sediments deposited in a shallow sea in the vicinity of the coast (littoral), or in a few hundred meters of water (neritic sediments) are wholly unlike those which settled further away from the coast in approximately 1,000 meters of water (the so-called bathyal sediments) or at an even greater depth (abyssal sediments).

Littoral or neritic sediments cannot form in more than a few hundred meters of water, but we sometimes come across a sequence of neritic sediment totalling a few thousand meters in thickness, and it is clear that this must

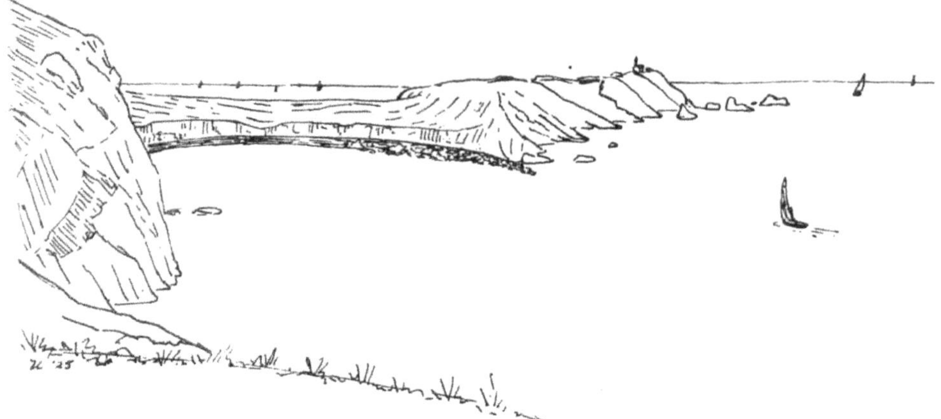

Fig. 13. The erosive action of the sea, near Camaret, Crozon-peninsula, in the neighborhood of Brest. The lower central part consists of Pre-Cambrian schists; to the left and right Lower Ordovician sandstones are exposed, forming an anticline.

have been brought about by an accumulation of successive deposits during a slow subsidence of the sea floor. The name geosyncline has been given to these subsiding furrows of the terrestrial crust, into which sediments accumulated to abnormally thick sequences. A schematic view of a geosynclinal

Fig. 14. Transgression of Cambrian quartzites (1) on Pre-Cambrian Torridon sandstones (2). Quinag near Inchnadamff, in the North-western Highlands of Scotland. (Compare with fig. 20).

cross-section is shown in block 2 (fig. 15). The downward movement of these formations must be attributed to the influence of deep-seated forces, and the same forces will at a given moment put a stop to the subsidence. A growing compression of the crust then causes the geosyncline to be thrown into spasms and its contents to be crumpled and folded. Block

3 depicts a more advanced stage. The whole zone will then tend to rise, and from the elongated belt emerges a mountain range which rises upwards very slowly, though with unremitting persistence, frequently to a height of several thousands of meters.

The landscape reproduced in fig. 17 borders on the Caledonian geosyncline in the north-western part of the Scottish Highlands. To the left of it are seen Cambrian strata (these were deposited normally on an older Pre-Cam-

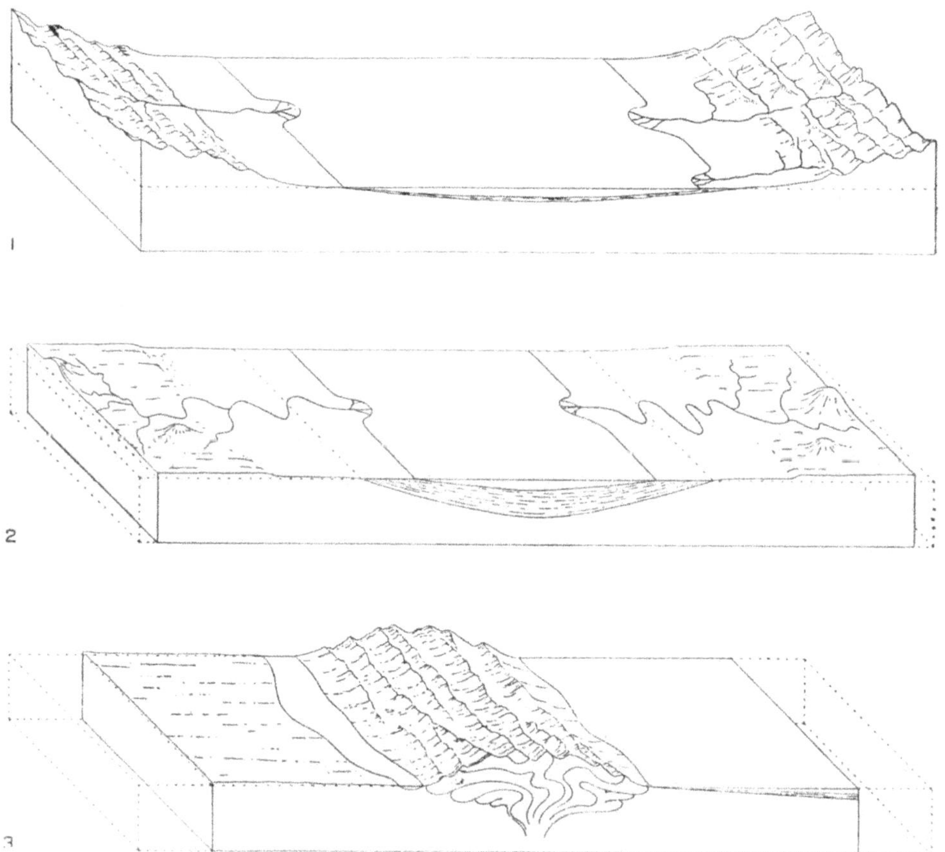

Fig. 15. Diagrammatic representation of a geological cycle.

brian basement, consisting in this case of Lewisian gneiss). To the right, however, a large "nappe" of Lewisian gneiss is clearly seen to have been overthrust towards the left, i.e. over the edge of the geosyncline.

The external forces embark upon their destructive action as soon as the folded chain emerges. A new area of subsidence and sedimentation is formed elsewhere (this is indicated schematically towards the right of block 3, fig. 15); a new geological cycle has opened. The vertical proportions in fig. 15 are exaggerated, but these rhythms will be dealt with again in Chapter IV (fig. 49).

A natural section such as that in fig. 18 provides the key to the reconstruction of a geological cycle. The strata in this consist of Devonian

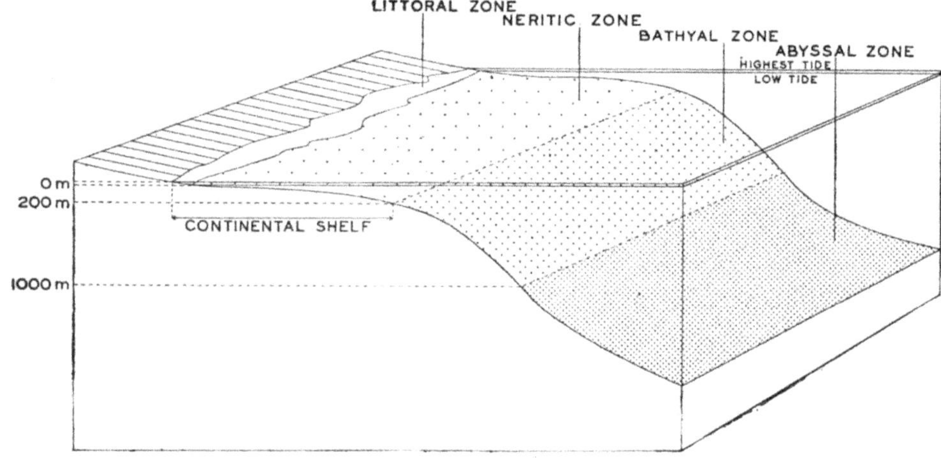

Fig. 16. Schematic block-diagram of different facies types.

Old Red sandstone resting unconformably on steeply dipping Silurian graywackes and shales. A plane of unconformity has a long history. The graywackes in fig. 18 were deposited normally, i.e. just about horizontally, during the Silurian, and were subsequently steeply folded and elevated above sea-level. Old Red Sandstone later settled on top of these eroded

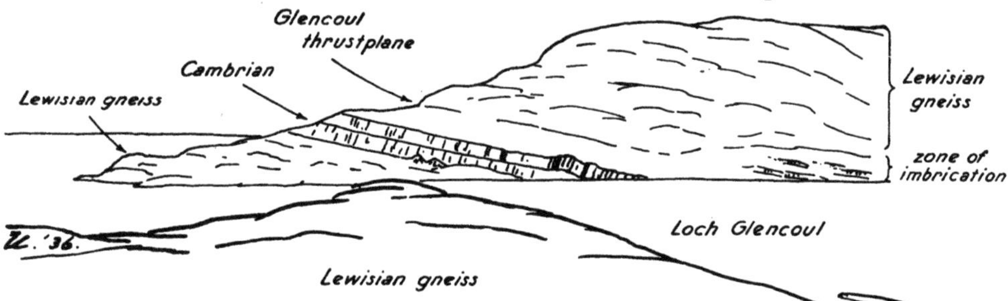

Fig. 17. Glencoul overthrust. Assynt Mountains, North-West Scotland (Compare with fig. 20).

and peneplained marine strata. The period of folding of the latter corresponds with that in which a geosyncline was compressed over large areas in Scandinavia, Scotland and Ireland. This last zone was later raised as a mountain-chain (Plate 1 indicates this belt with green lines). The actual Old Red Sandstone was supplied by detritus from these chains, and was later likewise worn away by erosion in extensive areas. Trough-shaped depressions, elongated in the trend of the Caledonian mountains, formed here and there (fig. 31), and the Old Red escaped erosion only in these depressions. Some of the troughs are bounded by faults (fig. 20), in which we once more find the trend of the old Caledonian geosyncline. These faults

Fig. 18. Old Red (1) resting unconformably on steeply dipping Silurian graywackes and shales (2). Siccar Point near Dunbar, Scotland. (Compare with fig. 20).

are observed along the surface over a considerable distance (fig. 19). Indeed, the relation between the scenery and structural history of an area is finely and instructively illustrated by the whole of the morphology of Scotland. A diagram of some of the most salient geological and "geomorphological" features of Scotland will be found in fig. 20.

In a few cases geosynclinal subsidence and folding are known to have occurred several times in the same region. One of these complicated areas is depicted in fig. 21. In the Ardennes, namely — near Fépin, along the banks of the river Meuse — the Lower Paleozoic deposits (Revinian) of an ancient Caledonian geosyncline were strongly folded and covered unconformably by Lower Devonian (Gedinnian) strata, beginning with a basal conglomerate. This younger sequence of the so-called Variscian geosyncline was later refolded, and the plane of unconformity was also strongly undulated.

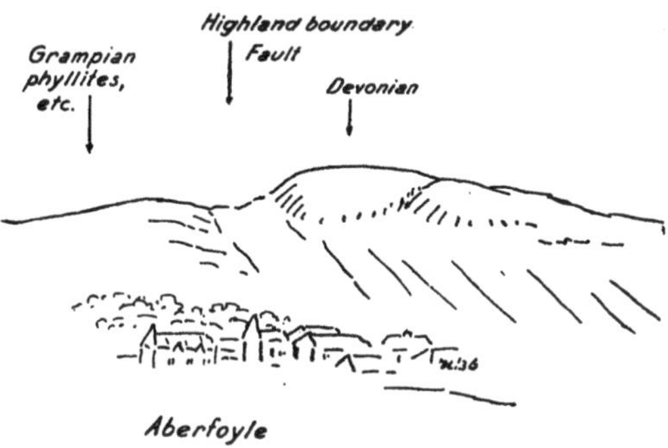

Fig. 19. The Highland boundary fault near Aberfoyle. (Compare with fig. 20).

The object of the preceding remarks, especially of the rough sketches in fig. 15, is to given an idea of the constant transformations that are altering the face of the earth. It should at once be added, however, that one of the most spectacular features of these transformations is periodicity.

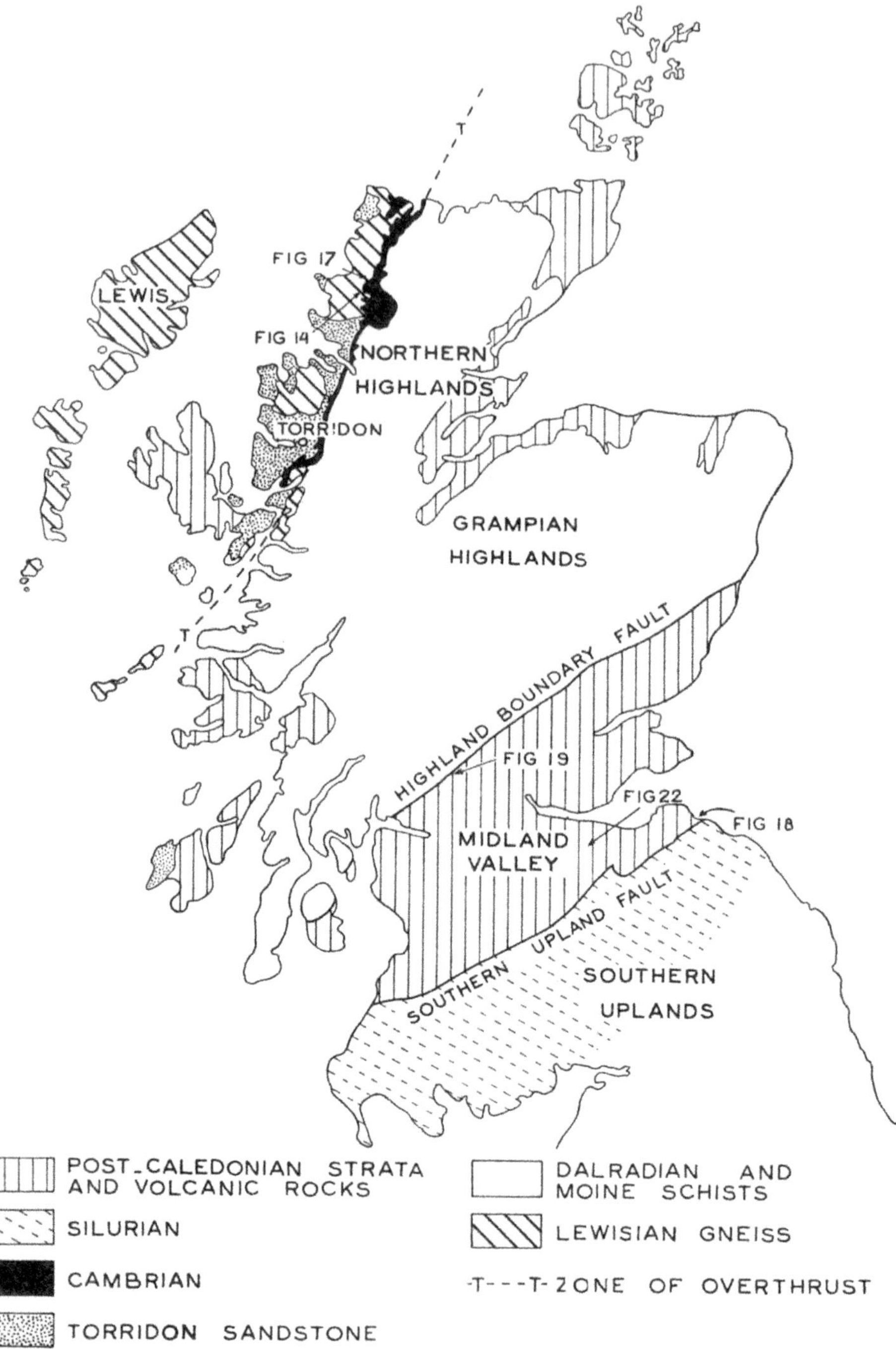

Fig. 20. Geological diagram of Scotland, showing the localities of Fig. 14, 17, 18, 19, and 22.

The geologist comes across periodicity in many of the pages which he is so arduously deciphering, — in the sequence of the strata, for instance, and their contents of former organisms (stratigraphy). He observes it elsewhere, in the pulse of the deep-seated forces that bring about subsidence first in one area and then in another, culminating in the folding of geosynclines; and again — in the intrusion of liquid melts or "magma" rising from some deeper part of the earth's interior; in the rhythmical invasion of the continents by epicontinental seas and the subsequent retreat of the latter (transgression and regression); he reconstructs the pulsation of the climates and the rhythmical evolution of Life.

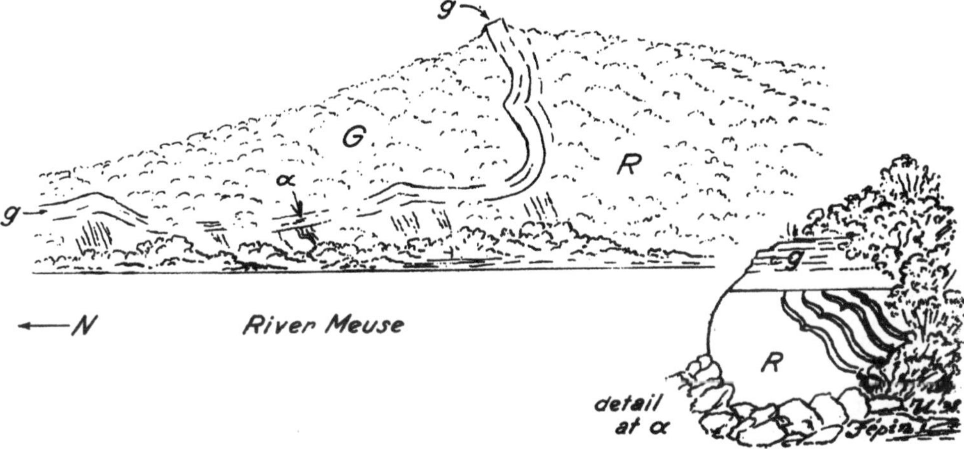

Fig. 21. The folded unconformity near Fépin in the Ardennes, a result of repeated geosynclinal subsidence and folding. G, Lower Devonian with basal conglomerate (g); R, Revinian.

Barrell's outstanding publication of 1917, which every geologist ought to read and re-read, opens with similar thoughts: "Nature vibrates with rhythms, climatic and diastrophic, those finding stratigraphic expression ranging in period from the rapid oscillation of surface waters, recorded in ripple-marks, to those long deferred stirrings of the deep-imprisoned titans which have divided the earth's history into periods and eras. The flight of time is measured by the weaving of composite rhythms — day and night, calm and storm, summer and winter, birth and death — such as these are sensed in the brief life of man. But the career of the earth recedes into a remoteness against which these lesser cycles are as unavailing for the measurement of that abyss of time as would be for human history the beating of an insect's wing. We must seek out, then, the nature of those longer rhythms whose very existence was unknown until man by the light of science sought to understand the earth".

Periodicity is the fundamental conception upon which the following pages have been built. The sequence of geosynclines and folded mountain-chains will be analysed in Chapter II. Chapter III will discuss another conspicuous feature of continental structure — the basins. The deeper structure of the earth's crust will be dealt with in Chapter IV, which moreover, will attempt to show the relation between tectonic cycles and the closely associated magmatic phenomena. Chapters II–VII deal chiefly with the history of the continents and the oscillations of the sea-level. The problems arising from the ocean-floors, will subsequently be examined in Chapter VIII. Chapter IX discusses the intricate problems of abnormal climates (the ice-ages). Chapter X attempts to trace the presence of a certain rhythm in the evolution of Life, especially in the evolution of the flora. Chapter XI deals with the linear patterns of the earth and the problem of a supposed displacement of the poles, and Chapter XII ends with a brief review of all the preceding phenomena which form *the pulse of the earth.*

References

AHRENDS, L. H. *Determination of the Age of Minerals by means of the Radioactivity of Rubidium* (Nature, no. 3983, vol. 157, 1946).
BARRELL, J. *The strength of the Earth's Crust* (Journ. of Geology, vols. 22 and 23, 1914 and 1915).
BARRELL, J. *Rhythms and the measurement of geologic time* (Bull. Geol. Soc. of America, vol 28, 1917).
BOK, B. J. *Galactic dynamics and the cosmic time-scale* (The Observatory LIX, Nr. 142, 1936).
BOK, B. J. and BOK, P. F. *The Milky way* (Philadelphia, sec. ed. 1946).
CHANDRASEKAR, S. *Galactic evidences for the time-scale of the Universe* (Science, vol. 99, no. 2564, pp. 133–136, 1944).
DALY, R. A. *Architecture of the Earth* (1938).
DALY, R. A. *Meteorites and an earth-model* (Bull. Geol. Soc. America, 54, 1943).
EDDINGTON, A. S. *The rotation of the galaxy* (Oxf. Univ. Press 1930).
EDDINGTON, A. S. *The expanding Universe* (Cambridge 1933).
ESCHER, B. G. *Moon and Earth* (Proceed. Kon. Acad. van Wetensch. Amsterdam, vol. 42, 1939).
ESCHER, B. G. *Het probleem der caldeira's en der maancircussen* (Geologie en Mijnbouw 1, 1940).
EVANS, R. D., GOODMAN, C., KEEVIL, N. B., LANE, A. C. and URRY, W. D. *Intercalibration and comparison in two laboratories of measurements incident to the determination of the geological ages of rocks* (Physical Review 55, 1939).
EVANS, R. D. and GOODMAN, C. *Alpha-Helium method of determining geological ages* (The Physical Review 65, 1944).
GOLDSCHMIDT, V. M. *Geochemie* (Handwörterb. d. Naturwissensch. 2e ed. 1933).
GOODMAN, C. *Geological applications of nuclear Physics* (Journ. of Applied Physics 13, 1942).
GUTENBERG, B. *Physics of the Earth* (vol. VII. Internal constitution of the earth 1939).
HOLMES, A. *The Age of the Earth* (Th. Nelson & Sons, London 1937).
HOLMES, A. *Principles of Physical Geology* (Nelson, Edinburgh, 1945).
HOLMES, A. *The geological time-scale* (To be published in Trans. Geol. Soc. Glasgow, 20, 1946).
HOLMES, A. *An estimate of the age of the earth* (Nature, 157, no. 3995, 1946).
HURLEY, P. M., and GOODMAN, C. *Helium retention in common rock minerals* (Bull. Geolog. Soc. of America 52, 1941).
JEANS, J. *Astronomy and Cosmogony* (Cambridge 1928).
JEANS, J. *Through space and time* (Cambridge Univ. Press 1934).
JEFFREYS, H. *The Earth. Its origin, history and physical constitution* (Sec. Ed. Cambridge Univ. Press 1929).
JEFFREYS, H. *The density distribution in the inner planets* (Monthl. Notices R. Astron. Soc. Geoph. Suppl. 3, 1937).
KUHN, W. und RITTMANN, A. *Uber den Zustand des Erdinnern* (Geolog. Rundschau 32, 1941).
LYTLETON, R. A. *The origin of the solar system* (Monthly Not. R. Astron. Soc. 96, 1936).
MOHOROVIČIĆ, S. *Das Erdinnere* (Zeitschr. f. angew. Geoph. 1, 1925).
MOHOROVIČIĆ, S. *Uber Nachbeben und über die Konstitution des Erd- und Mond-innern* (Geol. Beiträge zur Geophysik 17, 1937).
PLASKETT, J. S. *The dimensions and structure of the galaxy* (Oxf. Univ. Press. 1935).
PLASKETT, J. S. *Reports of the Committee on the measurement of Geologic Time* (Nation Research Council, Washington, 1936–1939).

RUSSELL, H. N. *The Solar System and its origin* (Edit. MacMillan Co, 1935).
RUSSELL, H. N. DUGAN, R. S., and STEWART, J. Q. *The solar system* (Ginn and Co., Boston, 1946).
SHAPLEY, H. *On the astronomical dating of the Earth's crust* (Am. Journ. Sci. 243, 1945).
SITTER, W. DE, *Kosmos* (A course of six lectures on the development of our insight into the structure of the universe delivered for the Lowell Institute in Boston, in November 1931, Cambridge Mass, 1932).
SITTER, W. DE, *Het uitdijend heelal* (Werken v.h. Genootsch. v. Natuur-, Genees- en Heelkunde. Amsterdam, 2e ser., deel 14, 1934).
SLICHTER, L. B. *Cooling of the Earth* (Bull Geol. Soc. America, 52, 1941).
TERMIER, P. *La Joie de Connaître* (1926).
UMBGROVE, J. H. F. *Periodicity in terrestrial processes* (Americ. Journal of Science 238, 1940).
URRY, W. D. *Ages by the Helium Method*, I *Keweenawan* (Bull. Geolog. Soc. of America 46, 1935), II *Post-Keweenawan* (Ibidem 47, 1936).
URRY, W. D. and HOLMES, A. *Age determination of Carboniferous Basic Rocks of Shropshire and Colonsay* (Geolog. Magazine 78, 1941).

Chapter II

MOUNTAIN-CHAINS

Introduction

"The profound revolutions, marked by folding, magmatic invasion and regional metamorphism were relatively brief periods closing long eras marked by diastrophic quiet and low continental relief". (J. Barrell)

The first critical analysis of the earth's structural history was made by Ed. Suess, who published a monumental book on this subject: *The Face of the Earth.* Suess was not only the right man for such a task, but he also undertook it at the right moment. The amount of data had already become so voluminous by then that an attempt could be made, by grouping them systematically, to analyse the surface of the entire earth. It would be impossible for a single man to perform such a task to-day, since a judicial examination of the enormous amount of world-literature would require more than a life-time. However, a series of regional treatises makes it much easier for us, at present, to find our bearings in this varied mass of publications and a further asset is, that the study of them requires far less of our time. Such synoptic works make it possible to form an idea of the whole region in question, without entering into a critical examination of all original data and fundamental details. In this manner it is possible to obtain a survey of all that appears to be known at present of the structural history of the continents and to base an opinion on its general aspects. This method had already been adopted previously by Born but he grouped his pictures regionally, i.e. continent-wise. We, on the other hand, have chosen a purely chronological sequence. Nevertheless, our two methods agree fully in one respect, i.e. the plain facts are stated separately, without being influenced by any pre-assumed hypothesis, and these are left to speak for themselves as they accumulate. The data have lastly been combined into a synoptic picture (Pl. 5), serving at the same time as a chronological analysis of continental structural history. At the end of the chapter the reader will find a list of the literature consulted. A summary and discussion of the various data will appear in the Appendix.

The following historical analysis will begin with the Cambrian, for the ages of the complicated structures of the Pre-Cambrian basement have only been determined in a few cases, and the study of their intricate patterns can therefore still be said to be in its infancy.

The intensive folding of the Ardennes had long been known to have occurred at a much earlier date than the tectonic activity of such mountains as the Alps, the Pyrenees and the Himalayas. The Upper Paleozoic chains were called the Hercynian Mountains, or — as this first name has now become obsolete [1]) — the Variscian (or Variscan). It had also been realized for some time that the final, intensive folding in Norway and Scotland occurred even earlier. The mountains in these last two areas were called the Caledonides. Nevertheless, as more and more facts came to light it became clear that the Lower Paleozoic or "Caledonian" belts had not all originated during the same period. Thus the Caledonides of North America, to mention but a few examples, were folded towards the close of the Ordovician, while those in Scotland and Ireland underwent a similar process some 30 million years later, towards the end of the Silurian (Gotlandian). We speak of a Taconic epoch of compression in the first case, and of an Erian in the second (fig. 22). Several different periods of origin are also observed in Variscian and

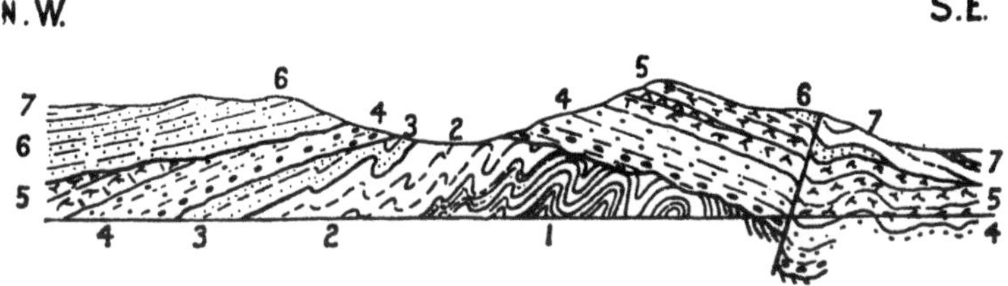

Fig. 22. Diagrammatic section across the Pentland Hills near Edinburgh (After W. H. Laurie), showing Erian and Mid-Devonian epochs of movement; 1 and 2 Silurian; 3 Downtonian; 4 and 5 Lower Old Red; 6 Upper Old Red; 7 Carboniferous. (Compare with fig. 17).

more recent mountains. A long list of epochs of compression has thus gradually been disclosed to us, the most important of which are listed in Table II.

Another important point is that the epochs are distributed over the whole world. A comparison between said phases and their distribution (See Pl. 1 – 4) will at once render clear that one movement was of particular importance in one group of mountains, while in another chain the influence of a different diastrophic epoch was especially strong. It would not always be easy, however, to state concisely which epochs made themselves felt in Caledonian chains [2]), and which other ones in Variscian [3]), for Caledonian

[1]) The name Hercynian is confusing in as much as it is commonly used to indicate a special direction in mountain-chains and faults of western Europe.

[2]) Though the oldest epoch in a so-called "Caledonian" belt is as a rule the Taconic, and the youngest the Erian or Mid Devonian (fig. 19) far more ancient Cambrian phases seem to occur in Asia. The oldest movements in Australia (Cambrian and Ordovician) were likewise regarded by Andrews as Caledonian epochs, and the youngest epochs in this "Caledonian" sector were observed to occur towards the close of the Devonian (the Bretonic epoch appearing elsewhere as the oldest in Variscian folded chains).

[3]) For while the Variscian era of folding ends in most cases with a Pfalzian, and more often an even younger epoch, "Variscian" mountain-building in the Dobrutcha, the Falkland Islands and the Cape Mountains appear to close with the Lower Cimmerian phase, whose influence may be described as more or less evident in many Cenozoic and Mesozoic chains.

	Formations		Principal Epochs of compression	
	— Pleistocene			
Tertiary		. . . .	Wallachian (= Pasadenian)	Alpine epochs
	Pliocene	Sicilian		
		Astian		
		Piacentian		
		. . . .	Rhodanic	
	Miocene	Pontian		
		. . . .	Attic	
		Sarmatian		
		Tortonian		
		. . . .	Late Styrian	
		Helvetian		
		. . . .	Early Styrian	
		Burdigalian		
		Aquitanian		
		. . . .	Savian	
	Oligocene	. . . .	Pyrenean	
	Eocene			
Mesozoic		. . . .	Laramide	
	Upper Cretaceous	Senonian		
		. . . .	Subhercynian	Mesozoic epochs
		Emscherian		
		Turonian		
		Cenomanian		
		. . . .	Austrian (Oregonian)	
	Lower Cretaceous	Gault		
		Neocomian		
		. . . .	Late Cimmerian (Nevadian)	
		Wealden		
	Jurassic	Tithonian		
		. . . .		
	Triassic	Rhetian		
		. . . .	Early Cimmerian (Palisade)	Mesozoic epochs / Variscian epochs
		Norian		
		Karnian		
		Ladinian		
		Anisian		
		Skytian		
Paleozoic		. . . .	Pfalzian	Variscian epochs
	Permian	Thuringian		
		Saxonian		
		. . . .	Saalian (Appalachian)	
		Artinskian		
	Carboniferous	Stephanian		
		. . . .	Asturian (Arbuckle)	
		Westphalian		
		Namurian		
		. . . .	Sudetic (Wichita)	
		Lower Carboniferous		
		. . . .	Bretonic (Acadian)	Variscian epochs / Caledonian epochs
	Devonian	Famennian		
		. . . .		
		Frasnian		
		Givetian		
		. . . .	Mid-Devonian	Caledonian epochs
		Eifelian		
		Coblentzian		
		Gedinnian		
		. . . .	Erian	
	Silurian (= Gothlandian)	Downtonian (incl. Upper Ludlow)		
		. . . .	Ardennian	
		Lower Ludlow		
		Wenlock		
		Tarannon		
		Llandovery		
		. . . .	Taconic	
	Ordovician	. . . .	Ordovician	
		. . . .	Sardic	
	Cambrian	. . . .	Cambrian	
		. . . .	Late Pre-Cambrian	

and Variscian epochs overlap. Moreover, the same applies to Variscian and younger convulsions.

Stille classified the post-Variscian Mountains as "Alpine" formations, I have divided them into two large groups, for the final phase is now known to have occurred during the Mesozoic — that is before the major movements in the Alps — not only in the Sierra Nevada of North America, but also in many other areas, e.g. in the eastern and south-eastern part of Asia. These belts have consequently been grouped separately (Pl. 3). The Cenozoic chains, beginning with the Laramide, will be found in Plate 4 (reference has also been made to Mesozoic movements where these are known to occur in their history).

The different belts can thus be divided into four large groups: the Early Paleozoic (Caledonides), Late Paleozoic (Variscides), Mesozoic, and Cenozoic (Alpine *sensu stricto*). The structure of the continents will be reviewed accordingly in Plate 5 [1]).

The accompanying table lists the diverse epochs of compression. A more diagrammatic view of their place on the time-scale will be found in Table II.

We will now review the more general aspects of the data discussed in the Appendix [2]).

Epochs of compression

Our historical analysis of the continents begins some 500 million years ago, for too little is known of events prior to that time. The significance of such a statement will be realized if we pause to consider that the earth is thought to be approximately 2,000 to 3,000 million years old, for this means that we only begin to find sufficient evidence of the past at a time when 3/4 to 5/6 of the earth's structural history had already elapsed. In other words, only one fourth — or possible even less — of this history can be said to be legible. So little is known of the preceding era — the Pre Cambrian — that we prefer to leave it alone altogether. This Pre-Cambrian era, however, not only covers the greater part of time, but the Pre-Cambrian areas also occupy the largest part of the continents. It should furthermore be noted that the continents, including probably large areas beyond them, had already passed through an extremely lengthy and complicated evolutionary stage before the Cambrian, and that hardly anything is known of this part of their development. Not only will a glance at the structural map in Plate 5 show that Pre-Cambrian formations occupy extensive areas, but the Pre-Cam-

1) These four groups can be subdivided into several others if classified according to the period of the formation of the belt, i.e. the youngest epoch of compression appearing in their history. The Caledonian Mountains can thus be subdivided into the Taconic and late Caledonian; the Variscian into the Sudetic, Asturian and Saalian; the Mesozoic into the early Cimmerian, late Cimmerian and Subhercynian; the Alpine into the Laramide, Pyrenean, Savian, Styrian, Attic and Wallachian. These groups would obscure the main outlines if drawn into a single structural map with separate notations, and we have therefore only indicated the four principal groups in Plate 5. Other details will be found in Plates 1—4 and in the Appendix.

2) The maps do not show the tectonic character of the structures. Mountains with intensive overthrusts, such as the Alps, appear side by side with other less intensely folded mountains, e.g. the Rockies and chains of the Jura type. It will be obvious that the coloured lines only indicate the general direction of the strike in a schematic manner.

brian basement is also known to crop out in almost all the larger Paleozoic, Mesozoic and Tertiary Mountains, proving clearly to what extent the younger part of the earth's history is engraved across the existing records of earlier ages.

We often read of the accretion of the continents, meaning that younger geosynclinal belts and folded chains are arranged consecutively around a Pre-Cambrian nucleus. In several cases these zones can be said to be younger the greater their distance from a nucleus (we will deal with this matter presently). There are indeed areas which demonstrate the outward displacement of zones of folding. It would not be right, however, to conclude from the above that a number of small blocks had existed in the Cambrian, which have grown larger and larger periodically, nor that the continents have consequently grown larger than at any previous date. On the contrary, it would be more plausible to assume that the continents had at one time been more extensive than they are at present. Not only have parts of the continents foundered below sea-level since Pre-Cambrian time, but they have even done so until quite recently, and their subsidence occasionally attained great depths! The present continents are mere fragments of one-time larger blocks. The same applies to the so-called "growth-zones". The Caledonides of Europe, North America and Asia originally covered wider areas than their actual fragments do to-day. The latter are bounded and cut off by the frontal chains of Variscian and later mountains. The Mesozoic geosynclines of Asia, North and South America still contain the vestiges of the former Variscian basement upon which they were formed. Most of the Alpine chains of Europe and Asia are situated in a zone which had already been folded during Variscian time.

It cannot be denied, however, that there are cases in which the younger the age of the geosyncline, the greater the distance between the frontal chains and a Pre-Cambrian nucleus. Here, too, a glance at Pl.5 will be enough to show that this applies to Asia, Australia, and partly to other continents. Certain Pre-Cambrian shields seem to form centres, around which originated a series of geosynclinal belts, spreading in ever-widening arcs. From Australia the Caledonian, Variscian, Mesozoic and, finally, Tertiary zones migrate east-and north-eastwards in a centrifugal sequence. An identical centrifugal displacement is observed around the Angara Shield in Asia, though with some complications. For the chains were compelled to bend around the Tarim and Ordos Massifs, including other Pre-Cambrian nuclei, which had long since been rigid blocks, and analogous centrifugal waves spread from the Kara-Sea and Tsuktschen massifs. The phenomenon of centrifugal migrations can be discerned to some extent in Europe, viz. in a southerly direction. These examples seem to imply that certain centrifugal impulses radiate from specific Pre-Cambrian areas. We might describe them as dynamic centers of tectonic activity (this is what Leuchs and Bailey Willis called them). Nevertheless, there can be no question of a "law" in this case, for it should at once be noted that such dynamic waves have never emanated from other Pre-Cambrian areas — neither from India, nor from Africa (with the possible exception of its extreme southern part), nor from the northern, western or southern part of Australia, nor from the eastern part of South America.

Leuchs therefore quite rightly described the block of India as a passive element in the structure of the Asiatic continent. It acted as an obstruction, against which Variscian, and subsequent Tertiary tectonic waves were shattered, and was responsible for the narrowing of the Tertiary zone of folding and the particularly intensive overthrusting and elevation of the Himalayas.

However, it would be a mistake to suppose that the formation and migration of various zones of folding were due exclusively to the mysterious activity of these "dynamic centers". For before passing on to an examination of their remarkable sequence through space and time, another equally important circumstance should be noted, viz. the world-wide occurrence of epochs of compression. Data have frequently shown that simultaneous movements occurred on all continents. One phase was of course of greater importance in one place than another, manifesting itself more intensely in one area than elsewhere. But the growing amount of stratigraphic studies make it increasingly evident that the terrestrial crust was subjected to a periodically alternating increase and decrease of compression. This concurrence between epochs of folding is not only found in geosynclines, but also in other areas beyond them — in basins and regions where folding had already occurred at a previous date, and where less accentuated movements accompanied the folding of the geosyncline (shifting of blocks, faulting, sliding, and at times even folding). I feel there is overwhelming evidence that the movements are the expression of a common, world-wide active and deep-seated cause.

The term "world-wide" should not be interpreted to mean that all phases can be observed everywhere, or that these movements occur in all areas. On the contrary, the special constellation of the basement may cause two geosynclines to form side by side, one of which will be folded to an intense degree during a given epoch of compression, while the other will be hardly influenced by it. A striking illustration of this is found in the Cordilleras along the west-coast of North America (fig. 23). The following are found side by side in this area: (1) the geosyncline of the Rocky Mountains, which was formed in the Pre-Cambrian and influenced by Laramide folding; (2) the geosyncline of the Sierra Nevada (Triassic and Jurassic, Nevadian folding); and (3) a geosyncline extending further west on a basement which was folded during the Nevadian revolution: the geosyncline of the Coast Ranges (Cretaceous and Lower Tertiary, Miocene folding).

The same phenomenon can be observed in the Paleozoic Wichita geosyncline of Central-America. The Wichitas and Criner Hills, both of which are situated in this belt, were influenced during the Wichita and the Arbuckle epoch, but the Ardmore basin in the immediate neighbourhood of the geosyncline, including the Arbuckle chains, were only affected by the latter.

A certain similarity is apparent in the linear shift of specific epochs of folding in mountain-chains, e.g. in the Dobrutcha, Crimea, Caucasus; the Donets and Ammodetic Mountains; and the Caledonides and Variscides on either side of the Atlantic Ocean.

These general aspects will help us with the further examination of the "dynamic centers". In which direction does that dynamic activity displace itself? The investigator would fain detect some regularity in its migration through space and time, some law, which at times we imagine can be formulated, though later it transpires that nature is more complicated than we supposed it to be. The sequence of the folded systems seems quite regular in Asia and the eastern part of Australia, but a far less simple state of affairs prevails in Europe. The Caledonides have been preserved along the north-western margin of the Baltic Shield, and perhaps originally bent around in a south-easterly direction. The Variscides, on the other hand, were not only partly formed in a Caledonian basement (fig. 21), but the front of these folded chains intersects the trend of the Caledonides in Ireland and England (Pl. 5). This is also true of the younger folded chains in North America. The latter intersect the trend of the Variscian Marathon-Ouachita Mountains, and these in turn cross the Wichita-Arbuckle geosyncline.

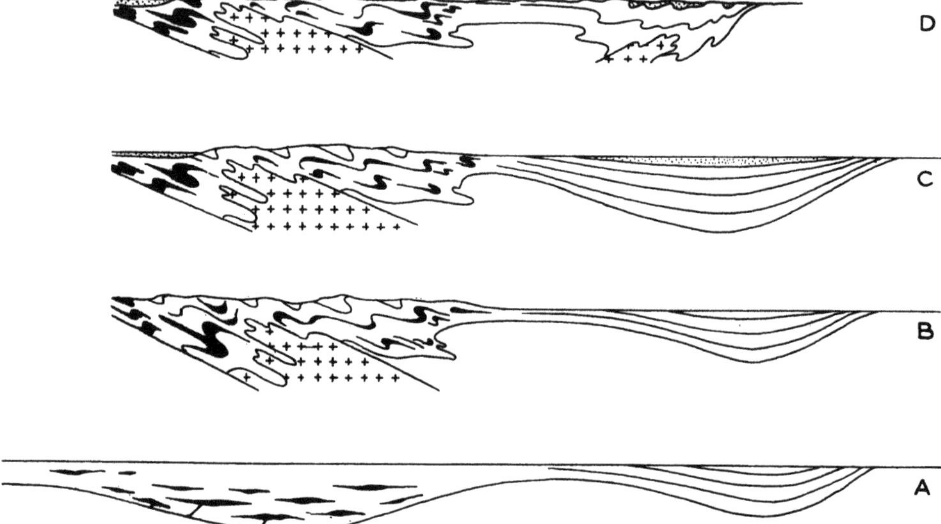

Fig. 23. Structural evolution of Rocky Mountains, Sierra Nevada and Coast Ranges. (After H. Stille). A, Nevadian geosyncline, showing basic volcanic rocks, and geosyncline of Rocky Mountains in Jurassic times. B, Nevadian (= Upper Cimmerian) folding and intrusion of batholiths in Nevadian geosyncline. C, continuation of subsidence in Rockies geosyncline during Cretaceous time. D, Laramide folding of the Rocky Mountains and subsidence of Coast Ranges geosyncline.

Besides, the idea of centrifugal migration breaks down entirely in America. For the younger Rocky Mountains extend on the eastern side of the older Sierra Nevada, along the Canadian Shield, and the even more recent formations of the Coast Ranges emerge on the other side of the Sierra Nevada. Such basins as Pugget Trough and California Valley formed again on the Canadian side of the Coast Ranges. These examples make it quite clear that the centrifugal displacement of zones of folding confines itself to definite areas. Even in those regions where a centrifugal migration of zones occurs in the major systems, the following phenomenon can be observed (e.g. in the Appalachians, and the European Variscides): a young trough of sedimentation originates on the continental side and is added as a last element to the folded chain. Moreover, intramontane troughs originate almost simultaneously. These phenomena will be discussed at greater length in the next chapter, in which attention will likewise be paid to the numerous basins and trough-shaped depressions on the continents.

Mountain-building

Not all epochs of compression observed in the structure of a mountain are followed by mountain-building elevation of any importance. The Eocene "flysch" of the Alps furnishes ample proof that some folded chains were raised and subjected to the influence of intensive denudation after an Upper Cretaceous phase. The Miocene "molasse" was the product of the sub-aerial denudation of ranges which had been elevated following an Oligocene epoch. However, intensive compression still persisted subsequently, during the Pliocene, and led to the severe overthrusting of the Helvetian nappes (fig. 24). Nevertheless, it was not until the close of the

Fig. 24. Cretaceous strata of the Säntis-"nappe" which was overthrust on Tertiary "molasse", near Kräzerli, northern Swiss Alps. The foreground is covered by moraine.

Pliocene that the Alps finally rose as a huge and lofty range of mountains. Similar conditions prevail in other Alpine mountains and older folded chains [1]).

Besides, many examples show that the process of elevation may subsequently be repeated in a folded and peneplained system. This second elevation, or "rejuvenating" movement, of the already considerably

[1]) A few examples will illustrate this. Though several epochs are known to occur in the Caledonian belt of Spitsbergen, Scandinavia and Scotland, mountain-building — i.e. a considerable rise above sea-level accompanied by erosion — is known to have taken place towards the close of the Silurian, succeeding the Ardennian phase in Spitsbergen and probably

peneplained older chains, which continues from the late Tertiary up to our own day, has brought about the present aspect of the Rocky Mountains, including that of the Sierra Nevada of North America and the South American Andes, and must be held responsible for the disparity between the deeply eroded structures of the latter, with their many intrusions of batholiths, and those of other mountains, in which the last principal epoch of folding and subsequent elevation was of a more recent date (cf. the Alps) [1].

All available data show that the largest expanse of mountain-chains since Cambrian time was occupied by the Variscides (Pl. 2), especially towards the Upper Carboniferous (post-Sudetic and pre-Mid Permian). The second largest area has been occupied by the Alpine system since the Pleistocene (Pl. 4). The Caledonides (Taconic and Silurian), extending over large tracts, especially in the continents of the northern hemisphere (Pl. 1), were less extensive than the Alpine belts. The Mesozoic Mountains covered the smallest areas, and surround particularly the Pacific (Pl. 3). The Variscides and Alpine chains also show a circum-Pacific arrangement, but these belts likewise occupied large areas in an E. W. direction, along the southern side of the northern continents. Mountains have a very strong influence on atmospheric circulation, and this influence must therefore have been exceptionally severe in the Upper Carboniferous and Pleistocene (this is illustrated by the schematic graph in Table II).

Small arrows in Plates 1, 2 and 4 indicate the possible trend of the Caledonides, Variscides, and Alpine systems, whose chains are intersected

---

the Ardennian or Erian in England and Scandinavia. The same occurred in the North American Caledonides after the Taconic revolution. The most important period of mountain-building in the central part of the European Variscides was posterior to the Sudetic epoch. In the northern, coal-bearing sector of Europe, however, and in Asturia, the most important mountain-building followed the Asturian phase, and an even later epoch — namely the Saalian, — in the Pyrenees and the Ural Mountains. The period of Variscian mountain-building varies in different parts of America. The Arbuckle-Wichita geosyncline was elevated after the Arbuckle period of compression and had already been peneplained to a considerable extent before being overrun by the Ouachitas during the Permian (Appalachian epoch). A large part of the Appalachians had already become an area of erosion after the Bretonic epoch, though the marginal deep was folded and overthrust only during the Appalachian phase. In South America the final folding and subsequent elevation occurred towards the close of the Paleozoic. In Asia, Variscian folding around the Angara Shield took place in extensive areas at a relatively early date (Bretonic and Sudetic), and large mountains must consequently already have existed as early as the Upper Carboniferous. On the other hand, the Variscian zones in Taimyr, the Ural and the southern and south-eastern sector of Asia appear to have originated as late as the Permian. The Australian Variscides, too, only originated as a large unite during the Permian, though one zone in the north-east may have formed at an earlier date. The final epoch of the formation of the Australian Caledonides dates from the Upper Devonian. The Nevadian folding of the geosyncline of the Sierra Nevada was more than likely succeeded by mountain-building, so that detritus was transported from this area both eastward and westward, where it aided the sedimentation in the geosyncline of the Rocky Mountains. Opinions differ as regards the previous existence of large Lower Cretaceous mountains. Crickmay was particularly doubtful in this respect. (Waters and Hedberg p. 24). It can probably also be assumed that no mountains were formed in the Rocky Mountains after the Laramide epoch. Hedberg and Waters even speak of a "low swampy plain across which the rivers from the down roots of the Nevadian Mountains wandered and laid down extensive flood deposits".

[1] I cannot, therefore, endorse Gerth's view that the characteristic features of the South American Cordilleras can only be explained by a westward drift of the continent.

by the coasts of the Atlantic. This question will be discussed extensively in the Appendix and will moreover be dealt with in Chapter VIII, which will include an examination of the problems of the oceanic sectors.

Subsided blocks

The presence of detritus in some folded chains can only be explained if the previous existence is assumed of areas of denudation which have since foundered deeply below sea-level. The most important submerged areas in maps 1—4 appear east of the Appalachians ("Appalachia"), west of the Sierra Nevada ("Cascadia"), west of Spitsbergen and Scotland (the "North Atlantic Continent" or "Scandia"), and, finally, south-east of Asia and east of Australia ("Melanesia"). These will now be treated accordingly.

Boesch is of the opinion that the coastal area of Nova Scotia represents the only visible portion of the hinterland of the Appalachian geosyncline which is commonly known as Appalachia or Nova Scotica (Pl. 2). It is impossible to estimate the size of this hinterland, but there certainly must have existed a mountainous area large enough to supply the Appalachian geosyncline with sediments. For not only was the foreland (consisting of the Laurentian Plateau, an extension of the Canadian Shield) covered by the sea, but Barrell was able to reconstruct an enormous Devonian delta in the Appalachians, and its structure shows that all the material came from the east (fig. 25). The very valuable seismic observations of Ewing and his collaborators reveal that the eastern part of Appalachia now lies buried from 3,000 to 4,000 meters below the shelf (fig. 26). This area would seem to have foundered more as a result of warping than of faulting. An important part of the shelf-sediments probably consist of Triassic and Jurassic strata. The region of the Appalachian Mountains can therefore probably be said to have remained above sea-level during these periods, and to have supplied the eastern area, which was then subsiding, with sediments. These deposits, together with Cretaceous and Tertiary strata, built up a considerable part of the shelf. If "Appalachia" can be assumed to have extended some 200 miles beyond the present coast (a conclusion arrived at both by Barrell and Schuchert), thick Paleozoic sediments — the erosion products of Appalachia —ought at any rate to lie burried in the Atlantic east of the shelf.

Some geologists believe that a land-area south of the Ouachitas supplied the homonymous geosyncline with detritus in the same way as Appalachia had formerly supplied the Appalachian geosyncline with erosion products. This borderland was called Llanoria (Pl. 2). Van Waterschoot van der Gracht regards the Ouachitas as the mere frontal ranges of a far more important and larger system, most of which would at present lie beneath the Gulf Coast basin.

In the opinion of Stille a "borderland" generally has to be considered as part of a geosyncline that was folded during a previous epoch of compression and subsequently became a geanticlinal ridge.

The geosyncline and folded chains of the Sierra Nevada should not be regarded as extending along the edge of the Canadian Shield, as these formations verge upon a borderland known as Cascadia (Pl. 3). The existence of a former area in the west was also deduced by Brock from the facies of Paleozoic and Mesozoic sediments in Canada. That such a mountainous area

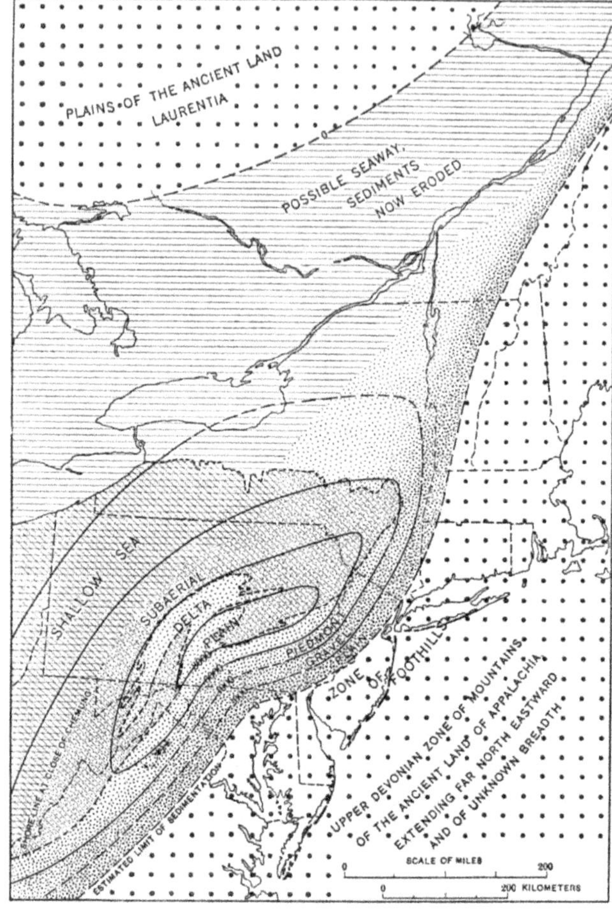

Fig. 25. Reconstruction of a Devonian delta of the Appalachian geosyncline (after J. Barrell). The contours show original thickness of sediments. The diagonal lines indicate the area where the deposits still occur. The small dots represent fresh-water and brackish-water deposits; the horizontal lines marine strata; and the large dots areas of denudation.

had existed in the west during the Mesozoic was a logical consequence, according to Born, of observations in this area, for the Mesozoic sediments of the geosyncline from which the Coast Ranges and the Sierra Nevada subsequently emerged, could not possibly have come from the east, since a wide zone, extending as far as — and including — the Rocky Mountains lay beneath the sea at that time. It seems highly improbable that the geosyncline of the Sierra Nevada would have been supplied with sediments from an area as far distant as the Canadian Shield, especially since the sediments

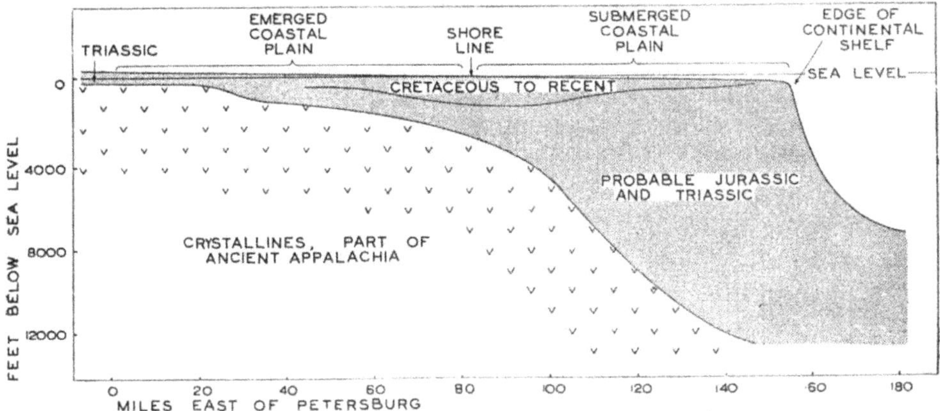

Fig. 26. Section across the Atlantic coastal plain of North America from Petersburg up to 180 miles East of this locality. Based on seismic refraction measurements of Ewing, Crary and Rutherford. (After B. L. Miller).

had to be transported over a submarine ridge between the Nevadian and the Rockies' geosynclines (see fig. 23).

Holtedahl pointed out that the Devonian and Tertiary sediments of Spitsbergen originated in an area further west, known as Scandia (Pl. 1). The latter must have vanished into the Atlantic at a time when the Tertiary strata of Spitsbergen were being folded. The Mesozoic strata provide another indication of the presence of an area of denudation in the west, according to Frebold, and both authors agree that this land cannot be identified as East Greenland, nor that it could have been connected with the same [1]). The only remaining part of Scandia, or the North Atlantic Continent, as it is sometimes called, may perhaps be the Pre-Cambrian area of the Hebrides and North-West Scotland, against which the Scottish Caledonides were overthrust (see fig. 17, 20 and Pl. 5). But the existence of a Pre-Cambrian block west of Spitsbergen and extending as far as Scotland need not imply that it occupied the whole North Atlantic [2]).

That a former mountainous area existed along the West Coast of Africa can be deduced from the distribution of the sediments in the Congo basin, which grow steadily coarser towards the west [3]).

The Melanesian Islands represent mere fragments of an extensive land-area The presence of Mesozoic sediments in New Caledonia can only mean that important areas of erosion, all of which have since disappeared, were present in the immediate neighbourhood. Besides, it might be asked whether it would be possible to indicate at present some area of denudation which could

1) Frebold, 1935, p. 177.
2) See e.g. Holtedahl, 1920, p. 18, fig. 12.
3) Veatch, 1935, p. 57, 156.

formerly have supplied the Tertiary sediments of Viti Levu with erosion products. The same applies to all those cases in which large quantities of pre-Tertiary and Tertairy detritus are found on small islands. We find it difficult to picture the formation of these rocks under the prevailing distribution of land and sea. These facts, including other data, (such as petrographical) have lead to the wide-spread assumption that the above area should be regarded as a continent, or, in other words, as a sialic block, large parts of which have subsided to a considerable depth. Geologists often refer to it as the Melanesian Continent (Pl. 4).

Fig. 27 shows the eastern and southern boundaries of Melanesia. The sub-

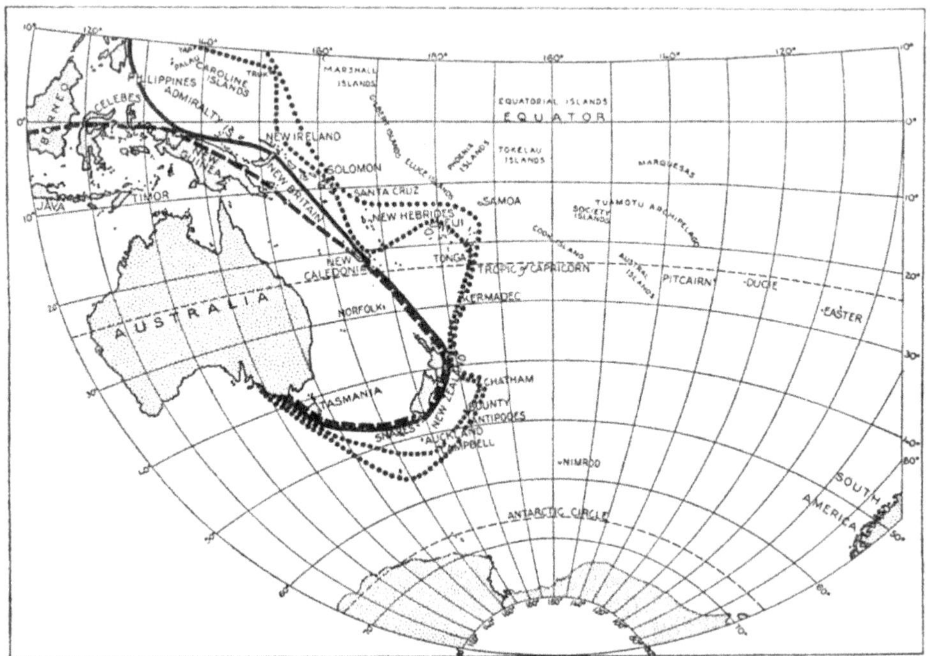

Fig. 27. Melanesia and the Pacific basin (From H. Ladd): ... probable boundary of former Melanesian continent; – – – outlines area showing metamorphic or plutonic rocks; — — — outlines area within Paleozoic rocks are present; ——— outlines area showing Mesozoic rocks.

marine relief of this block is extremely irregular at present. Many islands and submarine ridges show a linear or arcuate arrangement. Several authors believe that the preceding characteristics, including the long deep-sea troughs, prove that important crustal folding has also affected the submerged part of Melanesia. The deepest deep-sea trough (with a depth of more than 10,000 meters east of the Philippines) is found in this area, though even in this case "with islands containing rocks of definitely continental types such as granite, syenite, serpentine and schists on both sides of it". Faulting probably constituted another important feature of the foundering, or submersion, of large parts of the Melanesian Continent. Many geologists in distant parts of this region arrived independently at a similar conclusion [1]. Chubb [2]) describes the region as "an area of continental land, that was gradually folded, fracture and submerged in Mesozoic and Tertiary times".

The exact period of the "foundering" has not been ascertained in all cases but there are indications that in one area — in the south-eastern marginal zone — important faulting occurred during the Upper Tertiary, while in Viti Levu, to mention but one example, major movements were observed along the faults towards the close of the Tertiary, and these are thought to have accompanied the foundering of large parts of Melanesia [3].

The south-eastern boundary of Melanesia consists of a linear belt of volcanoes, and it is highly probable that the accumulation of these magmatic products is closely associated with a deep fault plane. This belt can be followed from New-Zealand to the neighbourhood of Samoa, and constitutes an important line of demarcation of volcanic rocks, known as the andesite-line. The subsided continent, as compared to the remaining Pacific sector east of Melanesia. is now known as an andesite zone from a petrographical point of view and will be discussed more fully in Chapters VII and VIII.

We will confine ourselves to these examples. A few less evident cases (e.g. Choco and Burckhardtland along the west coast of South America) will be discussed in the Appendix. And in Chapter VI the problem of subsiding borderlands will come up for discussion in connection with other phenomena of the continental margin.

It is difficult to say what influence was responsible for the subsidence of the blocks, but in a few areas the periods in which subsidence began, indicate that it can probably be associated with events in neighbouring geosynclinal zones. Appalachia began in fact to subside during the Triassic, i.e. after the mountain-building of the neighbouring Appalachian geosyncline. Cascadia subsided after the Upper Jurassic folding of the Sierra Nevada. The submersion of Melanesia seems to be related to the Laramide and younger periods of disturbance in south-eastern Asia. Scandia probably subsided during the Upper Tertiary, and contemporaneous folding (including faulting accompanied by volcanism) is known to occur in Spitsbergen.

References

Publications of general importance.

BORN, A. *Der geologische Aufbau der Erde* (Handb. d. Geophysik, Bd. II, 1932).

BORN, A. *Über Werden und Zerfall von Kontinentalschollen* (Fortschritte der Geologie und Palaeontologie, Bd. X, 1933).

BUCHER, W. H. *The deformation of the Earth's Crust* (Princeton University Press, 1933).

KOSSMAT, FR. *Paläogeographie und Tektonik* (1936).

SCHAFFER, F. X. *Lehrbuch der Geologie* (III, Geologische Länderkunde, Regionale Geologie 1941).

STILLE, H. *Grundfragen der Vergleichenden Tektonik* (1924).

STILLE, H. *Bemerkungen betreffend die „sardische" Faltung* (Zeitschr. Deutsch. Geol. Gesellsch. 91, 1939).

SUESS, Ed. *The face of the Earth* (*Das Antlitz der Erde*), transl. Sollas, vol. I—IV, 1904–1909).

---

[1]) Ladd, 1934, p. 51; see also Bryan, 1944.

[2]) Chubb, 1934, p. 289. See also Schuchert 1926, p. 101.

[3]) Ladd, 1934, p. 51. Bryan speaks of the dismemberment of the Australasian continent since early Tertiary times.

Europe, Spitsbergen and Greenland.
BAILEY, E. B. and HOLTEDAHL, O. *Northwest Europe, Caledonides* (Reg. Geol. d. Erde, Bd. 2. II, 1938).
BUBNOFF, S. VON, *Geologie von Europa* (Geol. d. Erde 1, 2, 1926–1936).
CADISCH, *Das werden der Alpen im Spiegel der Vorlandssedimentation* (Geol. Rundschau 19, 1928).
COLLET, L. W. *The structure of the Alps* (Arnold, London, 1927).
CORNELIUS, H. P. *Zur Vorgeschichte der Alpenfaltung* (Geol. Rundschau 16, 1925).
GROENLAND 1939. *Tagung der Naturf. Gesellsch. Schaffhausen* (XVI, 1940).
FREBOLD, H. *Geologie von Spitsbergen, Bäreninsel, etc.* (Geol. d. Erde, 1935).
HOLTEDAHL, O. *Paleogeography and diastrophism in the Atlantic Arctic Region during Paleozoic time* (Americ. Journ. Sci., 49, 1920).
JONES, O. T. *On the evolution of a geosyncline* (Quart. Journ. Geol. Soc. London, 94, 1938).
OWIN, A. K. *Outline of the Geological History of Spitsbergen* (Skrifter om Svalbard og Ishavet no. 78, Oslo 1940).
SEYDLITZ, W. v. *Diskordanz und Orogenese der Gebirge am Mittelmeer* (1931).
SCHWINNER, R. *Mikroseismische Bodenunruhe und Gebirgsbau* (Zeitschr. f. Geophysik, 9, 1933).
TEICHERT, C. *Geology of Greenland* (Geology of North America, Geol. d. Erde, I, 1939).
WATERSCHOOT VAN DER GRACHT, W. A. J. M. VAN, *A structural outline of the Variscian front and its foreland from South Central England to Eastern Westphalia and Hessen* (C. R. 2e Congr. Stratigr. Carb. Heerlen 1938).
WATERSCHOOT VAN DER GRACHT, W. A. J. M. VAN, *The Paleozoic geography and environment in Northwestern Europe as compared to North America* (Ibidem p. 1387–1429).

North- and Central-America.
BARRELL, J. *Upper Devonian delta of the Appalachian geosyncline* (Americ. Journal of Sci. 27, 1914).
BOESCH, H. H. *Zur Geologie des östlichen Nordamerika* (Eclog. Geol. Helvet. vol. 32, 1939).
BROCK, R. W. *Structure of the pacific region of Canada* (Transact. P. Pacific Sci. Congr. Australia, 1923, vol. 1).
EWING, M.; CRARY, A. P., RUTHERFORD, H. N. and MILLER, B. L. *Geophysical investigations in the emerged and submerged atlantic coastal plain* (Bull. Geol. Soc. America, vol. 48, 1937).
GERTH, H. *Die Fortsetzung der Venezuelanische Kordilleren in den Antillenbogen* (Geol. Rundschau 31, 1940).
KEIJZER, F. G. *Outline of the Geology of the Eastern part of the province of Oriente, Cuba, with notes on the Geology of other parts of the Island* (Acad. Thesis, Utrecht, 1945). With references to other recent publications on the geology of Cuba.
KING, P. B. *An outline of the Structural Geology of the United States* (Guide Book 28, XVI Int. Geol. Congres 1933).
LAFFERTY, R. C. *Central basin of Appalachian geosyncline* (Bull. Am. Assoc. Petrol. Geol. 25, 49).
LIVINGSTON, W. *Observations on the structure of Bermuda* (The Geogr. Journ. 104, 1944).
LONGWELL, CH. R. e.a., *Tectonic map of the United States* (1944).
REED, R. D. and HOLLISTER, J. S. *Structural History of California* (Bull. Americ. Assoc. of Petrol. Geol. 1936).
RUEDEMANN, R. and BALK, R. *Geology of North America* (Geol. d. Erde, vol. I, 1939).
RUTTEN, L. *Oude Land- en Zeeverbindingen in Midden Amerika en West Indië* (Tijdschr. Kon. Ned. Aardrijksk. Genootsch. 1934).
RUTTEN, L. *Geology of Isla de Pinos* (Proceed. Kon. Akad. Wetensch. Amsterdam, 37, 1934).
RUTTEN, L. *Über den Antillenbogen* (Proc. Kon. Acad. v. Wet. Amsterdam 38, 1935).
SAPPER, K. *Mittelamerika* (Handb. Reg. Geol. 8, 1937).
SENN, A. *Paleogene of Barbados and its bearing on history and structure of the Antillean Caribbean region* (Bull. Americ. Assoc. of Petrol. Geol. 1940).
SCHUCHERT CH. *Historical Geology of Antillen-Caribbean Region* (1935). *Historical geology of North America* vol. II, III.
STILLE, H. *Einführung in den Bau Amerikas* (1940).
WATERS, A. C., and HEDBERG, H. D. *The North American Cordillera and the Caribbean Region* (Reg. Geol. d. Erde 3, IVa, 1939).
WATERSCHOOT VAN DER GRACHT, W. A. J. M. VAN, *The Permo-Carboniferous Orogeny in the South Central United States* (Verh. Kon. Acad. v. Wet. Amsterdam 27, 1931).
WEYL, R. *Zum Bau des Antillenbogens* (Zentrallbl. Miner. etc. Abt. B, 1940).

South-America and Antarctica.
BAKER, H. A. *Final Report on geological investigations in the Falkland Islands* (1923).

GERTH, H. *Geologie Südamerikas* (Geol. d. Erde I, 1932, II, 1935, III, 1941).
GERTH, H. *Die Kordilleren von Südamerika* (Reg. Geol. d. Erde, Bd. 3, IVb, 1939).
NORDENSKJÖLD, O. *Antarctica* (Handb. Reg. Geol. 1913).
OLIVEIRA, A. J. DE, LEONARDO, O. H. *Geologia de Brassil* (Rio de Janeiro, 1943).
STEINMANN, G. *Zum Bau der östlichen Pazifik* (Geol. Rundschau 20, 1929).
STILLE, H. *Die Entwicklung des Amerikanischen Kordilleren-systems* (Sitzungsber. Preuss. Akad. Wiss. XV, 1936).
STILLE, H. *Einführung in den Bau Amerikas* (1940).
TAYLOR, G. *Antarctica* (Reg. Geol. d. Erde 1, VIII, 1940).
WADE, A. *The Geology of the Antarctic Continent* (Proc. R. Soc. Queensland, 52, 1941).
WILCKENS, O. *Der Sudantillenbogen* (Geol. Rundschau 24, 1933).

Africa and Arabia.
Carte géologique internationale de la Terre. (publ. Preuss. geol. Landesanstalt, Feuille 62, 63, 68, 69, Afrique).
DU TOIT, A. *Geology of South Africa* (Sec. ed. 1939).
HENNIG, E. *Afrika* (Reg. Geol. d. Erde, IV, 1938).
HEYBROEK, F. *La Géologie d'une partie du Liban Sud* (Leidsche Geol. Mededeelingen 12, 1942).
KRENKEL, E. *Geologie Afrikas* (Geol. d. Erde, 1–3, 1925–1938).
LAMARE, P. *Structure géologique de l'Arabie* (Ed. Béranger, Paris, 1936).
ROBERT, M. *Le Congo physique* (2° Ed. 1942).

Asia.
ANDREWS, C. W. *A description of Christmas Island, Indian Ocean* (Geogr. Journ., 13, 1899).
ARGAND, E. *La tectonique de l'Asie* (C. R. XIIIe Congr. Geol. Intern. 1922).
ARNI, P. *Relations entre la structure régionale et les gisements minéraux et pétrolifères d'Anatolie* (Bull. de l'Inst. et de Recherches Minières de Turquie 2, 1939).
BAIER, E. *Ein Beitrag zum Thema Zwischengebirge* (Zentralbl. f. Miner. Abt. B. 1938).
BAIER, E. *Das Iranische Binnenland* (N. Jahrb. f. Min. Abt. B. 1940).
BLANCKENHORN, M. *Syriën, Arabiën und Mesopotamiën* (Handb. Reg. Geol. 1914).
BROUWER, H. A. *The Geology of the Netherlands East Indies* (1925).
BROUWER, H. A. *Geological expedition to the Lesser Sunda Islands* (I, II, 1940; III, 1941; IV, 1942).
CHHIBBER, H. L. *The Geology of Burma* (1934).
COTTER, G. DE P. *The Indian Peninsula and Ceylon* (Reg. Geol. d. Erde I, VI, 1938).
FURON, R. *Géologie du Plateau Iranien* (Mém. Museum Nat. Hist. Nat. 1941).
Geological atlas of Eastern Asia (Tokyo Geogr. Soc. 1929).
Geological atlas of the Malay archipelago (Tokyo Geogr. Soc. 1932).
Geological map of the Union of Soviet Soc. Rep. 1 : 5.000.000 (1937).
GREGORY, J. W. *The structure of Asia* (1929).
HANZAWA, S. *Topography and Geology of the Riukiu Islands* (Sci. Rep. Tohoku Imp. Univ. 17, 1933).
HANZAWA, S. *Geolog. History of the Riukiu Islands* (Proc. Imp. University 11, 1935).
HAYASAKA, I. and TAKAHASHI, H. *An outline of the Geology and mineral resources of Taiwan* (1929).
HEIM, A. and GANSSER, A. *Central Himalaya* (Geolog. Observ. Denkschr. d. Schweiz. Naturf. Ges. 73, 1, 1939).
HEIM, A. and HIRSCHI, H. A. *Section of the Mountain Ranges of North-Western Siam* (Eclog. Geol. Helvetiae 32, 1939).
KANEHARA, N. *The Geology and mineral resources of the Japanese Empire* (1926).
KOBAYASHI, A. *A sketch of Korean Geology* (Americ. Journ. Sci. 26, 1933).
LEE, J. S. *The Geology of China* (1937).
LEUCHS, K. *Geologie von Asien* (I, 1, 1935, I, 2, 1937).
LEUCHS, K. *Geologische Entwicklung von Anatoliën* (Leipziger Vierteljahrschr. für Südosteuropa 2–2, 1938).
OBRUTSCHEW, W. A. *Geologie von Siberiën* (Fortschr. d. Geol. und Pal. 15, 1926).
OSWALD, F. *Armenien* (Handb. Reg. Geol. 1912).
OZAWA, Y. *The post-paleozoic and late mesozoic earth movements in the inner zone of Japan* (Koto Commemoration volume 1925).
PASCOE, F. H. *Geological map of India and adjacent countries* (5th. ed. 1931).
PHILIPPSON, K. *Kleinasien* (Hand. Reg. Geol. 1918).
RUTTEN, L. M. R. *Voordrachten over de Geologie van Nederlandsch Oost Indie* (1937).
SCHROEDER, J. W. *Essai sur la structure de l'Iran* (Eclog. Geol. Helvetiae, 37, 1944).

Scrivenor, J. B. *The Geology of Malaya* (1931).
Stahl, A. F. von, *Persien* (Handb. Reg. Geol. 1915).
Stahl, A. F. von, *Kaukasus* (Handb. Reg. Geol. 1923).
Umbgrove, J. H. F. *Geological History of the East Indies* (Bull. Americ. Assoc. Petrol. Geol. 22, 1938).
Weller, J. M. *Outline of Chinese Geology* (Bull. Am. Assoc. Petrol. Geol., 28, 1944).
Willis, B. *Geological observations in the Philippine Archipelago* (Bull. Nat. Res. Council of the Philippine Islands, 13, 1937).
Yabe, H. *Larger geotectonics of the island arc of Japan* (Proc. Imp. Acad. Japan 5, 1929).
Yabe, H., and Hanzawa, S. *Geological History of the Island Taiwan* (Proc. Imp. Akad. Tokyo, 6, 1930).

Australia, New Zealand and Melanesia.

Andrews, E. C. *The structural History of Australia during the Paleozoic* (Journ. and Proc. of the Royal Soc. of New S. Wales 71, 1938).
Benson, W. N. *The structural features of the margin of Australasia* (Transact. N. Zealand Inst. 55, 1924).
Bryan, W. H. *The relationship of the Australian continent to the Pacific Ocean, now and in the past* (Journ. and Proceed. of the R. Soc. of New S. Wales 78, 1944).
Chubb, L. J. *The structure of the Pacific Basin* (Geol. Magaz. 71, 1934).
David, T. W. E. *Explanatory notes to accompany a new geological map of Australia* (1932).
Hoffmeister, J. E. *Geology of Eua, Tonga* (Bernice P. Bishop Museum, Bull. 96, 1932).
Hoffmeister, J. E. and Ladd, H. *Geology of Lau, Fiji.* (Bernice P. Bishop Museum, Bull. 181, 1945).
Ladd, H. *Geology of Viti Levu* (Bernice P. Bishop Museum, Bull. 119, 1934).
Marshall, P. *Oceania* (Handb. Reg. Geol. VII, 2, 1911).
Marshall, P. *Stability of lands in the S. W. Pacific* (Rep. of the Australian and New Zealand Assoc. for the Advanc. of Sci. 21, 1933).
Mawson, D. *The Geology of the New Hebrides* (Proc. Linn. Soc. Nw. S. Wales 30, 1905).
Mawson, D. *Marcquarie Island; Its Geography and Geology* (Scient. Rep. Australian Antarctic Exp. 1911–1914, Sec. A. vol. 5, 1943).
Schuchert, Ch. *The problem of continental fracturing and diastrophism in Oceanica* (Americ. Journ. Sci. 47, 1916).
Stearns, H. T. *Geology of the Samoan Islands* (Bull. Geol. Soc. America, 55, 1944).
Wilckens, O. *Stratigraphie und Bau von Neu Kaledonien* (Geol. Rundschau 16, 1925).

CHAPTER III

BASINS AND TROUGHS

"The breaking up of the terrestrial globe, this it is we witness".
(ED. SUESS)

Introduction

Folded mountain-chains — those majestic features of the earth — have always attracted both the tourist and the geologist. There is another feature however, which, though less stimulating and even monotonous, can be said to be of primary importance in the earth's aspect. We mean such formations as troughs and basin-shaped depressions, containing more or less folded strata or even sediments which have remained almost undisturbed. As soon as we begin to examine these structures more closely, it transpires that they present heterogeneous characteristics, revealing different shapes, sizes, structures, depths, tectonic frames, periods of formation, and a varying distribution of facies and duration of development. In studying the earth's structural history, special attention will have to be paid — as in the case of the folded chains discussed in Chapter II — to the location and time of origin of the basins, without losing sight, however, of those other diverse characteristics which have just been mentioned.

The shape of these formations ranges from long troughs and elliptical depressions to practically circular basins. The thickness of the accumulated sediments varies between a few hundred meters and several thousands in often uniform shallow facies. Some basins are consequently not unlike geosynclines both as regards their form and the distribution of the facies of the sediments. It is in fact difficult to draw a sharp line between basins, troughs and geosynclines, or to formulate some definition which would distinguish clearly between them as separate physiographic units. The extremes are different enough; but the intermediate types cause difficulties. Thus we find one author calling a certain type of formation a geosyncline, while another refers to it as a basin [1]). Basins and troughs are often bounded by faults and

1) In 1923 Schuchert introduced a classification of geosynclines which he divided into mono-, para-, poly- and mesogeosynclines. At a later date Stille used the names ortho-geosyncline and parageosyncline the latter name, however, being a homonym. The orthogeosyncline is divided in a central eugeosyncline and a marginal strip named miogeosyncline. A new classification of geosynclines was proposed by Marshall Kay. In a preliminary paper of the year 1944 he introduced the names mio-, eu-, delta-, auto-, and taphrogeosynclines and from a correspondence with Professor Kay I learned that three more American types, each with a separate name, are in the making. The aim of the segregation of so many specific

overthrusts, and this circumstance makes it difficult to distinguish even between a trough and a graben. To make things still more complicated, names such as "stable and unstable shelf" have since been introduced!

We not propose to dwell upon all these different names, but will describe the features of each type of basin as it comes up for discussion and will then choose the name which seems to fit it best.

A study of continental basins and troughs has induced me to classify them into two groups, each group being subdivided as follows into two types (fig. 28):

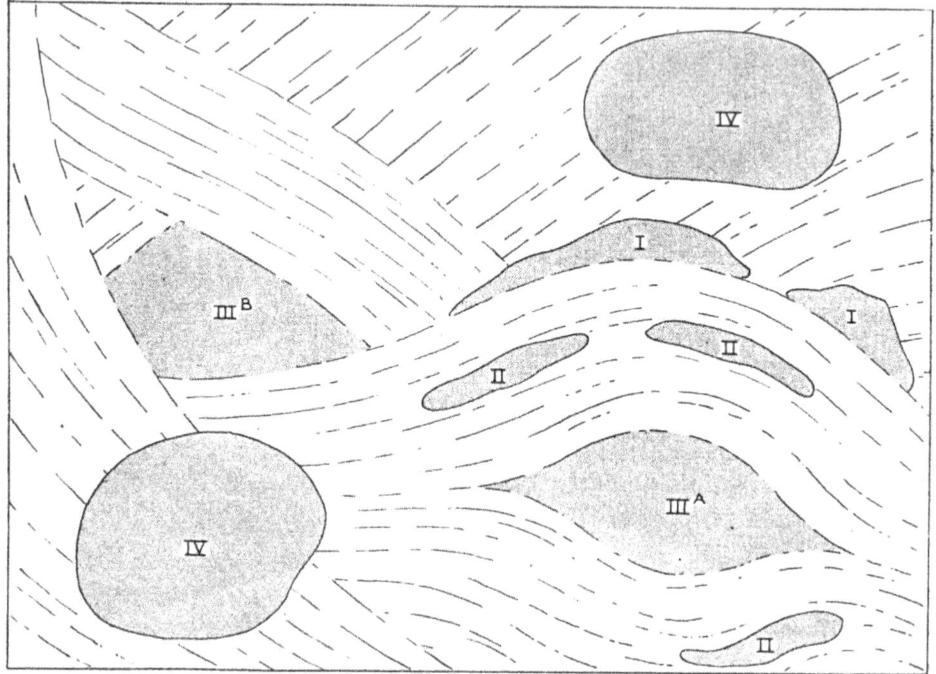

Fig. 28. Schematic review of four types of basins. The trends of folded structures are indicated by lines. I marginal deep; II intramontane trough; III nuclear basin, (A) with isochronous frame, (B) with anisochronous frame; IV discordant basin.

---

types is to sharpen description and to stimulate a more critical consideration of several tectonic, stratigraphic and magmatic characteristics of the subsiding areas and their surroundings. Obviously such a classification will partly overlap the classification used in our Chapter III, which is founded on different aspects of these structures and with a different valuation of their appearance in space and time. As an example it should be mentioned that the basin of Illinois (no. 4 on map 6) is classified as an autogeosyncline by Kay, and the Eocene Gulf Coast Basin (no. 7 on map 6) as a paralia-geosyncline. However both are classifi d as type IV, discordant basin, in the present book, though with three different notations regarding their time of origin. "Eu-geosynclines" fall out of the scope of the present chapter.

For a further discussion of different types of geosynclines the reader is referred to the Appendix, p. 342.

Group 1.

Type I	marginal deep.
Type II	intramontane trough.

Group 2.

Type III	nuclear basin.
	(a) with an isochronous frame.
	(b) with an anisochronous frame.
Type IV	discordant basin.

It is impossible to distinguish accurately between both types in each group. To the following review of continental types will be added a discussion and a classification of deep-sea basins and troughs. At the end of the chapter attention will be drawn to the chronological relations to other phenomena in the earth's structural history.

Continental basins and troughs

The extensive Variscian mountains of Western Europe emerged towards the close of the Lower Carboniferous (following the Sudetic epoch of compression). A relatively small area of sedimentation later formed along the northern margin of the chains. The latter can be traced from Ireland to Westphalia over England, and through Belgium and Limburg. This represents the subsiding trough in which the paralic and coal-bearing Upper Carboniferous was deposited, accumulating to a thickness of approximately 4,000 meters in Westphalia, and to as much as 7,000 meters in Upper Silesia. A considerable amount of detritus from the Variscian mountains gathered in this trough, which constituted a marginal deep of the European Variscides. Marine Devonian and Lower Carboniferous had already been deposited in this area previously, but the creation of the marginal deep only dates from Upper Carboniferous times. Its contents were then folded towards the close of the Upper Carboniferous, and during this same epoch (which was probably the Asturian), blocks of the older Variscian ranges in the south were overthrust towards the north as far as, and over, the sediments of the fore-deep (fig. 29). We will not enter into details at this point, nor will we describe the important role which the Brabant Massif played during these events. We refer the reader for these questions to the references at the end of this Chapter, especially to the publications of Van Waterschoot van der Gracht. The fore-deep should not be imagined to be a deep-sea basin in this case. On the contrary. The enormous supply of detritus from the south kept pace with the

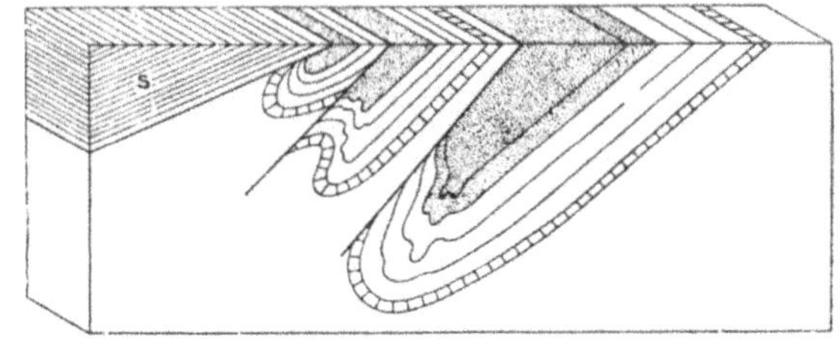

Fig. 29. Diagrammatic block of the folded and overthrust marginal trough of the European Variscides. S. Lower Paleozoic.

subsidence, producing a swampy zone which was intermittently flooded by a shallow sea. It was not, therefore, a deep in the morphological sense of the word, but in the structural or tectonic sense, as Stille formulated it. The mechanism of the subsidence and sedimentation of this Upper Carboniferous "mio-geosyncline" was the object of a special study by Pruvost.

While discussing epochs of compression, the principal phase in the Appalachians was shown to be the Bretonic (Acadian). It involved the eastern and most intensely folded and metamorphosed zone with its manifold intrusions of large batholiths. A marginal trough was formed along the wester edge of the newly elevated mountain range following this Acadian revolution. The marginal deep, which was filled with Carboniferous and Permian strata, was folded by the so-called Appalachian (Saalian) phase during the Permian. This is the classical Appalachian geosyncline of former authors, for which Schuchert introduced the name "monogeosyncline". As a tectonic unit, it formed the counterpart of the marginal deep of the European Variscides mentioned above [1]).

These two examples refer to Variscian marginal deeps. We will now give a brief summary of marginal deeps of the Alpine belt.

A Miocene trough was formed along the northern front of the Alpine chains after the intensive folding of the Oligocene. A sequence of approximately 2,500 to 3,000 meters of "molasse" was deposited into this trough, and the latter was overthrust towards the end of the Miocene (fig. 24). An identical formation — the Guadalquivir trough, which countains Lower Tertiary and Neogene sediments — can be observed in front of the Betic Cordillera. Its Oligocene "flysch" is folded; the Burdigalian and younger strata lie undisturbed. An analogous deep is the Neogene Siwalik trough, which extends along the southern margin of the Himalayas (fig. 96). On the northern side of the Alpine chains lies the depression of Karakum. The drainage basins of the Tigris and Euphrates and the Persian Gulf form the marginal deep of the Zagros and Iran chains [2]).

---

[1]) A glance at the distribution of Variscian epochs in Plate 2 will immediately make it clear that the southernmost Variscides in Asia were folded during the more recent Appalachian epoch, and that the Sudetic was the principal phase in the northern Variscides. A centrifugal displacement of epochs of folding (cf. also the same phenomenon in Australia) is what we observe here. This will be explained in the Appendix. We are as yet unable to say for certain whether we may speak in this case of a marginal deep, which would have formed on the southern side of the Asiatic Variscides, and would have been folded during the Appalachian epoch, but it looks as if we might be justified in doing so.

[2]) Von Bubnoff also regards the area between the Carpathian Mountains and the Podolian Massif, including the zone of subsidence north of the Dobrutcha, Crimea and Caucasus, as a marginal deep (Europa 1, p. 291). This same author also looks upon the South-Russian basin as a marginal deep — in this case of the Donetz ranges, namely (ibid. p. 208).

To the above must be added the huge coal basin of Kusnezk, which continues towards the South-Schenka basin. The basin of Kusnezk lies between the Caledonian Kusnezk Alatau and the Variscian Salair Mountains. It began to subside during the Lower Devonian, and 6,200 meters of neritic and paralic Upper Paleozoic strata were subsequently deposited in the basin (Obrutchew even speaks of 7,500 meters of coal-bearing sediments). The contents of the basin were influenced by rather intensive Sudetic and Upper Paleozoic folding (Leuchs 1, p. 155, fig. 48), and during this process even the Cambrian strata of the Caledonian Kusnezki — and the Variscian Salair on its other side — were overthrust over the edge of the basin. The

Recapitulating, we may say that a mariginal deep can be characterized as a subsiding trough of sedimentation, which originates along the edge of a folded chain — or eu-geosyncline 1) — shortly after folding and mountain-building occurred in this zone.

This trough faces the continental Shield in the European, North American and Asiatic Variscides, and the same applies to the Guadalquivir, Molasse and Karakum troughs of the Alpine ranges. The Mesopotamian and Siwalik troughs, however, are found on the other side of the Alpine belt, and an identical situation can be observed in the Asiatic Caledonides. We propose to give the formations a neutral name, for they can be called "foredeep" in one case, and "hinterdeep" in another, while at times both terms may apply — it all depends on the way an investigator looks at it. We shall

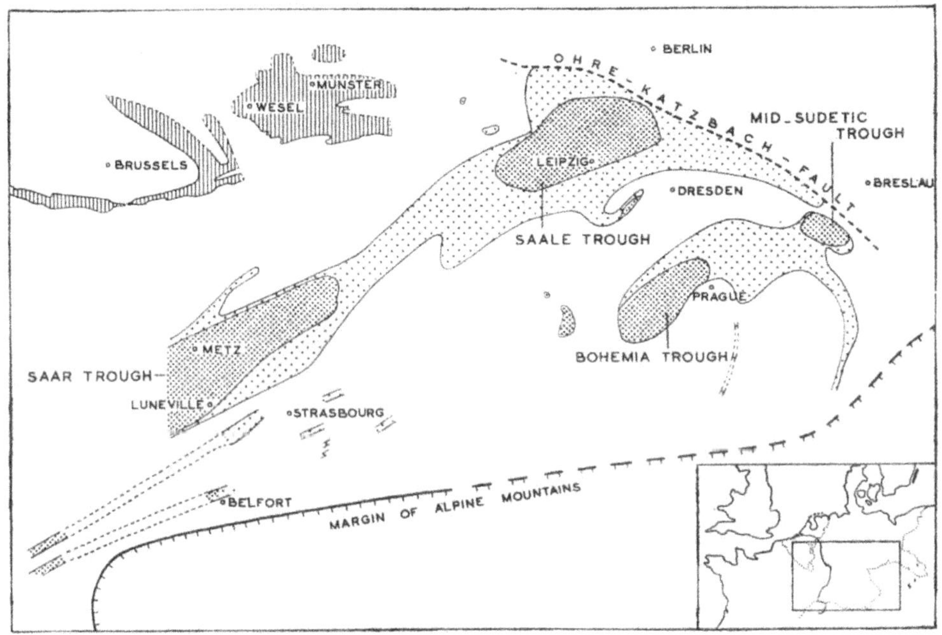

Fig. 30. Intramontane troughs in the Upper Paleozoic Variscides of Europe. Where subsidence started in Upper Carboniferous time areas are closely dotted, where it began in Permian times areas are widely dotted; marginal deep of Variscides is indicated by vertical shading.

subsidence was resumed in the Jurassic, following the Lower Cimmerian epoch, when 2,000 meters of Jurassic sediments were once more unconformably deposited upon the Paleozoic series.

The situation of the basin on the southern margin of the Asiatic Caledonides resembles that of a Caledonian marginal deep. Its subsidence began in the Devonian. One remarkable thing, however, is that it is now bounded on the other side by the Variscian Salair Mountains, which were folded during the Sudetic epoch (this basin represented the marginal deep of the Salair Mountains subsequently to this Sudetic epoch; it was folded towards the close of the Paleozoic). Another point which should be noticed is that this subsidence was repeated a third time in the Jurassic, resulting in the formation of a trough of the intramontane type.

1) See note 1) on p. 43, and the Appendix pp. 342.

therefore speak of them as "marginal deeps". Stille, too, refers to them as "Saumtiefe". As stated above, the Appalachians form an example of what Schuchert called a "monogeosyncline".

Geosynclinal migration was discussed in the preceding chapter, and we might be induced to speak of a repeated migration of geosynclines. We want to restrict the application of the term "marginal deep", however, to the youngest and final marginal trough in the history of a given mountain-belt.

One notable feature is that this marginal deep does in fact originate along the margin of the older folded chain, whereas the migrating geosynclines of

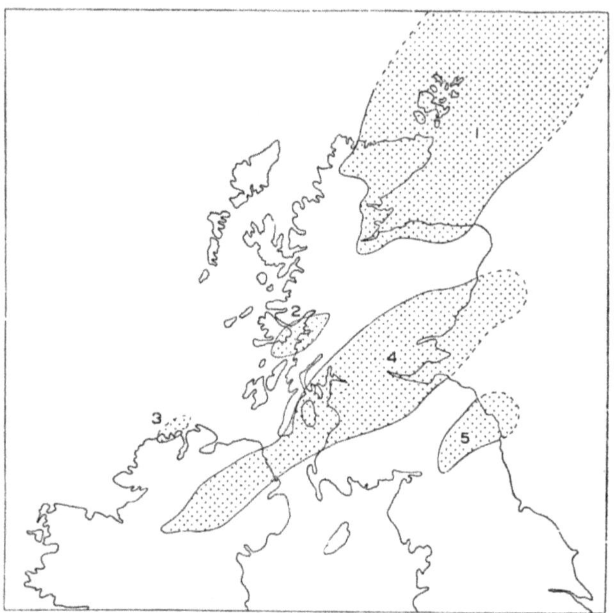

Fig. 31. Intramontane Old Red basins of Great Britain. 1. Lake Orcadie; 2. Lake Lorne; 3. Lake Fanad; 4. Lake Caledonia; 5. Lake Cheviot.

the earlier history of the belt are formed in most cases partly in the area of the preceding zone of folding and sedimentation.

Another notable feature is that the marginal deep is in many cases situated on the side facing the continental Shield, thereby opposing the general tendency of centrifugal development.

If a marginal deep were to be pictured without its contents, its shape would be that of a small and typical trough with a depth of a few thousand meters. The marginal deeps of Kusnezk and Karakum may be cited as exceptions. The latter have a more limited, basin-shaped appearance. This may merely be a secondary phenomenon, due to extensive overthrusting of the marginal areas. Thus the marginal deep of the European Variscides is also entirely "obscured" by overthrusting from the south in the western part of the north of France [1]) (see fig. 29).

[1]) See the tectonic map of Van Waterschoot van der Gracht, 1938.

Troughs of sedimentation also formed within the European Variscides shortly after the Sudetic folding of this zone. Special attention was paid to these formations by Born, and the problem of their origin was later studied by Stille and von Bubnoff. These troughs have long-drawn shapes, with more or less elliptical and at times irregular contours. Their longest axes extend parallel with the trend of the Variscides, and ridge-shaped "geanticlinal" elevations separate them (fig. 30).

The contents of these furrows, consisting of terrigenous strata, which are in turn composed for the most part of detritus from the geanticlinal zones of the Variscides, kept up with the gradual subsidence of the bottom, and

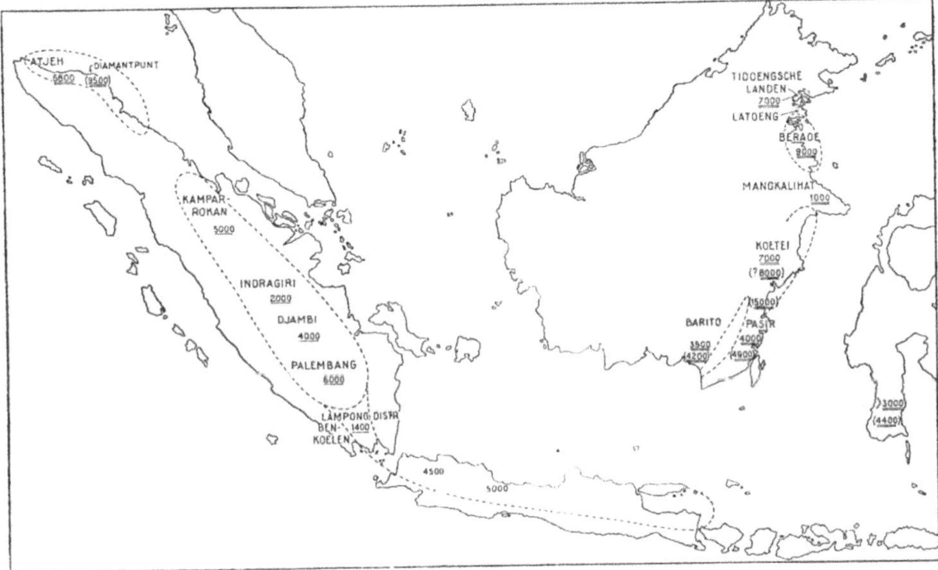

Fig. 32. Idiogeosynclines in the western part of the East Indies. The figures between brackets indicate the thickness of the whole Tertiary, in meters. The geosynclinal subsidence began during the Miocene wherever the figures are underlined and during the Eocene where double underlined.

became very thick in some cases, accumulating to as much as 6,000 meters in the trough of the Saar, 2,500 meters in that of the Saale, and 5,500 meters in the Mid-Sudetic trough. Still, precise observations have shown that the pace of this subsidence was not always the same, that the troughs were steadily enlarged in the direction of their longest axis, that they also broadened, and that there was a distinct relation between volcanic phenomen a hiatus in the sedimentation, and warping in the strata, especially between the middle and upper part of the Lower Permian. The beginning of subsidence varies between the lower part of the Upper Carboniferous and the upper part of the Lower Permian. Another striking feature is that many troughs (e.g. the Saar trough) originated on the boundary of a crystalline and sedimentary area. Thus in several cases these furrows may in a certain sense be said to represent marginal deeps, and it is in fact impossible to distinguish clearly between Type I (marginal deep) and Type II (intramontane trough).

The intramontane troughs of the Caledonides of Scotland and Ireland, which were also formed in the direction of the trend of the mountains, were filled with "Old Red". These troughs are known to us as "Lakes" Orcadia, Caledonia Cheviot, Fanad and Lorne (fig. 31) [1]).

There are many examples of intramontane troughs in Alpine zones of folding. For instance, the Maracaibo, Orinoco and Magdalena troughs in South America, which should be regarded as such, and the Pugget trough, Ventura basin and California Valley in North America. Other examples of intramontane troughs are the "ovas" in Asia Minor. Mention should also be made of those basins in the East Indian Archipelago and Burma to which I formerly referred to as idiogeosynclines (fig. 32). In Burma, Sumatra and Java these formations are bounded on one side by a pre-Tertiary area and on the other partly by a Miocene zone of folding (Arakan Yoma, Barisan, South-Java) [2]).

A sharp division between marginal deep and intramontane troughs cannot be based on the facies of the sediments, nor can the intensity of movements contribute anything towards such a division. For the fact is that the facies depends on the distance between the level of sedimentation and the corresponding level of the sea. The idiogeosynclines of the East Indies, unlike the intramontane troughs of the European Variscides, provide examples of

---

[1]) The so-called basin of Gorlowsk situated West of the Salair chains furnishes an example of an intramontane trough in the Asiatic Variscides (Leuchs, 1 p. 154, fig. 45; p. 156, fig. 52). Neritic to terrestrial Upper Paleozoic sediments from the contents of the trough which was rather intensely folded towards the end of the Permian, when many faults and overthrusts within the trough itself, as well as overthrusting of the framing chains in the direction of the trough came into existence, so that Leuchs even speaks of a "grabenförmige Einsenkung".

[2]) They cannot, however be regarded as intramontane troughs of this zone, for they originated partly during the Miocene (but before the Miocene folding), and, indeed, partly during the Eocene. The Miocene area of compression extends to the islands west of Sumatra and as far as the island-series formed by Timor, Tanimber, Kei and Ceram (fig. 120). The zone of Laramide folding of the archipelago was probably a much wider one, and it is not improbable that it also extends over a large portion of Sumatra and Java, for Cretaceous rocks are still found in the formations of the Barisan Mountains, which were folded during the pre-Tertiary, and in the exposures of pre-Tertiary rocks in Java (1938, p. 27, 28). The idiogeosynclines of Burma, Atjeh, Djambi, and Java may have a Laramide basement, or they may have originated on the edge of such a basement, but whether we call them marginal deeps or intramontane troughs is just another question of names.

This is equally true of Pugget trough and California Valley, and the idio-geosynclinal troughs of Northern and Southern New-Guinea, the South of Celebes and East-Borneo. The situation becomes even more complex in this last area, for a recent deep-sea trough originated on its eastern side, as will be described later. We already mentioned the interesting Kusnezk basin when dealing with marginal deeps. This basin, which was situated on the boundary between two elements of varying ages, began by being a marginal deep of the Caledonides, then became a marginal deep of the Variscides, and finally resumed its subsiding movement as an intramontane basin after the Lower Cimmerian phase. Mention must still be made, in this connection of the frequent appearance of longitudinal fault zones and graben, occurring mostly in the bordering geanticlinal areas, but striking at all events in the direction of the trend of the mountains. Examples are: the central graben of Spitsbergen, the rift valley in the Variscian Tianchan, the "graben" in the Chilean Andes (Gerth, 1939, p. 40), and the "longitudinal valley" in Sumatra.

troughs with a chiefly marine and paralic sedimentation. The frequent appearance of eruptive rocks in the intramontane troughs of Germany, and the complete, or almost complete absence of these rocks in a typical marginal deep would seem to constitute an important characteristic, but should in no case be generalized into a law, for volcanic products (viz. diabase,

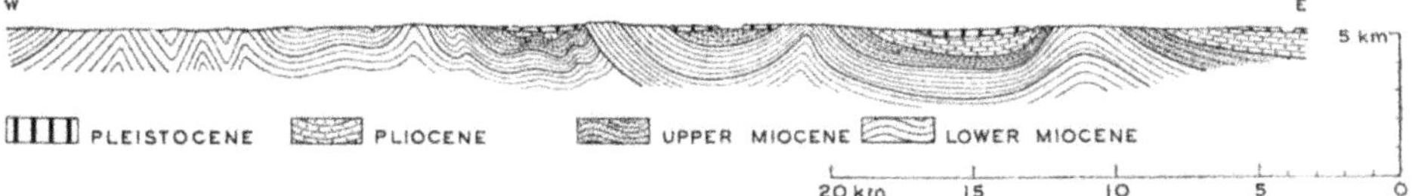

Fig. 33. Section across the geosynclinal basin of eastern Borneo. (After Jetzler).

melaphyr and basalt) do in fact occur in the Kusnezk basin a marginal deep of the Asiatic Caledonides. The following may be said as regards the third characteristic. The contents of the intramontane troughs of the European Variscides are hardly folded in comparison with the marginal deep of that zone, but we again find examples of a moderate folding of intramontane troughs in the East Indian idiogeosynclines (fig. 33).

We still have to consider the location of the troughs. The name intramontane in itself would seem to distinguish this type of trough from a marginal deep, but we saw that it is in some cases difficult to distinguish clearly between the two.

One more type of basin can be found in Europe, viz. the Hungarian, or Pannonian basin. Its basement was very probably folded during a Variscian epoch, and the basin more than likely formed a rigid nucleus around which

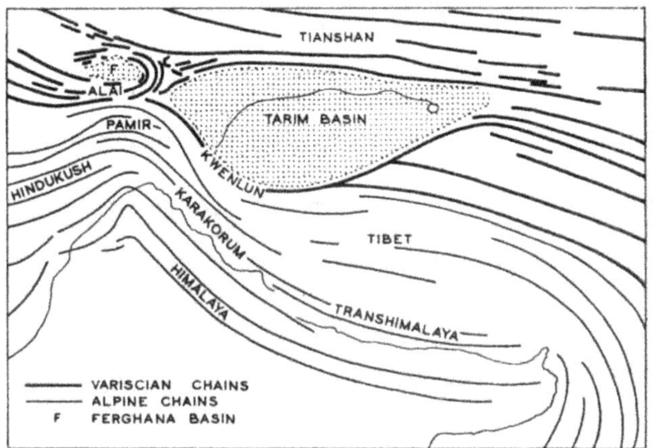

Fig. 34. Schematic illustration of the Ferghana and Tarim basins and the surrounding mountain-chains.

the Carpathian Mountains grouped themselves in the north and east, and the Dinarides in the south and west. The Pannonian Massif itself, however, subsided, forming a basin which was filled with Lower Miocene up to Pleistocene strata, and these were subjected to folding that adapted itself to the strike of the Alps in the west, and that of the Carpathians in the east.

The Pre-Cambrian nuclei of Ferghana, Tarim and Ordos played an important part during the Variscian folding in Asia. The chains were pressed against, and arranged around the nuclei (fig. 34). However, the massifs later showed a tendency to subside, and became basins. Sedimentation in the basin of Tarim began during the Permian or Upper Carboniferous and continues up to this day. The neritic, but mainly the terrestrial deposits, are probably exceptionally thick, and show several unconformities and folding of a not very pronounced character.

Therefore, the most salient characteristic of type III is that it originates on a subsiding block which, situated in a zone of folding, formerly acted as a rigid nucleus and thus produced the fore-mentioned effect on the trend of the folded chains that frame it. This is why I called them "nuclear basins".

The situation of the large salt basin of Texas is a different one in some respects, for it subsided on a Pre-Cambrian basement, as the Llano Burnett uplift in the south-east, and the Colorado Plateau in the north-west show (see Plates 5 and 6). Its frame consists of the Variscian Marathon-Ouachita chains towards the south and east, and the older Amarillo-Wichita zone in the north. The former western confine of the basin is unknown, but it is now occupied by the Tertiary ranges of the Rocky Mountains. This salt basin (containing Permian deposits up to as much as 4,000 meters) may in a certain sense be compared to the Pannonian basin of Europe, but the frame of the latter, unlike that of the Texas salt basin, is isochronous, as it were (the chains originated during the same period of folding). The submarine Barent-Sea Massif also has an anisochronous frame. The thickness of the deposited sediments remains unknown, but Frebold pointed out that numerous Mesozoic transgressions have at any rate invaded this area.

The three types discussed so far are clearly connected with the mountain chains alongside of which, or within which the basins are situated, but the same is not immediately evident in the large group of remaining basins. We can begin by stating that the most conspicuous feature is that their boundaries intersect existing structures. They originate "right across everything", so to speak, and we will therefore call them discordant basins.

Table I gives a summary of the numerous basins which belong to this group. They will be found under the heading "Type IV", and their situation is shown by Plate 6. We will discuss only one example here — viz. the basin of Paris (fig. 35). The rest will be described in the Appendix. A glance at the geological map of France shows that the basin of Paris originated in Liassic times. The deepest part gradually moved from the south-east to the north-west, lying near Paris during the Oligocene, and near the Channel coast during the Miocene. The subsidence represents approximately 1,200 meters (boring of Ferrières) in the northern sector of the basin, while the Pre-Cambrian basement — on which rests the Lias — descends to a depth of 1,532 meters in the "Pays de Bray". Its continuation is found in Hampshire and Wight on the other side of the Channel, and here the sediments are even considerably thicker (the boring at Portsdown reached the Upper Triassic strata at a depth of 1,993 meters).

The anticlines striking west-north-west — south-south-east were formed during the Upper Cimmerian epoch. The Subhercynian phase is indicated by but faint movements, the Laramide revolution by faulting and the Pyrenean epoch by "epirogenetic" movements. Finally, a rather faint folding occurred during the Attic phase. It would be wrong to think of one, large basin-shaped depression. Not only did the deepest part move west-north-west, but the bowl-shaped habitus of the basin is the result of certain

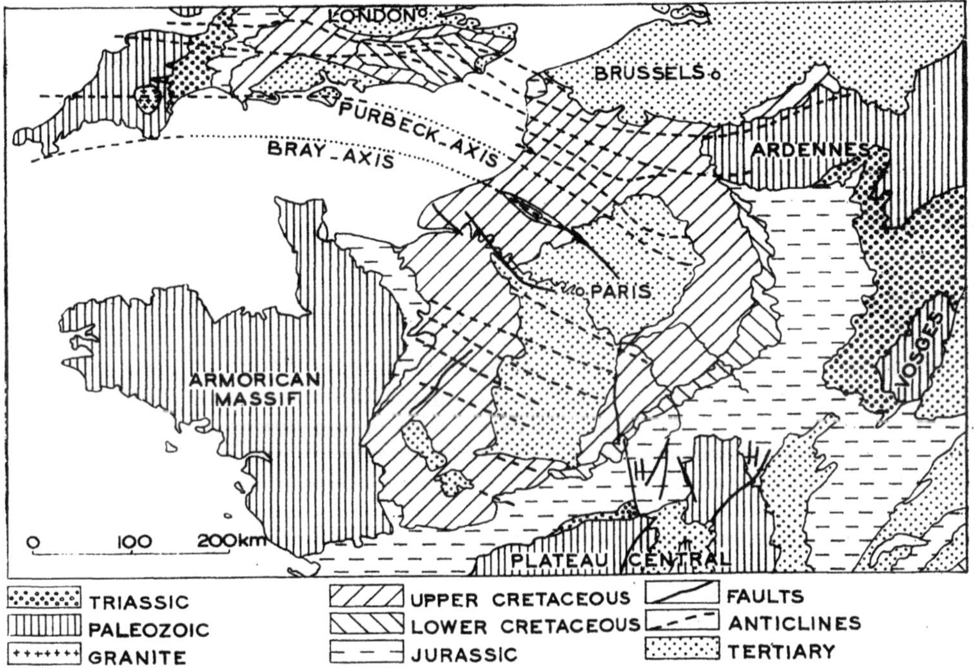

Fig. 35. Geological sketch map of the basin of Paris and its surroundings.

transgressions and of subsequent erosion, and does not seem to reflect the morphology of the basement. Lemoine drew attention to the fact that the liassic strata of a trough extending from Lorraine in the direction of the "Pays de Bray" are thicker than those found outside this area, and remarked that this feature also manifests itself in the foundation of the Mesozoic upper structure.

An examination of the manifold discordant basins (we refer to the Appendix for details) shows that these formation originated in heterogenous basements at different periods, and that their boundaries distinctly intersect those of pre-existing structures.

Yet it can be pointed out in a few cases that the original shape and boundaries of a basin are related to specific positive structural elements of their foundation. The original shape of the Liassic trough of the basin of Paris, for instance, the migration of its deepest part, and the strike of the anticlines, clearly reveal the posthumus activity of its Variscian foundation. Similarly, the north-south boundary between the North-Sea and the West-German Basin, or so-called axis of Erkelenz, and the Western boundary of the North-Sea Basin — the Pennine axis, in which positive movements occurred during the Saalian epoch — probably represent very ancient trends in the structure of the basement complex (see fig. 194). The entire arrangement of positive and negative elements in North America is apparently also the reflection of the Pre-Cambrian structure of the Canadian Shield. Africa may be cited as another example. A north-east — south west and a north-west — south-east strike (referred to respectively as the Somali and Erythrean trends) is inherent to the Pre-Cambrian structure of this continent which repeatedly manifests itself in one form or another, revealing itself also in the basins, their arrangement and boundaries.

These facts show that the original location of a discordant basin, if examined more closely, will probably reveal a certain "concordance" with some of the structural elements of its surroundings, and the devision of basins into nuclear and discordant basins consequently appears to be not so much a fundamental one, as one of degree.

Deep-sea basins

In their attempt to classify geosynclines, troughs and basins, some authors have excluded the deep-sea troughs and basins. To the present author such a procedure seems unreasonable. For whether a subsiding area appears at the surface as a depression or not depends merely on the relation between subsidence of the bottom and supply of detritus derived from the adjoining areas. And this relation may even vary during the evolution of a trough. The East Indian idiogeosynclines furnish a clear illustration. The subsidence of these intramontane troughs begins in continental regions. Thus e.g. the geosynclinal series begins with fluviatile-terrestrial sedimentation in the Barito basin of S.E. Borneo and in S. Celebes (fig. 32). In other places the first deposits consist of neritic sediments of a Miocene transgressive and epicontinental sea, which extended far beyond the area of subsidence, as e.g. in Sumatra (fig. 132). After some time the subsidence surpassed the rate of sedimentation. Hence the strata represent a deeper hemipelagic facies and accordingly the morphology of the area was a submarine trough. Still later, in the Pliocene, a much shallower facies developed, ending in the area becoming land. Obviously the subsidence had become slower and was surpassed by the rate of sedimentation. The submarine trough had become a swampy land surface.

A few submarine basins have been mentioned among the examples given in Table I and the Appendix. These include the Persian Gulf (an example of a marginal trough), the Gulf of Martaban and Madura Straits (representing the intramontane type). The Barent-Sea and Kara-Sea stand as examples of nuclear basins, and the Black-Sea and Caspian-Sea have been cited as discordant basins. All these cases, with the exception of the Caspian and Black-Sea, deal with shallow seas, and as most of the continental basins and troughs were filled with neritic, paralic and terrestrial sediments, a comparison between these formations and the above submarine furrows is obviously warranted.

If deep-sea basins are to be classified according to our four types, their characteristics would have to comply with the following conditions. To begin with, they would have to correspond with the features of continental basins, both as regards their location and the time of their formation. And, in the second place, the morphology and the surroundings would have to satisfy the same conditions. Finally, the facies of the deposits in these depressions would have their counterpart in fossil basins on the continents.

The submarine relief of the East Indian Archipelago, which has become so well-known since the Snellius Expedition and Kuenen's excellent mor-

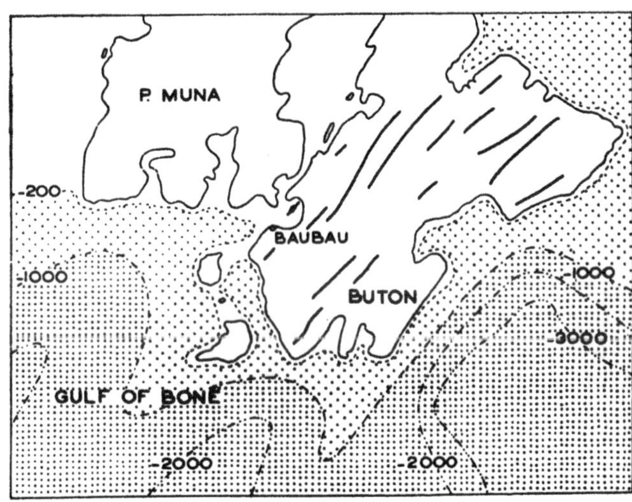

Fig. 36. Upper Tertiary anticlines in the island Buton, East Indies.

phological analysis of this area, would seem a suitable subject for the discussion of submarine-basins.

The present deep-sea relief of the East Indies must have been formed in the recent geological past. I treated this question at greater length in "The Geological History of the East Indies", as well as in a later paper of the year 1938, and will therefore confine myself here to a few of its salient features. I want to mention in the first place that the trend of the folded Miocene strata is intersected at an angle by the present boundaries of the deep basins in several places, from which it follows that the adjacent submarine relief must have originated at least after that Miocene folding. Molengraaff explained, for example, that the Amanuban mountain-chain, in Timor, is intersected at an angle of 12° by the coast. The view that the origin of the deep-sea, and the elevation of the series of islands between them, probably occurred simultaneously, is now generally accepted.

It is difficult to determine the exact time of the beginning of these submersions. It may perhaps be placed in the Pliocene. However, the Upper Neogene strata of the island Buton were folded with a trend which the neighboring Gulf of Bone intersects at right angles (fig. 36). The time of the formation of the basin can in this case consequently be fixed as uppermost Tertiary or Pleistocene. And if the submersion of these basins and the elevation of the intervening series of islands can be said to be closely connected, which seems probable, we are able to conclude that the most important part of the submerging movement must also have taken place in the Pleistocene in the case of the other basins. There can be no doubt that the rising movement of the islands occurred in the Pleistocene. This most recent movement, to mention but one example that could be supplemented by many others, brought parts of mountains in Central Ceram (which lay below sea-level during the sedimentation of the marine Pliocene strata in the Masiwat-Bobot graben) to a height of at least 3,000 meters above the sea. The recently elevated reef limestones and terraces give a particularly

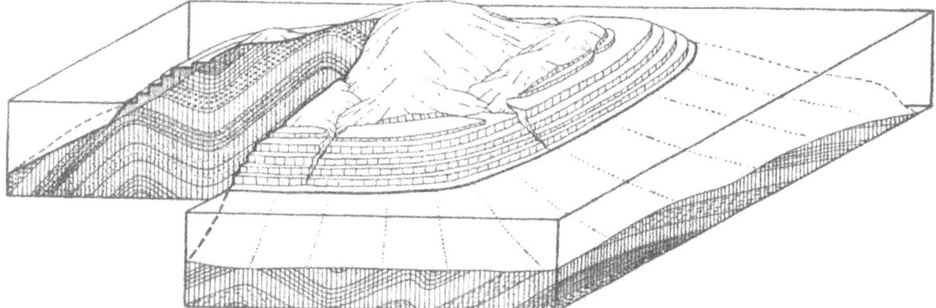

Fig. 37. Raised reef-terraces of the island Kissar, Southern Moluccas. (From Ph. H. Kuenen).

good idea of the amount and intermittent character of these recent movements. Fig. 125 gives some figures for part of the southern Moluccas The actual Pleistocene age of the elevated reef terraces has been verified in some cases. Fig. 37 shows the raised reef-terraces of the island Kissar.

Kuenen arrived at a similar conclusion in his geomorphological analysis of the bottom relief, and he exhaustively argues that the deep-sea basins originated recently as a result of the subsidence of "continental" (sialic) areas.

In the case of the East Indies we are dealing with a double system of island-arcs concave towards the Asiatic continent. On their convex external side the islands are bounded by a nearly continuous series of long and comparatively narrow deep-sea furrows. However, much broader irregularly shaped basins with relatively steep sides and flat, horizontal bottoms, are situated at the internal side of the island-arcs. It is not the place here to go into the problem of the origin of such a remarkable pattern of island-arcs and accompanying deep-sea basins (see Chapter VII). For the moment we will only point to the well known fact that a more or less similar arrangement of morphologic and structural elements characterizes the Western and Northern margins of the Pacific Basin proper.

The same view as regards the origin of the basins and troughs along the Eastern margin of Asia has also been held by Born, Leuchs, Lawson and others. Yabe pointed out that the entire region of Taiwan, the Riu-Kiu, Japanese and Kurile islands lying within the isobath of 720 meters was a land area which was connected with the Asiatic continent until Pleistocene or still more recent times [1]).

Leuchs and Born both assume that the arc of the Aleutians and the Tertiary chains of Alaska and Kamschatka surround a foundered pre-Tertiary "massif" part of which is at present buried beneath the sea. At

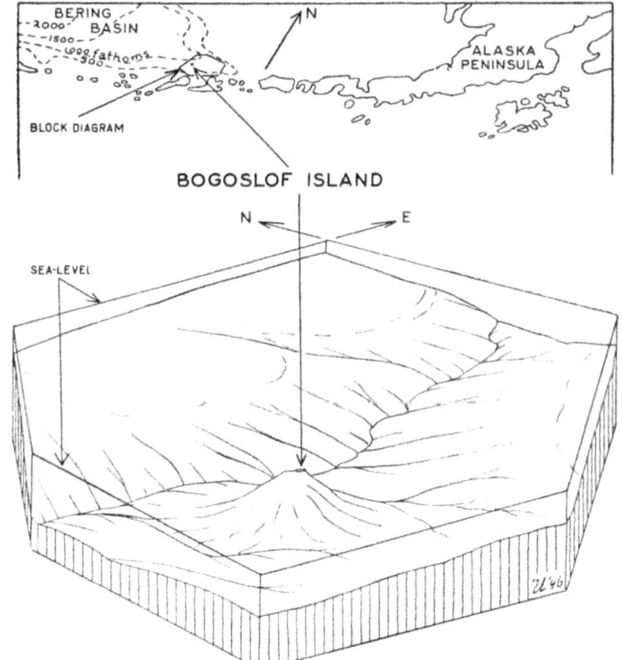

Fig. 38. Isometric block-diagram of the submarine relief near the island Bogoslof, Aleutians.

any rate, the chain of the Aleutians has an elongated deep-sea furrow along its convex side (the Aleutian trench) and an irregularly shaped deep-sea basin at its concave side (the Bering basin). Therefore probably a similar mode and time of origin may be supposed for the structural pattern of the Aleutians which in their principal elements show such a striking similarity to the arrangements of island-arcs, troughs and basins along the eastern and southeastern margin of Asia. Now the subsiding movement of the floor of the Bering basin was revealed by echo-soundings near Bogoslof,

---

[1]) YABE, H. *The latest land connection of the Japanese islands to the Asiatic continent.* (Proc. Imp. Acad. Japan, 5, 1929).

a small volcanic island in the central part of the Aleutian arc. Smith published a contour-chart of the region after which the block-diagram of fig. 38 was drawn.

It shows a drowned landscape. On the southern "plateau" is Bogoslof volcano, the summit of which only protrudes above the present level of the sea as a little volcanic ruin. The submarine stream pattern seems to be adapted to the presence of the volcano and therefore it appears to be of subaerial origin. Smith thinks that the submarine part of the volcano was also formed under subaerial conditions. If his interpretation is correct, it may be inferred at once from the isobaths that the region around Bogoslof was subjected to a submergence of at least 1,300 fathoms, approximately. Bogoslof is situated on a sea-bottom which forms the southern slope of the deep Bering basin at the concave side of the arc. Hence it would not be astonishing that Bogoslof and the surrounding sea-floor should have moved downward through a comparatively large distance since Pleistocene times. Neither is there anything astonishing in the revealing of a deeply drowned pattern of subaerial erosion on the sea-floor around Bogoslof.

Kuenen distinguishes between five groups of basins and troughs in the East Indies on morphological grounds (fig. 39). We will discuss four of them, though purposely in a different order. Kuenen's fifth group consists of grabens and other formations, to which we will not here give our attention. The dubious cases, indicated with a special notation in fig. 39, will likewise not be considered.

Kuenen's fourth group consists of the long furrows on the external side of the zone of Miocene folding (fig. 39), i.e. the trough west of Nias and the Mentawei-Islands, the trough south of Java (which has an occasional depth of 7,000 meters), and the Timor trough. These troughs concur with marginal deeps of the Miocene zone of folding as regards the time of their formation, their "tectonic" location and general morphological features (Plate 8).

It will be pointed out, in Chapter VII (p.188), that genetically two different types of marginal deeps have to be distinguished. Those accompanying a double arc originated in a fundamentally different way from a marginal deep in front of a single arc.

His third group includes not only considerably shallower formations such as e.g. the Mentawei trough, the trough along the south coast of Java, the Sawu-Sea and the Wetar deep, but also the Weber deep, which has an exceptional depth. They differ morphologically from the preceding group in that they present a less oblong and therefore a more basin-shaped appearance. As regards the location of this group, the Mentawei trough and Sawu-Sea have both subsided in the Miocene zone of folding (fig. 120), and as the submarine ridge south of Java can be said to belong in a morphological and gravimetrical respect to the Miocene belt of folding extending from Mentawei towards Timor (see fig. 120 and Plate 8), the same may probably be assumed of the basin situated immediately south of Java. This conclusion also holds good for the Wetar and Weber deeps, since these formations are bounded on one side by the Miocene zone which can be followed from Timor via the Tanimber and Kei Islands towards Ceram and on the other by the submarine continuation of the Miocene zone of the Lesser Sunda Islands. These troughs may consequently be said to belong to the intramontane type as regards their situation, morphology and formation.

It is doubtful whether the Flores Sea ought to be regarded as a marginal

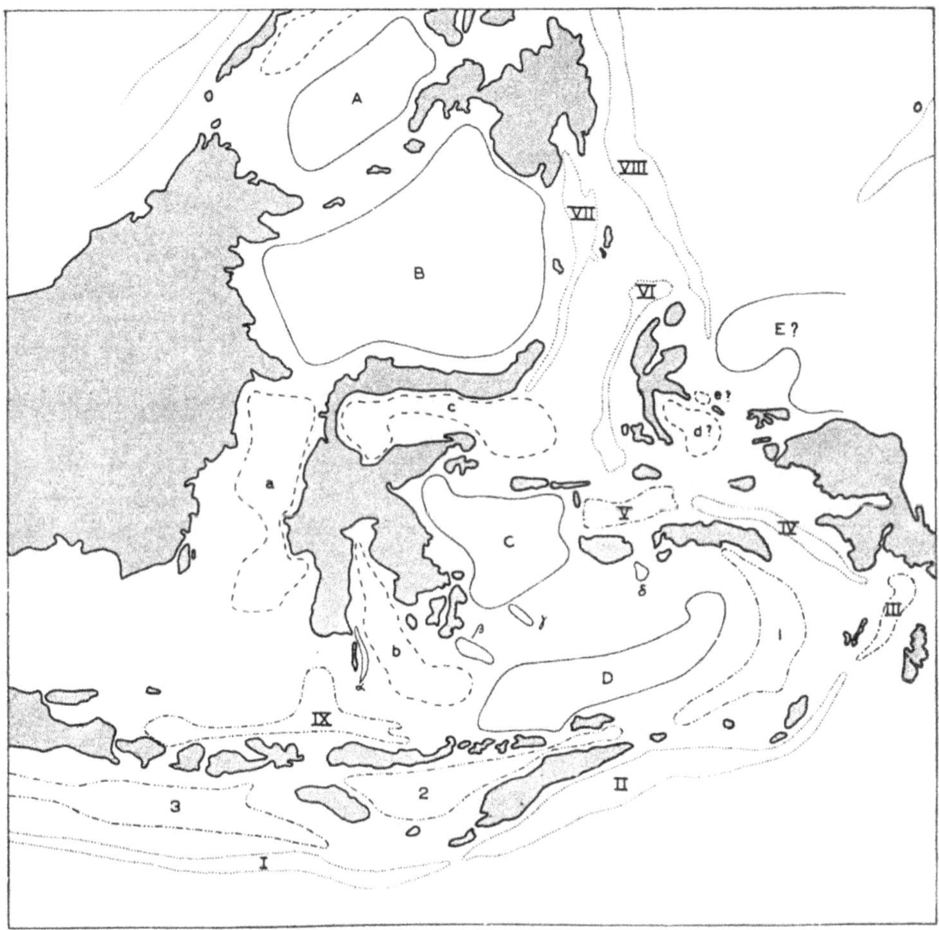

Fig. 39. Types of submarine basins and troughs in the East Indies (from Ph. H. Kuenen). I—IX marginal deeps; 1–3 troughs of the intramontane type; A—D nuclear basins; a—c discordant basins. I Java trough; II—VII Timor-Ceram and Molucca troughs; VIII Mindanao trough; IX trough of the Flores Sea. 1. Weber deep; 2. Sawu Sea; 3. Java-Mentawei trough. A Sulu Sea, B Celebes Sea, C and D Banda Basins. a Makassar strait; b Gulf of Bone; c Gulf of Tomini.

deep or as an intramontane trough. Kuenen included it in his fourth group. If this interpretation be correct, we would find a marginal deep on either side of the Miocene zone of folding, i.e. the Java-Timor trough on its southern side, and the Flores trough north of it. The Siwalik and Karakum troughs are placed similarly in respect to the Himalayas. (For another interpretation of the Flores deep the reader is referred to Chapter VII, p. 194).

Kuenen's first group is composed of basins with relatively steep sides and flat, horizontal bottoms, (the Banda basins, the Celebes-Sea and the Sulu-Sea). Dr. Fr. Weber pointed out that the southern Banda-Sea, or at least the greater part of it, had lain above sea-level during the Mesozoic. The folded chains group themselves around the actual Banda-Sea, and the latter thus closely resembles such nuclear basins as the Pannonian, Tarim, Ordos and Kara-Sea basins as far as the time of formation and the morphological and tectonic characteristics are concerned. (A different view concerning the origin of these basins is put forward in chapter VII, p. 194).

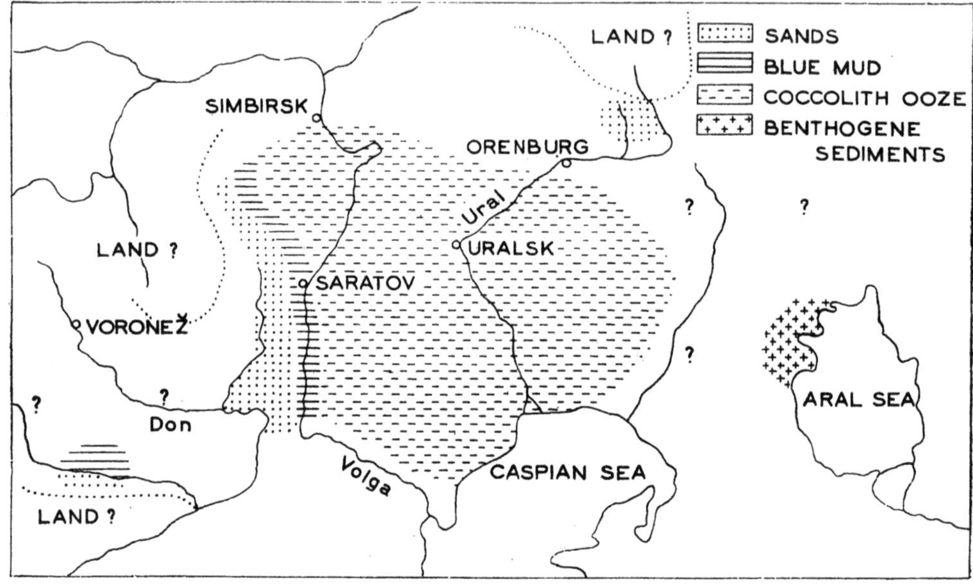

Fig. 40. Distribution of the abyssal coccolith-ooze in the Caspian basin during Maastrichtian times (after S. von Bubnoff).

The Macassar, Bone and Tomini basins make up Kuenen's second group. These formations have a somewhat shallower flat-bottomed cros-section, and a long and irregular shape. The Upper Tertiary chains are intersected by the steep coast of the Macassar and Bone basins (fig. 36) [1]).

These basins (at any rate the first two) are similar to those discussed as discordant basins under our Type IV.

It seems most remarkable, therefore, that Kuenen's division of deep-sea furrows into four groups should correspond with our own classification of four types of continental basins. We discussed Kuenen's types in a different order because we wished to emphasize the similarity between the results.

[1]) The recent formation of the Gulf of Tomini was pointed out in my publication of the year 1939.

One more question remains to be examined, namely whether basins or troughs can be found on the continents, which — like the present East-Indian deep-sea basins — were filled with sediments of bathyal and even abyssal facies. This question can be answered in the affirmative, for the Indian Archipelago itself contains examples of such fossil troughs, e.g. on Timor. One or more troughs with bathyal and abyssal sediments existed in Timor during the Mesozoic. A paleogeographic reconstruction of these troughs is not possible at the moment for too little is known of the complicated structure of this island, but the distribution of the facies types leaves no doubt that deep-sea deposits originated in the vicinity of areas of neritic sedimentation. Thus, too, we know that troughs containing bathyal and perhaps even abyssal sediments existed in the complicated geosyncline of the Alps during Mesozoic times. The cherty to siliceous limestone and black shales of the Ouachitas and the Marathon Mountains may possibly be indicative of pelagic conditions and sedimentation in deep water. And, lastly, mention must be made of the Caspian basin (fig. 40), a large portion of which was filled during Maastrichtian times by coccolith and globigerina ooze, such as now accumulates at a depth of from 2,000 to 3,000 meters according to the last oceanographic data.

We have enumerated a few examples of fossil deep-sea troughs and basins, and most of the present deep-sea basins of the East Indies may consequently safely be classified according to our system. The morphological aspect of an area of subsidence merely depends on the existing relation between the speed of the subsidence and the quantity of detritus supplied. The idiogeosynclines in East-Sumatra and North-Java (fig. 32) were filled with thick deposits because of the huge amount of detritus supplied by the extensive area of Sunda-land, which lay above sea-level at that time. The position of the Mentawei trough is a less favorable one, and this furrow is therefore not filled to a high level. The Weber deep is a very deep trough, as exceptionally few deposits accumulated within it.

We mentioned already that rising movements of the surrounding areas probably accompanied the subsidence of the basins. Nor do conditions in the Indian Archipelago differ, in this respect, from those prevailing in the surroundings of the basins on the continents.

Further examples of deep-sea basins will be found in the Appendix, in which reference is made to the basins of Eastern Asia and those of the Mediterranean, West-Indies and Southern Antilles. All these formations are situated inside continental folded chains and loops of continental islands, or immediately alongside of them. A discussion of their relation to the origin of the island-arcs will be found in Chapter VII. Besides these, the bottom relief of some oceanic areas proper also reveals basins and ridges of a type similar to that observed on the continents. These will be discussed in Chapter VIII.

Deep-sea basins are very numerous at present. It is obvious, however, that a comparison with the past has to restrict itself to the present boundaries of the continents. Leuchs observed quite rightly that many of the marginal deep-sea troughs lie at such a distance from an area of erosion of any importance (e.g. along the arc-shaped islands in Eastern Asia), that it is difficult to understand how they could be filled by a thick geosynclinal sequence of strata. Another point which deserves attention is that the abyssal sediments which are at present deposited within these troughs contain little or no calcium carbonate. This lack of lime has only been observed in a few exceptional cases in fossil troughs (e.g. in Timor).

Thus, though but a few examples are known of fossil basins which are similar to our present deep-sea basins, we should not forget that these formations are only met with in exceptional cases on the continents. In addition to this the relief of our continents is at the present time particularly high, and not only did the Alpine chains undergo a "rejuvenation" in the form of a rising movement, but even many older chains partook of this process. Moreover, if we stop to consider that the subsidence of a submarine area in many regions is accompanied by a rising movement of the surrounding land, it will at once be clear that vertical movements became very intense within the crust during the latest part of the earth's history and that it created a strong relief on the continents, and many deep basins below sea-level.

A comparison between the present relief of the floor of the sea and fossil troughs would be misleading, for the continents are the only areas that supply us with data of former periods. No one can tell whether the relief of the floor of the ocean had been more accentuated during certain periods than others. Deep-sea troughs cannot be said to have occurred very abundantly periodically; nor can the floor of the deep-sea be said to have flattened out in the intervening time. Still, it would be quite as unfounded to affirm that this certainly was not so. This is one of those typical points which are left open to conjecture.

Chronological relations with other phenomena

The most striking results of an examination of basins and troughs is that the origin and history of these formations are related to certain epochs of folding and mountain-building. This is not surprising in the case of marginal deeps and intramontane troughs, if we consider their situation. The nuclear basins, too, lie inside, or in between certain folded zones, and are surrounded by mountain-chains. It is obvious, therefore, that these basins must be subjected to the influence of processes deeper down in the earth when the latter occur underneath these particular orogenic belts. A problematical point is what happens in the substratum. We might be inclined to associate this question with changes and displacement of sub-crustal matter, but in doing so we would stray too far into the domain of speculation for the moment. We will return to this question in Chapter IV, p.85.

One remarkable feature, however, is that even discordant basins show chronological relations with folded belts, though these basins sometimes originate at a great distance from such zones. The subsiding movement of discordant basins, too, begins after a special epoch of folding, and in addition to this, certain phenomena in the history of these basins, such as unconformities, faulting, moderate folding and a repeated tendency to subside appear to be chronologically related to specific phases known in folded chains. One thing and another not only points to a deep internal terrestrial process but also shows that this process has a world-wide activity. Lastly, the greater part of the basins (as well as of the troughs) form after certain Variscian epochs (see Plate 6). A similar and exceptionally intense formation of basins succeeds certain Alpine phases. Those basins which are connected with Caledonian epochs as far as their origin is concerned, are undoubtedly far less numerous, and only a few can be associated with Lower and Upper Cimmerian phases (see Table I).

Can it be a mere accident that an identical sequence was arrived at in the case of periods of mountain-building. The given classification of basins can no doubt be improved upon. Some readers will perhaps feel inclined to change certain notations. However, I am of the opinion that nothing can alter the remarkable result that the periods of Variscian and Alpine mountain building, which were of far greater importance than the Caledonian and Cimmerian, corresponded with periods in which a considerably larger number of basins were formed (Table I).

It certainly is not a mere accident that the periodicity and intensity of mountain-building is related in point of time to the analogous periodicity and intensity of the formation of basins (Table II).

Epochs of folding, i.e. periods of increasing pressure in the earth's crust are succeeded by periods of decreasing compression, which also represent periods of mountain-building. As soon as the intensive compression ceases, the mountains not only begin to rise, but the negative elements, too — the basins — begin to take shape.

As pointed out previously, the shape of some discordant basins is known to be determined by certain structural elements of the basement. Nevertheless so few details are known of these basins that it is in most cases impossible to say what brought about their shape and location. (Cf. Chapter XI). On the other hand, there can be no doubt that they are the result of the whole structure of the foundation and the surrounding areas, and this is probably why subsequent phases of folding are manifested more strongly in one basin than in another. These factors are likewise responsible for the fact that the tendency of a contemporaneous basin to subside is continued for a greater length of time in one basin than in another. The same applies to geosynclines.

Generally speaking, those basins that are found in Pre-Cambrian Shields are wider than those originating within, or near, a later folded belt.

I hope that the above (including the Appendix, which discusses the various data on which this chapter is based) will have succeeded in showing that the origin of basins and troughs is related to periodic processes in the earth's interior, and particularly that periods of decreasing compression in the earth's crust are characterized both by a rising movement of the folded chains, and the beginning of local tendencies to subside, — in other words the beginning of the formation of new geosynclines and basins.

References

The works on regional geology, which are cited in the bibliography of Chapter II, should also be consulted.

BÄRTLING. R, *Transgressionen, Regressionen und Faziesverteilung in der Mittleren und Obere Kreide des Beckens von Münster* (Zeitschr. Deutsch. Geolog. Gesellsch. 72, 1920).

BORN, A. *Über jungpaläozoische Kontinentale Geosynclinalen Mitteleuropas* (Abh. Senckenberg Natur f. Gesellsch. Bd. 37, 1921).

BUBNOFF, S. VON, *Grundprobleme der Geologie* (1931).

COSTER, H. P. *The gravity field of the Western and Central Mediterranean, with new bathymetric chart* (Acad. Thesis, Utrecht, 1945).

ESCHER, B. G. *Beschouwingen over het opvullingsmechanisme van diepzee-slenken* (Verh. Geol. Mijnbouwk. Genootschap Geol. Ser. Dl. III, 1916).

GRABAU, A. W. *Migration of geosynclines* (Bull. Geol. Soc. China, 3, 1924).

HESS, H. H. *Geological interpretation of data collected on cruise of U.S.S. Barracuda in the West Indies* (Preliminary Report. Transact. Americ. Geophysical Union 1937).

HESS, H. H. *A new bathymetric chart of the Caribbean area* (Advanced report of the commission on continental and oceanic structure 1939).

HOLTEDAHL, O. *Geologische Karte der Arktis mit angrenzenden Gebieten* ("Arktis" 3, 1930).

KAHRS, E. *Zur Palaeogeographie der Oberkreide in Rheinland und Westphalen* (N. Jahrb. Geol. u. Pal. Beil. Bnd. 58, 1927).

KAY, G. H. *Development of the Northern Alleghany synclinorium and adjoining regions* (Bull. Geolog. Soc. America, 53, 1942).

KAY, G. M. *Geosynctines in continental development* (Science, 98, 1944).

KUENEN, PH. H. *Geological interpretation of the bathymetrical results* (The Snellius Expedition, Vol. 5, part. 1, 1935).

LAWSON, A. C. *Insular arcs, foredeeps, e c.* (Bull. Geol. Soc. America 43, 1932).

LEMOINE, P. *Considerations sur la structure d'ensemble du bassin de Paris* (Livre Jubilaire Centenaire de la Soc. Géol. de France, Tome II, 1930).

LEMOINE, P.; HUMERY, R. et SOYER, R. *Les forages profonds du bassin de Paris. La nappe artésienne des sables verts* (Mem. Muséum Nat. Hist. Nat. 1939).

LEUCHS, K. *Tiefseegräben und Geosynclinalen* (N. Jahrb. Min. etc. Beil. Bnd. 58B, 1927).

MOLENGRAAFF, G. A. F. *On recent crustal movements in the island of Timor and their bearing on the geological history of the East Indian Archipelago*(Procee d. Kon. Acad. v. Wetensch. Amsterdam, 1912).

NORIN, E. *The Tarim Basin and its border regions* (Region. Geolog. der Erde, 2. IVb, 1941).

PRUVOST, P. *Sédimentation et subsidence* (Livre Jubilaire, Centenaire de la Soc. Géol. de France, Tome II, 1930).

PUTNAM, W. C. *Geomorphology of the Ventura region, California.* (Bull. Geolog. Soc. America, 53, 1942).

REED, R. D. *Southern California as a structural type* (Bull. Americ. Assoc. of Petrol. Geol. 21, 1937).

SCHUCHERT, CH. *Sites and nature of the North American Geosynclines* (Bull. Geol. Soc. America 34, 1923).

SCUPIN, H. *Palaeogeographie* (1940).

SMITH, P. A. *Submarine topography of Bogoslof* (The Georg. Review, 27, 1937).

STEVENS, CH. *Le relief de la Belgique* (Mém. de l'Instit. Geolog. de Louvain, XII, 1938).

STILLE, H. *Alte und junge Saumtiefen* (Nachr. K. Gesellschaft der Wiss. Göttingen. Math. Phys. Klasse, 1919).

STILLE, H. *Die Oberkarbonisch-Altdyadischen Sedimentationsräume Mittel-Europas in ihrer Abhängigkeit von der Variscischen Tektonik* (Congr. de Stratigr. Carbonifère. Heerlen 1927, Ed. 1928).

TEICHERT, C. *The Mesozoic transgressions in Western Australia* (The Australian Journal of Science, Vol. 2, 1939).

TEICHERT, C. *Marine Jurassic of East Indian affinities at Broome, North-Western Australia* (Journ. of the R. Soc. of Western Australia, 26, 1934–40).

UMBGROVE, J. H. F. *Verschillende typen van tertiaire geosynclinalen in den Indischen Archipel* (Leidsche Geol. Mededeelingen VI, 1933).

UMBGROVE, J. H. F. *De pretertiaire historie van den Indischen Archipel* (Leidsche Geol. Mededeelingen VII, 1935).

UMBGROVE, J. H. F. *On the time of origin of the submarine relief in the East Indies* (C. R. Congr. Internation. de Géographie, Amsterdam 1938).
UMBGROVE, J. H. F. *Atolls and barrier reefs of the Togian islands, North Celebes* (Leidsche Geolog. Mededeelingen. XI. 1939).
UMBGROVE. J. H. F. *Periodical events in the North Sea Basin* (Geological Magazine, 82, 1945).
VEATCH, A. C. *Evolution of the Congo Basin* (Geolog. Soc. of America, Memoir. 3.1935).
WATERSCHOOT VAN DER GRACHT, W. A. J. M. VAN, *Lateral movements on the Alpine Foreland of North Western Europe* (Proceed. Kon. Acad. v. Wet. Amsterdam, vol. 41, 1938).

CHAPTER IV

CRUST AND SUBSTRATUM

"Pas de synthèse tectonique sans vision d'un continu à trois dimensions en train de se déformer". (E. ARGAND)

Introduction

Chapter I gave a brief outline of those rocks which can probably be said to build up the earth's crust. This important question will now have to be examined somewhat more closely. It should first be noted, however, that opinions differ considerably as regards the composition of the terrestrial crust though all agree on a few fundamental points.

An imaginary cross-section of a continent would probably present more or less the following aspect (fig. 41). On top we would find a veneer of sedimentary strata, totalling as much as 15 km in the geosynclines, or possibly more, but thinning out to zero at points where their foundation is exposed. Beneath this veneer would stretch a zone of acid rocks, consisting mostly of granite and gneiss, and descending to a depth from 15 to 35 km below the cover of sediments. The thickness of the crust in the continental regions is thought to vary between 40 and 70 km. Gutenberg's latest review shows that comparatively smaller figures apply to such regions, for instance, as New-Zealand and the north-eastern part of Japan (30 km). The largest values found are in the Sierra Nevada, California and the Alps (60–70 km). The lower side of the crust is known to form an important primary discontinuity — "the Mohorovičić discontinuity". There is no one to-day who doubts that the lower part of the continental crust is composed of basic rocks with an occasional thickness of 10–25 km. The lower boundary of the crystalline crust is within the basic material.

Opinions disagree as regards the nature of these basic rocks, which are piezo-gabbro or olivine basalt according to some authors (Daly, Kennedy), and peridotite according to others (Holmes, Hess, Eskola and others). The most important point, however, as far as we are concerned, is that this lower part of the crystalline crust is now universally assumed to consist of basic to ultrabasic rocks, the so-called sima. The chemical composition of the sima differs but slightly from that of the substratum upon which rests the crust. The lower layer of the crust is indeed regarded as a layer of crystalline sima. Some writers, including Daly, assumed that amorphous sima — the "vitreous substratum", an overheated and very viscous basaltic liquid under high pressure — follows beneath the Mohorovičić discontinuity [1]).

To avoid any misunderstanding, I want to emphasize that the terms "sima" and "sial" (which were introduced by Suess) are used in the petrographical sense, and not in Wegener's physical sense, such as rigid and elastic sial and viscous sima.

Some authors are of the opinion that one, or possibly more than one layer, with a thickness of about 25 km, extends between these two parts of the earth's crust. There are many conflicting views as regards the composition of this intermediate layer, some assuming it to consist of granite or diorite, and others of granodiorite or granulite, and even of amphibolite, olivine-free basalt or gabbro.

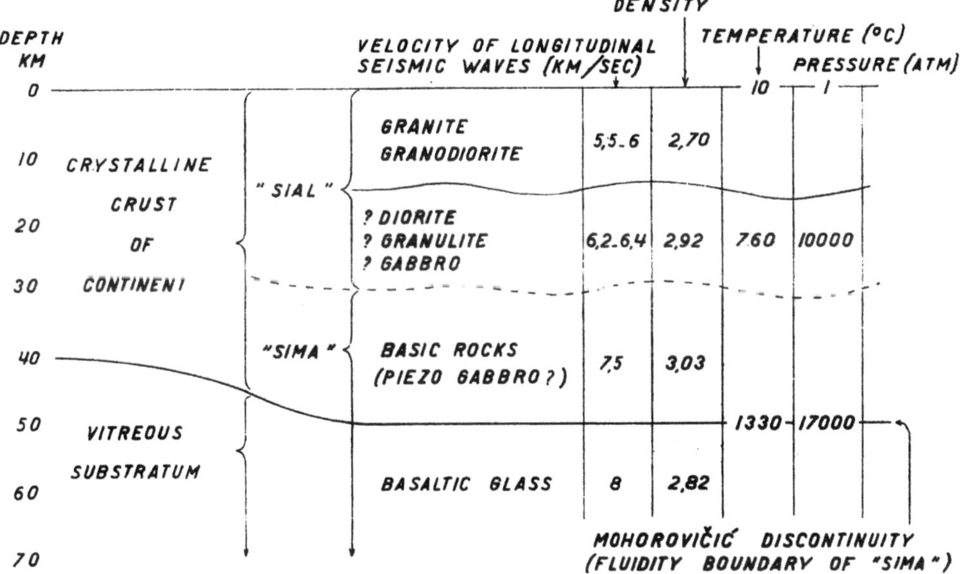

Fig. 41. Diagram of a continental cross-section.

Fig. 41 contains a few data on petrography, density, pressure, temperature and the velocity of longitudinal waves in the terrestrial crust. As pointed out in the first chapter, the general opinion is that a considerably thinner layer of sial exists under the Atlantic and Indian Ocean and that no such layer is found under the Pacific east of the andesite line (fig. 42). The nature of the sialic sheet of the Atlantic forms another point of uncertainty. One possibility is that the upper, granitic layer of the continents as well as one or more of the intermediate layers, are thinning out towards the oceans. Another supposition is that one of these layers is entirely absent from the Atlantic

[1]) In 1942, however, Daly wrote:.. "it appears probable that glassy basalt cannot now be world-wide and continuous: that here and there, within broad segments, the ancient energy-charged layer has crystallized because of the slow cooling of our planet". It seems very difficult to give a definition of the earth's crust in such places!

sector [1]). We will return to this question in Chapter VI. Though the floor of the Pacific differs in volcanologic, petrographic, magnetic and seismic respect from a continent, the same cannot be said of the floors of the Atlantic and Indian Ocean and the surrounding continents [2]).

The crystalline crust of the Atlantic, where seismic data (together with the evidence found in the granitic inclusions observed in the lavas of some volcanic islands [3]) have revealed the presence of a thin layer of sial, must be quite as thick and possibly even thicker than that of a continental

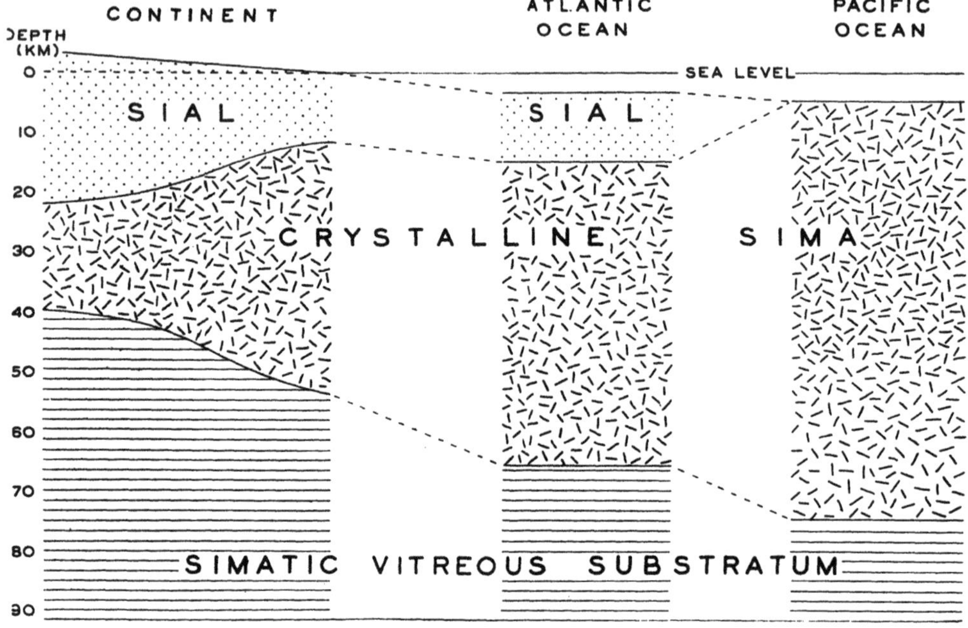

Fig. 42. Schematic sections through the earth's crust.

sector. Under the Pacific, where a sialic layer appears to be absent, the rigid crust, consisting entirely of crystalline sima, is thicker than anywhere else (fig. 42). Daly's view is based on the assumption that the thickness of

1) Vening Meinesz pointed out that the gravimetric profiles show a sudden increase of the positive anomalies (of 30—100 milligals) when passing from the shelf to deep water. After a discussion of the available seismic and gravimetric data, he comes to the conclusion that the most probable explanation is ... "to assume the granitic layer only to be present in great thickness in the continents and to end rather suddenly at the edge of the shelf or, if present in some parts of the oceans to be thin there. The next layer seems to continue under the oceans, perhaps somewhat thinner under the deepest parts of the Atlantic and Indian oceans and certainly thinner under the central part of the Pacific east of the andesite line; it may even be absent in this last area but in this case we must assume a still deeper crustal layer to be present there of a smaller density than the layer below it. The principal part of the gravimetric shelf-profiles as well as of the seismic data could thus be explained".

2) Gutenberg 1939, p. 302—320.

3) Ascension e.g., including probably Tristan da Cunha. Plutonic rocks, together with sediments of various ages and metamorphic rocks are also known to occur in some of the Cape Verde Islands and the Canaries.

the crust depends upon the production of heat by radioactive minerals, the sial being considered to be more radioactive than the basic rocks of the sima [1]. The terrestrial crust will consequently be thinner under a sialic continent, when in a state of thermal equilibrium, than under an ocean, where the sial is either thin or non-existent, and the solid crust will have to be even thinner under the additional quantity of sial of the "root" of a folded mountain-chain [2]. The lower side of the crystalline crust should at the same time be regarded as the upper limit of melting of the sima, since the temperature at this depth is equal to the melting point of the rocks under the prevailing temperature. The results obtained so far indicate that the temperature from radioactive minerals in the crust is insufficient to melt the rocks anywhere in the crust. "This suggests that extensive melting of the crust is only likely to occur in geosynclinal areas where the crust is thickened, or in some other abnormal circumstances" [3]. One of these abnormal circumstances is the formation of a crustal down-buckle or mountain-root. In such an area the Mohorovičić discontinuity will shift upwards, as shown in fig. 47, 49, 51, 123 and 130.

Rittmann's model of the structure of the earth's crust was based on arguments similar to those of Daly, and their view was lately confirmed by investigations of an entirely different kind by Vening Meinesz. A study of gravity fields over the Hawaian Archipelago and the Madeira area has shown, amongst other things, that there cannot be any important difference in the crustal rigidity and thickness beneath both these areas [4], and that the positive anomalies of the isostasy strongly indicate the existence of a rigid crust, acting as an elastic sheet that is bent by the load of the volcanic islands which formed on top of it.

It was claimed by Molengraaff in 1916 that a subsiding movement of volcanic islands in the Pacific would be brought about by slow isostatic downward movements, the subsidence continuing until the islands had reached the level of the simatic ocean-bottom, which he regarded, as Wegener did, as a viscous liquid. Molengraaf thus sought to explain the ultimate formation of thick coral reefs, and the creation, in conformance with Darwin's theory, of a barrier reef, and, finally, an atoll, from a fringing reef by the gradual subsidence of the volcanic foundation. However, no strong downward movement is apparent in many of the Pacific islands, as has since been shown by observations along their coasts. Sub-recent negative shifts of the shoreline with a height of approximately 6 meters occur in a number of these localities, and the world-wide distribution of this phenomenon proves that the sea-level has fallen over the whole world in comparatively recent times. The discovery of an equal amount of negative

---

[1] α-rays are emitted from the nucleus of a radioactive element at a constant rate. After an α-ray has lost its energy by absorption, it settles down as a helium atom. The major source of terrestrial radioactive heat is produced by absorption of the α-rays; β-rays and γ-rays contribute only about 11 percent of the heat (Evans and Goodman 1941).

[2] Daly, 1938, p. 35.

[3] Bullard, op. cit. 1945, p. 35.

[4] The figures by Vening Meinesz as regards the thickness of the crust (25 to 45 km) are much lower than those of Daly mentioned above.

shift on the simatic Pacific islands shows that the subsidence of the latter (i.e. assuming that they were really subsiding) has at any rate ceased. Vening Meinesz' geophysical investigations show that the subsidence of volcanic islands in the Pacific is confined to the down-bending of the simatic floor of the ocean, and that the latter acts as an elastic sheet, and not as a viscous liquid, as Wegener and Molengraaff supposed.

Magmatic clans

The conception of the earth's crust at which we arrived, above, explains several striking volcanological and petrographical phenomena. Only basic magma can reach the surface where the sialic layer is lacking, and this material — through differentiation — can furnish trachyte to phonolite (the-so-called atlantic rocks), but no quartz-bearing acid rocks, and therefore certainly no granite. If the basic magma of the simatic substratum suddenly invades a continental crust by a so-called abyssolithic injection into faults, the only material to extrude at the surface will again be basic material.

The question therefore arises how the acid rock-tribe comes into being. It has commonly been supposed that a great variety of igneous rocks may originate from one and the same parental magma by processes of normal, i.e. gravitational differentiation. All the volcanic and plutonic rocks have thus been considered as products derived from three types of magmatic melts, called atlantic, pacific and mediterranean respectively. The atlantic clan of consanguinous rocks evolved from basic parental material with the composition of an olivine-basalt. In the foregoing pages we assumed it to be present in the form of a world-embracing layer below the crust. Becke, who introduced the terms atlantic and pacific provinces, chose these names because examples of the first are known from certain islands in the Atlantic Ocean, whereas more acid and calc-alkaline igneous rocks characterize most of the volcanic belts surrounding the Pacific Ocean.

Without exception, however, the volcanoes within the Pacific Basin proper are composed of atlantic lavas. The name mediterranean clan was introduced by Niggli for a group of igneous rocks in which potash is a highly characteristic chemical constituent, sanidine and leucite being the two most characteristic minerals.

The mineralogical differences are shown by Rittmann's diagram of fig. 43. The geographic distribution of the magmatic clans or "provinces" is exhibited by Plate 7.

In order to explain the origin of the pacific rocks Kennedy distinguished two basaltic layers in the earth's crust. According to this theory the sialic part of the crust rests on rocks of tholeiitic composition and this layer in turn rests on material with the composition of olivine-basalt. The tholeiitic material would give rise to pacific differentiation products, whereas the atlantic lavas would derive from a melt of olivine-basaltic material from the lower layer. It would, however, be unreasonable to consider one of the layers to have originated as a differentation-product of the other. This objection invalidates the hypothesis of Kennedy. A much more plausible theory is advocated by Backlund, van Bemmelen, Fenner, Holmes, Rittmann,

Walker, Wegmann and many others. According to their opinion only one kind of parental basic magma exists, having an olivine-basaltic composition.

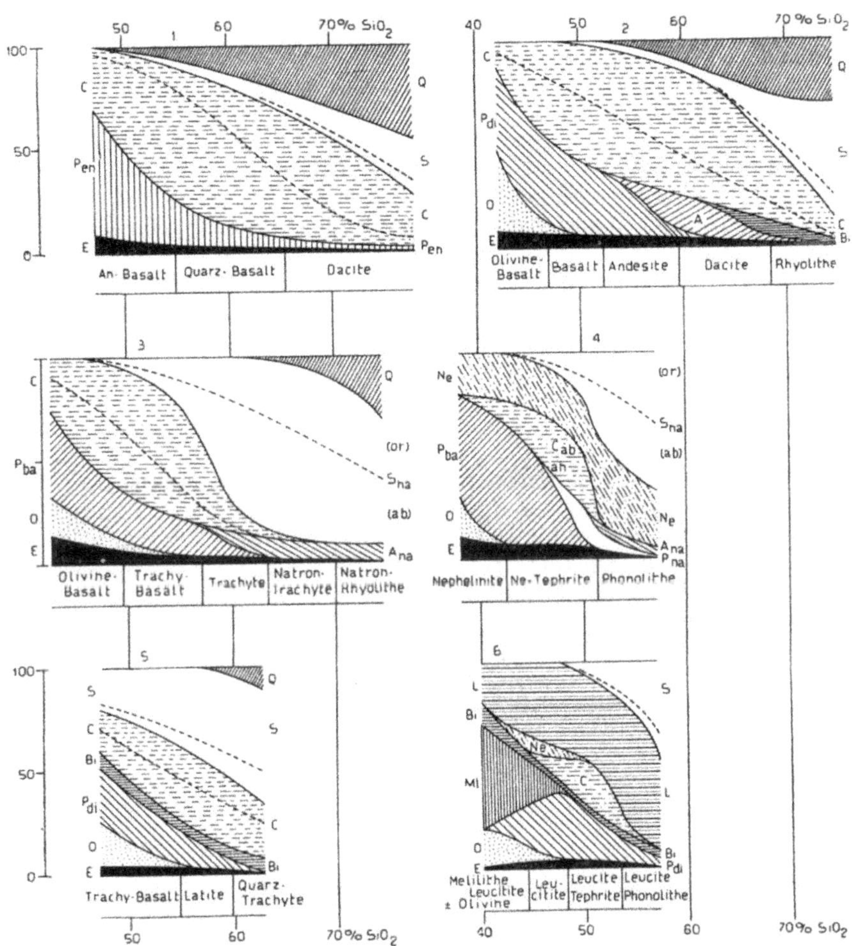

Fig. 43. Mineralogical composition of volcanic rocks belonging to different igneous provinces Horizontal axis: mineralogical types. Vertical axis: percentage of SiO_2.
Types of magmatic provinces: 1 strongly pacific, 2 slightly pacific, 3 slightly atlantic, 4 strongly atlantic, 5 slightly mediterranean, 6 strongly mediterranean.
Q quartz, S sanidine, S_{na} natron-sanidine, C plagioclase, Ne nepheline, L leucite, P_{en} enstatite and pigeonite, P_{di} diopside-augite, P_{ba} titane-augite, P_{na} aegirine-augite, A hornblende, A_{na} natron-amphibole. Bi biotite, O olivine, M melilithe, E ore and apatite. (After Rittmann).

The tholeiitic tribe originates from the olivine-basalt magma as a result of its acidification by melting and assimilation of parts of the sialic crust.

Pacific rock suites might originate as primary or juvenile differentiates from a parental magma of gabbroidal composition as was supposed by Bowen.

But it seems very probable that acid melts are frequently formed in a very different way.

One suggested process is that the ascending magma becomes altered into an acid melt by a process of melting and assimilation of pre-existing sialic crust-rocks. The new liquid has a composition intermediate between alkali-basalt and "pure" sial. It is called a syntectic magma.

Another process consists of the ionic or gaseous transfer of elements or mineral-components. And it seems highly probable that emanations from the substratum may convert continental crustal-rock into igneous rocks of a fundamentally different composition. Such a process is called *migmatization* and it may ultimately lead to *granitization* of pre-existing rocks of all kinds.

"Those who regard granites as having been largely formed from the pre-existing rocks recognize that they passed through a stage when part of the material was mobile or fluid. The partially fluid mash has been styled *migma* by Reinhard, to distinguish it from a mash consisting of incompletely crystallized magma. Moreover, magma may be generated from migma either by the attainment of complete fluidity or, at any stage, by the squeezing out of the fluid portion" [1]).

By the action on a large scale of one or both of these two processes pre-existing plutonic rocks and sedimentary strata such as schists, graywackes, quartzites, gradually become metasomatically altered into granite-bodies of large dimensions. Probably most, if not all, of the granite batholiths are not of primary igneous origin but originated as secondary or palingenetic products of granitization. The crustal rocks are enriched by alkali-silicates and

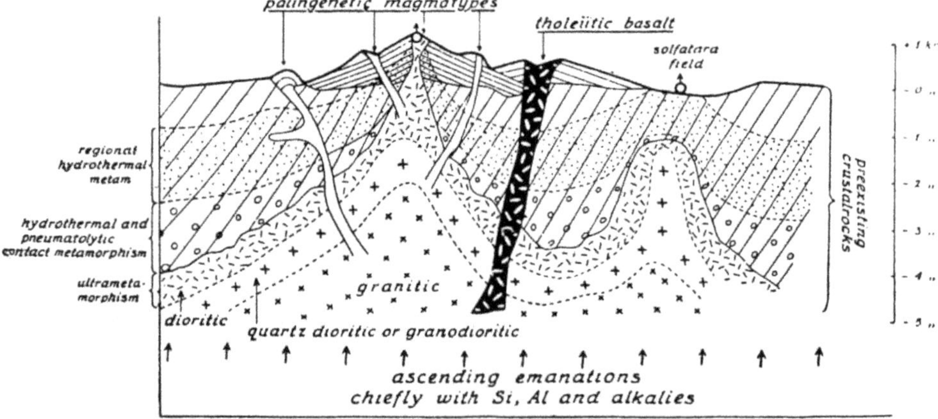

Fig. 44. Schematic representation of the rising migmatite front in the volcanic inner arc of the East Indies. (From R. W. van Bemmelen).

alkali-aluminates supplied from the ascending emanations. A process of progressive metamorphism — *migmatization* — converts them into calc-alkali rocks, the ultimate product being a granite body. Gradually the process

1) This is quoted from a paper by A. Holmes (1945), who published an excellent review of the historical succession of opinions and controversies on the natural history of granite.

migrates upward. Backlund and Wegmann speak of a rising *migmatite front.* Ever higher levels of the crust are invaded by the secondary (palingenetic) granite-melt. Immediately surrounding the granite mass is a zone of ultra-metamorphism, in upward direction passing into a zone of pneumatolitic and hydrothermal metamorphism, and still further by an aureole of crust-rocks that were influenced by hydrothermal processes only. A schematic representation is given in fig. 44, which represents Van Bemmelen's interpretation of his observations of plutonic and volcanic phenomena in Sumatra and Java. The distribution of the East Indian magmatic provinces and their relationship to the structural history of this remarkable island-festoon will be considered at some length in Chapter VII.

According to Holmes the geochemical relationships involved in the granitization of the country rocks can be briefly summarized in the formula: "granite = pre-existing rock plus added material (A) introduced by and abstracted from the incoming emanations (A + x), *minus* displaced material (B) driven forward with the outgoing emanations (B + x)". In the course of the notable investigations of the Newry igneous complex in Ireland Doris L. Reynolds has proved [1] "that the minimum introductions (A) were sodium, calcium and silicon; while, after several intermediate exchanges that are traced in detail, the displaced materials (B) eventually carried forward consisted of aluminium, iron, magnesium, potassium, hydrogen, titanium, phosphorus and manganese. The latter, together with some remaining sodium, calcium and silicon, became fixed in adjacent bands of hornfels which were thereby basified and transformed into rocks chemically equivalent to certain varieties of quartz-diorite. From these results and other relevant evidence, Dr. Reynolds concludes:

(a) that the introduced material (A + x) cannot have been an ordinary magma, since x has left no recognizable traces in the rocks;

(b) that the basic material migrating from a region of granitization, besides enriching the surrounding aureole in biotite and other minerals, was probably also responsible for the igneous-looking basic and ultrabasic rocks that overlie the granitic rocks of many plutonic complexes; and

(c) that before a given mass of country rock was actually granitized, it passed through a preliminary stage of basification".

The origin of the igneous rocks belonging to the mediterranean or potash clan is ascribed by Daly, Rittmann and Van Bemmelen to a process of calcification of a primary basaltic magma. Magma-sclerosis as Rittmann calls it, was described by him in a monograph on the structure and history of Mount Vesuvius. During an eruption blocks from Tertiary, Cretaceous, and Trassic strata underlying the volcano, are ejected. They are found in great quantities among the tuffs of Monte Somma (fig. 45). The fragments derived from Tertiary strata show hardly if any magmatic influence. Blocks of Cretaceous limestone, too, are mostly unaltered. Exceptionally they are recrystallized into crystalline limestone. They have been in contact with the magma only during the short time of their ascent in the craterpipe during an eruption. In strong contrast the blocks of Triassic dolomite and limestone

[1] Quoted from A. Holmes, op cit. 1945.

are throughly influenced by chemical processes. Many pneumatolitic silicate-minerals developed in them and not seldom the rock is entirely metamorphosed into silicate rock. The Triassic dolomites must have been subjected to a close and long-enduring contact with magmatic emanations. Therefore it is concluded that the upper part of the magma chamber has reached the

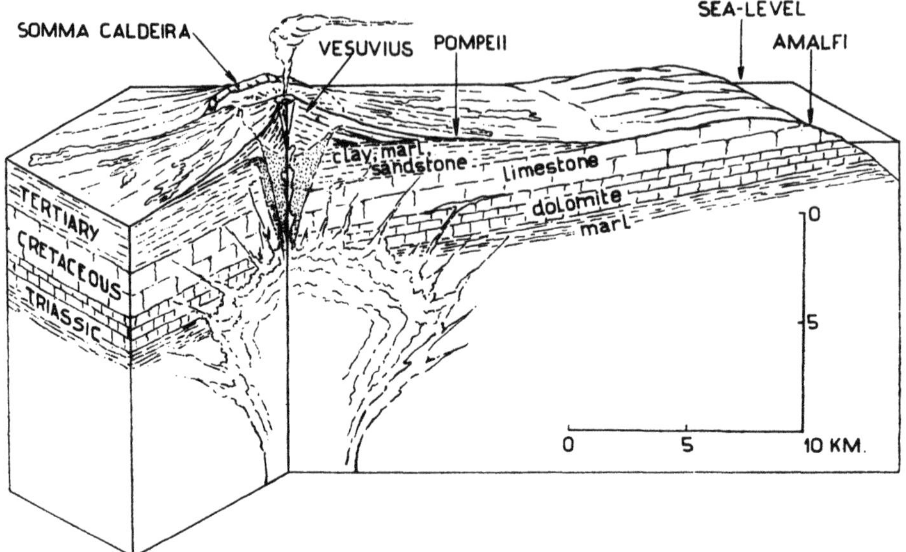

Fig. 45. Schematic block-diagram of the Somma-Vesuvius volcano.

Triassic strata and is gradually assimilating them. The Mesozoic strata crop out at the surface in the peninsula of Sorrento and Amalfi. So their thickness and the angle of their northwestward dip are known. From these data the depth of the top of the magma chamber below Vesuvius is estimated at approximately 5 kilometers below sea-level (fig. 45). From a closer examination of the lavas and tuffs of Somma and Vesuvius it appears that the composition of the magma was considerably modified in the course of time. The oldest eruption products of Somma [1]) are trachytes followed by vicoites and leucite-basanites and leucite-tephrites. And the still more recent eruption products of the Vesuvius consist of leucitites abounding in plagioclases (i.e. leucite-tephrite rich in leucite). The gradual transformation of the magma is ascribed to the progressive assimilation of Triassic dolomite and limestone.

Assimilation of limestone is considered to have caused the hydrothermal

[1]) The first eruptions of the Somma volcano began in the Pleistocene or shortly afterwards (according to Rittmann approximately 12000 years ago). The present Somma caldera originated during a large destructive eruption of the year 79. At that time Pompeii was covered by leucitic pumice, while Herculanum was buried under mud-flows. After some time the still active Vesuvius volcano grew up in the centre of the Somma caldera.

escape of soda. Consequently the potash content of the residual melt became comparatively enriched.

The extinct Laziale volcano, SE of Rome (fig. 46) probably[1]) may be

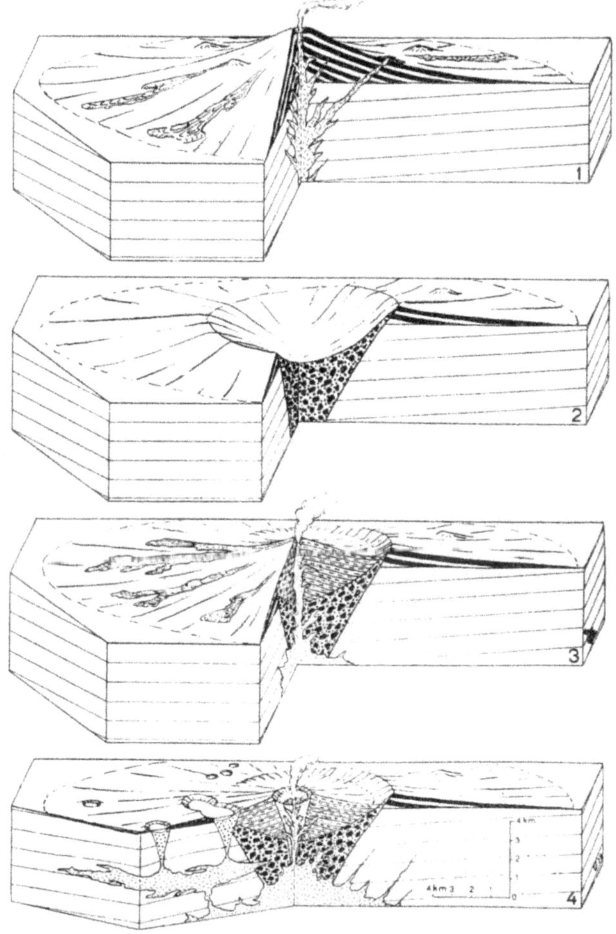

Fig. 46. Schematic representation of the history of the Laziale volcano, S.E. of Rome, Italy.

mentioned as another example of a volcano displaying an increasing calcification of its lavas.

The frequent occurrence of alkali rocks among the products of Italian volcanoes induced Niggli to introduce the name mediterranean province.

---

[1]) During a preliminary exploration in the so-called Albano mountains, SE of Rome, it appeared to the present author that the most recent lava flows of the Laziale volcano countain abundant leucite, together with numerous inclusions of dolomitic limestone blocks.

The relationship of the mediterranean magma clan to certain structural areas of the East Indies will be discussed in Chapter VII (p. 194).

The theory that the alkaline rocks may originate through the interaction of calc-alkali magma with calcareous rocks has been vigorously combated by several petrologists. Other theories were put forward by Bowen, Evans, Shand, and Holmes. Perhaps the truth is that the origin of alkaline rocks cannot be explained by one single generalizing theory. In any case the theory of magma-sclerosis seems to break down entirely for the leucitites and other mediterranean rock-types of the Bunyaruguru district of Uganda. For according to Holmes, who investigated the rock-suites from Uganda, no signs of limestone assimilation have been detected. And the distribution of the volcanic rocks in the area is independent of that of the dolomite and limestone. He suggests that the alkaline rocks may be due to a more intensive action of magmatic emanations.

Another question of geological interest is the remarkable relationship of certain igneous rocks to special epochs in the evolution of a folded mountain-chain and it is to this phenomenon of the so-called magmatic cycles that the next section will be devoted.

Tectonic and magmatic cycles

We now return once more to the question of a geological cycle, a diagrammatic view of which will be found in fig. 15. Should the terrestrial crust be drawn into this figure, it would become clear that a thick downward bulge of acid basement rocks must have penetrated the underlying, denser basic rocks during the process of crustal folding. A root of lighter rocks must then have formed under the folded geosyncline, either symmetrically and perpendicular or asymmetrically and oblique (fig. 47). In the last case the riding and overthrusting part of the crust caused the contents of the geosyncline to become squeezed unilaterally (B in fig. 47). The phenomenon is known very well from the Alps. In both cases, however, the resulting sial-root is due to the increasing compression of the earth's crust as a whole. Moreover this appears clearly to be true from the world-wide contemporaneity of the principal epochs of compression (as was set forth in Chapter II). It ought to be possible to determine the existence of such a root by means of gravimetrical observations, as the same would constitute an anomaly of the isostatic equilibrium.

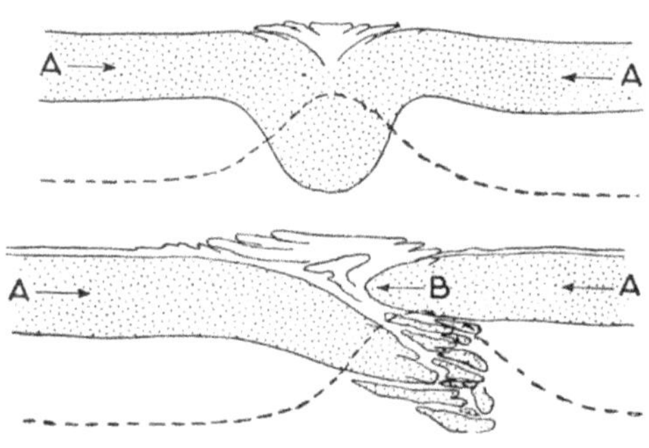

Fig. 47. Symmetrical and asymmetrical types of mountain-roots.

This sounds obvious, but, remarkably enough, the isostatic anomalies were in fact observed first, and from these was derived the existence of a sialic root. There can be no doubt, as Griggs remarked, that "one of the greatest contributions to the understanding of tectonics during the twentieth century has been Vening Meinesz' discovery of the great bands of gravity deficiency in the East and West-Indies" [1]). Vening Meinesz discovered abnormally large isostatic anomalies during his gravity expeditions in the East Indies in 1929 and 1930. One of the principal features of these deviations is a narrow belt of strong negative anomalies (fig. 116 and Plate 8), which was found to continue for about 5,000 miles. Anomalies are expressed in milligals, i.e. the third decimal — or, roughly, millionth parts — of gravity [2]). Anomalies of more than 50 milligal are but seldom observed, but this strip shows anomalies of more than 100, and occasionally of even more than 200 milligal. An analogous band was discovered in the West-Indies by the expeditions of Hess, Browne, Vening Meinesz, Hoskinson and Ewing (fig. 116). These remarkable strips of negative anomalies are undoubtedly due to an abnormal distribution of crustal material beneath the strip. Vening Meinesz' explanation — the only one that seems to cover all the facts — is that a great isoclinal protuberance of the crust was downfolded into the substratum. This led him to suppose "that the crust under the action of great horizontal stresses is buckling inwards, and that the folding and overthrusting of the surface layers, as found by the geologists, are an accompanying feature of this great phenomenon [3]). It would be impossible for us, at this point, to enter into an examination of the details of either the accompanying strips of positive anomalies, or their relation to the structural history — and seismic and geomorphological features — of the explored areas. These questions are dealt with extensively in Chapter VII.

We are at present particularly concerned with those phenomena in general which may be said to be associated with a subsiding and subsequently buckling crust. One characteristic phenomenon is that basic to ultrabasic rocks are found in the axis of a geosyncline, but that no acid plutonic rocks are observed at this stage of the development. Conversely, acid intrusions enter the geosynclinal belt during and after its folding. Thus an outpouring of large flows of lava is known to have occurred in the Jurassic Franciscan beds of the Sierra Nevada geosyncline, and these flows are actually found as intercalations between the strata. Diabase, gabbro and ultrabasic intrusions (serpentine), and, lastly, volcanic products, the latter mingling with the marine sediments, accompanied these outflows. Intensive folding ensued, during which the numerous batholithic intrusions of granodiorite, quartz-diorite and quartz-monzonite were emplaced [4]). Kossmat described a similar sequence in the Alpine, Variscian, Caledonian and older folded belts of Europe and in the Australian Caledonides. Stille grouped some European and American examples systematically in a table (fig. 48). The Alpine, Variscian, and Caledonian Mountains of Europe had already been grouped schematically at a previous date by both Stille and Lotze. This last diagram has been reproduced in Table II (see the last column at the

---

[1]) Griggs, 1939, p. 614—616.

[2]) Normal gravity amounts to 987.049 gal at the equator, and to a thousand times as many milligals.

[3]) Vening Meinesz 1933, p. 372.

[4]) Waters and Hedberg, Reg. Geol. der Erde, 1939.

extreme right). A great many examples will also be found in regional treatises [1]).

This remarkable sequence of magmatic rocks can also be derived without the aid of strained hypothetical constructions from the previously sketched structure of the earth's crust if we assume that a zone was first subjected

	EUROPE NORTH OF THE ALPS	ALPS	NEW ENGLAND	WESTERN N. AMERICA S. NEVADA	ANDES	CENTRAL AMERICA ANTILLES
PLEISTOCENE AND TERTIARY		o o o o		o o o o o o o o	o o o o o o o o	o o o o o o o o
CRETACEOUS		~+~+~+~+~ v v		~+~+~ ~+~+~	~+~+~+~+~ v v v v	~+~+~+~+~ ~+~+~ v v v v
JURASSIC		v v v v v v v		~+~+~+~+~ v v v v	v v v v v v v v	v v ?
TRIASSIC		v		v v v v v v v o	v o o o o	
PERMIAN	o o o o	o o o o		o v v v ~+~+~+~+~ v v v v	~+~+~+~+~ v v	
CARBONIFEROUS	o ~+~+~+~+~ v ~+~+~	~+~+~+~+~		v v v v v v v v	v v	
DEVONIAN	v v v v v v v v	v v v v v v v v	~+~+~+~+~ v v v v	v v v v		
SILURIAN	v v v v	v v v v	v v v v	v v v v		
ORDOVICIAN			~+~+~+~+~ v v v v v v v v	v v v v		
CAMBRIAN			v v	v		

o POSTOROGENIC VOLCANISM
\+ OROGENIC ACID INTRUSIONS
v PREOROGENIC BASIC EXTRUSIONS
~ EPOCH OF FOLDING

Fig. 48. Relations between magmatic phenomena and epochs of folding in different regions. Compare with Table II (After H. Stille).

to geosynclinal subsidence, and that it subsequently buckled as Vening Meinesz has suggested. The following considerations [2]) require but one assumption, i.e. that an upper layer of acid rocks rests on top of a layer of

[1]) With the aid of data on radioactive determinations of age, Kuenen recently showed that, pending confirmatory data, the relatively short periods of intrusion, which were separated by longer periods of quiescence, can similarly be determined in Pre-Cambrian times, and concluded that "the ultimate cause must lie below the crust and must be world-embracing". (Kuenen 1941, p. 337).

[2]) More or less similar considerations were published by A. Rittmann (Zur Thermodynamik der Orogenese, Geolog. Rundschau, 33, 1942), without mentioning, however, the paper of the year 1939 by the present author.

basic rocks. The intermediate layer will be left out to make things simpler. The block-diagrams of fig. 49 show that the crystalline crust consists of a sialic layer of approximately 20 km and this is seen to lie on a basic layer of equal thickness. It will be obvious that the ensuing reflections would hold good even if other thicknesses were concerned. In the meantime, I would like to make it clear that fig. 49 and 51 should be regarded as rough and entirely schematic sketches of the events. Moreover, anyone who has attempted to make such a drawing will have realized for himself that many dubious and difficult points arise in the course of such an undertaking.

When an area of subsidence originates in the earth's crust, sediments may accumulate at the surface in abnormal "geosynclinal" thickness. The lower part of the crystalline crust enters an area of higher temperature and pressure and subsides to the fluidity boundary. If that part of the crust were to melt, it would only be able to furnish basic to ultrabasic magma similar to the basic substratum already present beneath it (block I).

The strength of the crust will be exceeded when the downward movement reaches a certain level, and disruption will follow. The basic magma will then immediately (since it is subjected to high pressure) fill the fissures and faults in the crust. As the bending of the crust is strongest in the axis of the geosyncline, basic magma in this part will have the greatest chance of reaching the surface along faults. Moreover, that portion of the crystalline crust may be considered to have thinned as a result of subsidence, which caused the fluidity boundary to shift to a higher level (block I).

Once the sialic crust has been bent so far that it collapses, one of the consequences will be that it will buckle inwards into the deeper layers of the earth's crust (block II). The result will be:

(1) Folding of the contents of the geosyncline, together with the basic rocks in it.

(2) A local thickening of the sialic crust, resulting in the formation of a "root" of sialic material, penetrating into an area of progressively higher temperature and pressure.

(3) When the material of the root begins to disintegrate and invade the surface strata as batholithic intrusions, plutonism and volcanism cannot but be strongly acid ("pacific") in character, since the batholiths assimilate more and more sialic material from the root, and eventually from the sediments of the already folded geosyncline (acidification). The ten characteristics of batholiths, and the explanation which Daly gave of them [1]), fit the diagrammatic explanation given here.

(4) As soon as the compressive forces in the sialic crust decrease, the "root", in order to resettle into isostatic equilibrium, will have a strong tendency to rise. This effect is heightened by the expansion of the sialic root, which, through fusing, grows specifically lighter. This explains how a mountain-chain is formed from a folded geosyncline. More and more deeply situated batholiths are exposed by denudation at this stage (block III).

(5) The upward movement will continue till the isostatic equilibrium has been re-established. The sima will only be able to crystallize beneath

---

[1]) Daly, 1938, p. 169—173.

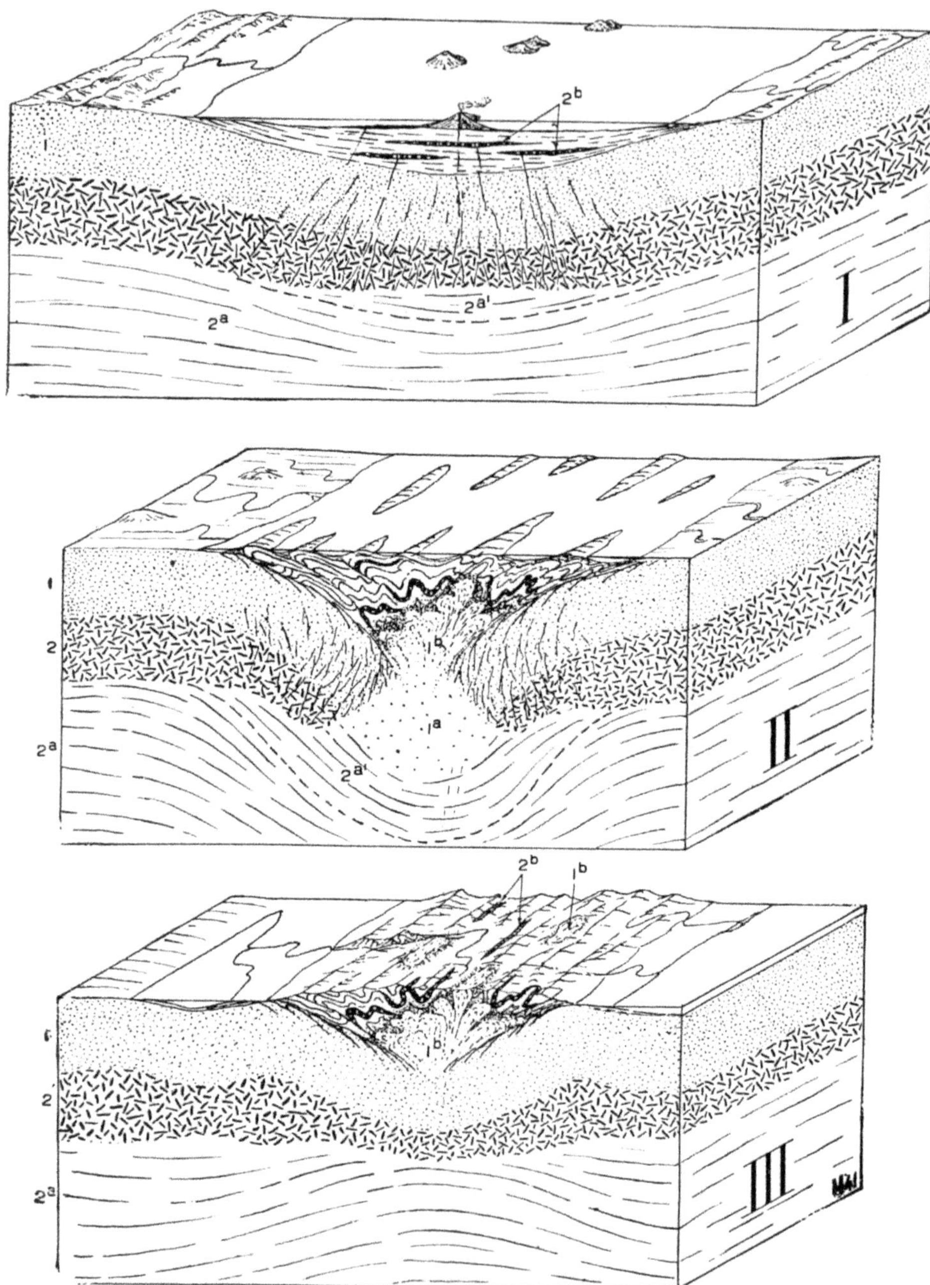

Fig. 49. Schematic and tentative block diagrams illustrating the relation between tectonic and magmatic cycles. 1 and 2 crystalline crust; 1 sialic layer; 1a molten part of 1; 1b acid batholithic intrusions and metamorphic aureole; 2 basic layer; 2a basic substratum; $2a^1$ molten part of 2; 2b basic abyssolothic injections, extrusions and volcanism.

the upper sialic layer to a thickness corresponding approximately with that of the sialic layer after part of the sialic root, stiuated at an abnormal depth, has spread out and disappeared (i.e. the crystallized sima will be thinner as the sial grows thicker) [1]). As an accompanying phenomenon of the rising zone, sima will flow in from the neighbouring strips of the earth's crust. This process may perhaps explain why a depression or "marginal deep" originates on one side, or both sides of a mountain belt.

As pointed out in Chapter II, the "zonal migration" of geosynclines can be traced distinctly in the history of some continents from the Cambrian up to our own time. Though the exact nature of events in the substratum cannot be determined, the subsidence of a geosyncline may be stated to be the result of internal terrestrial processes, whereas the sedimentation may be described as a secondary, external effect.

Fig. 48 and Table II also include post-orogenic volcanism, which has been indicated with separate notations. This volcanism may perhaps be partly due to differentiation of the orogenic batholiths, but it is also partly related to the dome-shaped elevations which will be discussed in Chapter XI (p. 308).

A peridotite-layer in the substratum

Hess has pointed out that the axial part of a folded geosyncline is characterized by a zone of dunite or serpentinized peridotite (also called "ophiolite"). He considers, moreover, that in the island-arcs the peridotites were intruded only during the formation of the first great down-buckle.

Later epochs of compression implying a rejuvenated down-buckling of the sial-root, were not accompanied by a further series of peridotite intrusions. The complete absence of such ultramafic intrusions elsewhere clearly indicates

Fig. 50. Situation of a world-encircling peridotite layer. (From H. H. Hess).

the relationship between the development of the first down-buckle and the intrusion of peridotite from a primary dunitic magma-layer. Hess speaks of a peridotitic substratum and supposes it to be present at a depth of approximately 60 km (fig. 50). Regarding the mechanism and time of peridotite intrusion he wrote [2]):

"During the forcing of the bottom of the downbuckle into the upper part of the peridotitic substratum, sufficient stress is present to permit squeezing off of a product of partial fusion of the peridotite substratum (this is the hydrous peridotite magma). Under any other conditions this product of partial fusion would merely remain between the interstices of the grains

[1]) Vide antea, p. 69 and cf. Daly, 1938, p. 190, fig. 151.

[2]) H. H. Hess, 1938, p. 333.

of the peridotite substratum, because as a rule probably insufficient stress would be present to squeeze it off, or if squeezed off, it might not be able to penetrate the relatively plastic basaltic layer above it. The relatively rigid and strongly deformed down-buckle allows the hydrous peridotite magma to migrate up its vertical structures, and thus the products of this magma are formed over or near the axis of the downbuckle or in the belts on either side of the downbuckles".

"That the peridotite intrusions accompany only the first deformation may be accounted for by two possible factors: (1) that after the first squeezing off of a product of partial fusion insufficient low melting material is present to yield a second body of magma; or (2) that fusion of the bottom of the downbuckle after the first deformation forms an impermeable cap through which the ultramafic magma cannot be intruded. Either one or both of these possibilities may be operative".

We agree with the interpretation given by Hess with the exception of one point. It seems more plausible to suppose the peridotite layer to be present at a lower level below the crust [1]). Three reasons enduce me to prefer this supposition.

Firstly the presence of a layer of basaltic composition immediately below the crystalline crust can hardly be doubted. For it is in accordance with what is known about the world-wide distribution of basalt-outpourings.

Secondly, the absence of peridotite intrusions during later epochs of rejuvenation may be interpreted to mean that the first downbuckle probably invaded deeper into the substratum than it did in later stages. Probably the first was more rigid and capable of giving fissures that could tap the magma. Later ones would tap higher magmas.

In the third place, a deeper situation of the peridotite layer agrees better with the distribution of the fields of strong positive anomalies of isostasy in the East Indies. This point will be made clear in Chapter VII (p. 196).

If the depth to which the *first* downbuckle invaded were known, the depth of the peridotite-layer would be known too. Anyhow, we would like to insert it tentatively in fig. 41 approximately at the lower boundary of the scheme (Cf. also fig. 130).

Zonal migration of geosynclines. Continents and ocean-floors

The oldest cores of the continents are known to consist of intensely folded and migmatized belts. None of the oldest areas observed so far can be regarded as a primary, i.e. an unfolded fragment of the earth's original sialic crust. That part of the drawing in fig. 49 which represents the seemingly simple structure of the basement of a geosyncline, or of an area of folding, can undoubtedly be stated to be very complicated, for it passed through one or more geological and magmatic cycles. (fig. 51 therefore acts as a supplement to fig. 49, II). This fact, if born in mind, might induce us to

---

[1]) A similar opinion was expressed by Daly (Floor of the Oceans, 1942, p. 58).

suppose that the continents might have originated as a result of periodical folding, culminating in a consecutive thickening of a relatively thin sialic layer originally enveloping the entire earth. We then feel inclined to enquire whether two problems cannot be explained simultaneously, viz. (1) the formation of continental blocks (in which geosynclines and folded mountain-

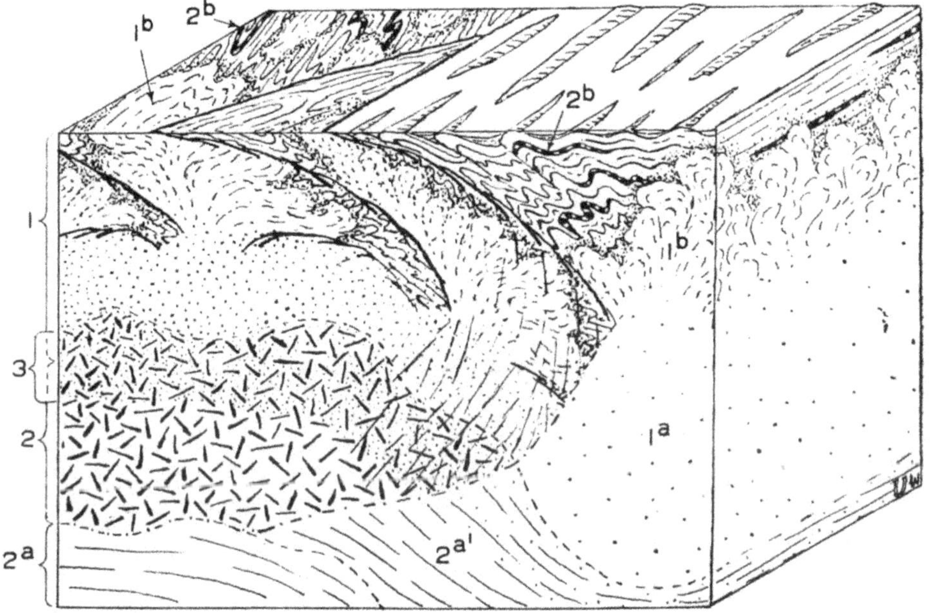

Fig. 51. Tentative illustration of four subsequent tectonic and magmatic cycles in a continental area, completing the left part of block II in fig. 49. For explanation of the figures see fig. 49 and 52.

chains originated subsequently, according to the complicated patterns discussed in Chapter II), (2) the formation of sial-free parts of the earth, such as the floor of the Pacific (fig. 68). These speculative reflections differ to a considerable extent from the hypotheses of Fisher, Wegener, Schwinner, Escher and others mentioned in Chapter VIII. For, unlike these authors, the opinion may be put forward that an upper and acid sialic layer which enveloped the whole earth during a very early stage of its history was indeed stretched over a basic lower layer. In the beginning the material of both layers must have originated from a primordial melt containing the combined elements of the diverse, now contrasting, crustal and deeper shells.

These entirely hypothetical considerations lead directly to the important problem of the origin of continents and oceans, but this question will be dealt with in Chapter VIII. However, I would like to draw attention to an important deduction. If we assume the primordial crust to consist of two layers — an upper acid and a lower basic layer, indicated as 1 and 2 respectively in fig. 52 — 1a would represent the acid melt formed as a result of buckling of the crust. This liquid, however, will probably not only move upwards in the shape of batholithic intrusions (1b), but will also flow out laterally in the direction of the arrows. If the molten sial (1a) and sima (2a and 2a[1]) recrystallize during a later stage, i.e. during the rising of the root, a zone containing a mixture of 1a and 2a will have to form in the crust. This may help to explain the formation of a layer such as 3, which is intermediate to the layers 1 and 2 both as regards its position and its petrographical composition (zone 3 has been indicated similarly in fig. 51).

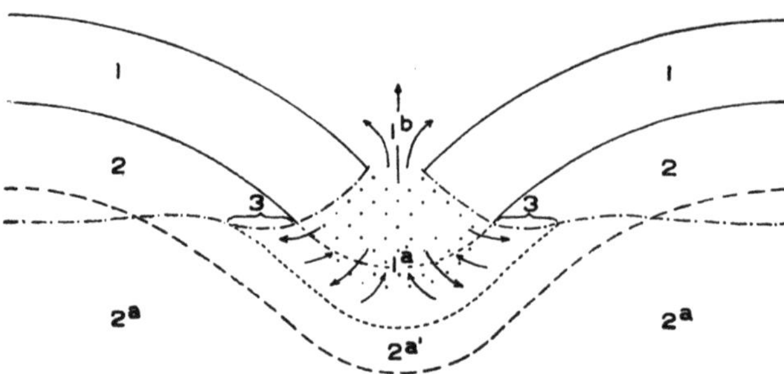

Fig. 52. The possible origin of an intermediate continental layer (3). For an explanation of other figures see fig. 49 and 51.

Volcanism in basins

The foundation of a subsiding basin, like the down-bending crust of a geosyncline, will be affected by fracturing and become injected with magma. The magmatic products intrude as sills, and flow out at the surface as lava flows, or mingle with the sediments as clasmatic products. This can be observed in many basins, and it would certainly be worth while to investigate to what depths the bottoms of geosynclines and basins subsided before the first series of basic rocks made their appearance. This would provide us with valuable data as regards the possible mechanical process in the earth's crust. Repeated reference is made in the descriptions of these basins to basalt, gabbro, diabase, melaphyr and other basic rocks, but no acid rocks have ever been observed so far as I know. It need hardly be said that those basins which are closely connected with zones of folding (the basins of the intramontane type e.g.) should be considered *cum grano salis*, for pacific rocks in their immediate vicinity invaded the crust before and at times even during their formation. It is certainly noteworthy that the marginal deep which originates subsequently to the principal epoch of folding of the geosyncline (cf. Chapter III) should, generally speaking, seem devoid of volcanism. This negative characteristic must have some connection with the deep crustal buckle which was not formed until a short while before and along which the basins and troughs are arranged. The question will be dealt with at some length in Chapter VII.

Fragmentation and growth of continents

"And so the earth is growing old". This sentence appears at the end of one of Stille's numerous papers on the structural history of the continents [1]).

[1]) Stille, Research and Progress, vol 1, no 1, 1935, p. 14.

In my opinion, however, the distinguished author to whom we owe so many valuable data and suggestions regarding periodical events in continental areas should not have closed his reflections with such an *accord en mineur*. In his interpretation of the face of the earth Stille is obviously impressed by the breaking down in later times of large stretches of the mighty belts of Caledonian and Variscian mountains. America has lost territory to the east and west in the foundered blocks of Appalachia and Cascadia. Africa, India and Australia, all of which paid heavy tribute to the ocean, are mere remnants of larger areas. They seem to illustrate a proces of continental fragmentation and decay. Along the eastern border of Asia a whole series of basin-shaped depressions originated by the collapse of a corresponding series of sialic masses. Rift-valleys opened like gaping wounds in the earth's skin. They nearly cross Europe from the Vettern graben in Sweden to the border of the Mediterranean and they break across Africa from the Red Sea to the Zambesi.

There is nothing to contradict these statements. However, in spite of these and other similar facts I cannot agree with Stille in his speaking of a *marasmus sinilis* of our planet. It may be true that part of the impressive chains of Variscian times are covered by the waters of the Atlantic. But we are living in a geological epoch which has its own mighty chains of Alpine grandeur. And we should not forget that in several districts the lofty chains of the Caledonides and Variscides ruthlessly dissected and destroyed structural units of previous periods. The Paleozoic earth had also its wounds. And are not we witnesses of the rejuvenation of many an old pattern into a fresh-looking expression? Constantly erosive forces arc wearing down the earth's surface trying to demolish its vigour into the dullness of continental peneplaines. So long, however, as the earth disposes of the youthful power to withstand the destructive work of denudation I cannot detect any symptoms of senility in the pulse of the earth.

If we were to adopt the metaphor of a human creature we could better compare the processes of the earth's interior to processes of metabolism in a living being. For what happens is a constant change of surface expression under the influence of periodical processes in the interior.

The surface of a continent is steadily lowered by the perpetual action of denudation. Loss of matter at the surface is partly compensated by isostatic uplift rejuvenating the general relief of the continents. So, the flotation of the continents remains a feature enduring through the aeons of geologic time. If, however, no fresh sialic matter were to be supplied to the continental bucklers, they would become ever thinner and thinner. At a final stage the areas occupied by the primordial continents would have degenerated into very thin and perfectly peneplained sial-flakes. There would remain no possibility for them to rise or sink. Some 9/10 of their original thickness would have been removed for ever, — assuming that sial has a mean specific gravity of 2,7 and sima 3,0 [1]).

Hence the question arises what processes may be available to counterbalance the constant lowering of continental relief by erosion and denudation. Transport of waste-products from the surrounding mountain-belts would account only for a certain retardation of the process of levelling. Probably

[1]) LAWSON, 1932, pp. 363 and 364.

therefore, addition of sialic matter must come from below. And, indeed such a process is not improbable in as much as the molten part of the sial-root under a mountain-chain (represented by 1a in block II of fig. 49) will spread laterally. On account of its light specific weight and high radioactive content it has an urge to migrate in an upward direction, melting its way into the crust. Convection currents may transport material from the root over long distances and, thus, far distant parts of the crust will be supplied by sialic matter. However, one or more zones of geanticlinal updoming generally originate in the vicinity of a belt of crust buckling (see Ch. VII, p. 191). These are places *par excellence* for the ascent of light and acid magmatic emanations and melts.

Obviously a sialic melt could be provided not only by the molten part of a mountain-root but also by any deeply subsided part of the crust. Sub-crustal sial-melts may also originate below the bottom of a continental basin, in the subsided part of a continental margin (cf. p. 120), and at the base of a deep-sea trough or of a subsided block like Appalachia.

In any case it seems clear that new sialic melts must be continually formed under certain parts of the crust. And the spreading, migration and subsequent ascent of acid melts due to periodical processes in the earth supplies the underside of the continents with the sialic material that they lost at the surface. So, far from growing old, the continents are being periodically rejuvenated.

Summary

The conception of the structure of the earth's crust, as advocated above, makes it clear that the presence of certain magmatic rocks must (assuming that a period of geosynclinal subsidence is followed by crustal buckling, as postulated by Vening Meinesz) be related to the phenomena of geosynclinal subsidence, folding and mountain-building. Thus, it is obvious, for instance, that only basic to ultrabasic rocks can occur in a geosyncline, and that intrusions of acid batholiths take place after the crust has buckled and the contents of the geosyncline been folded. All evidence shows that periods of increasing compression of the earth's crust alternate with periods of decreasing compression. During this last stage the sialic root (which was brought about by buckling of the crust) is given an opportunity to rise until the isostatic equilibrium has been restored. A mountain-chain, in a geographical sense, then comes into existence. This process has thickened the sialic layer in the meantime, and the continents may consequently have originated as a result of periodical folding and thickening of an originally thinner sialic layer enveloping the whole earth.

The accumulation of sediments in a geosyncline is a secondary and external effect of a subsiding crustal furrow, and this phenomenon, together with the alternating periods of increasing and decreasing compression, and the migration of geosynclines, must be attributed to some common and world-embracing, deep-seated cause, for all the phenomena are related chronologically.

The theory is summarised of a common origin of all the igneous rocks belonging to three magmatic tribes called atlantic, pacific, and mediterranean — from a parental material of olivine-basaltic composition which forms a world-encircling layer immediately below the crystalline crust. Most of the granite batholiths are thought to be of palingenetic or syntectic origin. The genesis of rocks belonging to the mediterranean clan by a process of assimilation of dolomite and limestone is described from Somma and Vesuvius. Other theories on the origin of alkaline rocks are mentioned. At any rate the example of the Bunayaruguru district in Uganda shows that a process of magma-sclerosis cannot have taken place in that area. For no signs of limestone assimilation have been detected and the distribution of the alkaline rocks is independent of that of dolomite and limestone in that area.

A peridotite-layer is probably present in the substratum below a layer of olivine-basalt. This explains why peridotite intrudes a mountain-root only during its first deep down-buckling, whereas later epochs of compression, implying a rejuvenated down-thrusting of the sial-root, are not accompanied by another series of peridotite intrusions.

The continents are rejuvenated periodically, receiving a supply of sialic melts derived from the spreading, migration and subsequent ascent of sialic melts from the molten part of mountain-roots and the subsided parts of the crust such as basins and submerged borderlands.

References

BACKLUND, H. S. *Der "Magma-aufstieg" in Faltengebirgen* (C. R. Soc. Geolog. Finland, 9, 1936).

BACKLUND, H. G. *The granitization problem* (Geolog. Magaz. 83, 1946).

BEMMELEN, R. W. VAN, *Igneous geology of the Karang Kobar region and its significance for the origin of the Malayan potash provinces* (Ingen. Nederl. Indië 4, 1937).

BEMMELEN, R. W. VAN, *On the origin of the Pacific Magma types in the volcanic Inner arc of the Soenda Mountain system* (Ingen. Nederl. Indië 5, 1938).

BUBNOFF, S. VON, *Grundprobleme der Geologie* (1931).

BUCHER, W. H. *The deformation of the Earth's Crust* (Princeton Univ. Press 1933).

BULLARD, E. C. *Heat flow in S. Africa* (Proceed. R. Soc. London A. 955, vol. 173, 1939).

BULLARD, E. C. *Thermal history of the Earth* (Nature, 156, 1945).

CLOOS, H. *Hebung — Spaltung — Vulkanismus* (Geol. Rundschau 30, 3A, 1939).

CLOOS, H. und RITTMANN, A. *Zur Einteilung und Benennung der Plutone* (Geolog. Rundschau 30, 1939).

DALY, R.A. *Architecture of the Earth* (1938).

DALY, R. A. *The roots of Volcanoes* (Trans. Amer. Geophysic. Union. Report General Assembly 1938).

DALY, R. A. *Igneous Rocks and the Depth of the Earth* (Mc. Graw-Hill, 1933).

DALY, R. A. *Volcanism and petrogenesis as illustrated in the Hawaiian Islands* (Bull. Geol. Soc. America, 55, 1944).

ESKOLA, P. *On the origin of granitic magmas* (Miner. u. Petrogr. Mitt. 42, 1932).

ESKOLA, P. *Wie ist die Anordnung der äusseren Erdsphären nach der Dichte Zustande gekommen?* (Geol. Rundschau 27, 1936).

FENNER, C. N. *A view of magmatic differentation* (Journ. of Geology 45, 1937).

GRIGGS, D. *A theory of mountain-building* (Americ. Journ. of Sci. 237, 1939).

GUTENBERG, B. *Internal constitution of the Earth* (Physics of the Earth VII, 1939).

GUTENBERG, B. *Seismological evidence for roots of Mountains* (Bull. Geolog. Soc. America, 54, 1943).

GUTENBERG, B. *Variations in physical properties within the earth's crustal layers* (American Journ. of Sci., 243 A, 1945).

HARKER, A. *Metamorphism, A study of the transformation of Rock masses* (London 1932).

HESS, H. H. *Primary peridotite magma* (Americ. Journ. Sci. 35, 1938).

HOLMES, A. *Contributions to the theory of magmatic cycles* (Geol. Magaz. 62, 1925; 63, 1926).

HOLMES, A. *Radioactivity and earth movements* (Trans. Geol. Soc. Glascow 18, 1928–'29).

HOLMES, A. *The origin of igneous rocks* (Geol. Magaz. 69, 1932).

HOLMES, A. and HARWOOD H. F. *The Petrology of the Volcanic Area of Bufumbira* (Geol. Surv. Uganda. Mem. 3, part 2, 1937).

HOLMES, A. *Leucitized Granite Xenoliths from the Potash-rich Lavas of Bunayaruguru, S. W. Uganda.* (Americ Journ. Sci. 243–A, 1945).

HOLMES, A. *Natural history of Granite* (Nature, 155, 1945).

JEFFREYS, H. *Types of isostatic adjustment* (Americ. Journ. Sci. 243–A, 1945).

KENNEDY, W. Q. *Trends of differentiation in basaltic magmas* (Americ. Journ. of Sci. 25, 1933).

KENNEDY, W. Q. and ANDERSON, E. M. *Crustal layers and the origin of magmas* (Bulletin volcanologique II, 3, 1938).

KOSSMAT, F. *Paläogeographie und Tektonik* (1936).

KOSSMAT, F. *Der ophiolitische Magmagrütel in den Kettengebirgen des Mediterranen Systems* (Sitzungsber. Preuss. Akad. d. Wissensch. 24, 1937).

KUENEN, PH. H. *The negative isostatic anomalies in the East Indies* (Leidsche Geol. Mededeelingen VIII, 1936).

KUENEN, PH. H. *Major geological cycles* (Proc. Nederl. Akad. v. Wetensch. Amsterdam XLIV, 1941).

KÜHN, W. und RITTMANN, A. *Über den Zustand des Erdinnern und seine Entstehung aus einem homogenen Urzustand* (Geolog. Rundschau 33, 1941).

LAWSON, A. C. *Insular arcs, foredeeps, and geosynclinal seas of the Asiatic coast* (Bull. Geol. Soc. America, 43, 1932).

MOLENGRAAFF, G. A. F. *The coral reef problem and isostasy* (Proc. Koninkl. Akad. v. Wetenschappen Amsterdam, 19, 1916).

NIGGLI, P., and BOGER, G. J. *Gesteins- und Mineralprovinzen* (Borntraeger, Berlin 1923).

NIGGLI, P. *Das problem der Granitbildung* (Schweiz. Min. Petr. Mitt. 22, 1942).

PETRASCHEK, W. E. *Gebirgsbildung, Vulkanismus und Metallogenese in den Balkaniden und Südkarpathen* (Fortschr. der Geologie und Palaeont, 14, heft 47, 1942).

RASTALL, R. H. *The Granite Problem* (Geolog. Mag. 82. 1945).

READ, H. H. *Meditations on granite* (Proceed. Geol. Assoc. 54, 1943; 55, 1944).

REYNOLDS, D. L. *The albite-schists of Antrim and their petrogenetic relationship to Caledonian orogenesis* (Proc. R. Irish Acad. 48, 1942).

REYNOLDS, D. L. *Granitization of Hornfelsed sediments in the Newry Granodiorite* (Proc. R. Irish Acad. 48. 1943).

REYNOLDS, D. L. *The south-western end of the Newry Igneous complex* (Quart. Journ. of the Geol. Soc. London 99, 1944).

RICHARDSON, W. A. *Problem of Batholithic Intrusion* (Geolog. Magazine 60, 1923).

RITTMANN, A. *Vulkane und ihre Tätigkeit* (1936).

RITTMANN, A. *Die Entstehung des Sials und die Herkunft der Vulkanischen Energie* (Geologie en Mijnbouw, 1939).

RITTMANN, A. *Die geologisch bedingte Evolution und Differentiation des Somma-Vesuvmagmas.* (Zeitschr. f. Vulkanologie 15, 1933).

RITTMANN, A. *Zur Thermodynamik der Orogonese* (Geolog. Rundschau 33, 1942).

RUTTEN, L. *On the age of the serpentines in Cuba* (Proc. Kon. Acad. v. Wetensch. Amsterdam, vol. 43, 1940).

SCHWAIBACK, M. *Vulkanismus und Senkung in den Kaledonischen Geosynklinale Mitteleuropas* Geolog. Rundschau 34, 1943).

SEDERHOLM, J. J. *Batholithes and the origin of the granitic magma* (14*er* Int. Geolog. Congress, Bd. I, 1933).

STILLE, H. *Zur Frage der Herkunft der Magmen* (Abh. Preuss. Akad. d. Wiss. Jahrg. 1939, 1940).

UMBGROVE, J. H. F. *Verschillende typen van tertiaire geosynclinalen in den Indischen Archipel* (Leidsche Geol. Mededeel. 6, 1933).

UMBGROVE, J. H. F. *The relation between magmatic cycles and orogenic epochs* (Geol. Magazine, vol. 76, 1939).

UMBGROVE, J. H. F. *Vulkanische verschijnselen in de omgeving van Napels* (Tijdschr. Kon. Nederl. Aardrijksk. Genootschap, 2e ser. 56, 1938).

UMBGROVE, J. H. F. *De Vulkaan van Rome* (Tijdschr. Kon. Nederl. Aardrijksk. Genootschap 74, 1940).

VENING MEINESZ, F. A. *The mechanism of mountain formation in geosynclinal belts* (Proceed. Kon. Acad. v. Wetensch. Amsterdam, vol. 36, 1933).

VENING MEINESZ, F. A. *Gravity over the Hawaian archipelago and over the Madeira area; conclusion abouth te earth's crust* (Proceed. Nederl. Acad. v. Wetensch. 44, 1941).

WEGMANN, C. E. *Zur Deutung der Migmatite* (Geolog. Rundschau 26, 1935).

WEGMANN, C. E. *Geological Investigations in Southern Greenland. Part I. On the structural Divisions of Southern Greenland* (Medd. om Gronland. 113, no. 2 1938, pp. 74–83 and 136–142).

CHAPTER V

OSCILLATIONS OF THE SEA-LEVEL

"We have to account for periodic transgressional seas and orogenesis upon an enormous scale and clearly associated with vertical continental movements". (J. JOLY)

Introduction

The sea has repeatedly invaded large portions of the continents and then subsequently retreated. This part of the earth's history is an extremely monotonous one, but this very monotony, this continuous recurrence of the same events, constitutes one of its most important aspects. Why does the sea invade the continents, and why does the water later withdraw to the oceanic receptacles, only to advance again afterwards? In other words, in geological terminology, — what is the origin of the periodic trans- and regressions?

Before considering the major periodic movements, however, we should first note that there are undoubtedly examples of trans- and regressions of a more limited, regional importance. In these cases it is but seldom possible to connect the relative shifts of the sea-level directly with one or more clearly discernible impulses.

In geosynclinal areas the phenomena may occur at a different rate as a result of their own and possibly opposed movements, and the negative or positive shifting of the sea-level may perhaps be quite counterbalanced. It is even conceivable that transgression may occur in a geosynclinal zone simultaneously with a negative phase, and vice versa. On the continents, too, the strand may advance or retreat over an area of greater or lessser extent as a result of different causes, but this need not necessarily indicate a world-wide movement.

A few such cases should be discussed before passing on to an examination of changes of world-wide importance.

Regional transgressions and regressions

The sea invaded the Baltic regions twice during the melting of the so-called Fennoscandian ice-sheet (fig. 53). The first to form as the ice-sheet retreated was the Baltic ice-lake, which occupied the southern part of the actual Baltic Sea (fig. 53 A). A strip of land in the south of Sweden became free as a result of the steadily increasing recession of the ice-front. This area was so low that it was flooded by the ocean, causing the Baltic fresh-water lake to be transformed into a sea — the Yoldia-Sea, which was

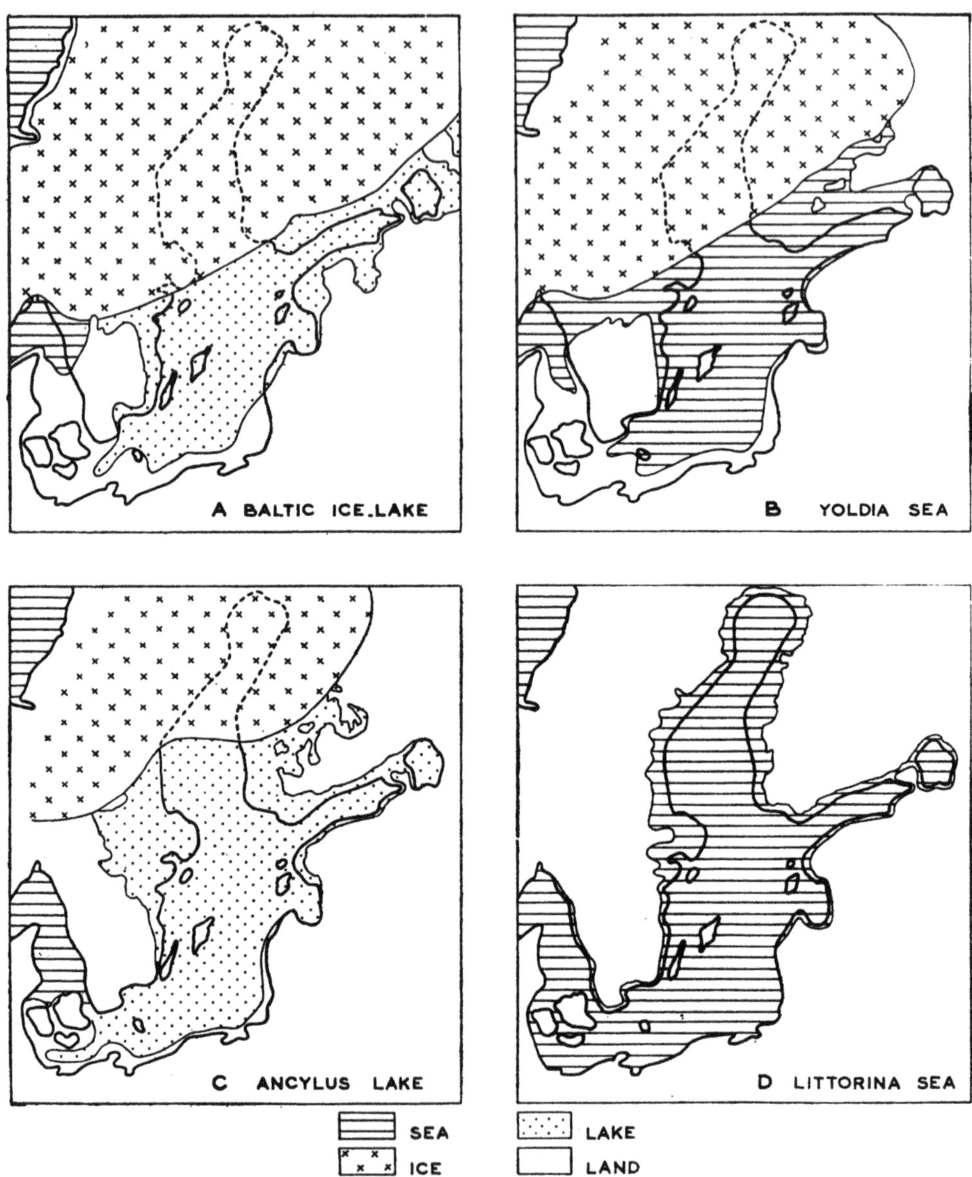

Fig. 53. Pleistocene history of the Baltic region.

connected with the Atlantic (fig. 53 B). The crust of South and Central Sweden was now rid of the load of ice which had previously been pressing it down and began to rise gradually. The connection with the Atlantic was consequently severed, and the Yoldia-Sea was soon replaced by a fresh-water body — the Ancylus lake (fig. 53 C). While the crust had been strongly basined under the load of land-ice, the marginal belt around the glaciated tract had been super-elevated. As pointed out, the land began to rise slowly after the ice had melted. On the other hand, the peripheral belt — which includes the North of Germany and Denmark — began to subside. This, combined with the rising of sea-level by the melting of ice, caused the con-

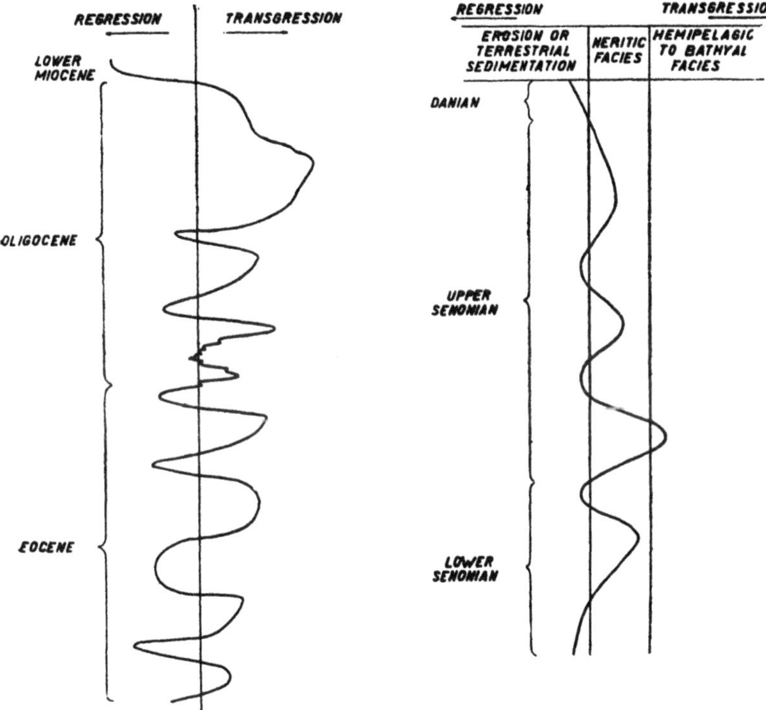

Fig. 54. Tertiary transgressions and regressions in the basin of Paris (After H. Stille).

Fig. 55. Transgressions and regressions in the Upper Cretaceous of Limburg, in the Netherlands.

nection with the waters of the Atlantic to be re-established. The Straits of the Sund, the Great Belt and Little Belt were opened, and the fresh-water of the Ancylus lake was replaced by salt water — the so-called Littorina-Sea (fig. 53 D), which was gradually transformed into the present Baltic-Sea.

I mentioned this example because it clearly illustrates how the local rhythm of the alternation of fresh and salt water can be derived from one constant, universal and non-periodic factor. The land movements are in fact a consequence of the retreat of the Fennoscandic inland-ice, while the rise of the sea-level is due to the melting of the Fennoscandic ice-cap and other ice masses. All this is brought about by one changing factor, viz. a world-wide increase in the average annual temperature.

This example clearly shows that the repeated advance and retreat of the sea in a restricted area need not necessarily be the reflection of a universally active cause that changes periodically. The curve in fig. 54 indicates the successive trans- and regressions which affected the basin of Paris. If we bear in mind that the basin has a complicated history (see Chapter III), it will become clear that the latter has also found expression in the curve. In many cases the origin of trans- and regressions remains a matter for conjecture. To what cause or causes must we for instance attribute the oscillations of the sea-level in South-Limburg during the Senonian (fig. 55)? We have to content ourselves with more or less vague surmises. In a very readable article, Born mentions examples of great differences in amplitude and time of such rhythmical movements. It seems as yet impossible to furnish a satisfactory explanation for all these phenomena. When considering a world-wide "rhythm", we should not forget that numerous local movements, which may have had a very different origin, have undoubtedly also played a part in several regions.

World-wide transgressions and regressions

Large transgressions, however, are undoubtedly known to have occurred at a time when there could be no question of a considerable change of sea-level under the influence of the melting and extension of vast masses of ice,

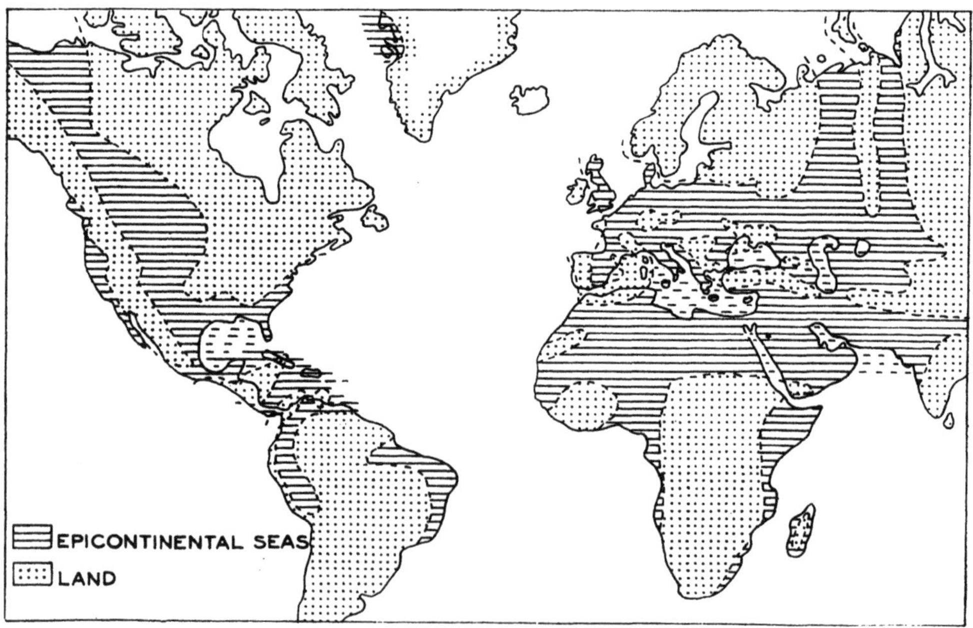

Fig. 56. Part of a world-map showing transgressive seas of the Upper Cretaceous.

the considerable expanse of which only proves that they cannot merely be explained by the movement of parts of a continent.

An example of this is furnished by the epicontinental seas of the Upper Cretaceous (fig. 56). The gradual advance of the sea into the European Continent can be closely followed. This great invasion of the sea was not only felt in Europe, but also on other continents. Conversely, a world-wide retreat of the sea prevails towards the close of the Mesozoic, and is in turn succeeded by a positive phase of the epicontinental seas in the Lower Tertiary. Suess was the first to show that these phenomena have a world-wide importance and that the periods of transgression were far longer than the relatively short periods of regression.

Suess, Haug, Stille and Grabau paid special attention to the chronological and regional distribution of the transgressions, and each in turn came to the conclusion that a great number of major transgressions took place each being separated by periods of widespread emergence of the continents. There was a rhythmic advance and retreat of the sea. We can therefore only conclude that the trans- and regressions on the continents must be ascribed solely to a world-embracing cause. Stille expressed the synchronism of the great trans- and regressions in his law of epirogenic synchronism, which Bucher formulated as follows: "In a large way the major movements of the strandline, posive and negative, have affected all continents in the same sense at the same time".

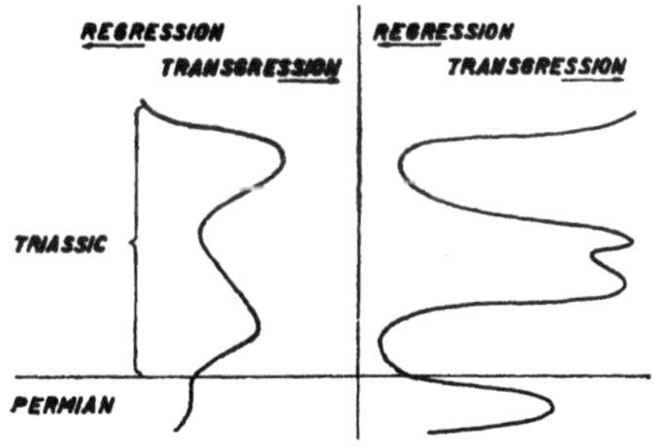

Fig. 57. Two curves showing the relative movement of the sea-level in Triassic times. The left-hand curve shows the world-wide "eustatic" movement, the second curve the oscillations of the sea-level in Germany (After H. Stille).

Once this rule is established statistically, the local and regional exceptions appear clearly. Thus Stille gives two curves for the Triassic, one representing the world-wide movement of the sea-level, and the second the strongly diverging and even opposed transgressions and regressions of the Triassic in Germany (fig. 57).

I discussed the problem of trans- and regressions in a paper of 1939 and will therefore deal with them very briefly at present. A study of the oscillating sea level caused me to draw the following conclusions:

(1) World-wide regressions are brought about either by a periodical elevation of the continents, or by periodical subsidence of the ocean-floors, or by simultaneous but opposed movements of both continents and ocean-floors. The last possibility appears to be the most probable one, and the same applies — *mutatis mutandis* — to the transgressions.

(2) Epochs of folding coincide with periods of world-wide regression, which were relatively short in comparison to the much longer intervening periods of transgression.

(3) Movements of the sea-level and folding are caused by subcrustal processes, which elapsed periodically, as it were, with a rhythmic cadence.

(4) Other phenomena (the magmatic cycles, the formation of basins and dome-shaped elevations, and the alternating decrease and increase in compression) are closely related to this "rhythm", and they, too, show that their origin must be sought deep within the earth, as pointed out in the preceding Chapter.

In Chapter IV, I explained that my conception of the structure of the earth's crust coincides with that of Daly and Vening Meinesz. This means that the earth is actually enveloped by a rigid crust. A cross-section of a continent shows that the crust is composed of sial and crystalline sima, and that a thinner sialic layer (though resting on a thicker layer of sima) is present under the Atlantic, and that a very thick layer of crystalline sima extends under the Pacific. Wegener's view that sima represents a viscous substratum in which continents would only drift as rigid blocks is unacceptable under the present day circumstances [1]). It may still conjectured whether the rigid and crystalline upper layer of sima enveloping the earth at present had at some time been considerably thinner or possibly non-existent (aside from the fact as to whether they were controlled by periodic radioactive processes, as Joly assumed, or whether another unknown internal process was responsible for them). At any rate, the periodicity of trans- and regressions shows that the differences in level between continents and ocean-floors have changed periodically!

The curve in Table II indicates the result as regards the relative movement of the level of the sea, showing the positive and negative shifting of the sea-level and its chronological relation to epochs of compression. This curve is entirely schematic. Besides, it is highly probable that certain trans- and regressions, which cannot be indicated for lack of data, had a greater amplitude than others.

A detailed study of the facies of the strata deposited on the continents by transgressive seas may produce evidence of the height of the sea-level at various periods above certain parts of the continents. But in that case local movements (e.g. of basin-shaped depressions) will also have to be taken into account, and the problem is thus far from being a simple one. The vast epicontinental extent of Upper Cretaceous strata, which are often compared to the present cocolith-ooze found at a depth of at least 1,000 meters, would lead one to suppose that an important shift of the strandline accompanied some transgressions. Kuenen assumed that a major transgression would result from an eustatic movement with an amplitude of some 40 meters. This amount, as he himself declared, should be regarded as a minimum value. We might add that it ought undoubtedly to be more. For Daly pointed out that if all the ice of the present ice-caps were to melt, this fact alone would cause the sea-level to rise some 40 to 50 meters [2]). We may safely say that the period in which the last world-wide regression reached its deepest level now lies behind us, but we are undoubtedly still far removed from the day when the prevailing transgression will have reached its maximum. During the maximum glaciation in the Pleistocene, the sea-level was approximately 100 meters lower than it is now, and the sea was already

---

[1]) See Chapter VIII, pp. 230 and 280. [2]) See Chapter IX p. 260.

in a new transgressional stage at that moment. Thus the maximum for a transgression must be fixed at at least 150 meters, plus an unknown amount, and the rise of the sea-level towards the maximum of the now prevailing transgressional phase will attain 50 meters above the present level, plus an unknown but undoubtedly considerable amount. Penck estimated that transgression were caused by a relative of the sea-level with an amplitude of 500 meters, and Joly evaluated the amount of the eustatic change between a deep regression and an extreme transgression at 1,200 meters.

Many other factors complicate this problem. For instance, the formation of the Mediterranean and the East and West Indian and Asiatic basins would have resulted in an eustatic lowering of the sea-level of approximately 60 to 80 meters according to Kuenen. The most important question concerns the depth to which the sea-level was depressed in distinct periods of intensive regression, in other words the extent of the change to which the distance between the surface of the continents and the ocean-floors was subjected during the pulsating rhythm of subcrustal processes. Joly was the only one who approached this question from the geophysical side and he arrived at an order of 1,000 meters. No one can foresee what further geophysical speculations may lead to eventually, but it is clear that this question has an extremely important bearing on various geological problems. For if the sea-level moved downwards far beyond its present position during certain periods, it would have had a strong influence on many phenomena. In this connection may be mentioned the origin of isthmian links, and broader transoceanic land-bridges. Further, all those physiographic factors that lead to glaciation would be considerably accentuated.

Obviously, not a single marine sediment could have been deposited on the continents or within shoal-water geosynclines during such a large emergence of the continents as may be expected to have preceded the late Paleozoic and the Pleistocene glaciation. As far as I have been able to ascertain, however, the actual facts do not confirm the idea of a world-wide regression of such extent that the surface of the sea would at some time have been lowered to a few thousand meters below its present level.

References

BARRELL, J. *The strength of the Earth's Crust* (Journal of Geology 22, 1914).

BARRELL, J. *Rhythms and the measurement of geclogic time* (Bull. Geol. Soc. of America 1917).

BORN, A. *Periodizität epirogener Krustenbewegungen* (Rep. 16th Intern. Geol. Congress Washington 1933, 1, 1936).

BUCHER, W. H. *The deformation of the Earth's Crust* (1933).

DALY, R. A. *The changing world of the ice age* (1934).

GRABAU, A. W. *Oscillation or pulsation* (Rep. 16th Intern. Geol. Congress Washington 1933, 1, 1936).

GRABAU, A. W. *Palaeozoic formations in the light of the pulsation theory* (1936).

GUTENBERG, B. *Changes in Sea level, postglacial uplift and mobility of the Earth's interior* (Bull. Geolog. Soc. America, 52, 1941).

HAUG, E. *Les géosynclinaux et les aires continentales; contribution à l'étude des transgressions et regressions* (Bull. Soc. Géol. de France, 3e ser. 18, 1900).

JOLY, J. *The surface history of the Earth* (Oxford Univ. Press. 1925).

KUENEN, PH. H. *Quantitative estimatons relating to eustatic movements* (Geologie en Mijnbouw 1, 1939).

KUENEN, PH. H. *Causes of eustatic movements* (Proceed. sixth Pac. Sci. Congress II, 1940).

MACHATSCHEK, F. *Zur Frage der eustatischen Strandverschiebungen* (Petermann's Geogr. Mitteil., 85, 1939).
MASAROVICH, A. N. *On the rhythm in the history of the Earth* (Bull. de la Soc. des Naturalistes de Moscou, 48, sect. géolog. 18, 1940).
PENCK, A. *Theorie der Bewegung der Strandlinie* (Sitzungsber. Preuss. Akad. Wiss. 1934).
SCHUCHERT, CH. *Correlation and chronology on the basis of paleogeography* (Bull. Geol. Soc. of America 26, 1916).
STILLE, H. *Studien über Meeres- and Bodenschwankungen* (Nachr. K. Ges. Wiss. Göttingen. Mat. -phys. Kl. 1922).
STILLE, H. *Grundfragen der Vergleichenden Tektonik* (1924).
SUESS, ED. *The face of the Earth* (*transl. Sollas*) 2, 1906.
UMBGROVE, J. H. F. *The amount of the maximal lowering of the sea-level in the Pleistocene* (Proc. Fourth Pacific Science Congress, vol. 2, 1930).
UMBGROVE, J. H. F. *On rhythms in the history of the Earth* (Geol. Magazine 76, 1939).

CHAPTER VI

THE CONTINENTAL MARGIN

"the shelves are quite generally marked by trenches, valleys, and embayments referable to rivers that formerly crossed them and which have not yet been concealed by sedimentation. These features imply that the continent was recently so deformed that these shelves were out of water, and that the rivers reached the true borders of the continental platforms. They equally imply a general movement toward continental submersion since, . . ." . . ."the region back from the coast was warped upwards, the streams being thereby rejuvenated and the conditions provided for the formation of the rapids of the infra-coastal tracts".

(TH. C. CHAMBERLIN)

Introduction

Almost everywhere the real edge of the continent is situated below sea-level. Generally a shallow platform borders the continent. Its edge lies about 200 meters below sea-level (Fig. 58) and an area of 10,000,000 square

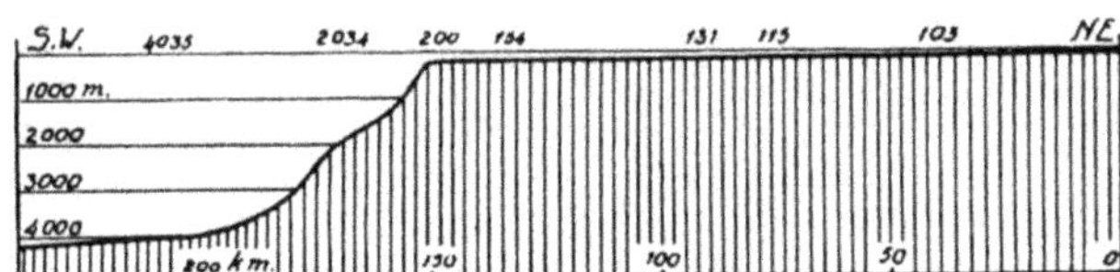

Fig. 58. Profile across the shelf of Southern France. (After Kossinna).

miles or 13% of the continents is thus covered by shallow shelf-seas. What is the reason of this world-wide uniformity of depth of the shelf-surface and its edge?

Echo-soundings and geological and geophysical explorations carried out during the last decade have, moreover, revealed many features hitherto unknown about the morphology and structure of the shelf and its outer slopes. From these observations it appears that the history of the shelf is a rather complicated one. Sedimentation, abrasion and denudation all played their rôle. The area was subjected to changes of sea-level and movements of the bottom. Wind-waves and tidal currents acted upon the sediments of the shelf. The influence of each of these and still more factors in the building of the submerged part of the continental margin is still an open question.

The outer slope is notched by great valley-like trenches descending toward the abyssal depths of the ocean. The origin of these huge gullies is another problem that challenges us. The outer slope of the Atlantic shelf of the United States — often called continental slope — drops from 200 to 2,400 meters and more in approximately 50 miles. In a few sections the slope is as steep as $7^{1}/_{2}^{\circ}$ (about $13^{0}/_{0}$) or 700 feet per statute miles [1]).

On the other hand the region inland from the coast offers a series of problems of its own. Wide-spread features of rejuvenation and a number of elevated marine terraces are characteristic of the marginal part of the continental land. The prevalence of these phenomena in all latitudes along the continental border is impressive and requires an explanation which extends beyond local interest. How can these features be understood so as to form a harmonious supplement to those of the submerged part of the continental margin?

Fig. 59. Shelf of the North Sea showing reconstruction of rivers debouching near the Dogger Bank. (After C. Reid).

These, in short, are some of the intricate and highly attractive problems to be considered in the following pages.

Generally the shallow platforms bordering the continents are classified in two categories, viz. the inner and the outer shelves. The first group comprises such regions as the shelf of the North Sea and the Sunda Sea. Bathymetric charts show them to be furrowed by river-like trenches (fig. 60). And indeed, in the two examples just mentioned their sub-aerial origin as rivers extending over these regions during low stands of the sea-level in Pleistocene times was clearly demonstrated by data furnishing converging evidence. Their course can be followed towards the debouchement of present-day rivers. Moreover, the frequent finding of remains of large vertebrates as far as the Dogger Bank (fig. 59) shows the North Sea to have been a land area in the near past.

In the case of the Sunda-shelf (fig. 60) the congruence of such animals as fresh-water fishes clearly proves e.g. the rivers of western Borneo and eastern-Sumatra to have been connected in sub-recent times. The shelf of the

1) Veatch and Smith 1939, p. 13

Barents Sea is considered to belong to the European continent. In this area Spitsbergen and Bear Island portray the nature of structural units which, over large stretches, were probably truncated and covered by the sea. So, between Spitsbergen and Scandinavia the inner shelf of the Barents Sea

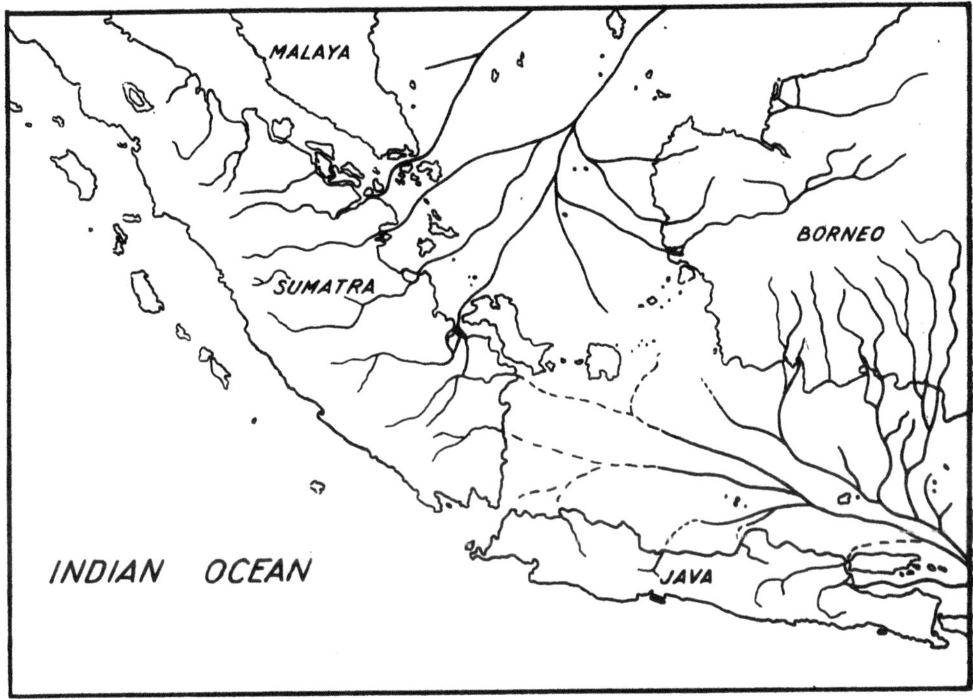

Fig. 60. Drowned rivers in the Java Sea and South China Sea.

gradually passes towards the outer shelf of north-western Europe. Here, too, the sea-bottom shows a drowned system of rivers. However, although highly interesting in many respects, the origin and history of the inner shelf regions will be left out of consideration here, since they do not belong to the marginal zone proper of the continents.

According to Kossinna the shelves all over the world cover 27,500,000 square kilometers, i.e. 7.6 percent of the surface of the oceans. The figure for the outer shelves is 9,900,000 square kilometers, i.e. 3.1 percent distributed as follows:

Atlantic Ocean	4.6 million km^2,	i.e.	5.6	percent of the sea-surface
Indian Ocean	2.4 million km^2,	,,	3.2	,, ,, ,, ,, ,,
Pacific Ocean	2.9 million km^2,	,,	1.7	,, ,, ,, ,, ,,

The origin of the outer shelves is the first problem to be considered.

The Surface of the Shelf

Waves laden with sediment and shingle attack the shore. Their continual bombardment undermines the coast. Gradually the sea encroaches upon the land, abrades and invades it. Undertow transports the abraded material seawards, and tidal currents assist in the submarine distribution of the sediments. The result is clearly demonstrated in the morphologic profile fig. 61, which shows the small shelf round the island St. Helena [1]). From the cliffed coast over some distance seaward the surface of the submarine bank consists of an abrasion-plane formed by the erosive action of the breakers. The more distant part of its surface, however, is built up by the accumulation of sediments derived largely from the land. From the edge of the shelf, situated here at a depth of 100 meters, the outer slope of the small shelf may be followed downwards to a depth of 700 meters. At that depth the slope diminishes; the reason for this seems obvious on that level the normal slope of the St. Helena volcano is reached, as is suggested by the broken line drawn in fig. 61.

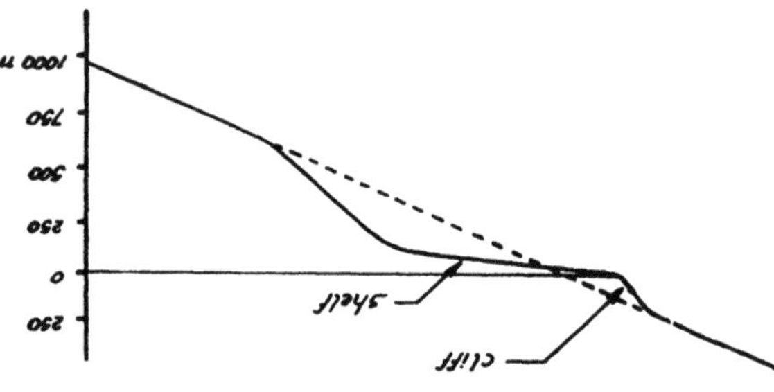

Fig. 61. Formation of a gradation-plane along the coast of Saint Helena.

The surface of the shelf therefore seems to be a plane of equilibrium [2]) between two antagonistic forces. One of these is the demolishing, erosive transporting action of the sea. The other is the accumulating action of the same medium, the sediments being transported, selected and deposited by wave-action, undertow, tidal currents and other off-shore sea-currents. The surface of the shelf expresses a balance between two opposing sets of factors which in short might be indicated by the terms submarine erosion and aggradation.

One of the first doubtful points is the depth of the abrasive action of the sea. Obviously the demolishing power of waves is dependent on the intensity of wave-motion. The motion of waves following one another in trains, is expressed by the formula $L = VT$ when L is the length of the wave (measure from crest to crest), V its velocity, and T the period of oscillation of the particles which are affected by successive waves.

Now, at the surface the radius of the circling movement of a revolving particle is equal to half the height of the wave. Proceeding downwards a similar circling movement is found but with a radius ever diminishing in a geometrical progression, while the depth increases in arithmetical

[1]) R. A. Daly. The Geology of Saint Helena Island. Proceed. of the American Acad. of Arts and Sci. 62, 1927. See also Ph. H. Kuenen, Geology of Coral Reefs. The Snellius Expedition, vol. V, part 2, 1938, p. 99, fig. 91.

[2]) Compare N. M. Fenneman. Development of the profile of equilibrium of the subaqueous shore terrace. Journ. of Geology 10, 1902. See also Johnson op. cit. 1919, pp. 225—228.

progression. The result is that towards deeper water the power of even big storm waves decreases rapidly.

Jenkins [1]) gives the following examples: Waves 90 meter long and 3 metres high are not uncommon with strong winds in the open ocean.

"At 10 metres depth the height of the subsurface trochoid would be 1.5 metres; at 20 metres depth 0.75 metres; at 50 metres only 9 centimetres; in 100 metres not quite 3 millimetres — that is hardly perceptible".

"An ocean storm-wave 600 feet long and 40 feet high from hollow to crest would have at a depth of 200 feet a subsurface trochoid with a height of 5 feet; at 4C0 feet (6/9 of the length) a trochoid of 7 or 8 inches only".

It is clear, therefore, that the abrasive action will hardly be able to be effective beyond the depth of average wave-base. Accordingly marine planation by abrasion at wave-base seems to be restricted to a narrow strip, parallel to the coast, in the litoral zone. The more so since material tumbling down from the cliffed coast forms a barrier sheltering the coast from further abrasion for an appreciable lapse of time. It would seem, therefore, that the sea can invade the land by abrasive action over a large extent only when it is accompanied by a simultaneous positive shift of the coast-line.

Some authors, however, are of the opinion that the formation of the entire shelf-surface is controlled exclusively by the demolishing action of waves and currents. According to their theory the shelf is considered to be a feature which originated mainly by abrasion. Jessen is a modern advocate of the origin of the shelf by abrasion. In order to explain the frequent occurrence of the shelf-edge at a depth of 200 meters he put forward the following theory. Firstly he supposes the abrasive forces to be effective down to a depth of 60 meters. Secondly he thinks that — as a result of sediments being transported from the land oceanwards — a worldwide rise of the sea-level of 30 meters should have taken place since the beginning of the Pleistocene. Finally, a glacial lowering of sea-level would account for the remaining amount of 100 meters. In this hypothesis Jessen confuses the movement of the water which under normal conditions is able to displace sand-grains over the surface of the shelf, with an abrasive action at wave-base. The amount of 30 meters is chosen to suit the hypothesis, and is not based on well founded estimates. Moreover, granted even for a moment that these figures were correct, a glacial lowering of sea-level to an amount of 100 meters is inadequate to explain the features of the shelf-surface, as will be shown in the following section.

Novák, too, is convinced that the shelves were carved out of the continental blocks. He considers they were formed mainly by submarine

---

[1]) J. T. Jenkins. A Textbook of Oceangraphy (Constable and Co, London 1921), p. 108. Jenkins summarizes as follows: The orbits and velocities of the particles of water are deminished by *one-half* for each additional depth below the mid-height of the surface wave equal to *one-ninth* of a wave-length — i.e.:

depth in fractions of a wave-length below the mid-height of the surface wave $\left.\right\}$ $0 \frac{1}{9} \frac{2}{9} \frac{3}{9} \frac{4}{9} \frac{5}{9} \frac{8}{9}$, etc.

Proportionate velocities and diameters $\left.\right\}$ $1 \frac{1}{2} \frac{1}{4} \frac{1}{8} \frac{1}{16} \frac{1}{32} \frac{1}{256}$, etc.

peneplanation during a low stand of sea-level, but marine abrasion and subaerial planation are admitted by him as co-operating factors. Novák, however, believes that the duration of the Pleistocene stages of low sea-level did not last long enough for the shelf-surface to be formed entirely by these factors. He therefore thinks it inevitable to look back in geological time to a period when sea-level stood 100 m or so lower than at present for a time sufficiently long for the cycle of the present shelf to be completed. For that reason he places the formation of the present shelf in Maeotian-Pontian times! Moreover his hypothesis implies a subsequent rising of the sea during the Pliocene to some 200 meters above its present level, in order to explain the higher marine terraces which occur along the shore of the Mediterranean (Calabrian terrace). Finally, a world-wide lowering of the sea-level would have taken place during the Pleistocene, but was interfered with by glacial oscillations. For several reasons his hypothesis is untenable. Firstly, paleogeographic conditions at the end of the Pliocene are opposed to the view that sea-level was 200 meters above its present level. On the contrary, the upper Pliocene was a time of wide spread regression of the epicontinental seas, which were then confined to a few basin-shaped depressions.

Moreover the whole idea of the shelves being carved into the continental blocks — which was also the gist of Buchanan's hypothesis — is rendered out of date since the geophysical explorations carried out by Ewing, Crary and Rutherford. For they made it highly probable that the Atlantic shelf of North America is composed of an accumulation of sediments ranging from Triassic to recent times resting on the submerged part of the former continent and attaining a thickness of 4000 meter at a distance of 60 miles off the shore-line (fig. 26).

Similar results were obtained by Bullard for the shelf to the west of the English Channel. He also found the surface of the continental basement — consisting of crystalline rocks — to slope steadily downwards to abyssal depths on receding from the land. At the 100 fathom line the surface of the basement is over 8,000 feet below sea-level and covered by a prism of off-shore deposits (fig. 62).

So it appears at any rate that the continental shelf must not be regarded as a huge step incised in the continental block. Wegener's interpretation of the shelf, too, must be discarded for the same reasons. And this applies also to analogous views of Du Toit.

However, apart from the problem as to the role played by several factors in the formation of the shelf-surface, there is a question of primary importance, viz. at what depth would a shelf-surface and its edge be situated when formed entirely under present conditions? The process of grading of the shelf implies that there exists a certain hypsometric relation between sea-level and shelf-surface.

The example of the island Saint Helena (fig. 61) makes it probable that the edge of the plane of equilibrium — in other words the outer edge of the shelf — is formed at a depth of about 100 meters below the level of the sea. At least for the southern part of the Atlantic this does not seem an exceptional figure. The island of Ascension for example, shows exactly the same feature. "As usual", Daly wrote [1]), "the horizontal distance between the shore line and the 50-fathom isobath is considerably greater than that between the 50-fathom and 100-fathom isobaths. The break of slope at the outer edge of the coastal shelf is located close to the 50-fathom line".

In his paper on the morphogenesis of Ben Lomond Rode [2]) mentions that

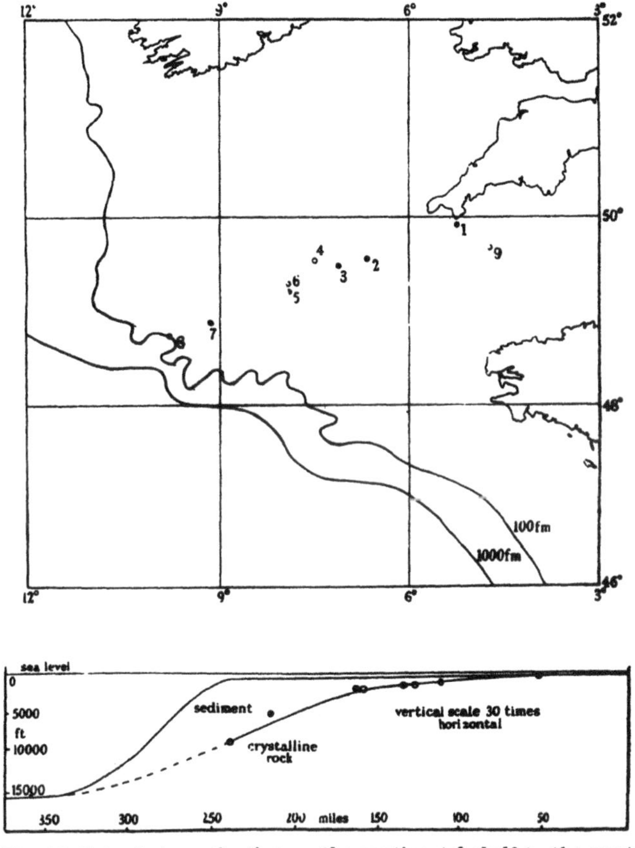

Fig. 62. Seismic investigation on the continental shelf to the west of the English Channel. The map shows the position of seismic stations. The two stations marked by a black point in the section give only an upper limit of the level of the crystalline rocks. (From C. G. Bullard and T. F. Gaskell).

along the Californian coast the submarine action of waves and currents extends down to a depth of at least 50 meters, but not surpassing 100 meters.

In his general description of the American shelves Shepard [3]) summarises:

1) R. A. Daly. The geology of Ascension Island. Proceed. of the Americ. Acad. of Arts and Sci. 60, 1925, p. 9.

2) K. Rode. Geomorphogenie des Ben Lomond. Zeitschr. f. Geomorphologie 5, 1936.

3) F. A. Shepard. Continental Shelf sediments in "Recent Marine Sediments", a Symposium 1939, p. 219.

"steepening beyond the shoal platforms is initiated at an average depth of about 65 fathoms". In some areas, however, the plane of balance between aggradation and erosion, decidedly occurs at a slighter depth. According to Stetson [1]) for example, "60 to 70 meters can probably be regarded as the limiting depth to which the storm waves of the Atlantic, when aided only by normal tidal currents, can agitate the bottom deposits."

From the statements mentioned above one might expect the break of slope on a shelf the surface of which is completely adjusted to present sea-level and which was formed entirely under the same conditions, to occur at the isobath of approximately 70 meters (or as lower limit 100 meters). A deeper figure may be regarded as a rare exception. It might locally be caused by very intense tidal scour [2]).

It is well known, however, that actually the edge of the continental shelf frequently occurs at a depth of 200 meters. Bourcart paid special attention to the morphological features of the shelf. Again he found that nearly everywhere a break of slope may be noticed near the isobath of 200 meters. Respecting the eastern border of the Atlantic he mentions only two exceptions. Along the coast of Angola and off the Ivory Coast the isobath of 1500 to 2000 meters approaches the coast so nearly that no shelf at all is present. The Atlantic shelf off North America shows lower figures, ranging from 60 fathoms east of Chesapeake bay to 80-fathoms north of the Hudson and on Georges Bank (fig. 68). Hence, the problem has to be solved as to how the shelf-surface below the isobath of 70 meters (40 fathoms) came into being [3]).

The influence of changing sea-level

In order to explain the break of slope so frequently occurring at a depth of 200 meters the possible influence of eustatic changes of sea-level and movements of the bottom need to be examined. Undoubtedly changes of sea-level rank among the principal agencies which have affected the surface of the shelf in the near geological past.

One factor is the changing sea-level associated with transgressions and regressions caused by periodical processes in the earth's interior (See Chapter V). The last regressional stage of the sea not controlled by changes in the volume of ice-caps happened to occur towards the end of the Pliocene, simultaneously with the intensive folding and overthrusting which characterises so many mountain-chains belonging to the Alpine belt of folding. Since then a transgressional stage has begun. But at yet data are lacking which could show us the amount of the transgression since the regressional stage of the Pliocene. It may well be that it is rather of the order of ten or tens of

---

[1]) H. C. Stetson. Summary of sedimentary conditions on the continental shelf of the East Coast of the United States, in "Recent Marine Sediments" a Symposium, 1939, p. 237.

[2]) Stetson (ibid., p. 239) mentions that well sorted sands are found to a depth of 132 meters off Nauset.

[3]) "At the present time, it appears that the outer half of each broad continental shelf is too deep to represent a profile in equilibrium with the forces responsible for the building of the continental terrace" (Daly, The floor of the Oceans, 1942, p. 100).

meters than of a hundred or hundreds of meters. And of course the extent of the transgression at a certain stage of the Pleistocene must be less. Owing to lack of data this amount has to be left out of account.

Apart from the positive shift of the shore-line considered just now, the sea-level swung downwards and upwards four times during the immediate geological past under the influence of the growing and melting of the ice-caps. The amount of the lowering of sea-level at the time the ice-sheets reached maximum size, may be estimated at 75 meters, possibly as a lower limit 90 to 100 meters [1]).

A certain amount (as mentioned above an unknown though probably a small amount) has to be added in as much as these figures are based on estimates of the volumes of ice in excess of the present volume, which existed when the Pleistocene ice-sheets reached their maximum extension.

From these estimates one might perhaps feel justified in considering the shelf-surface between the isobaths of 70 (100) and 200 meters a plane of equilibrium adjusted to one or more stages of low sea-level in the Pleistocene. In doing so, however, the implicit assumption has to be made of a stable bottom. Possible this holds good in a few special cases, viz. in such oceanic islands as Ascension and Saint Helena. For in all probability these volcanic islands date from at least pre-Pleistocene times. According to Daly [2]) who summarized the different data bearing on this question, Saint Helena dates from pre-Glacial, and probably from pre-Pliocene time. Thus, e.g. he wrote [3]): "The stupendous cliffs of St. Helena seem to represent a duration of wave-attack which may be reasonably guessed as more than 2,000,000 years".

In fig. 63 the sea-level related to the strongest glaciation and the corresponding abrasion and shelf-surface has been drawn tentatively. It will be seen at once that the glacial shelf-body is entirely buried by the more recent shelf-formation and that its present morphology is the result of both bodies being fused together. In reality the situation must necessarily be still more complex because of the repeated alternations of glacial and preglacial sea-levels. In other regions present circumstances might be such as to cause the recent profile of equilibrium not

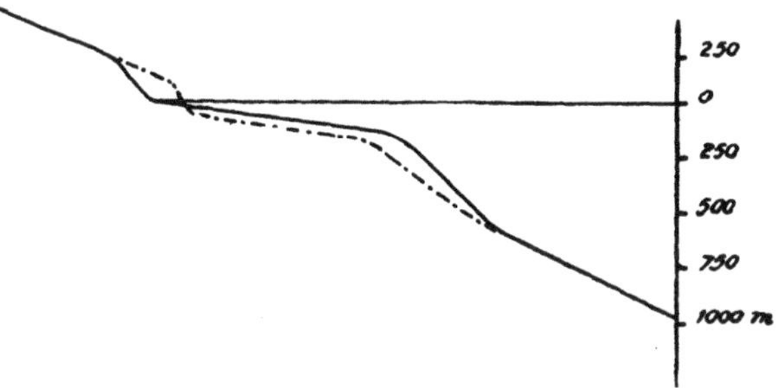

Fig. 63. Probable complications in the formation of the shelf of Saint Helena.

[1]) See R. A. Daly. op. cit. 1934, pp. 46—48 and R. A. Daly, submarine canyons, Americ. Journ. of Sci 31, 1936, p. 402; J. H. F. Umbgrove The amount of the maximal lowering of sea-level in the Pleistocene. Proceed. Fourth Pacific Sci. Congr. Java 1929, 2, p. 105—113; Ph. H. Kuenen. Geology of Coral Reefs. The Snellius Expedition 5, part. 2, 1933, p. 115.

[2]) Daly, op. cit. 1927, pp. 87—88.

[3]) R. A. Daly. The Geology of Ascension and St. Helena Islands. Geological Magazine 69, 1922, p. 156.

to cover the Pleistocene shelf-surface entirely. In that case one or more older surfaces might be exposed in the outer area of the shelf between the isobaths of 70 and 200 meters. However, the outer parts of the continental shelves present further problems worth of special attention.

Subsidence of the shelf-area

It has been mentioned already that the preceding considerations are based on the implicit assumption of stable conditions of the shelf-area in Pleistocene times. There are reasons, however, for saying that for the outer shelves of the continents such a surmise is not justified. The problem is much more complicated. One remarkable feature of the outer portion of the shelf is the frequent occurrence of coarse sediments, sometimes even called "rock bottom" (fig. 64). It was noticed on both sides of the Atlantic. On more than one occasion Shepard reported it from the shelf along the eastern coast of the United States. And Bourcart made a thorough study of this phenomenon in shelves of North Africa and southern Europe. Shepard reported it also from off the Californian coast. This feature is of great importance as it appears that the sediments on the continental shelf form an exception to the well-established rule of decreasing grain-size with increasing distance from the shore. The conclusion of Shepard and Cohee reads as follows [1]):

"The study of a large number of samples from the shelf off the Mid-Atlantic States has shown that there is the same absence of outward decreasing gradation of grain-size which had been suggested previously, on less substantial evidence, for the shelves of the world in general. The sediment is, for the most part quite out of harmony with present-day conditions and was evidently deposited during the Pleistocene states of lowered sea-level".

In a later paper Shepard [2]) advocated: "the concept of an older generation of sediments formed on the shelf when it was exposed to subaerial conditions during the last glacial epoch and left uncovered since the rise in sea-level,

1) Shepard and Cohee, op. cit. 1936, p. 457. See also Shepard, op. cit. 1932.

2) Shepard, op. cit. 1939, p. 227.

Twenhofel supports the idea that the coarse sediments were carried outwards together with the fine deposits during maximum conditions of transport under storm conditions and that afterwards during ordinary circumstances fine sediments buried the coarse ones near the shore (W. H. Twenhofel, Marine unconformities, marine conglomerates and thickness of strata. Bull. Americ. Assoc. of Petrol. Geol. 20, 1936, pp. 686—689) Shepard demonstrated the improbability of this assumption (Shepard, op. cit. 1939, pp. 227—228). There remains the question whether the deeper part of the outer shelf is free from recent sediments because the new level of equilibrium has not yet built itself outwards so far, or because of currents hampering sedimentation in this area. Kuenen suggested that absence of fine deposits at the outer margin of the shelf is due to the fact that this region acts as a brake on a large proportion of the energy of long waves coming in from the ocean and being cut down until they hardly stir the bottom, sweep over the shelf-edge and free it of its deposits (Ph. H. Kuenen. The cause of coarse deposits at the outer edge of the shelf. (Geologie en Mijnbouw 1939, pp. 36—39). It is an interesting task for future investigations to unravel what kind of movements of the water actually does occur in these areas of the outer shelf and to settle the open question whether appreciable amounts of fine deposits are swept away from these areas by active currents or whether the accumulation is so slow here that even without transportation elsewhere the area would morphologically and petrographically remain in a clear contrast to the shallow part of the shelf inside the isobath of 70 meters.

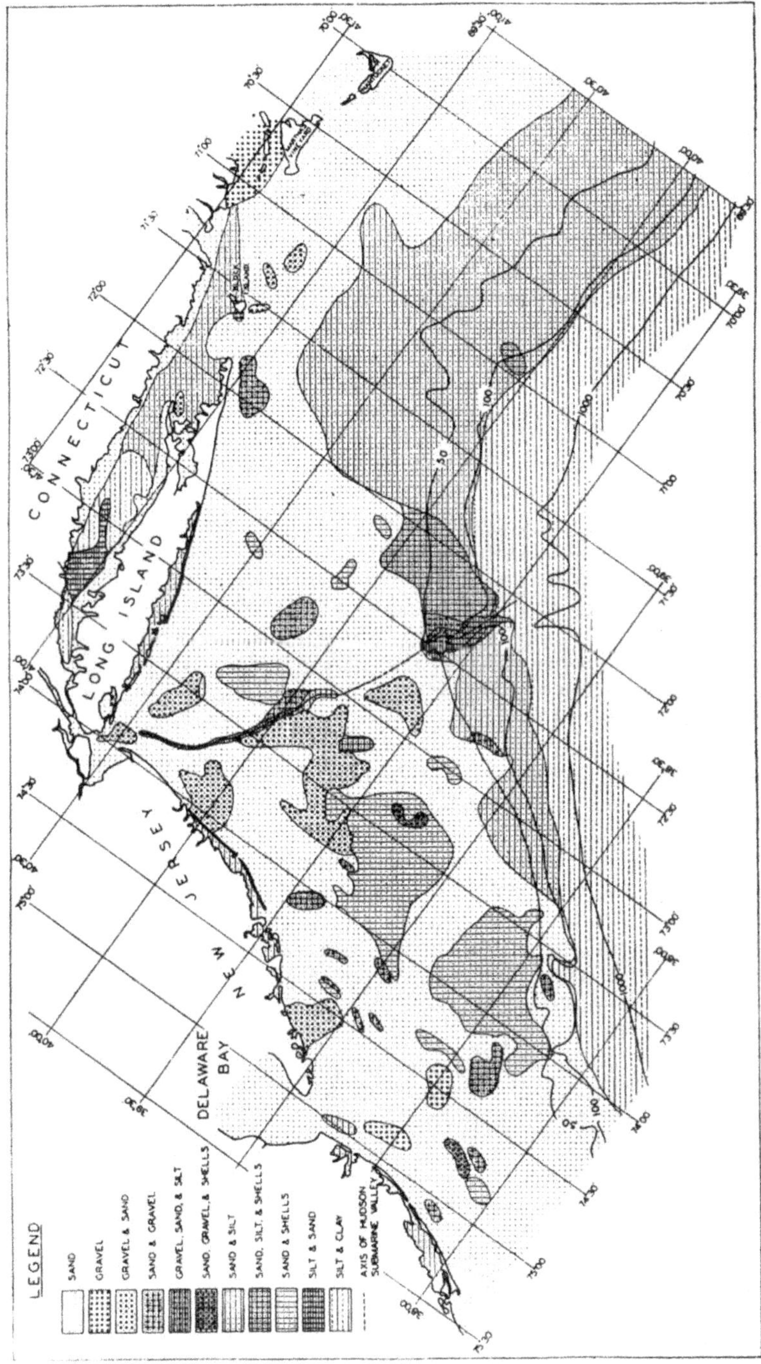

Fig. 64. Distribution of sediments on the continental shelf of the Mid-Atlantic States. (After Shepard and Cohee).

because the currents may have removed some of the finer portions of the sediments".

This quotation from Shepard leads us to the following remarks. A theory based on changing level of the sea only, might account for a Pleistocene plane of equilibrium situated at a deeper level than the 70-meter isobath [1]) and a shelf edge at about 200 meters. This would correspond to the possibility of subaerial erosion at a depth of (at maximum) 100 meters below present sea-level. Hence, if the level of the coarse sediments be situated below the isobath of about 100 meters it never could have been exposed to subaerial conditions unless it be assumed that — apart from a rise of sea-level — a simultaneous or subsequent downward movement of the bottom has also taken place! Conversely, if convincing arguments could be presented to show the formation of the coarse sediments to have taken place either in the litoral zone or under subaerial conditions, we would — for the areas under consideration — have to suppose a subsequent downward movement of the bottom superimposed on the changing sea-level.

Now, according to Bourcart the outer shelves, at least those of southern Europe and North Africa, which he studied in some detail, can have no other interpretation. Indeed, they provide good reasons for believing in the wide-spread influence of a subsidence of the bottom as an additional factor of importance. For he found the continental platform strewn with a variety of boulders of rather large dimensions. Often their diameter is 50 or 60 centimeters. The possibility of their being brought from the land towards their actual place by sea-currents is out of the question. Moreover, to many of them Bryozoa and other delicately branched organisms have attached themselves and the very fragility of these organisms prove the boulders not to have been displaced or even moved by submarine currents. A further asset of importance is their occurrence in regions — e.g. off the low coasts of southern France — which were unaffected by glaciers. In general the dimensions of the boulders surpass those of boulders carried to day by rivers in the coastal area. Apparently they were transported by more powerful rivers in the Pleistocene. For the reasons just enumerated the boulders have to be considered as orginally deposited in the litoral zone. And remarkably enough, in many places especially off the coast from Gibraltar towards Senegal the boulders were found to occur in elongated tracts of increasing depth, parallel to the coast. One between 0 and 50 meters, a second between the isobaths of 50 and 100 meters, a third between 100 and 200 meters, and off Azila an elongated accumulation was even met with at a depth of 400 meters below sea-level. So, the deepest must be considered much older than those which frequently were found down to 200 meters and which are supposed to be Mousterian age *sensu lato* by Bourcart. Apart from the 0 to 200 surface Bourcart distinguishes a second surface along the oceanic slopes of the shelf at a depth of 200 to 500 meters and a third between the isobaths of 500-1000 meters. Their distribution off the Atlantic coasts as studied by him [2]) is shown in the table IV. The two deeper surfaces show

[1]) Compare Stetson, op. cit. p. 238 and D. W. Johnson, op. cit. 1919, Chapter V.

[2]) Bourcart, op. cit. pp. 438—441.

a greater inclination than the upper surface of 0 to 200 meters. The accompanying illustrations fig. 65, show these features in a convincing manner. Different surfaces of the Atlantic shelf of North America will come up for discussion in the following section.

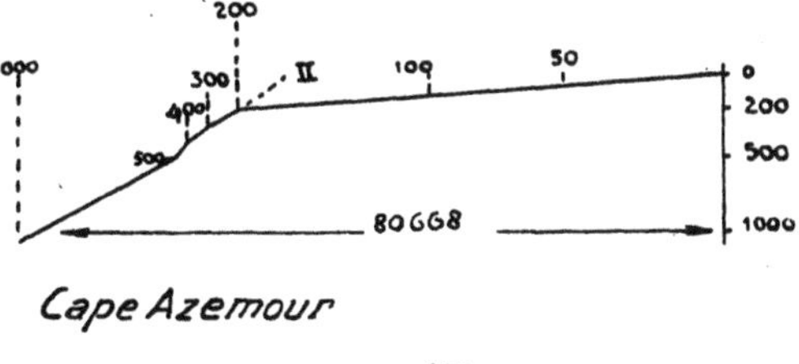

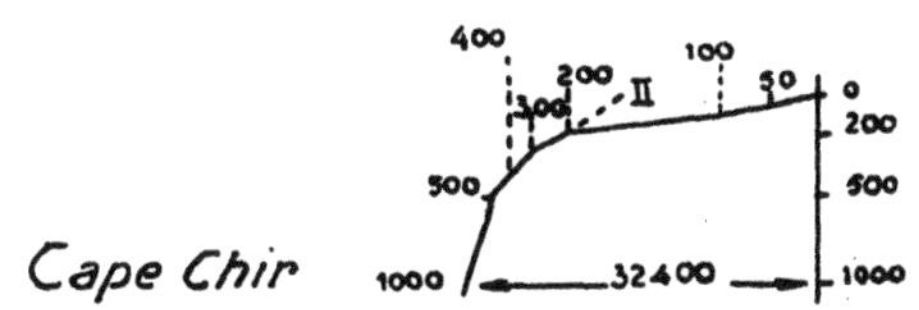

Fig. 65. Deeper surfaces on the outer slopes of the shelf. (After Bourcart).

An interpretation of the more remote part of the shelf region along the same theoretical lines would lead to the conviction that the original continental margin has sunk deeply below the level of the sea. And this is the view which has actually been advocated by many authors respecting several so called sunken borderlands of the continents as has been mentioned in Chapter II (p. 35).

The marginal flexure of the continents

In clear contrast to the positive shift of the shelf area the land-side of the continental margin shows wide-spread evidence of a movement in the opposite direction. Zones clearly demonstrating a rejuvenation of the relief are known along the greater part of the oceanic coasts under consideration. It is interesting to note that in their well-known textbook of Geology, Chamberlin and Salisbury wrote as early as 1906 [1]:

> "Nearly every coast is bordered by inlets which are almost invariably submerged valleys; but, followed inland, these inlets usually graduate into deep sluggish rivers, and these, farther inland, are very often replaced by rapids or falls, or at least by steepened gradients. When the continental borders are examined throughout their full extent in all latitudes, the prevalence of this phenomenon becomes impressive. The chief exceptions are the great rivers which drain interior basins through broad gaps in the elevated tracts that so generally border the continents, as the Mississippi, which issues through the great gap between the Appalachians and the mountains of Arkansas and Indian Territory, and the Amazon, that issues between the Parima and the Brazilian mountains. A critical study of the gradients of the normal coast-border river-channels, embracing at once the submerged portions, the inlet portions, and the high-gradient portions, indicated a warping rather than a simple uplifting and depression, such as is implied in the epeirogenic conception. This is an important factor in the alternative interpretation".

The monotonous succession of transgressions and regressions shows the difference in level between continents and ocean floors to have changed

[1]) Th. C. Chamberlin and R. D. Salisbury, Geology, vol. 3, p. 523.

Table IV

Different surfaces along the outer slope of the shelf, after Bourcart

	0-200 m	200-500 m	500-1000 m	deeper slope
Scoresby Sund, Greenland	narrow	well developed	absent	500–2000
From Blosseville to Angmassalik	well developed	,, ,,	,,	500–2000
S. E. Greenland	narrow	narrow		500–1500, 2000
Mayen island, S. coast	present	present	present	–
Iceland, N. coast	,,	,,	absent	–
,, , S. coast	,,	,,	,,	–
,, , S. E. coast	,,	absent	,,	200–1500
Faroer	,,	present	present	–
Rockall-bank	,,	,,	,,	–
N.W. Scotland (Orkneys, Shetlands, Hebrides)	very broad	absent	absent	200–1500; 2000 steep
W. Coast of Ireland (Procupine bank)	broad	well developed	,,	500–2000, steep
S. E. Coast of Ireland	,,	,, ,,	present	1000–4000 not steep
Western entrance of Channel	very broad	absent	absent	200–4000, steep
W. Coast of France (fosse de Cap Breton)	well developed	,,	,,	200–1200, 2000
N. Coast of Spain	narrow	,,	,,	200–3000, steep
Coast of Portugal (Sado, Tage, Nazaré canyons)	present	between Durro and Tague	–	–
W. off Gibraltar	,,	present	present	1000–deep-sea, not steep
Morocco	narrow	,,	absent	400–> 1500, steep
Mauritania, up to "Cap Blanc"	rather broad	,,	,,	–
,, , S. of "Cap Blanc"	well developed	absent	,,	200–great depth, steep
,, , N. of "Cap Vert"	very narrow	–	–	steep
,, , between "Cap Blanc" and "Cap Vert"	,, ,,	slightly developed	slightly developed	not steep
Between Dakar and Morovia	broad	present	present	–
Liberia	narrow	narrow	absent	500–2000 steep
Ivory coast	absent	absent	,,	isobath of – 1500 near coast
Gold coast	present	,,	,,	200–5000, steep
Dahomey-Nigeria	narrow	,,	,,	200–2000, steep
Niger delta	present	present	–	–
Congo (Congo canyon)	,,	,,	present	–
Angola	absent	absent	absent	isobath of 1500–2000 near coast
S. W. Africa	present	present	present (narrow)	not steep

periodically (See Chapter IV, p. 83). However, apart from a periodical rejuvenation of the continental and oceanic sectors as a whole, warping or tilting seems to have taken place spasmodically along the continental border. This caused a periodical submergence of what formerly was the margin of the continent and a simultaneous bowing up of a marginal tract parallel to the newly formed coast-line.

Bourcart, who has given some instructive examples in favour of this theory speaks of a marginal flexure of the continents which periodically grows steeper, in the way suggested by fig. 66 and 182.

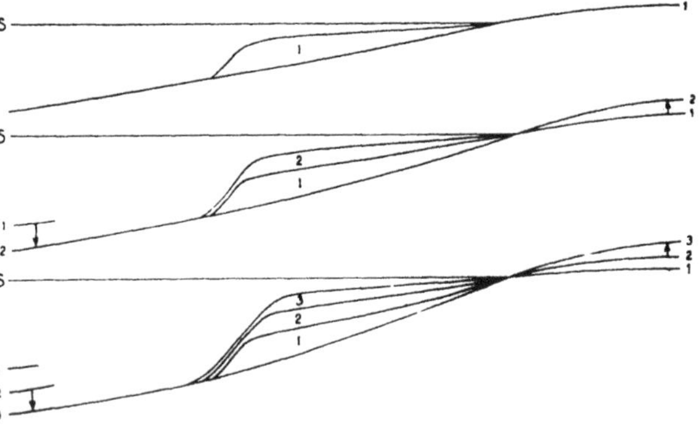

Fig. 66. Schematic representation of the movements along the continental flexure, supposing a stable hinge-line at sea-level.

Similar ideas were published a few years later by Jessen who devoted a voluminous memoir to this subject. His treatise gives much additional evidence of the widespread occurrence of upward movements of the present continental edges — resulting in a rejuvenation of the relief along the continental border — and a downward movement of a zone parallel to it, now submerged below the level of the sea.

Rejuvenation of the relief is a characteristic feature of the recent geological past of the whole globe including continental and oceanic sectors. The facts assembled by Jessen offer some very welcome points of view regarding rejuvenated zones surrounding continental and submarine basins. But we will restrict ourselves at present to the borders of the continents. According to Jessen the thickenings of the continental margin may be classified according to three types, viz. (1) warping accompanied by faults, (2) updoming accompanied by faults and slight folding, (3) folding and overthrusting, accompanied by underthrust of the oceanfloor, giving origin to a deep-sea trough [1]).

It is clear that a subsequent landward migration of the hinge of the marginal flexure causes a shifting of the shore-line [2]). This phenomenon may perhaps account for the local absence of a marginal zone of elevation as for example along the Atlantic coast of southern France.

The notion of the continental flexure is also found in a publication by Veatch and Smith of the year 1939. Veatch found that the Tertiary and Cretaceous beds of the coastal plain near the mouth of the Congo have a thickness of over 3000 meters and rest on a Cretaceous peneplained surface sloping seawards. The uplifted part of the peneplain has been eroded to a new peneplain in about Mid-Miocene times. This Mid-Miocene peneplain was uplifted and warped by Mid- to Late Miocene disturbances.

1) Illustrations are given by Jessen in his plate 30, fig. 79.

2) cf. Bourcart and Jessen, p. 30, fig. 78, p. 470.

Probably, however, successive post-Miocene uplifts and warpings have tended to form several additional partially developed peneplains. And these involve "probably not only upwarping landward but accompanying downwarping seaward" (1933, p. 16).

The result is: "a steepening of the continental slope initiated by the warping of the Miocene peneplain, both by subsequent marginal deposition and the upwarping responsible for the younger partially developed peneplains".

Veatch moreover pointed out that similar processes took place on the opposite side of the Atlantic in the eastern United States. His opinion is well illustrated by a sketch, reproduced in fig. 67.

Fig. 67. Sketch section of Atlantic coastal plain and continental shelf. (From Veatch and Smith)

However, the accurate charting of the Atlantic shelf of North America revealed some complicated features of great interest. One feature, as revealed by the excellent charts published by Veatch and Smith shows a gradual deepening of part of the shelf-surface from the south towards the north including the edge of the shelf.

The shelf broadens from 95 kilometers east of Cape Henry to 150 km near the Hudson channel and it remains approximately 125 to 135 km wide from off Long Island up to Georges Bank. The related characteristics of the surface and edge of the shelf are compiled in the table V. The gist of these data is summarized in the accompanying schematic block diagram fig. 68.

TABLE V. Topographic data of the Atlantic Shelf of the United States.

		From East of Cape Henry to East of Pocomoke Sound	From East of Pocomoke Sound to East of Atlantic City	From East of Atlantic City to South of Nantucket	Georges Bank
Distance in Kilometers	From shore to shelf edge	95	105–130	125–140	> 140
	From shore to isobath of 40 fathoms	90–85	85–110	110–80	–
	From isobath of 40 fathoms to shelf-edge	5–16	16–20	20–40	40–50
	From isobath of 40 fathoms to isobath of 60 fathoms	5–10	10–25	25–40	–
Depth in fathoms	"Franklin shore"	35–40	40–55	55–60	–
	Shelf edge	60	80	80 ("Nichols Shore")	90–100

Obviously the outer part of the shelf shows a downward tilt to the north-northeast. This phenomenon is revealed (1) by the lowering of the shelf edge from 60 fathoms down to 80 fathoms. (2) by the deepening of a zone of narrowed isobaths called "Franklin Shore" — from 40 fathoms to 60 fathoms over the same distance, i.e. between the southern limit of the charted area east of Cape Henry to south of Nantucket. The downward northerly tilt was clearly recognized by Smith who considered it as "probably the residual tilt due to incomplete recuperation from ice-loads" [1]). The Franklin Shore is regarded by Smith "as the result of a stillstand in the sea level during one of the intermediate glaciations". I cannot agree with this interpretation of the Franklin Shore. Both the shelf edge and the Franklin Shore show the same amount of notherly tilt, viz. 20 fathoms, and both diverge from the shore in a northern direction. Therefore both were probably in existence when the

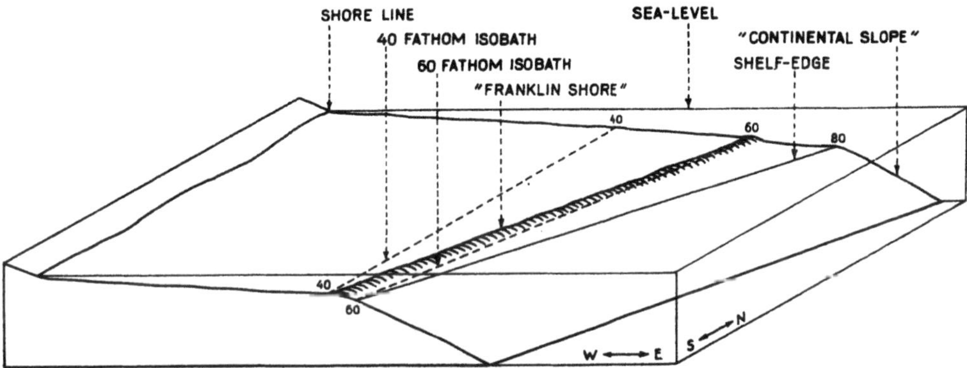

Fig. 68. Schematic block-diagram of the Atlantic shelf of the United States from east of Cape Henry to east of Atlantic City.

bottom movements happened to occur. However, remarkably enough the distance between the present shore-line and the 40 fathom isobath remains approximately constant (80-90 km). For obvious reasons the area occupied by the Hudson delta forms an exception (it is indicated by the figure 110 km in table V). As a tentative explanation of these features I suggest that the region east of the Franklin Shore is an older shelf surface which was brought downward by a movement along the continental flexure. After this movement a new shelf-surface formed adjusted to the then prevailing conditions of submarine erosion and aggradation. This surface is still to be seen in the region between the isobath of 40 fathoms and the Franklin Shore which may be considered as the seaward end of the new surface. Then a northerly tilt of the shelf-area occurred, probably due to ice-load and the subsequent beginning of a partial isostatic restoration of the region due to vanishing of the ice-masses, as suggested by Smith. Finally, the shelf-surface between the present shore line and the isobath of 40 fathoms formed as a surface adjusted to conditions prevailing at present. Summarizing, our suggestion will make

[1]) Veatch and Smith, 1939, p. 44.

clear that the morphology of the Atlantic shelf of North America shows the influence of two different sets of movements, one being a spasmodic seaward tilt due to a movement along the continental flexure, the other being an additional notherly tilt due to local processes of loading and unloading of a Pleistocene ice-mass. The age of the respective movements cannot be fixed exactly for lack of data. Nor is a correlation of the surfaces to those postulated by Bourcart attempted. We may say only that if the northerly tilt be due to the load of ice then the most recent ice-mass was the Wisconsin ice-cap which is known to have advanced as far as Long Island. Now the Wisconsin Ice Age was at its maximum approximately 50,000 years ago and the recession of the ice-cap started approximately 30,000 years ago. Hence, if the northerly tilt was caused by the Wisconsin ice-mass, then the shelf-surface between the 40 fathom-isobath and the Franklin Shore would date at least from pre-Wisconsin time. And this would hold good a *fortiori* for the lower surface of the shelf between the Franklin-Shore and the shelf-edge. On the other hand the shelf surface between the shore and the isobath of 40 fathoms would have been modelled during post-Wisconsin time, i.e. during the last 30,000 years.

Marine coast-terraces

The most recent part of the movement of the coastal tract is demonstrated by the elevation of marine terraces along the continental border. In "The Changing World of the Ice-Age" Daly has given a clear review of the intricate problems inherent to the age and correlation of the marine coast-terraces. Although locally marine terraces were found at exceptional heights [1]) the majority of the coastal tracts show four terraces, the uppermost one — called Sicilian in the Mediterranean — occurring at an altitude of 100 meters above sea-level. In such far distant regions as the eastern border of the Pacific, New Zealand, South Africa, Chile, Alaska and Japan the present altitude of marine terraces has been tentatively correlated by various authors with the well known scheme of four marine terraces occurring in the Mediterranean.

It would be a too naive idea to surmise eustatic emergence causing a constant height of the terraces all over the world. For as a result of the shallowing of the ocean during glacial stages the coast-terraces would not now be of exactly uniform height. Moreover, even the melting of all the land-ice now existing would rise the sea-level only about 40 meters. Hence the present altitude of the Sicilian terrace cannot be explained by eustatic movements alone. The same holds good for the Milazzian terrace which stands at 55 to 60 meters above present sea-level [2]).

[1]) See note [2]), viz. the part on page 115.

[2]) Instead of ascribing these higher terraces to elevation of the land Novàk thinks "of a part of the difference as being probably due to sinking of the sea-level going on through the greater part of the Pleistocene, of course if we abstract from the glacial oscillations" (op. cit. p. 15). This suggested world-wide lowering of sea-level in the Pleistocene would have been preceded by a slow rising of the sea during the greater part of the Pliocene. And again this eustatic movement would have been preceded by a low position of the sea surface during the earliest Pliocene or latest Miocene (Maotian and Pontian). As was set forth on p. 102 his theory is untenable. Moreover Novàk's estimate of a sea-level 200

The possibility of matching the marine terraces with the glacial stages was put forward by Daly who pointed to the fact that the fossil molluscs of the 30-meter Tyrrhenian terrace lived in water considerably warmer than the present Mediterranean. He therefore suggested the correlation of the Tyrrhenian with the Yarmouth (Mindel-Riss) interglacial stage, which is known to have been the longest and warmest Pleistocene stage. Possibly the climate of the whole world was then so mild that little land-ice was left and the sea-level was raised some 30 meters above its present level i.e. to the height of the Tyrrhenian terrace. If this assumption be accepted as correct it follows that an elevation of the coast must have taken place in post-Sicilian, and even in post-Milazzian times. Now the Sicilian terrace is usually dated as pre-glacial which means that the sea-level of that time stood 40 meters higher than at present. According to this line of argument the total elevation of the Sicilian terrace can be estimated at 90 to 100 minus 40, i.e. at 50 to 60 meters. If we suppose the Milazzian terrace to have been formed under conditions approximately equal to the present so called "post-glacial" time, an elevation of 55 to 60 meters — the present altitude of the Milazzian terrace — must have taken place in post-Milazzian though pre-Tyrrhenian times. And this amount corresponds exactly with the amount of elevation of the Sicilian terrace!

However, everyone will feel that the problem is beset with many doubtful questions and numerous pitfalls [1]). For, in the first place the inter-regional correlation of terraces is a difficult and delicate procedure in itself. Secondly the fossil molluscs of the Sicilian terrace give evidence of a sea cooler than the present Mediterranean. On the other hand the fauna of the Milazzian terrace seems more in accordance with the idea of warmer conditions than are found in the Mediterranean waters of our days. Unless one would try to explain these features by supposing paleogeographic conditions to have been responsible for these facts, they present difficulties which do not fit in the above given reconstruction of events. Hence, future research may lead to quite different results!

A further difficulty is presented by the marine terraces of Oahu and Maui, Hawaii [2]). Stearns found them to occur at the same height which

meters above the present one at the beginning of the Pleistocene — based on the elevation to that height of terraces in Calabria — is to be rejected too. There are Pliocene sediments in Calabria at more than 1000 meters above sea-level, in Algeria at 500 meters, etc. I agree with Daly (1934, p. 156) that these exceptionally high figures are reasonably explained by calling upon their situation near the young and therefore rising mountain chains of the Alpine belt.

Zeuner holds that the interglacial sea-levels were successively lower, which would mean that in each interglacial stage deglaciation was less than in the preceding one. As his assumption is not acceptable in view of the climatic evidence of the interglacial deposits, he claims that the glacial fluctuations of sea-level were superimposed on a gradual lowering of sea-level due to some major cause, viz. a sinking of the bottom of the sea. (Zeuner, op. cit., p. 164, 248).

1) The reader interested in this intricate problem may be referred to Paterson's treatise of 1941 which tries to give a correlation of Pleistocene terraces all over the world. In his opinion variations of power-volume of the rivers are coincident over all the areas examined. These variations would result in three great cycles of sedimentation divisible into seven phases.

2) It was the situation of the marine terraces in the Hawaiian islands that particularly impressed Novàk and strengthened him in his opinion "that the arrangement of the marine terraces the world over can hardly be explained on any other supposition than that it was caused by sinking of sea-level during the greater part of the Pleistocene" Compare p. 114 footnote 2).

is characteristic of those found by Cooke in S. Carolina and they were correlated to the well known scheme of Dépéret for southern Europe and that of Krige for South Africa. The Hawaiian islands present a difficulty in so far as they form a comparatively small area which, moreover, has no sialic substratum — according to now prevailing ideas. Hence one ought to explain why in these islands the same coastal features are met with as in the marginal zone of the continents — notwithstanding the apparent differences just mentioned.

Another point of importance is that a quite different age determination of the most recent movements is to be deduced from the coastal profiles of Morocco (fig. 69). According to Lecomte the elevated marine shore and

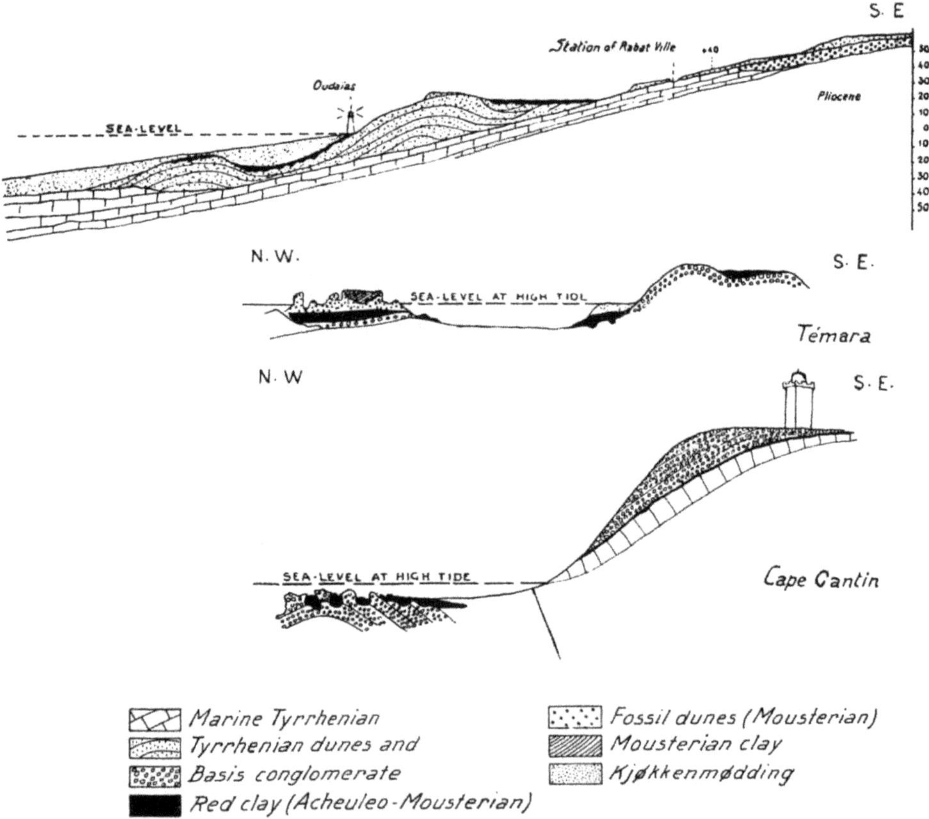

Fig. 69. The continental flexure along the coast of Morocco. (After Bourcart).

dune landscape, from Tanger towards Cape Cautin de Safi, is of Chellean age. It is, however, covered by an encrustation which, according to Antoine, has to be considered of Acheulean age and by a loamy soil of Mousterian times. Bourcart considers the fossil dunes and the elevated shoreline synchronous with the Tyrrhenian Strombus-beds. From fig 69 illustrating this part of the continental flexure of Morocco, it would necessarily follow that there the most recent movement has to be estimated as post-Mousterian i.e. post-Würm glaciation. From fig. 69 it may be seen that at the landward side the tilting movement attained about 50 meters at least. In passing it is interesting to note that, according to Bourcart, the deformation of the coastal tract of Morocco resulted in the formation of three zones which were more elevated than the two intervening "synclines". If Bourcart's interpretation of the profile and his age determinations are correct the question arises whether the age of the most recent movement of the continental flexure in Morocco is an exceptional case or not? At any rate one would like to know how it can be brought into harmony with the above conclusions regarding the time of elevation of the terraces in the Mediterranean.

Still another difficulty arises when an endeavour is made to correlate the movements of the shelf-area to those of the coastal tract. From the geological sections of the coast region of Morocco a synchronism of the most recent movements in both zones may be readily deduced. And according to Bourcart, the formation of the uppermost shelf-surface off large stretches of the European and African coasts would date from mid-Paleolithic times. The following table summarizes his opinion on the time of origin of the different shelf surfaces.

TABLE VI, Different shelf-surfaces, after Bourcart

Surface I,	0– 200 m isobaths:	Mid-Paleolithic (Acheulean-Mousterian), regressional stage between the transgression of the Tyrrhenian and the Normanno-Flanderian.
Surface II,	200– 500 m isobaths:	either regression of the upper-Pliocene or regression between Sicilian and Tyrrhenian.
Surface III,	500–1000 m isobaths:	regression of the Pontian.

Hence the last downward movement of the shelf-area would date from post-Mousterian time, i.e. after the Würm-glaciation.

If, however, a pre-Tyrrhenian elevation of the coastal tracts should prove to be the general rule, no synchronism would seem to be established between the movement of the shelf-area and the adjacent coastal strip. Obviously the determination of the ages of the Pleistocene movements is still far from being a settled question. We have to wait for many more data which only can be assembled by detailed geological surveys of the continental margin.

The geophysical side of the Problem

In the preceding sections an endeavor has been made to picture the events that seem characteristic of the continental margin in its superficial parts. More difficult is the problem of the processes underlying these phenomena in deeper layers of the earth.

In a schematic section through the continental margin of eastern Australia (fig. 70) Jenson expressed his opinion that a displacement of deep-seated material ought to take place from the down-faulted blocks of the shelf and the deep-sea towards the continent. Probably currents in the simatic substratum have to be considered as the primary agencies. Their action appears to involve the movements along the continental flexure — in this case the origin of the faults bordering the shelf of eastern Australia and the subsequent movements along these faults. The attempt to unravel the mutual connections of these phenomena means entering the domain of

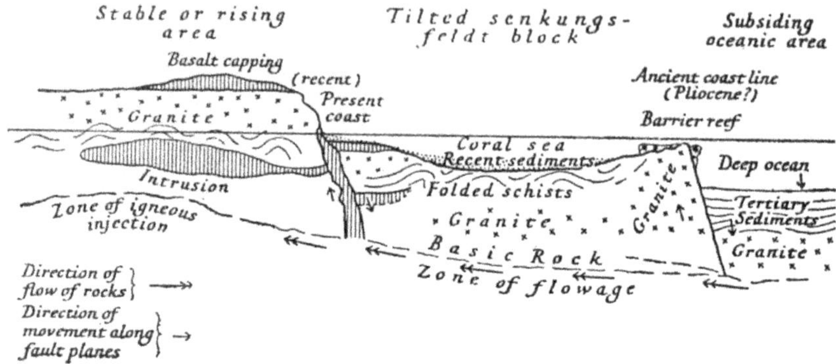

Fig. 70. Geological section across eastern Australia and the shelf of the Great Barrier reef. (After Jensen).

mainly theoretical speculations. Let us summarize the diverse points which form the geophysical side of the problems discussed in the preceding pages.

The theory of continental flexure implies the periodical occurrence along the continental margin of movements of a rather small though uniform amount causing a subsidence of the shelf areas and an elevation of the adjacent coastal tracts. The questions arises whether such processes can be accounted for from a geophysical point of view. A further question is by what mechanism isostatic equilibrium is re-established in the deeper parts of the down-bent or down-faulted portions of the earth's crust. *Mutatis mutandis* the same problem arises for the periodically elevated part of the continental margin.

In his discussion of the features of the continental border Chamberlin [1]) supposed the ocean basins to have sunk and to have crowded somewhat upon the land sectors. Hence at their junction the sea-bottom would have tended to sink and at the same time to have been pushed under the land, "while the latter tended to rise relatively, and perhaps even to spread above toward the ocean-basin". According to his opinion the submerged portions of the shelf-area would be carried obliquely out of the water during epochs of a strong world-wide compression of the outer shell of the earth. After such an epoch of compression accompanied by the bowing up of the

[1]) Chamberlin and Salisbury. Geology 3, p. 526.

continental border a slow reverse movement would set in and this would again submerge the now strongly eroded coastal region.

Similarly an alternating sinking and elevation of the ocean-bottoms as a whole, accompanied by simultaneous epochs of upheaval and downward movement of the continental border is ascribed by Jessen [1]) to hypothetical processes in a presumed sheet of "salsima" which is supposed by him to underly the sial of the ocean-bottom and to terminate at the sialic thickenings of the continents.

The facts known at present do not seem in favour of these hypotheses. In general the shelf-area is intermittently subjected to a downward movement, whereas the continental side is only elevated periodically. The rejuvenated relief of the continental border is worn down by sub-aerial erosion in the course of time and the denudation products are transported seawards where most of it accumulates in the shelf-area.

Gravity observations at sea carried out at right angles to the coast, generally show an increase in the isostatic anomalies of 30 to 100 milligals, when passing from the shelf to deep water. This rather sudden jump in the anomalies mostly causes positive values at the ocean side of the profiles (fig. 71). In all 26 complete profiles have been obtained by Vening Meinesz

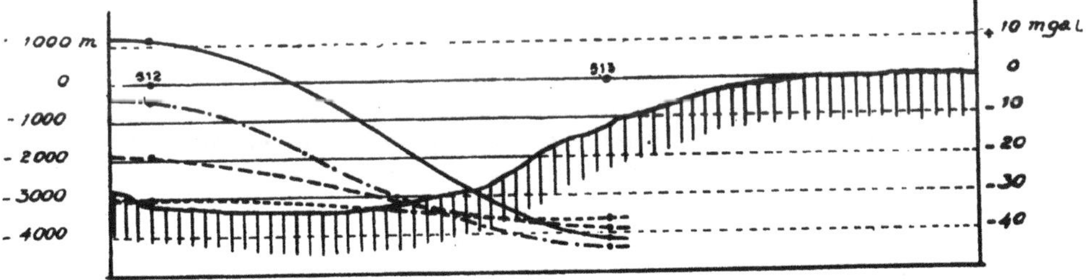

Fig. 71. Gravity observations across the shelf southeast of the Canary Islands according to different methods of isostatic reduction. Horizontal scale 1 : 3,000,000. Vertical scale 1 : 300,000. The dotted line represents the anomalies for T = 20 km and R = 0; the broken line for T = 30 km and R = 0; the point-dot line for T = 30 km and R = 116.2 km; and the full drawn line for T = 30 km and R = = 232.4 km. T represents the thickness of the crust. R represents the radius of the bending of the crust according to different assumptions of regional compensation. R = 0 represents the case of local compensation according to Heiskanen's assumptions. (After Vening Meinesz).

on board submarines of the Royal Dutch Navy; three at the end of the Channel, one near Lisbon, four along the coast of W. Africa, two near the mouth of Chesapeake bay (east coast of the United States), four between Panama and San Francisco (one incomplete), six off the east-coast of S. America, two off the west-coast of Australia, one off the south coast of Ceylon and one incomplete near Socotra.

From a study of several methods of isostatic reduction of the gravity stations Vening Meinesz concludes that the most probable explanation of the data is to assume one or more of the sialic layers of the earth's crust to thin out rather suddenly at the continental margin. His solution is in agreement with the generally accepted idea — based on seismological and

[1]) O. Jessen, op. cit. pp. 144—148 and plate 25.

petrographic data — of the presence of a thick granitic layer in the continents and a rather sudden thining out at the continental margin.

A much thinner sialic layer is thought to present under the Atlantic and Indian Oceans. And probably this layer is totally absent in some other oceanic sectors including the North Polar Basin and a large part of the Pacific Ocean inside the so-called andesite line (see pages 67, 165 and 228). The question why these upper layers of the earth's crust are distributed in such a remarkable manner involves the whole problem of the origin of continents and ocean floors. It forms one of the most fundamental and baffling problems of geology, which will be treated in Chapter VIII.

However, possibly the peculiar distribution of sialic and simatic masses in the border zone of continental and oceanic areas causes the intermittent action of convection currents and these, in turn, cause the periodical movements along the continental flexure. In order to explain why the submerged parts do not emerge during a period of rest of the convection system one has to admit that sialic material of the deeply submerged lower part of the continental margin is transported continentward. It thus supplies new sialic material to the continent from below, whereas erosion on the surface tends to the opposite effect.

Finally, if the intermittent submergence of the shelf-area has to be considered as a world-wide phenomenon continuing from primeval times we would have to conclude that a complete series of sediments lies buried in the deposits of the shelf-body. Chamberlin and Salisbury already laid emphasis on this aspect of the situation. However, from the few facts at hand such a sequence seems only partly to be realised. For example, the shelf-sediments accumulated on the sunken borderland Appalachia seem to cover a sequence ranging from Mesozoic to recent times (fig. 26). It appears that Appalachia began to subside during the Triassic, i.e. after the mountain building of the neighboring Appalachian geosyncline. Similar time-relations between the beginning of the submergence of a borderland and certain tectonic phenomena in adjacent geosynclinal belts were mentioned in a previous section (p. 39). In addition it may be mentioned that the subsidence of the present shelf-area of southwestern Europe cannot possibly have begun earlier than after the Variscian chains of Europe had been formed. Hence, the shelves that existed in these regions previous to those the origin of which can be dated approximately, must necessarely have been situated farther off the present continental margin. What was the origin, situation and history of these presumed shelves of a very remote past is a problem that cannot be attacked at the moment for lack of reliable data.

Submarine valleys

The continental margin shows a geomorphological feature of high interest in the wide spread occurrence of submarine canyon-like valleys extending from the shelf downwards to two or three thousand meters. It is not yet possible to furnish a well-founded classification of these furrows on the upper surface and outward slopes of the shelf since our knowledge of the submarine relief is still far from complete. However, the following

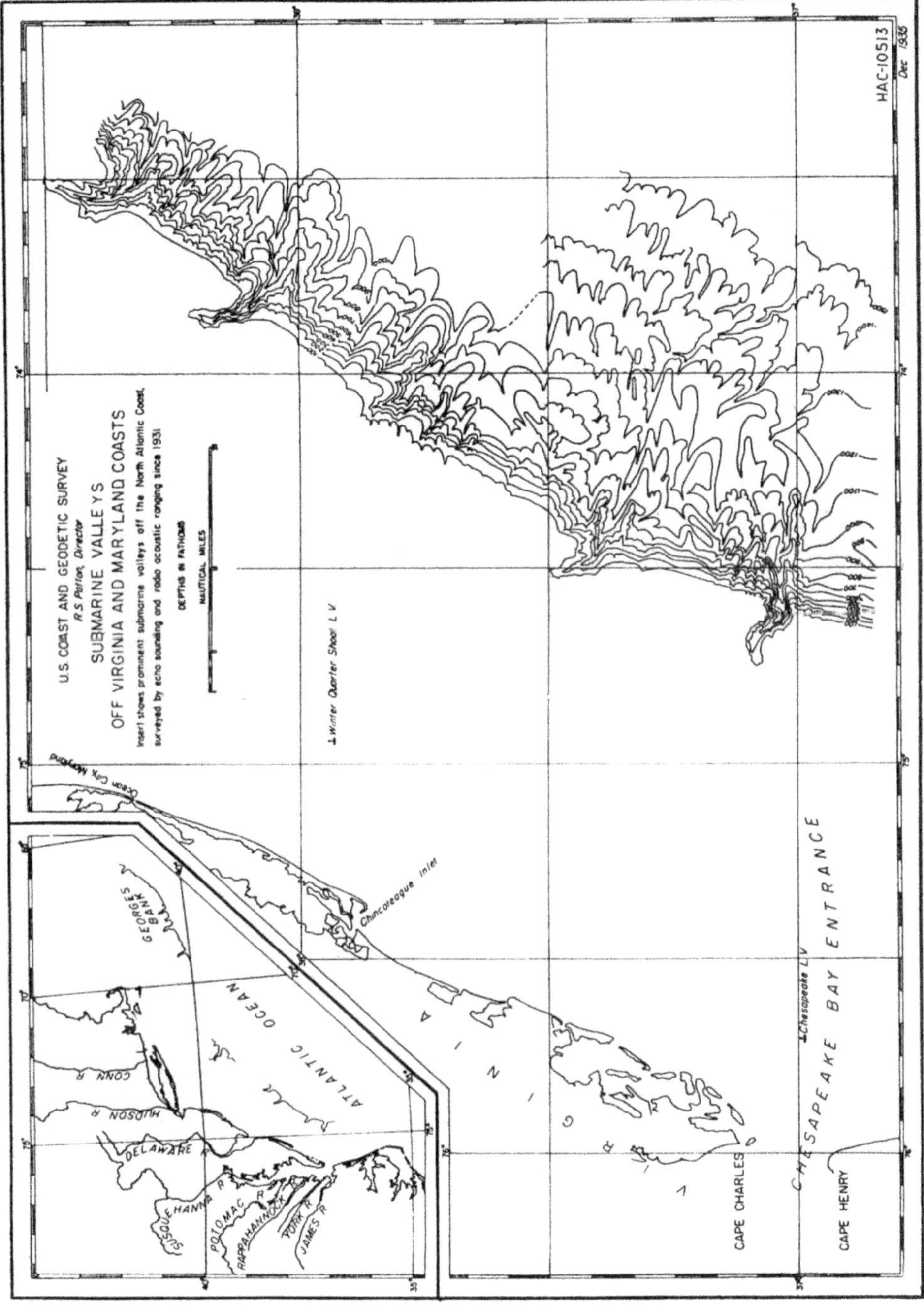

Fig. 72. Submarine valleys at the shelf-edge off the coast of Virginia and Maryland. (From P. A. Smith).

three types may be fairly well distinguished and tentatively classified.

(1) As a first group submarine gorges originating near the edge of the shelf and running downwards to great depth have to be mentioned. In many places they were found crowded together in great numbers (fig. 72). Their course is in general uniformly parallel, and perpendicular to the edge of the shelf. They are cut back only a short distance into the platform of the shelf, their headward extensions being seldom more than 5 to 10 miles. As an additional feature these gorges only exceptionally show branching of appreciable extent and have comparatively few tributary gorges. Respecting steepness and depth they are comparable to large subaerial canyons. Cretaceous and Tertiary fossils dredged by Stetson in 1936 from the walls of some of these deep submerged marginal notches of the shelf bordering eastern North America seem to prove that at least the upper part of the canyons were cut in Late Tertiary or Pleistocene time. According to Shepard and Beard [1]) the average gradient observed on river valleys in the Appalachians, Wasatch Mountains, Sierra Nevada and Coast Ranges was 6.07 percent, as compared with the average gradient of 6.54 percent for the oceanic canyons. And then they continue: "of course another choice of land canyons might have given very different results. The four mountain ranges show a difference in gradients [2]) Land valleys are well know to show decreasing gradients along their length. The submarine canyons have the same tendency, as is shown by the averages of 11.62; 6.63 and 4.76 for the head, center and end portions. Furthermore it

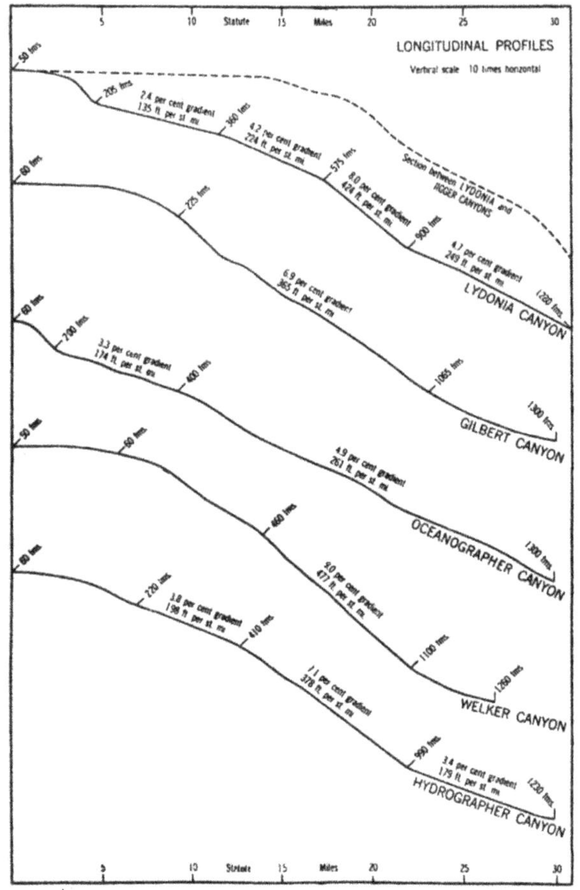

Fig. 73. Longitudinal profiles of some canyons of the eastern United States. (From Veatch and Smith).

1) Shepard and Beard, op. cit. 1938, pp. 450, 451.

2) Viz. Appalachian Mountains 2.05, Wasatch Mountains 4,06; Sierra Nevada 9.03 and Coast Ranges 7,3 (averages).

is true of both land valleys and submarine canyons that most of the large rivers have much lower gradients than the small."

Fig. 73 shows a series of longitudinal canyon profiles of Georges Bank according to a more recent publication by Veatch and Smith [1]. Veatch pointed out that similar features, including the maximal depth of the canyon below the rim, exist in subaerial regions [2].

The same author was of the opinion that the high average slope, the steepness of the gradients of the streams of the American continental slope and the probable uniformity of the Cretaceous and Tertiary beds of the region are responsible for the characteristic tendency of the valleys to be so uniformly parallel [3].

(2) A few gorges of the type mentioned just now extend across the continental platform, their much shallower headward extensions reaching the

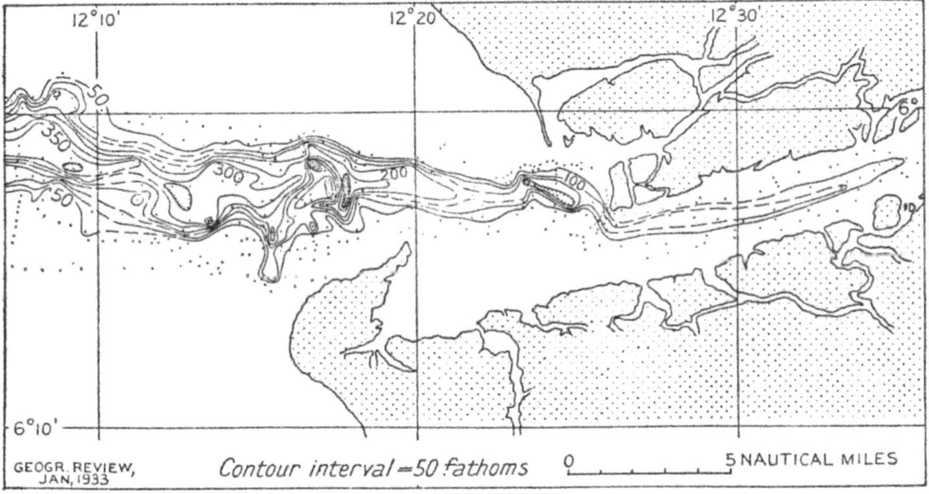

Fig. 74. Head of the Congo canyon. Each dot represents a sounding. (After F. P. Shepard).

vicinity of the shore near or at the debouchement of a large river (fig. 74). According to Bourcart the gradient of the longitudinal profile of these gorges rather surpasses the gradient of a normal river profile in its lower or even median course. He mentions the following percentage figures: Congo 19.7; Indus 9; both the Hudson gorge and "Gouf de Cap Breton" 2; Sado 2.5; Tage 3.3 These figures are, however, widely diverging from those given by Shepard and Beard, viz.: the submarine slopes of the Congo the Indus, and the Ganges having, according to them gradients of 1.1, 1.0 and 0.8 percent respectively. These figures seem more trustworthy as may be seen from fig. 75, which shows a longitudinal profile of the Congo Canyon. The canyon is incised at the deepest as much as 1480 meters (820 fathoms)

1) Veatch and Smith 1939, p. 18, fig. 6.
2) Ibidem, p. 20.
3) Ibidem, pp. 16, 17.

below the general slope of the shelf. The Hudson canyon is cut 1120 meters below the rim near the edge of the shelf. These are not extraordinary figures

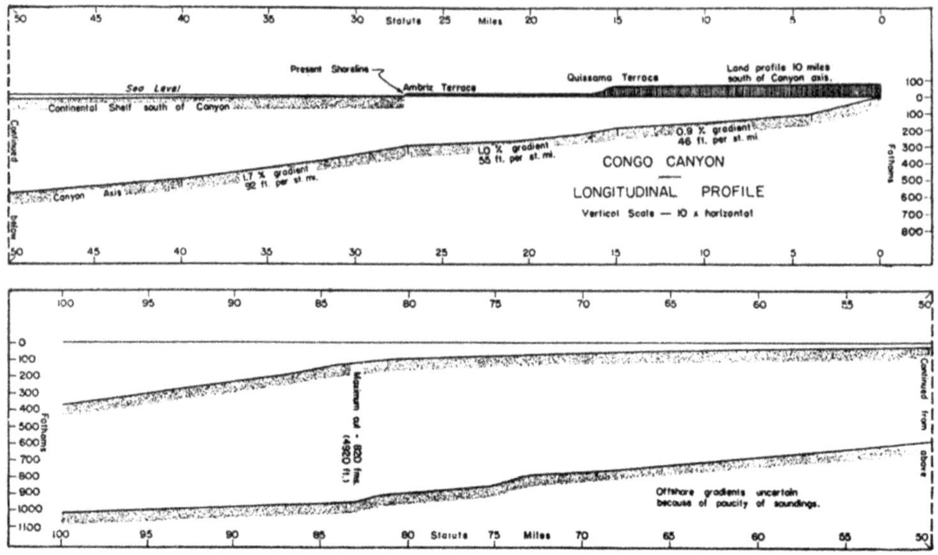

Fig. 75. Longitudinal profile of Congo Canyon. (From Veatch and Smith).

if compared to the maximal depth of 1830 m incision below the rim of the Colorado canyon and 2370 meters of the Hell's canyon of Snake River in Oregon and Idaho [1]). Two cross-sections through the Hudson canyon are shown in fig. 76 [2]). Veatch regards the Congo valley as of post-glacial origin. According to this interpretation the main erosion of what is now the elevated interior Congo basin took place when the Congo river had a much more northerly course and reached the sea several hundred miles north of the present mouth. He asserts that

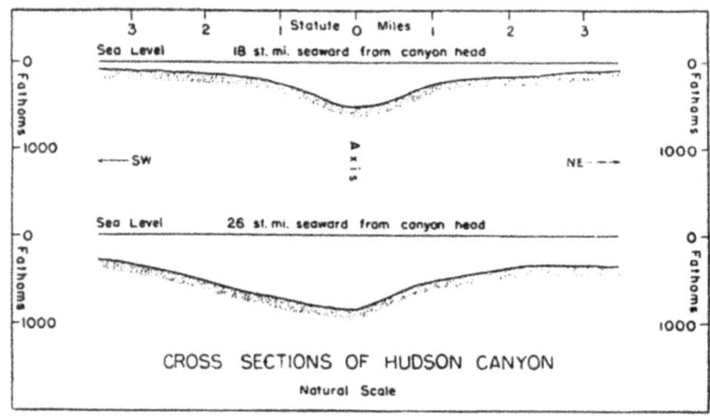

Fig. 76. Cross-section of Hudson canyon. (From P. R. Smith).

"faulting along the western part of this original course of the river caused the water to pond in the interior basin forming a great lake about 400 to 500 miles in diameter" (1939, p. 25). The water of the lake flowed over a

---

[1]) Veatch and Smith, 1939, p. 15.

[2]) Ibidem, p. 20, fig. 8.

low point of the surrounding rim and started the Congo on its present course to the sea. Now the time of this lake is fixed by the abundant human artifacts found beneath the silts of the lake — and regarded as Mousterian, some of them even as Aurignacian and Solutrean. Hence, the present course of the Congo seems of post-Mousterian age. Most authors regarded the origin of the submarine canyons of groups (1) and (2) as contemporaneous. And since it appears that the gorges are cut in Cretaceous and Tertiary beds, it seems certain that they are at least younger than Pliocene. This conclusion applies to the Congo canyon as well as to those off the eastern coast of the United States [3]).

(3) Another type of submarine canyon was revealed by sonic soundings off the coast of California (fig. 77). They are characterized by a dendritic river-like pattern, deeply incised in the surface of the continental shelf (fig. 77) and thence continuing towards the great depth of the Pacific Ocean.

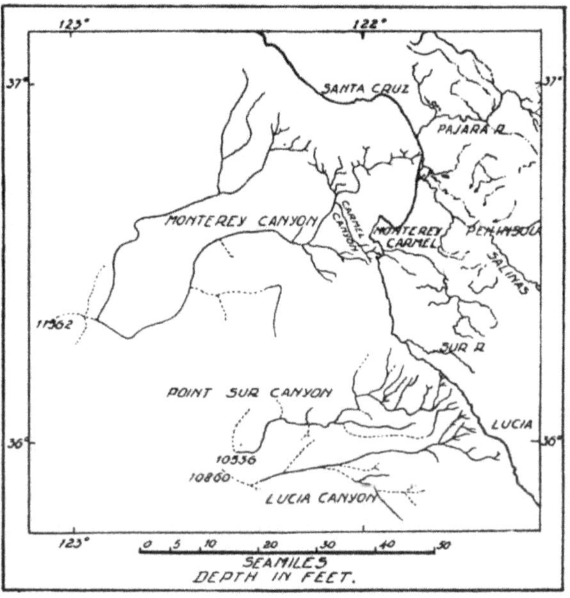

Fig. 77. Dendritic pattern of the Monterey- and Lucia-canyons off the coast of California. (After F. P. Shepard).

The submarine gullies mentioned under (1) and (2) are similar in many respects. It will be seen that the Californian gorges are a special case to be considered separately.

It is to these submarine notches incised in the surface of the continental shelves and running down along their outer slopes towards the abyssal regions of the oceans that the following considerations will be devoted. From the very beginning, however, a few submarine features which by some authors have been introduced into the literature of "submarine canyons" have to be eliminated from our discussion. For it would be erroneous to classify all elongated submarine depressions under the same head and to assume their history to be genetically the same.

For example, a structure which will be left out of consideration here is the submarine depression occurring southwest of the delta of the Mississippi (fig. 78). It looks like a scar formed by the sliding down of a part of the shelf, possibly

[1]) Although Smith argues that the prominent canyons are probably not all contemporaneous, he still agrees that all were probably incised during the Pleistocene (Veatch and Smith, p. 43).

along faults. Shepard is of the opinion that its origin was connected with the many submarine saltdomes found in the vicinity of the edge of the shelf.

The submarine trough-like bottom configuration in the Bahamas

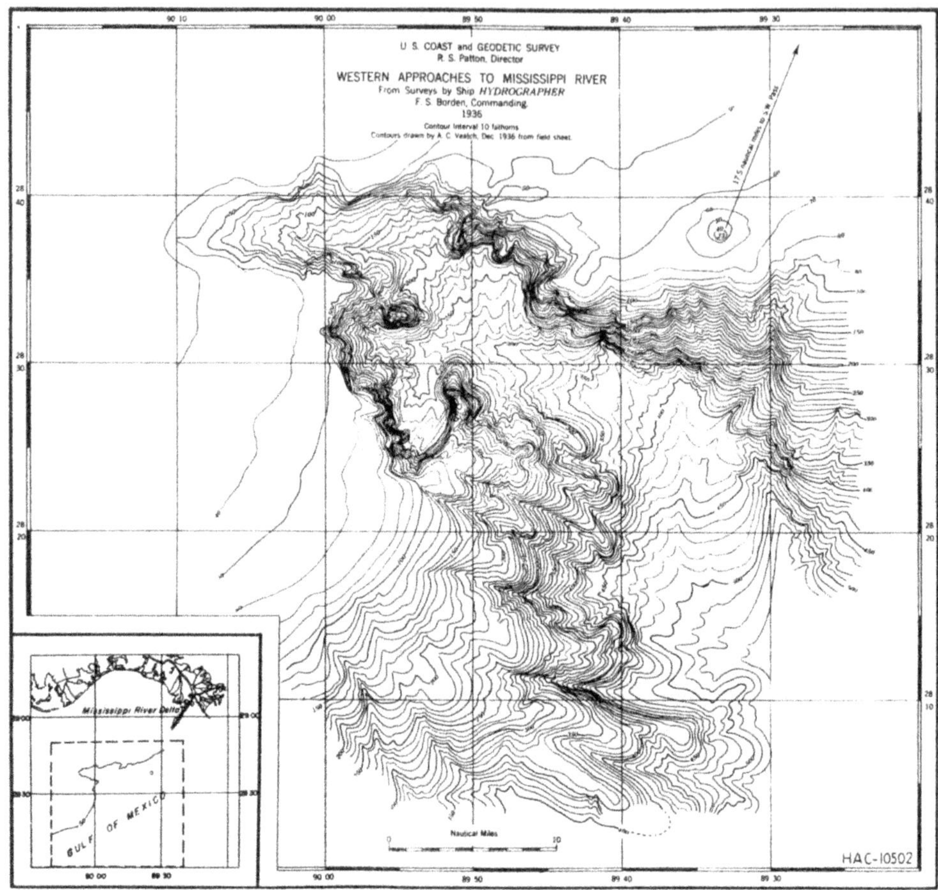

Fig. 78. Submarine morphology SW of the Mississippi delta. (After P. R. Smith).

described by Hess, will also be left out of consideration here. Whatever its origin may be the morphological features seem clearly contrasted with those of the gorges, incised in the continental shelves, which come up for discussion in the following pages.

According to Shepard fig. 79 shows a "graben-"type of submarine valley off the coast of Japan.

The same author published a special article which is mainly devoted to submarine glacial troughs. The remarkable South-Norwegian depression (fig. 80) is also classified by Shepard among the morphological features belonging to the group of glacial phenomena. The characteristic difference of cross-sections as compared to a submarine canyon of the "river-type", is demonstrated by fig. 81.

A further case which has to be discarded from the problem of the submarine canyons is the submarine relief which was revealed round the island Bogoslof (fig. 38). Smith who published a contour-chart of the surroundings of this little volcanic island in the Aleutians was probably quite right in arguing that the relief of the bottom originated under subaerial conditions and that it was subsequently drowned to great depths (cf. p. 58). In the case of the relief surrounding Bogoslof, however, there is again no shallow outer shelf platform. All the problems connected with the presence of the outer shelf are simply lacking and Bogoslof therefore will be left out of consideration. To include the case of Bogoslof in a discussion on the general problem of submarine gorges along the surface and edge of the shelf — as has been done by some authors — would be to compare unequal situations [1]).

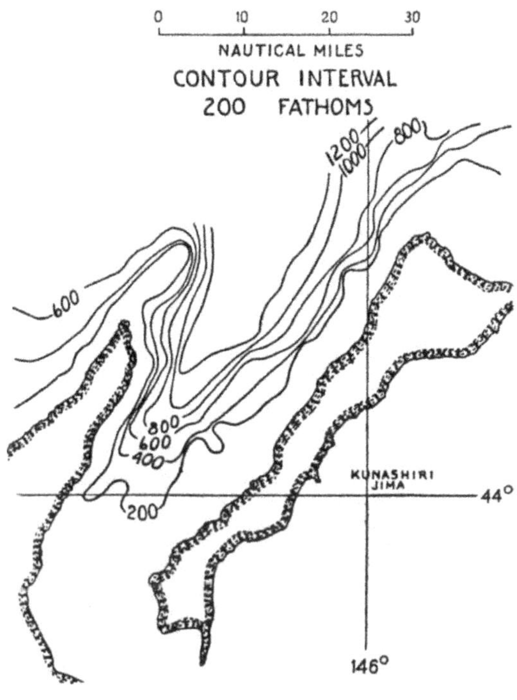

Fig. 79. A "graben" type of submarine valley off the coast of Japan. (From F. P. Shepard).

After these preliminary remarks the three groups mentioned before, will now be discussed.

(1) The notched shelf edge.

Submarine gorges of the type mentioned under (1) were charted in great abundance along the outer slope of the shelf of Georges Bank and from Cape Hatteras to the Grand Banks (fig. 82). These gorges have been traced to a depth of more than 8000 feet. They display, however, the remarkable feature that their headward extension is seldom cut back into the surface of the shelf more than 5 to 12 miles. The exceptions will be mentioned later on. In the shallower parts of the shelf (less than 50 fathoms) all trace of them is lost (fig. 82 and 83). Their head and side-walls are steep; v-shaped in cross-section, and from 2 up to 6 miles in width.

Hypotheses involving a tectonic origin were put forward by some authors

[1]) It is for the same reason that the submarine valleys of Oahu will be left out of discussion. The more so as I was unable to study detailed charts of this region. Compare also note [2]) on page 115.

including A. Wegener, J. W. Gregory and A. C. Lawson. Although a few canyons display peculiarities which may find an interpretation by structural

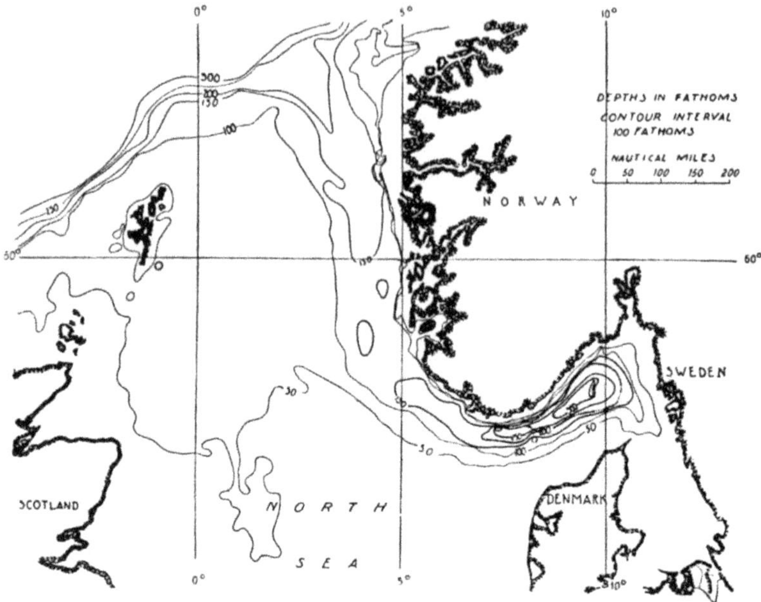

Fig. 80. The South Norwegian submarine trough showing a deep depression. (From F. P. Shepard)

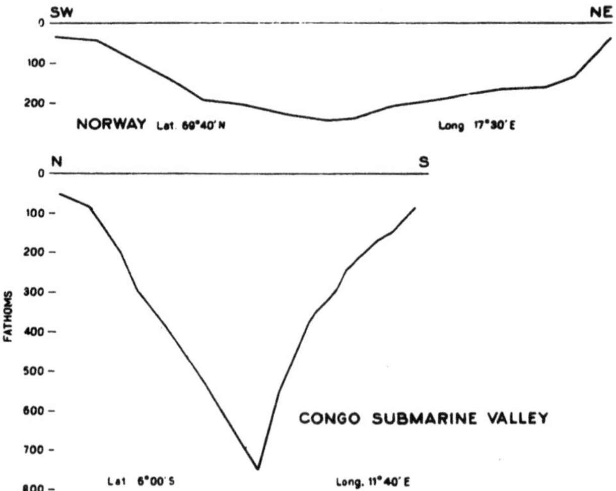

Fig. 81. Transverse profiles showing the contrast between a submarine valley of the "river" type (lower) and glacial trough (upper). (From F. P. Shepard).

processes, the great majority of submarine notches exhibit characteristics which cannot be explained by tectonic hypotheses.

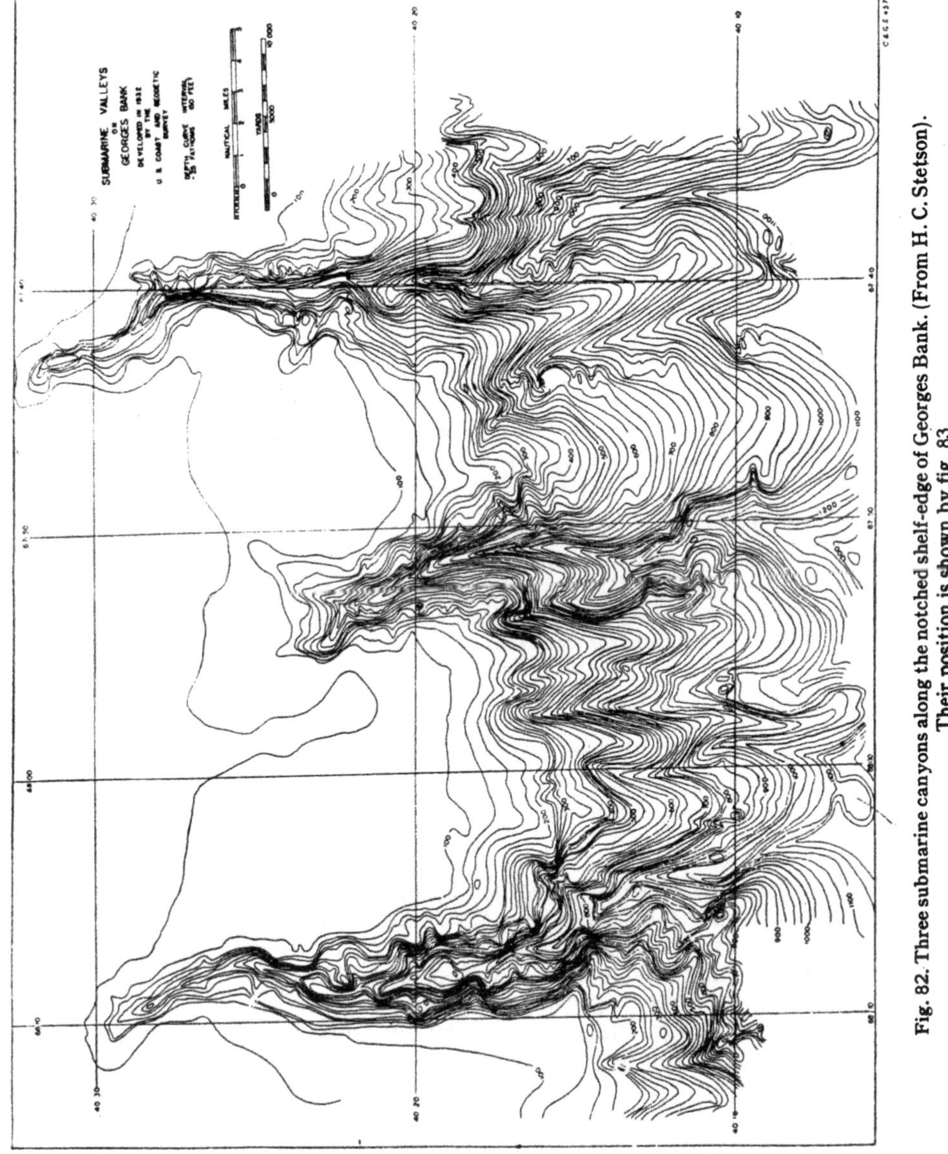

Fig. 82. Three submarine canyons along the notched shelf-edge of Georges Bank. (From H. C. Stetson). Their position is shown by fig. 83.

The two major groups of hypotheses are those involving a subaerial origin and those involving submarine agencies. The preceding sections on the origin of the shelf lead to a decision as to which of these two great groups of

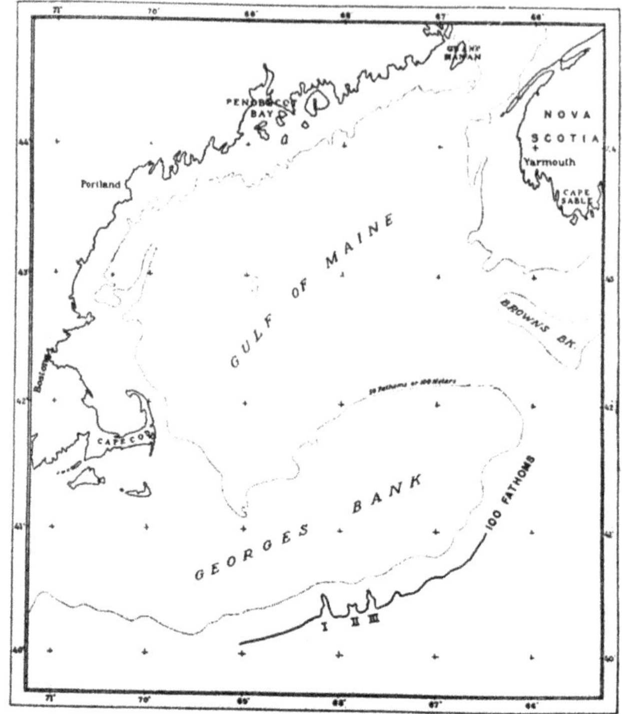

Fig. 83. Position of the three submarine canyons illustrated by fig. 82. (From H. C. Stetson).

antagonistic theories is to be preferred. The principal lines of thought of both of them will therefore be explained presently.

Theories of subaerial origin

In his article on the geology of Georges Bank Stetson [1] wrote: "the abundance of the canyons themselves precludes the possibility that they are all caused by streams crossing the shelf. Upward of forty large ones are found on the southern face in addition to many minor ones. So closely spaced and so similar in size are they that it is difficult to conceive that the necessary number of rivers flowing out of a hinterland could have existed side by side, even before the existence of the Gulf of Maine. Stream piracy would certainly have enlarged certain ones at the expense of the others".

And then Stetson continues: "The writer is indebted to Kirk Bryan for a suggestion which might explain this situation. It is not necessary to appeal to some ancient river, with its headwaters in the hinterland, to provide the means for cutting these canyons. In a plateau country, starting upon the face of a scarp, canyons of the order of magnitude represented here can

[1]) Stetson, op. cit. 1936, p. 353.

grow by headward erosion through the agency of groundwater sapping. Many formed by this means are well shown on the United States Geological Survey's topographic sheet of the Bright Angel quadrangle, Coconino Country, Arizona" ....etc...... "The same conditions on a smaller scale are well illustrated by a scarp near Cameron, Arizona". [1])

Indeed the similarity is very striking and many authors have tried to explain the submarine gullies by suggesting a subaerial origin and a subsequent submergence of the area. Consequently the continuation of the above quotation of Stetson's reads as follows:

"Elevation of the continental margin would provide ideal conditions for erosion of this sort. If Georges Bank now stood about 7000 feet above sea-level, its southern face would look not unlike the scarp near Cameron".

Here, however, lurks the unsurmountable difficulty viz. the improbability of an oscillation of thousands of meters either of the continental margin or of the sea-level since Pleistocene times. According to the usual estimates, Pleistocene changes of sea-level would account only for an oscillation of about 100 meters [2]). And even Shepard who once claimed a glacial lowering of sea-level of some 3000 feet by suggesting enormous ice-caps about four miles thick over the Arctic and Antarctic regions — was unable to account for the lower 3000 to 5000 feet of the submarine notches [3]).

Hess and MacClintock even went so far as to put forward a hypothesis which holds that under the influence of some stellar body the earth's rotation might have been decreased causing a depression of the sea-level in low latitudes and raising it in high latitudes. A sudden change in the speed of rotation is, however, highly improbable from a mechanical and astronomical point of view. Such a happening would moreover bring about formidable changes in other processes on the surface of the earth. Nothing of the kind is known to geologists.

On the other hand a theory surmising that the continental margin could have been uplifted to a height of 3000 or more meters above its present level and that subsequently it could have been submerged by a similar amount has to be rejected on very sound arguments. Elevation of the shelf-area accompanied by comparative stability of the land would undoubtedly have resulted in a very characteristic drainage system of the dammed-up rivers at the land side. Again no effect of the kind is to be noticed in the present coastal areas of the continents. Moreover, if rivers be assumed to have cut the many trenches, then — as Daly [4]) remarked — "the master rivers of the continent must have simultaneously induced physiographic changes (either cutting broad valleys with floors now far below sea-level or else making local, voluminous deposits of sediment) on a much larger scale and visible

[1]) Some other illustrations have been given by Veatch 1939, Plates 3 and 4.

[2]) The same conclusion was reached by Veatch (1939, p. 77): "— the wave-formed surfaces and other shoreline phenomena now exposed around the Atlantic do not, in so far as known, show differences in level that would warrant the inference of earth movements in the Atlantic area of more than a few hundred feet in Pleistocene time".

[3]) Veatch also postulated "in human time — Mousterian or later — a recession of the Atlantic waters to a shore line located 10,000 feet or more below the present sea level, and the return of these waters to the present level less than 10,000 years ago" (1939, p. 30, see also p. 41).

[4]) Daly, 1936, p. 405.

in the existing morphology and structure of the continental land. No drastic effects of the kind have been found".

Remarkably enough difficulties of this short have been overlooked even by a few very serious workers. Bourcart and Francis-Boeuf for example are of the opinion that during the low levels of the Pleistocene when the shelf-surface was largely situated at or above sea-level, the upper part of the canyons were incised. Applying his theory of an intermittent down-bending of the shelf-area one would have to ascribe the formation of the deeper parts of the canyons to a more remote past. This theory would be adequate if the depth of the gullies at the edge of the shelf were a negligible factor. However, near the shelf-edge (in the region of the 200 meter isobath of the shelf-surface) most of the gullies show a depth of more than 1000 meters. In order to account for an erosion-base at that level one would have to admit the hypothesis of an oscillation of at least over 1200 meters!

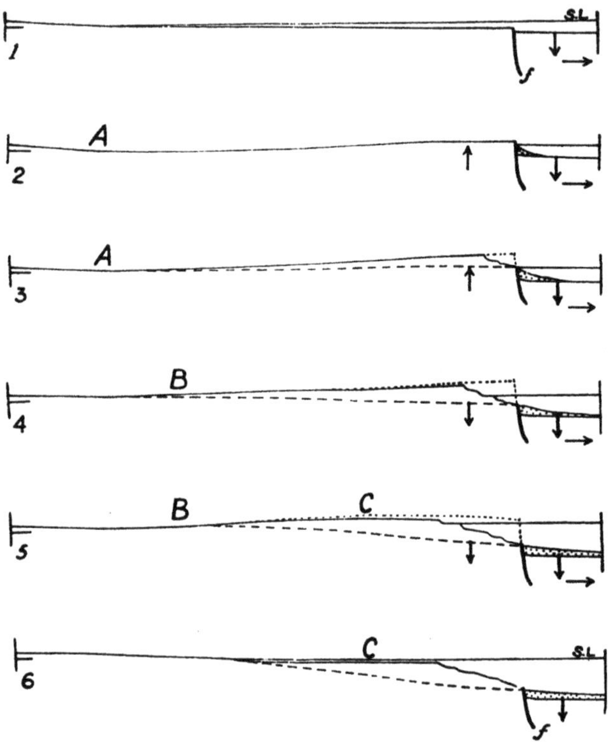

A great marginal up-arching followed by a coastal downwarping of still greater magnitude is indeed postulated by Du Toit.

His scheme — fig. 84 — involves both faulting and warping with trends more or less parallel to the continental margin. In his opinion a process of periodic tension, due to continental drift and operating from the Late Mesozoic onwards, was responsible for the origin of one or more coastal faults accompanying the downwarping of the continental rim along the Atlantic and Indian oceans. The same process of thinning of the oceanic sial-layer would cause the continental side of the fault block to bend upwards "in accordance with isostatic and paramorphic principles" while the rivers maintained and deepened their courses forming subaerial canyons. In order to explain the present submarine gorges a coastal up-arching of some 4000 feet is postulated. Then: "continued or renewed tension causes further sagging of the ocean

Fig. 84. Hypothetical stages in the evolution of a submarine canyon; the broken line indicates the floor of the canyon. (From Du Toit).

floor with downdragging of the rim". The result would be a great marginal down warping of the furrowed coast, the drowning of the gorges, and finally the formation of a marine abrasion plane. This plane would constitute the continental shelf which in most cases would have been cut right across the drowned gorges that mostly became severed from their subaerial parts. All that has been set forth in the previous sections on the origin of the shelf is in strong contradiction to such a theory. For, the characteristics of the shelf-surface only allow for a subsidence of the shelf combined with changes of sea-level to a total amount of some 300 meters since Pleistocene time, certainly not for an amount more than four times as large. Nor do the marine coast-terraces fit in such a theory.

We are, therefore, compelled to look for submarine agencies which might have caused the remarkable phenomena under discussion. However, before entering into this aspect of the problem, it may be remarked that some other difficulties which were inherent in older theories are explained very nicely by Bourcart's theory of the continental flexure. In the first place his view accounts for the occurrence of a break of slope at 500 and 1000 meters and perhaps at deeper levels below the sea. In the mean time it explains the steepening of the outer margin of the shelf and finally it might perhaps explain why the average gradient of the submarine canyon-profile is steeper than those generally found in rivers on land.

Theories involving submarine agencies

The problem, however, of the origin of these submarine trenches remains untouched by Bourcart's considerations. Therefore, in order to explain these submarine furrows attention must be given to submarine agencies as a possible cause of notching of the edge of the shelf. Several factors have been advocated by various authors. The late D. Johnson [1]) brought them under the following headings: submarine landslides, submarine currents, subterranean river outlets, foundering of subterranean caverns, solution along faults either by uprising subterranean waters or by downfiltering marine waters, non-deposition along faults due to uprising subterranean waters and finally — sapping by submerged artesian springs (the latter hypothesis being put forward by Johnson himself).

It is clear from such an *embarras de choix* that the problem of the notched shelf-edges is far from being solved. Indeed: "Fifty years of discussion, and much new knowledge of the characteristics of the canyons, have not, however, brought agreement among geologists respecting the genesis of these forms. Many hypotheses have been advanced, but none has yet attained the standing of a generally accepted theory" [2]). We may refer anyone interested in the variegated diversity of these theories to Johnson's review and the more recent literature cited at the end of the chapter.

An exception, however, will be made in respect of Daly's theory in which appeal has been made to the widespread influence of turbidity currents. For

[1]) D. Johnson, op. cit. 1938, 1939. See also "submarine canyons" in Geograph. Review 27, 1937.

[2]) D. Johnson, 1938, p. 112.

even his opponents will avow that it represents a fine example of a well considered working hypothesis. And, moreover, it has so far received more attention from other investigators than has any rival hypothesis involving a submarine origin of the trenches. According to Daly waves beating upon the loose sediments of the continental shelf during epochs of a lowered sea-level of the Pleistocene, charged the sea-water with much sediment held in suspension. These turbid waters tended to flow down over the shelf-edge as density-currents which carved the trenches. During Kuenen's laboratory-experiments concerning Daly's hypothesis I more than once witnessed the fascinating view of a suspension running down into the deeper water of the tank along the artificial "canyon" previously carved by him in the outer slope of the shelf-model (fig. 85). It is, however, a crucial point whether currents of muddy seawater of glacial times were able to erode shelf-notches to a depth of 3000 and more feet. Daly and Kuenen are hopeful in so far as they think that further experiments and field data will bring a full confirmation of the theory. However, if the inner walls of the canyons are composed of solid rocks, instead of loose sediments, it seems hard to believe in Pleistocene density or turbity currents as the agency which was competent to notch the continental shelves. Now, according to Stetson the walls of the submarine canyons of Georges Bank actually consist partly of strongly lithified rocks and the same is held true by Bourcart regarding some trenches off the coast of Portugal and southern France. Both groups of data are worthy of further examination and we have to wait for many more facts which can only be gained by future investigations.

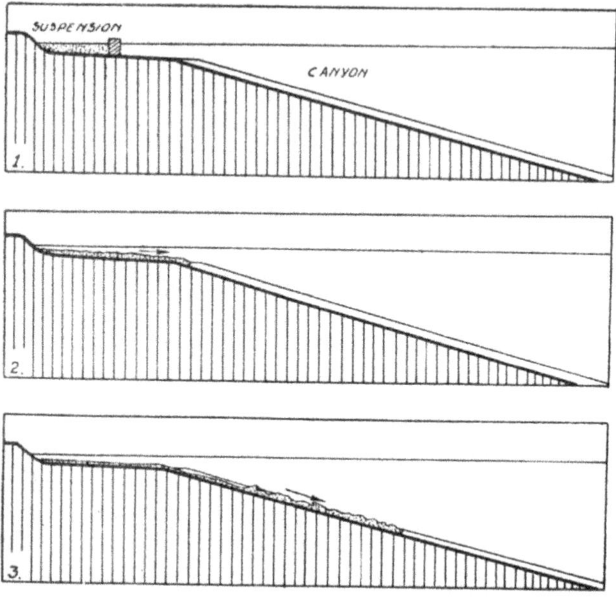

Fig. 85. Kuenen's experiments showing density currents running down the slope of a shelf-model (After Kuenen).

A different theory was offered by Bucher. He agrees that the cause of the submarine erosion must lie in movements of the ocean water. He suggests, however, that the action of seismic sea waves (tsunamis) has to be regarded as the principal factor capable of etching valleys along the continental slopes.

Bucher pointed out that the Upper Tertiary period of rejuvenation affected also such regions as the Mid-Atlantic Rise as is shown by elevated deposits of Upper Tertiary age on most of the oceanic islands of the Atlantic. This rejuvenating movement was accompanied by strong seismic disturbances. And these would have produced innumerable "tsunamis" that acted like the trigger effect. The highly seismic character of the Mid-Atlantic Rise would show that the period of rejuvenation is still going on.

(2) Gorges extending over the surface of the shelf. The question of the origin of the shallow headward prolongations over the shelf-surface, shown by a few submarine canyons stands apart from the genesis of the deep notches running from the shelf-edge towards the abyssal region of the ocean. Rather exceptionally a submarine valley can be traced upwards towards the mouth of a present river. The Congo canyon provides a well-known example (fig 74), whereas the head of the Indus canyon is situated 20 kilometers off the coast formed by the delta river after which it is named. It crosses the broad shelf at the edge of which its depth is about 1000 meters (fig. 86).

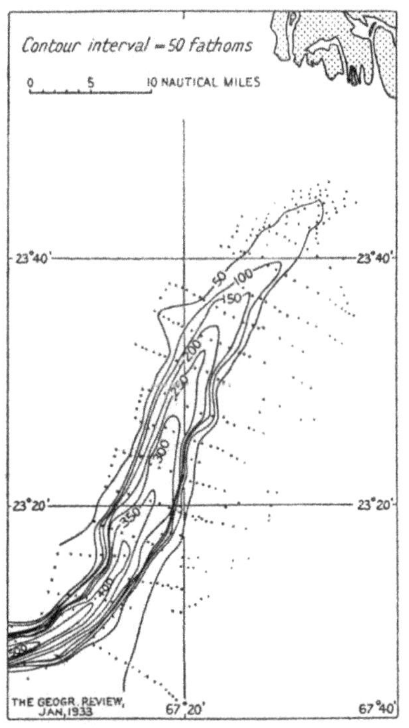

Fig. 86. The submarine valley off the Indus delta. Each dot represents a sounding. (From F. P. Shepard).

As a third example of a submarine gorge extending across the shelf fig. 87 shows the Hudson canyon. Out of the mouth of the river leads a shallow and rather broad submarine valley connecting New York Harbor with a submarine canyon which extends down the slope beyond. The outer gorge has been cut into the shelf platform twice as far as the neighboring canyons and those of George Bank, viz. 18 miles. And instead of being straight the course of the gorge is s-shaped. The shallower valley on the platform dies out near the margin of the shelf and the canyon heads about six miles northeast of the end of the shallow valley. Traces of an ancient Hudson delta are found near the continental edge. According to Shepard [1]: "tracing the canyon down the slope it is shown to make a double bend and come into line with the inner valley for most of its length. This suggests the possibility that the two were formerly connected, but that they were separated as a result of a complex series of sea-level oscillations which allowed the filling in of the head of the canyon and the production of a new head as a result of a diversion of the inner valley".

[1]) P. Shepard. Investigations of submarine topography. Transact. Americ. Geophys. Union 1937. See also Veatch and Smith 1939, pp. 41—48 and fig. 13.

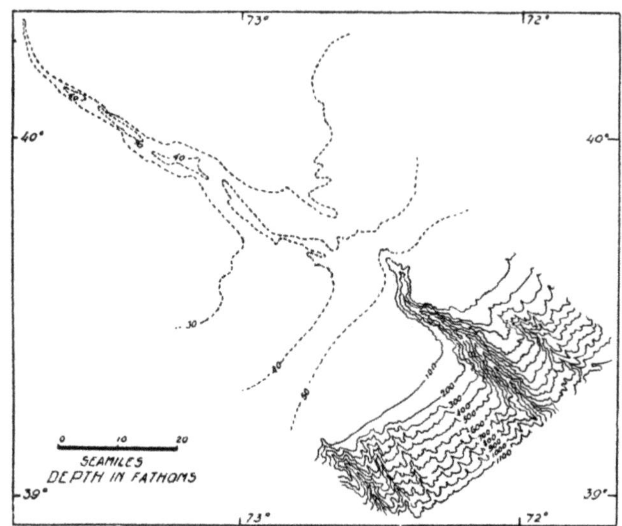

Fig. 87. Submarine valley of the Hudson. (From Kuenen).

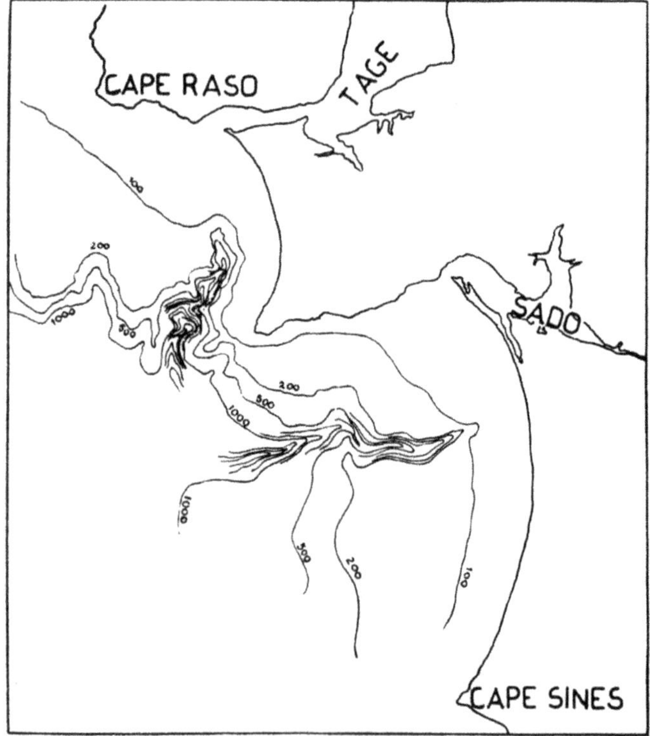

Fig. 88. Submarine canyons of the Tage and Sado, off the Portugese Coast (After Bourcart).

Other submarine canyons crossing the shelf are known from several other shelves. Fig. 88 illustrates the canyons off the Tage and Sado rivers and other examples are given in fig. 89, the Nazaré canyon, and fig. 90, the

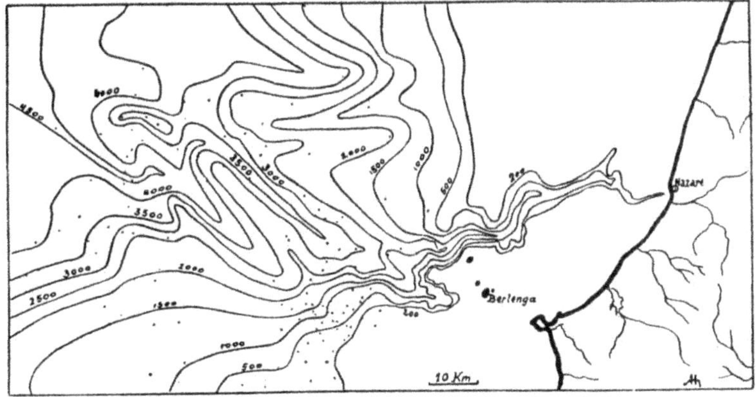

Fig. 89. Nazaré canyon (After Freire de Andrade).

gouffre du Cap Breton. Freire de Andrade is of the opinion that the canyon off Nazaré has to be considered as a graben with a southwest-northeast

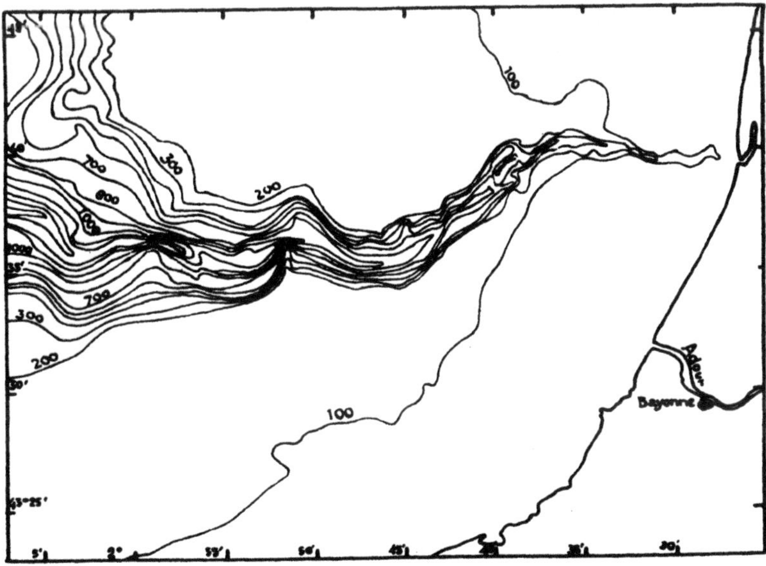

Fig. 90. "gouffre du Cap Breton" off the Atlantic Coast of southern France (After Bourcart).

trend. Even in this special case it seems open to doubt whether tectonic factors were of paramount importance in the formation of the submarine relief. The same holds good for the canyon off Cape Breton, a gorge for which a tectonic origin was claimed by some authors.

A few students of the problem have assumed that many more submarine gorges, which now appear to be incised only a short distance into the shelf, must once have had a greater headward extension. If these supposed extensions did exist, they must have been obliterated by marine planation and filled during certain epochs of the Pleistocene. And only if they were dependent on a large stream like the Congo river the obliterated channels were partly cleared in "post-glacial" times or a new channel was formed. Obviously during the modelling of the shelf-surface submarine agencies such as density and other currents have been at work. These currents may have held open gullies which originated at a low stand of the Pleistocene sea-level from consequent streams, whose shortened remains still exist. Moreover, ancient submarine trenches, which during previous stages of the Pleistocene were partly or wholly filled with loose sediments may then have been cleared out wholly or partly by such agencies. Some of the rivers debouching near the headward extensions of these gullies, including the Missisippi, Fraser, Indus (fig. 86), Ganges and Niger, are now building deltas. Hence during the formation of the gullies the oceanographic conditions were apparently different from present conditions.

(3) S u b m a r i n e c a n y o n s s h o w i n g a d e n d r i t i c r i v e r - l i k e p a t t e r n.

The great system of canyons off the coast of California was classed as a separate group on account of its being characterized by a strongly dendritic pattern.

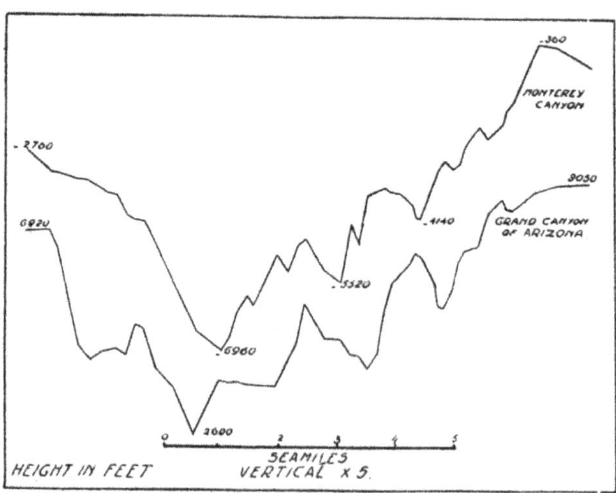

Fig. 91. Section of the Monterey submarine canyon compared with the Grand Canyon (After Shepard).

Fig. 77 illustrates its river-like appearance and fig. 91 the canyon-like type of cross-section when compared to the Grand-Canyon. These sections should also be compared to those of the Hudson Canyon shown by fig. 76. Probably the submarine topography of the Californian canyons may indeed be considered as a subsided marginal block of the continent and a drowned system of canyons originally formed under subaerial conditions. As a matter of fact Maxson [1]) argued that the coast of northern California may probably be regarded as a fault-coast.

Daly [2]) thought it "quite possible that a few trenches, including perhaps more than one off the coast of California, were actually river-cut far back in geological time, then drowned, afterwards more or less completely filled with sediment, and finally, recently, re-excavated by the same process which caused the majority of the open furrows now being discovered by soundings".

Investigations by Emery and Shepard give strong support to the theory of the subaerial origin of these canyons. Rock was obtained from 132 localities. Great quantities of fragments were obtained in a single haul. The fragments were often of large size, angular and with fresh fractures. These features suggest that at any rate a great deal of the samples come from the bedrock outcropping at the sampling locality [3]). The submarine topography much resembles the structural features of California. Several faults could be traced in the surface of the shelf. The rocks on the sea-floor include sedimentary, igneous, and metamorphic types. Triassic, Jurassic, Cretaceous, Eocene, Miocene and Pliocene are all represented and their distribution on the sea-floor "conforms to the paleogeographic maps given by Reed and Reed and Hollister, though changes are necessary for the Middle Miocene and the Pliocene maps...." On the wall of one canyon consolidated Pliocene sediments were found, indicating that at least this canyon was formed in post-Pliocene time [4]). At any rate this conclusion holds good for the upper part of the canyon.

Summary and Conclusions

(1) From the available data it appears that the gradation-plane of the surface of the shelf, so far as it is adjusted to present conditions, is formed as a surface gradually deepening from the coast to a depth of about 70 meters (40 fathoms) below sea-level. This figure is based on observations on the Atlantic and Pacific shelves of North America. For two islands in the southern Atlantic, Saint Helena and Ascension, a slightly higher figure — nearly 100 meters — seems to be the most probable estimate.

Generally, however, the break of slope which is mostly called the edge of

[1]) J. H. Maxson. Structural relationship of the coast and continental margin of California. Bull. Geol. Soc. America 44, 1933, p. 152.

[2]) Daly, op. cit. 1936, p. 405. On p. 419 he writes: "Here and there canyons and other types of valleys, cut by ordinary rivers in pre-Glacial time, have been drowned by strong, local subsidence of continental borders. Such old, rock-hewn valleys, if they had not been quite filled with fine-grained detritus, would naturally have attracted the water loaded with sediment during the Glacial epochs. The resulting, localized currents would have been likely to remove some of the filling of each of the old trenches, and thus revived the open-valley form more or less completely. Illustrations of this speculative process might be looked for particularly along the coasts of California. Japan and other uneasy parts of the continental borders". (See also Daly, 1942, p. 127).

[3]) Emery and Shepard, 1945, p. 431.

[4]) Shepard and Emery, 1941, p. 448.

the shelf is situated lower, often near the isobath of 200 meters. The question therefore, arises how the shelf-surface below the isobath of 70 meters originated. In order to explain this situation the possible influence of eustatic changes of sea-level and movements of the bottom have to be examined.

(2) At first sight it would seem that the problem may be solved in a simple way by surmising the deeper part of the shelf-surface to have been graded by submarine agencies during Pleistocene epochs of a lowered sea-level and to be adjusted to the existing hypsometric relation between sea-surface and shelf-surface. Although eustatic changes undoubtedly must be regarded as a factor of importance, the problem is more complicated. For, on both sides of the Atlantic the occurrence of coarse sediments and boulders was observed on the outer parts of the shelf and these features bear undoubted evidence of their being originally deposited in the littoral zone.

Therefore, apart from eustatic changes (which during the Pleistocene may have caused the deposition of littoral deposits at a depth of maximally about 100 meters below present sea-level) an additional widespread subsidence of the bottom in the order of 100 meters should be assumed. According to Bourcart the formation of the surface under discussion would date from Mousterian times. The same author distinguishes a second surface occurring along the oceanic slopes of the shelf at a depth of 200 to 500 meters, and a third between the isobaths of 500 and 1000 meters, as noticed in many regions off the Atlantic coasts studied by him. Apparently, in the course of time the original continental margin has sunken deeply below the level of the sea. A prism of sediments (for the greater part erosion products from the continents) accumulated on it. Such a situation was revealed by the seismic refraction measurements of Ewing, Crary and Rutherford in the case of the sunken borderland Appalachia, which is now covered by a pile of shelf deposits ranging from Triassic to recent times and attaining a thickness of 12000 feet at a distance of 60 miles off the present coastline.

(3) In clear contrast to the positive shift of the shelf-area, the land-side of the continental margin shows widespread evidence of a bottom-movement in the opposite direction, a zone parallel to the oceanic coast being characterized by rejuvenation of the relief. Chamberlin and Salisbury and in recent times especially Jessen and Bourcart paid attention to this phenomenon.

(4) It was the last-named author who in 1938 proposed the theory of the continental flexure. Spasmodically a warping or tilting movement seems to have taken place along the continental border causing a periodical submergence of what formerly was the margin of the continent and a simultaneous bowing up of a marginal tract parallel to the newly formed coast-line.

(5) The most recent movement of the coastal tract is demonstrated by the elevation of marine terraces along the oceanic coasts. The situation of the Sicilian terrace at 100 meters above sea-level clearly shows that apart from a eustatic shift (which would account only for a position of maximally 40–45 meters above the present level of the sea) its emergence must be ascribed (at least partly) to warping of the bottom. This must have amounted to at least 50 to 60 meters in post-Sicilian even post-Tyrrhenian times provided that the Tyrrhenian terrace (30 m above sea-level) can be correlated to the Yarmouth or Mindel-Riss interglacial stage.

Locally a more recent post-Mousterian tilting has been noticed.

(6) The periodical action of convection currents which are supposed to display themselves below the continental margin as a result of the peculiar distribution of sialic and simatic masses in the border zone of continental and oceanic areas, possibly accounts for the periodical movements along the continental flexure. Thus, the phenomena of the continental margin are possibly correlated to other periodic events occurring in the earth's crust and its substratum.

(7) The submarine canyon-like trenches which are incised in the outer slopes and — occasionally — in the surface of the shelf may tentatively be classified in three groups, viz. (a) slightly ramified notches in the shelf-edge running down towards great abyssal depth, (b) gorges of the same type which however, have headward extensions over the surface of the shelf, (c) submarine canyons showing a dendritic river-like pattern.

Near the shelf-edge the trenches of group (a) show a depth of 1000 meters and more. In order to account for an erosion base at that level one would have to admit an oscillation of at least 1000 meters either of the bottom or of the sea-level. However, such a hypothesis is incompatible with the data presently known about the geological history of the continental margin. Therefore theories involving a subaerial origin of the notched shelf-edge have to be abandoned. And hence their formation must be due to submarine agencies. The gullies of group (b) probably originated as oceanward extensions of large consequent rivers existing in Pleistocene and more recent times and debouching in notches of the shelf-edge. During the Pleistocene and more recent modelling of the upper shelf-surface the gullies probably were held open by submarine processes. The canyons of group (c) probably formed in pre-glacial times under subaerial conditions. Subsequently a subsidence of the bottom caused them to become drowned. Afterwards they were more or less completely filled with sediments and finally they were re-excavated by the same processes suggested for the digging of the furrows of group (a).

References

The works on regional geology which are cited in the bibliography at the end of chapter II should also be consulted.

BARRELL, J., *Upper Devonian delta of the Appalachian geosyncline* (Americ. Journ. of Sci. 27, 1914).

BOURCART, J. *La marge continentale* (Bull. de la Soc. Geolog. de France, 8, 1938).

BOWEN, N. L., CUSHMAN, J. A. and DICKERSON, R. E., *Shifting of the sea floor and coastlines* (Oxford Univ. Press. 1941).

BUCHANAN, J. H., *On the landslopes separating continents and ocean basins, especially those on the west coast of Africa* (The Scottish Geographical Magazine, 3, 1887).

BUCHER, W. H., *Submarine valleys and related geological problems of the North Atlantic* (Bull. Geol. Soc. America, 51, 1940).

BULLARD, E. C., *Submarine Geology* (Sci. Progress. no. 134, 1939).

BULLARD, E. C., *The Geophysical Study of Submarine Geology* (R. Inst. Great Britain Proc. 31. 1940).

BULLARD, E. C., and GASKELL, T. F., *Submarine seismic investigations* (Proceed. Royal Society of London, series A. no. 971, vol. 177, 1941).

CHAMBERLIN, TH. C. and SALISBURY, R. D., *Geology* (vol. 3, 1906).

DALY, R. A., *The Changing World of the Ice Age* (Yale Univers. Press, 1934).

DALY, R. A., *Origin of submarine "canyons"* (Americ. Journ. of Sci., 31, 1936).
DALY, R. A., *The floor of the Oceans* (Oxford Univ. Press, 1942).
DAVIS, W. H., *Submarine mock valleys* (Geogr. Review 24, 1934).
DESPOIS, J. *Les "oueds" sous-marins du Golfe de Gabès et les "canali" du Golfe de Venise* (Bull. de l'Assoc. de Géogr. Francais, no 102, 1937).
EMERY, O., and SHEPARD, F. P., *Lithology of the Sea floor off southern California* (Bull. Geolog. Soc. of America, 56, 1945).
EWING, M.; CRARY, A. P.; RUTHERFORD, H. N. and MILLER, B. L., *Geophysical investigations in the emerged and submerged Atlantic coastal plain* (Bull. Geolog Soc. America, vol. 48, 1937; EWING, M.; WOLLARD, G. P., and VINE, A. C., idem. 51, 1940).
FREIRE DE ANDRADE, C., *Os vales submarinos portugueses e.o. diastrofismo des Berlangas e da Estremadura* (Serv. geolog. de Portugal. Lisboa, 1937).
GREGORY, J. W., *A deep trench on the floor of the North-Sea* (The Geographical Journal 77, 1931).
HESS, H. H. and MACCLINTOCK, P., *Submerged valleys on continental slopes and changes of sea-level* (Science 83, 1936).
HOLTEDAHL, O., *The submarine relief off the Norwegian coast* (Det. Norske Videnskaps Akademii Oslo 1940).
HULL, E., *On the sub-oceanic physical features off the coast of Western Europe, including France, Spain and Portugal* (The Georg. Journal 13, 1899).
JESSEN, O., *Die Randschwellen der Kontinente* (Peterm. Geogr. Mitt. Erg. Heft 241, 1943).
JOHNSON, E., *Origin of submarine canyons.* (Journ of Geomorphology 1 and 2, 1938 and 1939).
JOHNSON, D. W., *Shore Processes and Shore-line development* (London, Chapman, and Hall 1919).
JOHNSON, D. W., *The New England-Acadian Shore line* (London, Chapman and Hall, 1925).
JONES, O. T., *Continental slopes and shelves* (The Geogr. Journal 97, 1941).
KOSSINNA, E., *Die Erdoberfläche* (Handb. d. Geoph. 2, 1933).
KUENEN, PH. H., *Experiments in connection with Daly's hypothesis on the formation of submarine canyons* (Leidsche Geologische Mededeelingen 8, 1937).
KUENEN, PH. H., *Onderzeesche canyons* (Tijdschr. Kon. Nederl. Aardrijksk. Gen. 55, 1938).
KUENEN, PH. H., *Density currents in connection with the problem of submarine canyons* (Geolog. Magazine 1938).
KUENEN, PH. H., *The cause of coarse deposits at the outer edge of the shelf* (Geolog. en Mijnbouw, 1, 1939).
LADD, H., *Geology of Viti Levu* (Bern. P. Bishop Museum Bull. 119, 1934).
LEET, L. D., *Status of geological and geophysical investigations on the Atlantic and Gulf Coastal Plain* (Bull. Geol. Soc. America 51, 1940).,
NOVAK, V. J., *On the origin of the continental shelf* (Mém. de la Soc. R. des lettres et du Sci. de Bohème, Classe des Sci. 1937, XVIII, p. 1–27, published 1938).
PATERSON, T. T., *On a world correlation of the Pleistocene* (Transact. R. Soc. of Edinburgh. vol. 60, part 2, 1941).
PEI WEN-CHUNG, *An attempted correlation of quaternary geology, paleontology and prehistory in Europe and China* (London Univ. Instit. of Archael. 1939, Occas. paper no. 2).
SCHUCHERT, CH., *The problem of continental fracturing and diastrophism in Oceanica* (Americ. Journ. of Sci, 47, 1916).,
SHEPARD, F. P., *Glacial troughs of the continental shelves* (Journ. of Geology 39, 1931).
SHEPARD, F. P., *Sediments of the continental shelves* (Bull. Geol. Soc. of America 43, 1932).
SHEPARD, F. P., *Submarine valleys* (The Geogr. Review 23, 1933).
SHEPARD, F. P., *American submarine canyons* (The Scottish Geogr. Mag. 50, 1934).
SHEPARD, F. P., *Canyons off the New England Coast* (Americ. Journ. of Sci. 27, 1934).
SHEPARD, F. P., *Submerged valleys on continental slopes and changes of sea-level* (Science, 83, 1936).
SHEPARD, F. P., *The underlying causes of submarine canyons* (Proceed. Nat. Acad. of Sci. 22, 1936).
SHEPARD, F. P., *"Salt" domes related to Mississippi submarine trough* (Bull. Geol. Soc. of America 48, 1937).
SHEPARD, F. P., *Daly's submarine canyon hypothesis* (Americ. Journ. of Science 33, 1937).
SHEPARD, F. P., *Submarine geology and Geophysics* (Nature, 146, 1940).
SHEPARD, F. P., and BEARD, C. N., *Submarine canyons distribution and longitudinal profiles* (The Geogr. Review 28, 1938).
SHEPARD, F. P. and COHEE, G. V., *Continental shelf sediments off the Mid-Atlantic States* (Bull. Geolog. Soc. of America, 48, 1936).
SHEPARD, F. P., TEEFTHEN, J. M. and COHEE, G. V., *Origin of Georges Bank* (Bull. Geol. Soc. of America 45, 1934).

SHEPARD, F. P., *Non depositional physiographic environments off the California Coast* (Bull. Geol. Soc. of America, 52. 1941).
SHEPARD, F. P. and EMERY, K. O., *Submarine Topography off the California coast, canyons and tectonic interpretation* (Geolog. Soc. America. Spec. Paper, 31, 1941).
SMITH, P. A., *Submarine valleys* (U. S. Coast and Geodetic Survey. Field Engin. Bull. 10, 1936).
SMITH, P. A., *Submarine topography of Bogoslof* (The Geogr. Review 27, 1937).
SMITH, P. A., *Atlantic submarine valleys of the United States* (The Geogr. Review 29, 1939).
SPENCER, J. W., *Submarine valleys off the American coast and in the North Atlantic* (Bull. Geolog. Soc. of America, 14, 1903).
STETSON, H. C., STEPHENSON, L. W. and CUSHMAN, J. A., *Geology and paleontology of the Georges Banks Canyons* (Bull. Geolog. Soc. of America, 47, 1936).
STETSON, H. C., *Current measurements in the Georges Banks Canyons* (Transact. Americ. Geoph. Union 1937).
STETSON, H. C., and SMITH, J. F., *Behaviour of suspension currents and mud slides on the continental slope* (Americ. Journ. of Sci. 35, 1938).,
STETSON, H. C., *Summary of sedimentary conditions on the continental shelf off the East Coast of the United States* (Recent Marine Sediments. A Symposium 1939).
TOIT, A. L. DU, *An hypothesis of submarine canyons* (Geolog. Magaz. 77, 1940).
UMBGROVE, J. H. F., *Origin of Continental shelves* (Bull. Americ. Assoc. Petrol. Geolog., 30, 1946).
VEATCH, A. C. and SMITH, P. A., *Atlantic submarine valleys of the U.S. and the Congo submarine valley* (Geolog. Soc. of America. Special Paper, 7, 1939).
VENING MEINESZ, F. A., *Gravity and the hypothesis of convection-currents in the earth* (Proceed. Kon. Akad. van Wetensch. Amsterdam, 37, 1939).
VENING MEINESZ, F. A., *Gravity over the continental edges* (Proceed. K. Akad. van Wetensch. Amsterdam. 44, 1941).
YABE, H. and TAYAMA, R., *Bottom Relief of the Seas bordering the Japanese Islands and Korean Peninsula* (Bull. of the Earthquake Research Institute, Tokyo Imp. Univers. 12. pt. 3. 1934).
ZEUNER, F. E., *The Pleistocene Period, Its climate, chronology and faunal successions* (The Ray Society, London, Monogr. 130, 1945).

CHAPTER VII

ISLAND-ARCS

"The inward buckling of the main crust, as has been assumed in the East Indies, has been a rather common occurrence in the history of the earth's crust". (F. A. VENING MEINESZ)

Introduction

Island-arcs are among the most challenging features of the earth's surface. Island-arcs border the eastern margin of Asia and appear in the West Indies and the so-called Southern Antilles which form a discontinuous festoon between South America and Antarctica. At the same time these arcuate structures represent one of the most intricate problems of structural and historical geology. When considering a geographical globe, one wonders how these remarkable island-festoons have developed into their actual shapes. A further question which arises is, why similar features are conspicuously absent from other areas such as e.g. the eastern border of the Pacific, the entire African continent, Europe, Greenland and Australia with the exception of its northern part. On the other hand a geological examination of the earth's surface shows that arcuate structures are by no means confined to certain border-regions of continental and oceanic areas. A single glance at Plates 1–5 shows that they form one of the most conspicuous features in the structural pattern of such continents as Asia and Europe. To some extent they also played an important part in the geological history of North and South America and eastern Australia. Again we are struck by the fact that they are lacking in some other continental areas, at least since Cambrian times, i.e. since the moment from which the surface history of the continents can be deciphered to some appreciable degree, thus far. The arcuate structures of the continents were discussed in Chapter II. It will be remembered that among the characteristics mentioned with some emphasis were the following two. Firstly, the fact that some of the arcuate belts are arranged consecutively around a central nucleus of respectable antiquity, spreading in ever widening arcs. Secondly, it may be stated that the younger the age of the belt the greater its distance from the Pre-Cambrian nucleus. Again, however, it had to be added at once that other continental areas do not reveal such a regular centrifugal configuration of tectonic belts. On the contrary, in some districts the sequence is irregular and non-concentric and may even display the reverse of the time-relations just mentioned. Elsewhere certain structural belts boldly intersect older zones. Once more we learn that the earth's crust is not built up by a single dynamic activity that can the expressed in a simple formula or "law".

Island-arcs are associated with several other remarkable features. They are the sites of conspicuous girdles of still active volcanism. Zones of strong seismic activity can be traced in their immediate vicinity. Deep-sea furrows are situated along their external side; deep basins appear at their concave side. Belts of large disturbances of isostatic equilibrium accompany them. The phenomenon of deep-focus earthquakes seems to be related to these zones. And, finally, terrestrial magnetism probably shows noticeable deviations from its normal values in these arcuate island-zones. However, once more it should be added at once that by no means all the characteristics enumerated so far are confined to these structures alone. It is true that the different types of basins and troughs accompanying island-arcs have their counterpart on the continents proper. And, to mention only one more striking feature, the structural zones of the East Indian festoon can be followed northwestward into similar structural zones in the Asiatic continent, viz. in Burma, whereas the Aleutian arc continues into the North American continent and the Kurile arc into Kamschatka. Accordingly, problems of the island-arcs are for evident reasons linked to those of their counterparts on the continents some of which have already been discussed in Chapter II. However, it will be seen that the island festoons have some characteristic features of their own justifying their being treated in a separate chapter.

The Problem

At this stage it is worth while to formulate in a few words the principal lines that will be followed in attacking the problematic origin of island-arcs.

At the outset the question why island-arcs occur in one district while they are conspicuously absent from others will be left out of consideration. Let us not begin by asking why the western border of the Pacific shows a series of island-festoons from the Aleutians to the East Indies, whereas no such arcs occur along its eastern border. Doubtless the presence of the Antillean arcs amidst the arc-less coasts of the Atlantic is a striking phenomenon. But to ask for an explanation of their absence from Europe, Africa and South America raises a problem which may well be treated apart from the question as to how the arcs originated in the districts where they actually exist. And it is to the last named question that the following sections are principally devoted. At the end it will be useful to return once more to the problem of their absence from other areas.

However, in order to avoid misunderstanding, the subject needs even further precision. It should be added that the problem treated in the following pages is restricted (1) to those single island-arcs which are associate with volcanism and show a marginal deep-sea trough in front of their convex side; (2) to double island-arcs, the inner one being a volcanic arc, the outer one a non-volcanic arc with a marginal deep in front of it. Hence *per definitionem* a single row of islands with a more or less arc-like shape, like those in front of the coast of British Columbia, fall outside the subject under discussion as the festoon is not associated with a deep-sea trough (fig. 92).

Fig. 92. Island-festoon in front of the coast of British Columbia

It is evident that the origin of single and double arcs is one of the principal problems to be attacked. We want to know how the row of volcanoes came into existence and why they are absent from the outer arc of a double festoon. Among the other points that come up for discussion are the origin of the marginal deep and other deep-sea troughs and basins which are associated with the island arcs. And the same holds good regarding the accompanying belts of strong seismic activity, zones and fields of isostatic anomalies, and certain features of structural geology, plutonism and volcanism. If possible the relations between these various groups of phenomena have to be revealed and an attempt at synthesis has to be made. A further question which deserves special attention is whether there is evidence for the theory that island-arcs differing in physiographic aspect represent at the same time different evolutionary stages in the history of a tectonic belt or not. Finally, the unsolved problems or questions for which no satisfactory explanation can be given, have to be outlined as sharply as possible.

The various groups of data will be considered in systematic order. It will be seen that, in doing so, gradually — step by step —, our insight into the genesis of the arcs changes from a chaotic assemblage of data into an attempt at a harmonious synthesis.

Physiographic Features

Older Theories

Island-arcs might be classified according to different principles. Some arcs are single, others double. Some have a slight curvature, others are more strongly arcuate. In the Aleutians the outer non-volcanic arc is only slightly developed along its eastern side, near the coast of Alaska.

On the other hand the outer festoon is much more strongly developed in the East Indies. It seems an obvious thing to suggest a genetic connection between the various types and to assume an evolutionary sequence from a simple stage towards a more complicated one.

According to Hobbs the evolution of an arc would "pass through a progressive series of changes marked by ever increasing curvature". It is moreover maintained that the succession of changes through which an arc passes is the result of a pressure caused by the progressive settlement of the ocean floor, the rock floor of the sea acting as a girder. In a youthful stage the arc would show a curvature of large radius and a low elevation above the surface of the sea. From this early stage the arc would pass through a series of movements in both horizontal and vertical directions. It is well known that Argand considered an island-arc like the East Indies to be a mountain-chain *in statu nascendi*. In a future epoch a mountain belt comparable to the Alps or the Himalayas would grow out of these embryonic structures. Then, once more the continent would have been enlarged at the cost of the oceanic areas. The idea of a youthful stage of the island festoons apparently finds support in the occurrence of a strongly accentuated relief of the surrounding sea-floor, the seismic activity in the vicinity of the arc, the presence of a belt of active volcanoes, etc. Moreover, Argand's reconstruction of the Alps at a previous stage of their complicated history showed them too as island-zones in a Mesozoic sea, fig. 93. This hypothesis

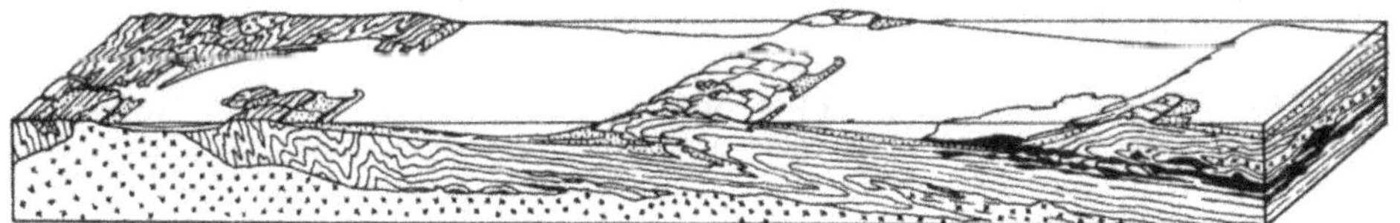

Fig. 93. Reconstruction of the Alpine geosyncline in the Mesozoic (From E. Argand).

is attractive by its combining so many different problems into a single synthesis. But, remarkably enough the island-festoons seem all to be situated inside the andesite line. A sialic crust even seems to reach as far as the festoon of the central volcanoes of the Aleutians. In the case of the Himalayas one might point to the large block of India pushing the chains towards the Asiatic continent. But what is the pushing power in such festoons as the Aleutian and the Mariana-Pelew arcs bordered as they are by the deep lying floor of the Pacific? It may well be that the comparison of the Alps in a Mesozoic stage with the island-festoons, especially with the East Indies, is misleading. The evidence offered by our knowledge of the structure and history of the island-arcs has to be examined without prejudice and then the question will again be taken up whether or not the different physiographic types of island-arcs can be arranged so as to represent a series of corresponding stages in an evolutionary sequence.

It is not my intention to mention all the theories bearing on the origin of island-arcs, which in the course of time have been presented by various authors. For we should then be compelled to treat and weigh separately more than thirty opinions of widely divergent character and importance. Many of these theories are out of date. There would, for instance be no sense in dwelling any longer on hypotheses involving large horizontal movements of island-festoons in the way suggested by Wegener, Du Toit, Wing Easton, Smit Sibinga a.o.

Horizontal movements and crustal shortening of geosynclinal belts are phenomena associated with epochs of compression. The movements are, however, not of the sort suggested by drift-hypotheses which envisage sialic blocks floating in a simatic syrup. The earth is surrounded by a world-embracing rigid crust and the tangential pressures in the crust have to be described and figured in a quite different manner. These aspects of the problem are dealt with in Chapter VIII. Accordingly these older theories may be left out of consideration for the present. Moreover, the reader who feels inclined towards a more detailed study of older views will find surveys of these theories in two papers, viz. one published in 1935 by Kuenen, and one shorter review by the present author in 1934 [1]). It is true that more than two dozen of the hypotheses considered in these publications bear especially on the East Indies, but it goes without saying that all theories presenting view-points of general importance may be found incorporated among them. Therefore I will restrict myself here in focussing the general attention only on those speculations that seem to me of special importance in the light of our present state of knowledge.

Deep-reaching thrustplanes and shear-zones

In this respect special relevance should be given here to the theories of Sollas, Lake and Lawson. The gist of Lake's theory can be summarized in a few lines. In his own words these read as follows [2]): "In 1908 Professor Sollas showed that many mountains and island-arcs are truly circular, so far as a large-sized globe will show, but he had no explanation to offer. We cannot infer from the shape that there must be a thrust plane at the base of each, but we can say with confidence that a thrust plane at the base would account for the shape and that no other explanation is so simple and complete."

"Moreover if there is a thrust plane at the base, it is easy from the form of the arc to determine the dip of that thrust plane at its outcrop, for the angle which the thrustplane makes with the surface is equal to the angle subtended at the centre of the Earth by the radius of the arc (fig. 94). The thickened portion of the line BT represents a thrustplane cutting through the outer crust of the earth. P is the pole of the arc made by the outcrop of the thrustplane. The angle ATB made by the thrustplane with the surface, is equal to the angle AOT subtended at the centre by the radius of the arc. All that is necessary is to determine the pole of the arc and measure the radius in degrees and minutes of a great circle."

---

1) Umbgrove, J. H. F. 1934, Chapter VII in Vening Meinesz Gravity Exped. at Sea II; and Kuenen 1935 in Snellius Exped. Bathymetric Results.

2) Lake, 1931, p. 150.

"If the mountain arc is not circular a negative inference is possible. There is not now a plane thrust-surface at its base. It is possible, however, that the range was formed upon a true thrust-plane and that there has been subsequent deformation. Or the surface along which movement took place may not have been a plane or the movement may have been of a different character altogether for there are other ways in which a mountain-range may be formed."

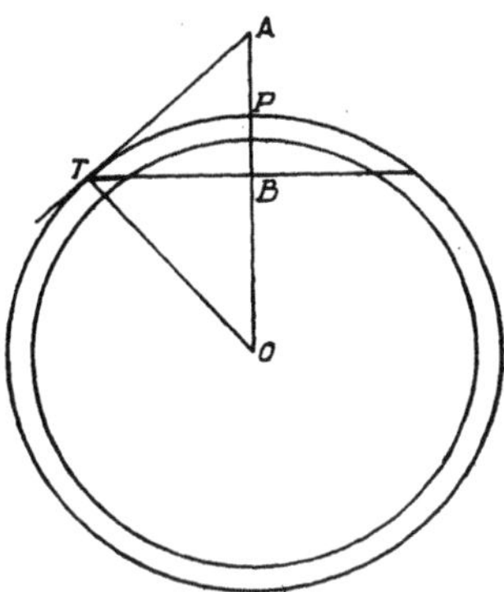

Fig. 94. Outcrop, angle and dip of a major thrustplane through the earth's crust (From P. L. Lake).

As a matter of fact the principal points of Lake's argumentation were already set forth by Sollas as early as the year 1903 [1]).

Lawson is one of the modern exponents of similar ideas about the association of island-arcs, large thrustplanes and seismic activity, with subsiding sea-floor along the concave side of the arc and an elongated deep-sea furrow along its convex side. We shall turn to his theoretical considerations in the section on deep-sea troughs (p. 170).

A difficulty which arises in the practical procedure of drawing the form of an arc is to choose the right place for the circular line on the globe or geographical map. Another point that is open to personal interpretation is the fixing of the terminal points of the arc. But let us proceed and consider some of Lake's further remarkable constructions — and conclusions — which are strikingly lucid and simple. Sollas had pointed out that the poles of his circular arcs lie upon the same great circle. Lake, surmising these arcs to coincide approximately to the outcrops of large thrust-planes, concludes "that the thrustplanes are all at right angles to the plane of the great circle that passes through their poles"! In order to examine this question more closely he proceeded as follows. After he had chosen the lines representing the arcs shown by fig. 95 the longitudes and latitudes of three points upon each line were determined, one at each end and one in the middle. "From these data the position of the pole of a circle passing

---

1) I feel that we should do justice to Sollas in this respect. And therefore a few passages from his paper may be quoted. After having dealt with some theoretical considerations, which for the present are no longer of significance to us, he wrote (1903, p. 187): "That the fractures by which at some stages re-adjustment has taken place have frequently been circular arcs is shown by observation, and there is much evidence to suggest that such fractures have frequently preceded as well as followed the folding-up of a mountain-chain". And regarding the arc of the Aleutians he said (1903, p, 180): "The numerous mighty volcanoes which characterize the region point to the existence of an extensive subterranean reservoir of lava, and to discontinuity of the earth's crust, in the form of a circular crack. It is difficult to look upon this part of the globe, to avoid the impression that we have before us the remains of a spherical dome or blister, which has broken down along circular and radial fractures, the islands standing over a circle, the coasts of the Kamchatkan sea marking irregular splits". Sollas had even recognized the association of earthquake zones with the arc-like structures. For he continues: "Deep-sea exists outside, not far removed from a region regarded by Prof. Milne as the origin of many of the earthquakes which shake the whole crust of the earth".

through the three points was computed. The positions of the poles are shown on the map, and in each case a circular arc — circular upon the sphere, not circular upon the map — has been drawn through the three selected points" [1]). The Japanese arc is not a good circular arc either geologically or topographically. Therefore the position of its pole can not be defined clearly. Hence, "the Japanese arc" should be left out of consideration for

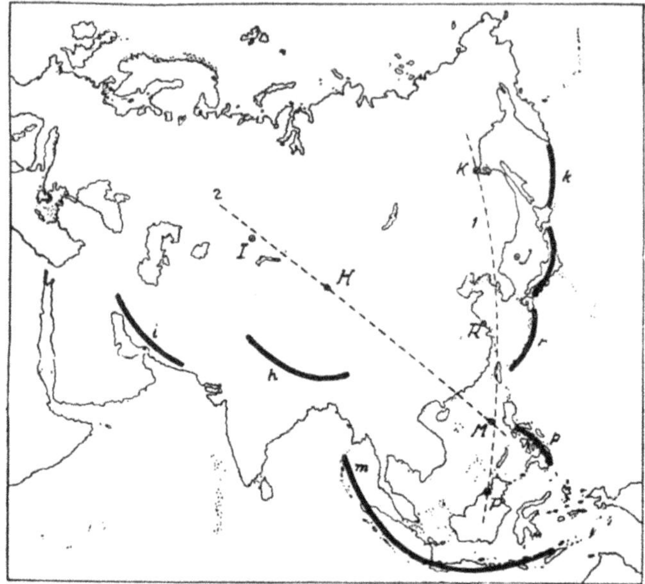

Fig. 95. Island-arcs of the western Pacific (After P. L. Lake).

the moment. The other four poles, however, lie so nearly on a great circle that it can hardly be a question of mere fortuity. This result is the more striking in as much as the arcuate lines were determined not on actual outcrops of the thrustplane but on various geological and topographic features which do not necessarily coincide exactly with the true outcrop. Accordingly Lake suggested that the Asiatic continent as a whole is being pushed over the floor of the Pacific upon a series of large thrustplanes and that the direction of the movement is at right angles to the great circle indicated by the line KP on the map fig. 95. More accurately the phenomenon has to be described as a continentward underthrusting of the Pacific area. It will be seen that Lawson laid special stress on this point and speaks of a landward creep of the Pacific sima. But there are still two other remarkable features in Lake's map. One concerns the construction of the poles of the East Indian, Himalayan and Iranian arcs. It appears that they lie upon a common great circle indicated by IM on fig. 95. And as a second feature of importance it was found that the pole of the East Indian

[1]) Lake, p. 152.

arc lies practically on both circles. "And this is what might be expected in the case of a corner arc with underthrusting from both sides."

A theory which has the merits of offering so many striking results is worthy of being examined in more detail. Lake did not speculate as to the possible cause either of the origin of the thrust plane or of the underthrusting. This part of the problem was taken up by Lawson. We will return to both in a later section (p. 170) and there too attention will be given to the Bonin-Pelew-Mariana arcs which, together with some other features, were left out of account by Lake.

A further questionable point is whether deep-reaching thrustplanes are known to us from older mountain-belts on the continent. Lake and Lawson are of opinion that the great boundary fault which is known to occur along the southern border of the Himalayas might be considered as an example of such a deep reaching surface. It generally lies between the Siwalik foot-hills and the pre-Siwalik rocks that constitute the Himalayan belt proper (fig. 96). And the main ranges were pushed up from north to south along this and several other thrust-planes. According to Lake the

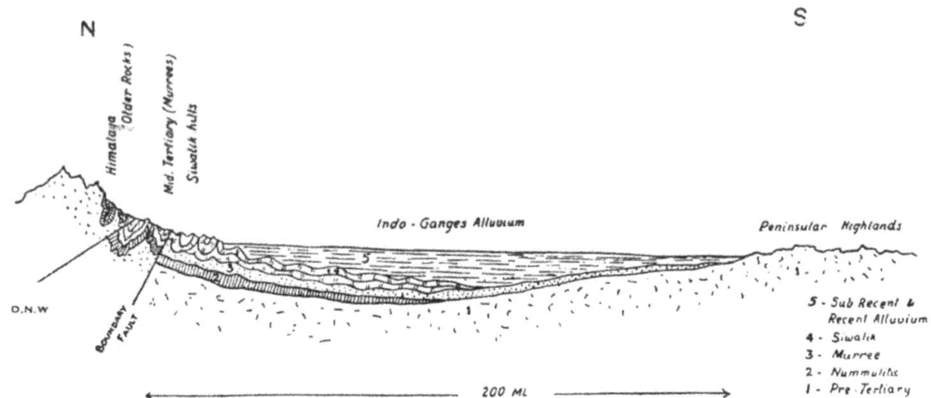

Fig. 96. Diagrammatic section across the Indo-Gangetic synclinorium. Vertical scale exaggerated. (From Wadia).

agreement of the dip of the main boundary fault with the amount of the angle of about 14° as inferred from the arc can hardly be an accidental coincidence. Reference was also made by Lake to the structure of the Carpathians. In most places the foreland formations are not only independent of those of the Carpathians, but they even run almost at right angles to the border of the mountain-belt where they are abruptly cut off [1]). Similarly the Carboniferous formations of the marginal trough along the northern front of the Variscides in northern France and Belgium has been over-ridden by older formations resting on a major thrustplane (fig. 29). As a further illustration fig. 17 shows one of the great thrustplanes upon which the Caledonides of northwestern Scotland were pushed over the

[1]) Lake, 1931, p. 149.

rocks of their foreland. Schematically the same phenomenon is illustrated by the geological sections through the Sierra Nevada geosyncline (fig. 23).

Personally I am not convinced that the marginal thrustplanes of the regions just mentioned should be considered as the outcrops of deep-reaching shear-planes similar to those that are supposed to border the western Pacific. Probably most of the continental examples — not all — are no more than superficial features resulting from the squeezing-out of the contents of a geosyncline over the rocks of its "foreland" (cf. fig. 15, 47, 49 and 51). In such a structure the dip of the thrustplane is largely due to the dip of the earth's crust originating from its downward buckling beneath the folded ranges. Kuenen is of a similar opinion. In his discussion of Lake's paper he wrote [1]: "Lake's thrusts and tectonic thrusts are as different from each other as stratification is from the layers of discontinuity in the earth's crust. They may of course develop into a normal thrust-plane at the surface, as in the case of the Himalayas, in the same way as for instance a river may change to a tidal estuary near its mouth". As a matter of fact, however, these examples formed the starting point of Lake's theory.

I quite agree with Kuenen's criticism. But apart from our doubts about the validity of some of the examples on the continents, Lake's theory remains a working hypothesis of high interest. Remarkably enough, deep-focus earthquakes were recognised in the same areas and their actual distribution induced some seismologists to construct large continentward dipping shear-zones (cf. fig. 101, 102) which reach a depth of more than 600 kilometers!

For the moment the existence of deep-reaching shear-zones connected with the location of island-arcs of the Pacific seems hardly open to doubt.

Now, obviously, a shear-zone in the substratum cannot possibly be a thrustplane of the type known from surface geology. For such a plane can not possibly persist in the plastic layers below the crust. However, if shear-phenomena happen to occur they appear to be bound to that special continentward dipping seismic zone. For the sake of clearness it may be called a *potential zone of shear.*

Moreover, it will be seen that the angle between the shear-zone and the earth's surface greatly surpasses the angle that might be expected from Lake's physiographic theory. So, *if* the curvature of the arc is due to a thrustplane this plane must have been restricted to the earth's crust. It must have had a gradient corresponding with the curvature of the arc and finally, it must have been disturbed and obliterated by crustal processes of later epochs. The only reason for adhering to such a *hypothesis ad hoc* is the simple explanation offered by Lake's hypothesis which, however, will have to be abandoned as soon as a more plausible explanation of the shape of the island-festoons has been given. Up to now this is not the case.

[1]) Kuenen, 1935, p. 92.

Earthquake zones

Classification

Jeffreys, after a thorough discussion of the available data, concluded in the year 1928 that the focal depth of the great majority of all tectonic earth-shocks does not surpass 35 kilometers. In the meantime Turner had argued as early as 1922, that in addition to earthquakes of the normal type, i.e. with foci in the earth's outer mantle, shocks might also originate at much greater depth. An analogous suggestion had been made by previous workers, though based neither on adequate data nor on scientifically sound determinations. However, in 1928, the year of publication of Jeffreys' conclusion, Wadati clearly showed that apart from the normal shocks, other earthquakes with focal depth of several hundreds of kilometers undoubtedly occur in the surroundings of Japan. Since then earthquakes with foci down to 700 kilometers below the earth's surface have been established in several other areas. Gutenberg and Richter distinguish three classes, viz. (1) normal shocks at depths not exceeding about 60 kilometers, (2) intermediate shocks at depths from 60 to 250 kilometers, and (3) deep shocks from 250–700 kilometers. The latter are named plutonic earthquakes by Macelwane. The classification into three groups should be regarded only as a preliminary attempt which is not always appropriate. Taking the evidence from all parts of the world together, it may be said that earthquakes are known to originate at practically all levels ranging from the surface down to depths of approximately 700 kilometers. No earthquake has been located with satisfactory reliability from a depth much in excess of 700 kilometers. This well-established phenomenon has been correlated by Bullen with three other changes of properties at a depth of approximately 700 km below the earth's surface, viz. (1) a rapid increase in density, (2) a rapid increase in velocity of seismic waves and (3) an increase in electrical conductivity. When the individual districts of seismic activity are considered separately, it appears, however, that there is no uniformity in the limits between the earthquake classes. The limits may vary regionally and nearly every district shows its own characteristic depth zones of seismic activity. Thus in the Japanese area deep foci are known almost continuously from 300 to 650 km, whereas in South America no shocks are known to occur at depths between 300 and 600 km. However, we should bear in mind that the considerations on shocks of the deeper classes are based on comparatively scanty data covering no more than 30 years at the most. It may be that the absence of shocks from certain depth zones is a feature of real significance. If so, we must leave an explanation of the phenomenon to the combined efforts of seismologists and other geophysicists.

The mechanism

A consideration of the technical foundations of the seismic results does not come within the scope of the present subject. One exception, however, has to be made, although in this case the subject will be treated concisely and suited to the general understanding. Seismologists agree that the cause of the deep-focus eartquakes is in no essential respect different from tectonic earthquakes with foci situated comparatively near the surface. This conclusion is of paramount importance to geological science.

The cause cannot, for example, be of an explosive character, due to sudden and local recrystallization of unstable mineral forms initiating explosive chemical reactions such as violent atomic readjustements, overheating, or overcooling accompanied by a change of volume. The large shear-waves observed in those earthquakes lave no doubt on that point.

For if an earthquake were of explosive origin, the initial movement in the surrounding rocks would be everywhere outward, and this would be so recorded in all the seismographs, regardless of their geographic site and distance from the epicentre. On the other hand, if the movement at the focus represents a sudden slipping displacement along a shear-plane the initial movements recorded will differ in a peculiar manner at different stations, some seismographs recording an initial movement toward the focus others away from it.

Conclusive proof is given by a comparison of two patterns of initial dilatations and compressions, one of an earthquake that originated in the surface layers, the other one of a deep-focus earthquake. Earthquakes of tectonc origin are known to cause a very characteristic distribution of the directions of the first shocks arriving at the surface. The direction of the first shock as recorded by the seismograph depends on the character of the event in the focal area. The first movement recorded by the apparatus will be upward, i.e. in the direction of wave propagation, provided the initial movement in the focus was of the same character. It is called compressive. The reverse type of shock in which the initial movement is directed towards the focus, is called a dilatation. The arrows in fig. 97 indicate the direction of the initial shocks caused by a tectonic earthquake in western Europe, each arrow representing a seismic station. The tectonic earthquake originated along a fault of approximately *NW–SE* direction. The movement of the block along the north side of the fault was directed towards the south-east. The southern block moved in the opposite direction, i.e. northwestward. It appears that if a movement along a fault-plane takes place the surroundings of the fault have to be divided into four quadrants,

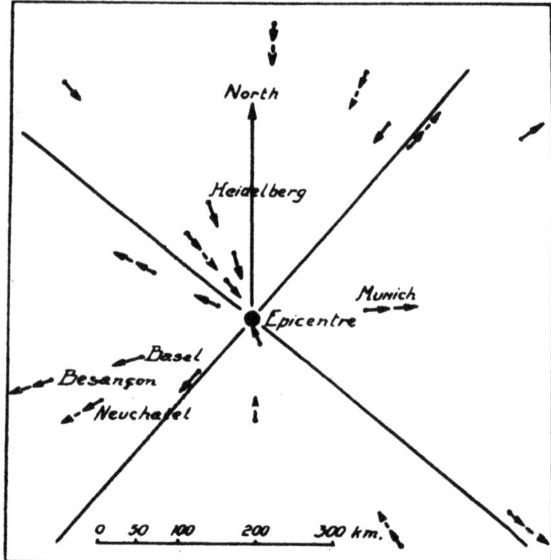

Fig. 97. Direction of initial motion of earthquake shocks originating along a NW-SE fault in Wurttemberg, on the 27th of June 1935 (After Hiller).

in each of which the initial shock has its own direction. As a result of the opposite movements of the blocks along the fault-plane the seismographs of the eastern and western quadrants shown by fig. 97 recorded an initial compression; in those of the northern and southern quadrants the initial shock was a dilatation.

Now the same effect has been noted in the identifications of deep-focus earthquakes. It is clear, however, that in the case of a deep focus the position of the system of quadrants (with their characterising compressions and dilatations) may vary endlessly. In a horizontal position of the fault-plane the surface features agree with those of fig. 97. If its position be vertical,

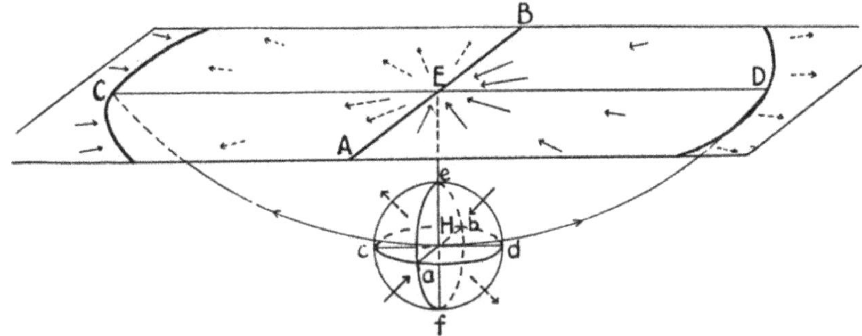

Fig. 98. Diagram to illustrate the distribution on the surface of initial dilatation (full drawn arrows) and compression (dashed arrows) of earthquake shocks deriving from a deep-focus earthquake at the hypocentre H, along a vertical fault plane cd-ef. E, epicentre. Further explanation in the text (After Honda).

cd — *ef* of fig. 98, the epicentral area will not show four quadrants but it will be divided into two areas of opposite character, according to the line *AB*. To the left of *AB* initial compression will be recorded. On the other hand the initial shock will be a dilatation in the area to the right of *AB*. An inspection of fig. 93 will show that still another effect is illustrated by that diagram. A seismic shock starting in the horizontal direction *Hd* gradually curves upward along the route *HD*. Shocks reaching the surface in an area situated outside of a circle with a radius *ED* come from below the surface *abcd*. It can readily be seen from fig. 98 that to the left of the nodal line *AB* they will cause initial dilatation or motion away from the focus, whereas to the right the initial motion will be of the compression type, i.e. directed towards the focus. Hence outside of the circle — the radius of which depends upon the depth of the hypocentre *H* — the seismic effect is the reverse of what is observed inside the circular region round the epicentre *E*.

Patterns of initial shocks agreeing with these two types of theoretically deduced systems have been recorded by Honda and Ishimoto in numerous examples of Japanese earthquakes. Fig. 99 shows an example of the type schematically explained by fig. 98.

The two examples that are offered, may serve as convincing illustrations of the important conclusion to which seismologists have arrived, viz. that both deep-focus and normal tectonic eartquakes originate by the same mechanism, viz. by shearing resulting from the sudden release of stresses accompanying a faulting movement.

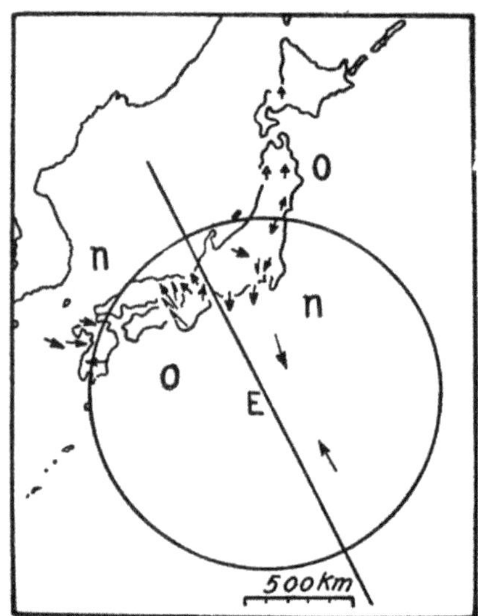

Fig. 99. Direction of initial motion of shocks deriving from a deep-focus earthquake. E, epicentre; N, initial movement directed towards focus; O, initial motion from the focus (After Honda)

Fig. 100. Epicentres of earthquakes in Hindukush (From S. W. Visser). For notations compare fig. 102.

Geographic distribution of deep-focus earthquakes

Notwithstanding the apparent incompleteness of our present knowledge the examination of the geographic distribution of intermediate and deep foci has brought to light some very remarkable features.

It appears that earthquakes of the intermediate class have been identified in all regions in which shocks of the normal class are of frequent occurrence. One district is of special interest viz. a comparatively small area in Hindukush (fig. 100). Repeatedly shocks from about 220 km have been recorded over a period of some 30 years.

Concerning the class of deep-focus earthquakes it may be said that they are known only from areas bounding the Pacific basin (fig. 101). A second feature of importance is the arrangement of foci in zones of ever increasing depth continentward. So they give the impression of originating along deep-reaching shear-zones, sloping inland from the Pacific border.

This idea was presented by Wadati for the first time in a map of the Japanese area, showing the contour lines for foci of equal depth (fig. 102). Fig. 103 without contour lines, reproduces a still more recent map — and hence containing still more data — by Gutenberg and Richter. The volcanic zones of Japan and its vicinity are shown in fig. 104.

For the present we will only stop to notice the parallelism of the volcanic belts to both the deep-sea troughs and the zone of normal shocks. The zone of intermediate shocks lies more or less under the islands with their superimposed rows of volcanoes. The possible relations of these various phenomena deserve special consideration in a later section.

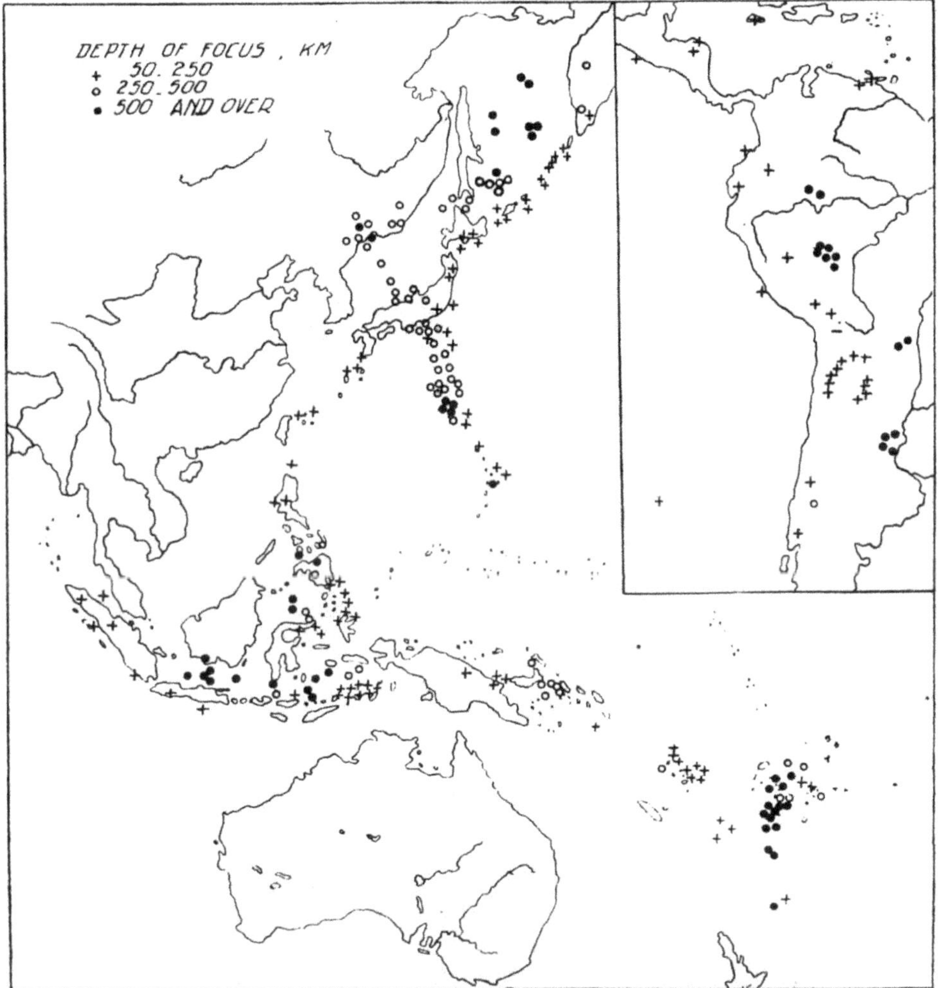

Fig. 101. Distribution of intermediate and deep earthquake-foci in the surroundings of the Pacific. (From Gutenberg and Richter).

Any one inspecting these map will notice the remarkable distribution of epicentres. It appears that the principal seismic activity is associated with the deep-sea troughs lying off the Japanese, Kurile- and Bonin islands. Foci of intermediate shocks lie further continentward, under Japan, and the class of deep foci still further away from the Pacific basin. It appears, moreover, that a zone of deep foci extends along the Bonin island towards the Marianas. And another zone may be seen to run parallel to the Kurile arc and the Riu-Kiu Islands.

More or less analogous results have been obtained from other districts.

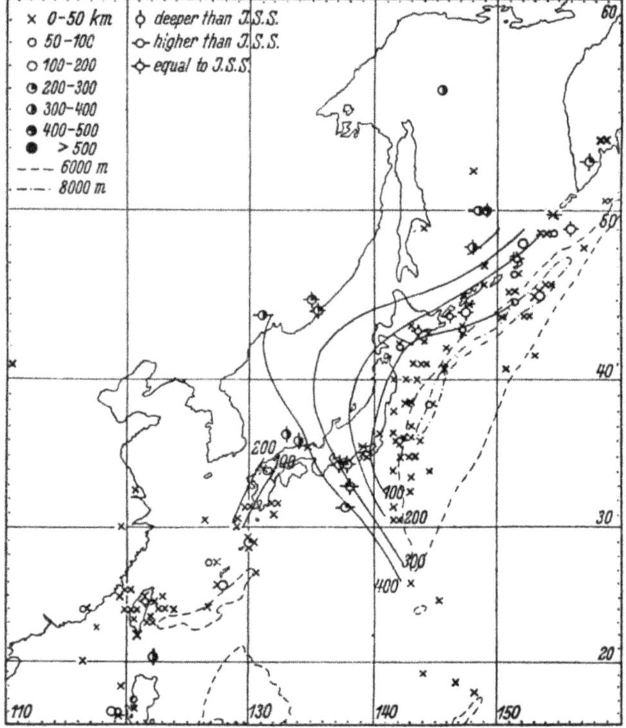

Fig. 102. Seismic isobaths in the Japanese area (From S. W. Visser).

Fig. 105 shows the distribution of deep foci along the Aleutian arc and their relation to the deep-sea trough in front of it [1]). Another district known for the frequent occurrence of deep-focus earthquakes extends from New Guinea and the Solomon islands towards the region of the New Hebrides and the Loyalty islands. Once more we see that it is bordered by deep-sea troughs such as the Tonga and Kermadec deeps (fig. 106). Again the same associations hold good in the Philippines, the East Indies and South America. The last named district deserves special attention because of the absence of island-arcs, which in the other districts appear to be associated with the occurrence of intermediate and deep foci along their concave side and with foci of the normal class in a zone extending continentward from the deep-sea trough on their convex side. In South America the epicentres again possibly lie on a plane sloping from the Pacific eastwards

[1]) This map is reproduced from a paper by Visser. For more recent data see the map of Gutenberg and Richter 1945, p. 623, fig. 3. The same holds good for fig. 107.

until it reaches a depth of more than 600 km east of the Andes, fig. 107.

The East Indies present a complicated geographic distribution of deep foci apparently due to the situation of the district along the southeastern margin of the Asiatic continent and in the vicinity of Australia. The region will be considered more closely in a later section. For the present, however,

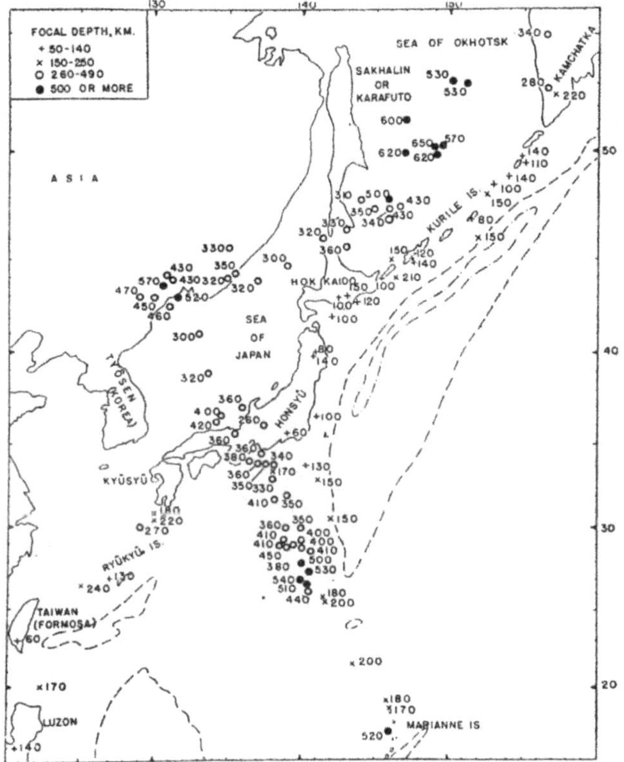

Fig. 103. Epicentres of earthquakes in the Japanese area (From Gutenberg and Richter).

it may be said that the less complicated area between Borneo and the deep Java trough again shows the same arrangement of foci along a continentward sloping zone (fig. 118).

A further important feature is the well-established absence of intermediate and deep shocks along the north-eastern border of the Pacific. Neither are deep foci known from the Antarctic and south-eastern Pacific. Remarkably enough deep-sea troughs of the type called marginal deeps in Chapter IV are also conspicuously lacking along the coasts of these regions!

Outside the Pacific border earthquakes originating at depths exceeding the thickness of the earth's crust are of the intermediate class. No true deep-focus shocks are known to be associated with large thrustplanes on the continents. This holds good even for the notable district of Hindukush (fig. 100).

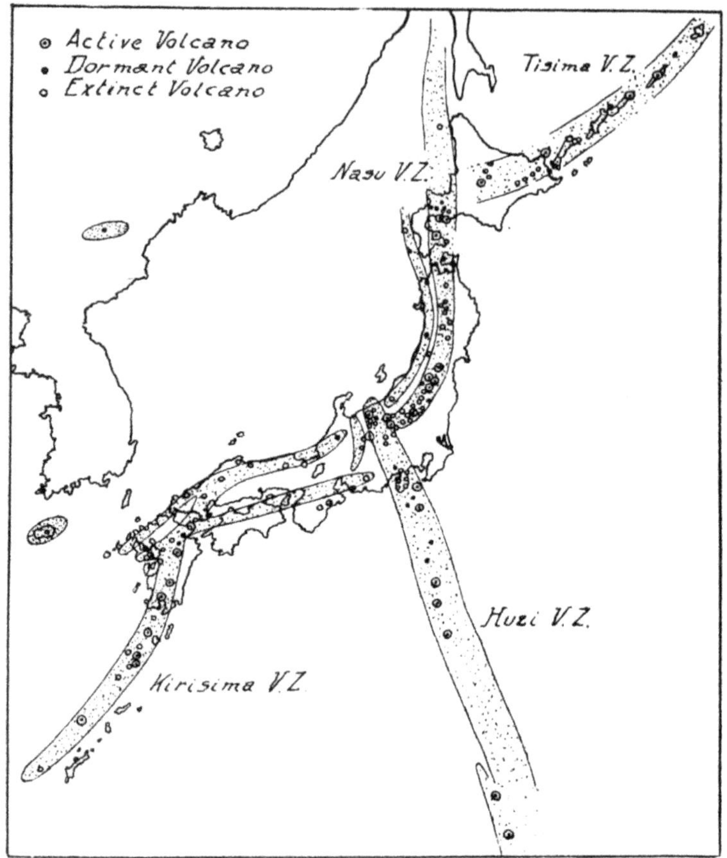

Fig. 104. Volcanic zones of Japan and vicinity (From K. Wadati).

Deep-reaching shear-zones and convection currents

It is clear that these seismic results imply several aspects of geological importance. One question is whether the notion of shearing zones down to a depth of several hundred kilometers may be reconciled with the geophysical theories of isostasy and of convection currents in the earth's interior. It appears from modern researches that the question may be answered in the affirmative. According to Griggs, experiments at high pressures showed in a convincing manner that "when a rock enters the region of plastic flow, it will not deform indefinitely, but will rupture if the deformation is carried far enough". Similar results were obtained by Bridgman and by Haskell.

Conclusion

The principal features mentioned in our short review of such seismological results as may be regarded of special importance to geology may be summarized as follows. Seismic data imply the existence of deep-reaching

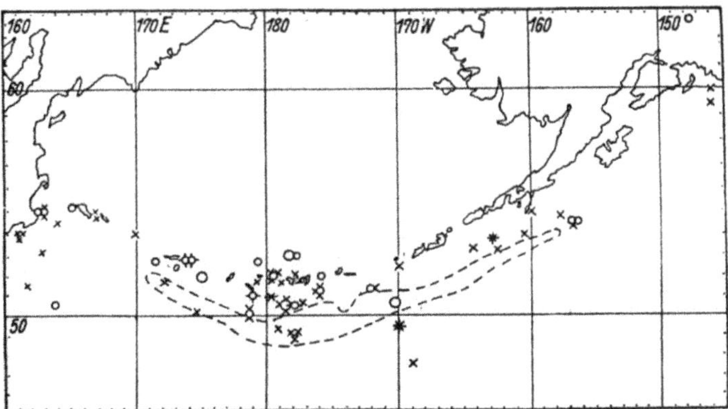

Fig. 105. Seismic activity in the vicinity of the Aleutian arc (From Visser). For notations compare fig. 102.

shear-zones along the Pacific border. And it seems hardly open to doubt that most of these shear-zones are intimately connected with the site of island-arcs.

It appears that in general the dip of the shear-zone is much greater than the amount of the dip that was calculated from the physiographic theory. Now, the theoretical construction of the curvature of the arc was rather arbitrary. In our present state of knowledge it would seem preferable to measure the curvature of the arc along the axis of the zone of negative anomalies of isostasy.

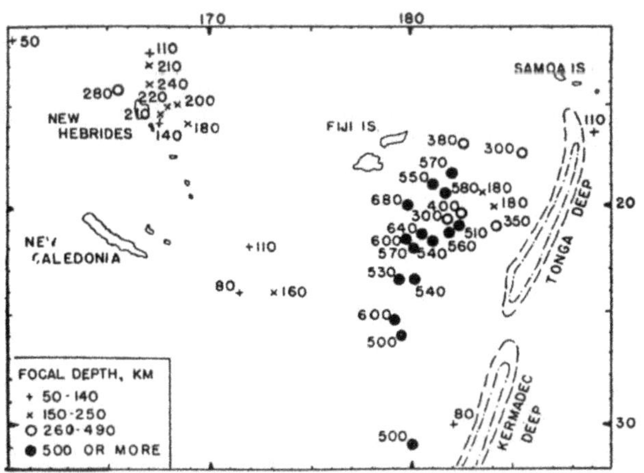

Fig. 106. Deep-focus earthquakes along the south-western border of the Pacific Basin (From Gutenberg and Richter).

One will notice, however, that the curvature may vary considerably in one and the same island-festoon (cf. Plate 8), but even the preliminary data show that the dip of the shear-zone varies also within comparatively short distances. More seismic data are needed in order to get a better idea of the shape and variable dip of the shear-zone and its relation to the degree of curvature of an island-arc or parts of it.

There remains however one problematic point viz., the discrepancy be-

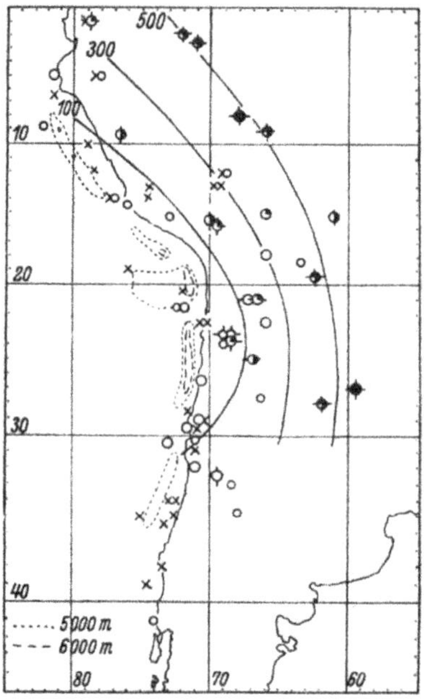

Fig. 107. Seismic isobaths in South America (From S. W. Visser).

tween the curvature of the arc and the angle the deep-reaching shear-zone makes with the earth's surface.

Possibly two sets of shear-zones formed simultaneously at the time of the origin of the arc, viz. one in the crust and another in the substratum.

The shear-zone in the crust may have had a gradient of some 15 degrees corresponding with the curvature of the arc. This shear-zone, however, was disturbed and partly obliterated by crustal processes of later epochs. On the other hand the potential deep-reaching shear-zone in the subcrustal matter was rejuvenated time and again. This tentative suggestion tries to make clear why it will be in vain to look for a correlation between the degree of curvature of the arc and the dip of the deep reaching shear-zone, whereas the latter shows a striking relation to the site of the arc.

Probably the potential shear-zone in the substratum (see p. 152) is caused by convection currents which are in some way related to the site of the island-arcs. It will be shown (p. 204) that island-arcs occur on the boundary of crustal areas of different composition. Hence it seems clear that the location of the island-arcs as well as the convection currents and the shear-zones all depend on the distribution of crustal matter.

Originally the present author [1]) thought the shear-phenomena would occur on the boundary of two systems of convection currents, a stronger system on the continentward side meeting a weaker one at the oceanic side, as suggested by Holmes [2]). In turn the shear-zone would seem the place *par excellence* for the origin of a crustal down-buckle. However, according to this theory one would expect a shear-zone in a nearly vertical position, which is not the case.

In a very recent theory Vening Meinesz (1946) made an attempt at an explanation of the obliquity of the shear-zone. He argues that owing to the higher temperature of the belt of down-buckling, a convection current will originate. It will show a rising column under the belt and a sinking column at the continentward side, viz., in the area of the deep-sea basin behind the arc. The largest values of shear are found in planes under 45° with the vertical at the four turning points of the system.

But he thinks that, probably, shear-phenomena will only take place

[1]) Umbgrove, 1945, p. 199.

[2]) A. Holmes Principles of Physical Geology 1945, p. 506, fig. 262.

under the influence of a trigger effect. And this effect is supposed to be caused by a rising movement of the zone of buckling.

Therefore shear-phenomena would occur especially in a zone dipping continentward at about 45° from the sialic mountain-root, which at the same time is revealed as a belt of strongly negative anomalies of gravity.

According to Vening Meinesz the curvature of the arc is the cause of the absence of another convection system of equal strength along the convex side of an island-arc. This is a weak point in his interesting hypothesis. For, the Kermadec and Tonga zone shows not an arcuate but a linear arrangement of volcanoes and deep-sea troughs. Therefore a symmetrical set of shear-zones on either side of the straight line might be expected according to the theory. However, they were found only on the west side (fig. 106). And in South America deep-focus earthquakes were found at the continentward side of structural elements of deep-sea and coastal districts which are decidedly concave towards the Pacific (fig. 107)!

Moreover Vening Meinesz suggests that the downward movement of such basins as the Banda Sea basin would be caused by the downward moving column of a convection system below the basin. However, the absence of a deep-sea basin in the area of the Java sea, where deep-focus earthquakes were found as well as in the Banda Sea (fig. 118) is not in accordance with his hypothesis. Another attempt at an explanation for the origin of the deep-sea basins will be put forward later on (pp. 194–195).

One thing will be clear from this short survey of recent theories viz., that many questions regarding the remarkable deep-reaching shear-zone still remain problematic. However, the relation between this zone and the zone of down-buckling seems unmistakable. We shall have to return to this point on pp. 185 and 204.

Terrestrial Magnetism

According to Visser anomalies of terrestrial magnetism might be regarded of interest to the present problem in two respects. In the first place the anomaly-fields reveal a rough relation to the regions in which deep-focus earthquakes occur. In the second place they appear to be in agreement with the theory of convection currents in the earth. However, both the sources of evidence are vague and the conclusions have the character of provisional hypotheses which stand in obvious need of confirmation by more detailed investigations.

The full-drawn strong lines of fig. 108 represent the "zero-meridian", being the great circle through the magnetic poles, and the magnetic equator. The two circles intersect in Central Africa and in the central part of the Pacific. The other curves represent positive and negative anomalies of the vertical component of terrestrial magnetism (expressed in 0,001 Gauss).

The principal areas of deep-focus earthquakes are indicated on the map by shading. It appears that they are confined to two great areas of positive deviations of the vertical component, viz. the large field which extends from the Asiatic continent into the western and southwestern part of the Pacific, and the part of another large field that extends as far as South America. However, no deep-focus earthquakes are known in another part of the same field that extends over the southern Atlantic and Indian Oceans.

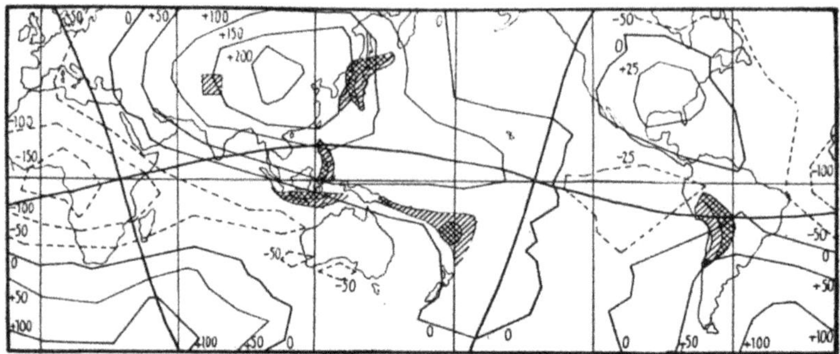

Fig. 108. Anomalies of vertical component of terrestrial magnetism (0,001 Gauss) and principal regions of deep-focus earthquakes (cross hatched) (From S. W. Visser).

A field of less intense positive anomalies may be seen to occur on the North American continent. A few indications of deep shocks have been recorded from the area of the south-eastern extension of this field in the Caribbean region. If these also were taken into account the arrangement of the principal deep-focus areas might appear to be roughly symmetrical with respect to the zero-meridian of the Pacific, as is the case with the fields of positive anomalies. It hardly needs saying that the relations suggested by Visser are not yet very convincing. He concluded: "If indeed the connection between anomalies of terrestrial magnetism and deep-focus earthquakes can be proved to be real, it is by no means a simple one, as appears when we compare the fairly capricious behaviour of the deep-focus areas of the southwest Pacific with the great magnetic anomaly of the same region."

Regarding the second question Visser presumes that convection-currents transport material with special magnetic properties from the deeper realms towards the more external parts of the earth's interior and, accordingly, cause positive anomalies of terrestrial magnetism at the surface. Again fig. 108 shows only a very rough agreement for some areas. The small area of negative anomalies in the Pacific might be mentioned as a weak point of Visser's theory. And we may stop to notice that the field of negative anomalies over the African continent is the reverse of what one would expect from the theory.

At any rate Visser has drawn the attention to a source of data which is worthy of detailed examination when more data on terrestrial magnetism become available.

Finally it should be mentioned here that in another paper, Visser reported that he had found some indications suggesting a relation between the magnetic anomalies in the East Indies and the gravity anomalies of the same region.

Volcanism and Plutonism

One of the two fundamental postulates, which were formulated in the introduction, pertained to the invariable association of the island-festoons with volcanism. As a general statement this point needs further precision. An arc of the single type is crowned by a majestic row of volcanoes, like those of the Pelew- and Mariana islands. In double arcs the inner one bears a crown of volcanoes, while in the outer arc volcanism is conspicuously lacking. Geological exploration, however, may reveal the presence of volcanic and plutonic rocks in the outer arc as well, but dating from more remote times.

Acid batholiths may occur in both of them. And according to Hess serpentines and other ultra-basic rocks are characteristic of the outer arc. It will be the aim of a separate section to deal with their mutual relations to the structural history of the whole region (p. 191). Such a procedure seems reasonable since a discussion of these features has to be preceded by a discussion of a few other sources of data. For the same reason, however, a treatment of deep-sea troughs and isostatic anomalies necessarily has to be preceded by a short enumeration of the volcanic and petrographic characteristics of the arcs.

The andesite line

In the first place we may recall that the true boundary of the Pacific Basin proper is generally represented by a hypothetical line, known as the andesite line (p. 64, 120, 228, and Plates 4 and 7). On the continental side of the line of the petrographic composition of volcanic rocks implies the probable existence of sialic material in the crust. On the other hand volcanic rocks from the Pacific Basin situated inside the boundary of the andesite line, invariably point to a simatic ocean floor, devoid of a cover of sial. Some doubt exists concerning the Albatross plateau off the coast of South America. Possibly a thin film of sialic matter may be present in that area. It is clear, however, that the andesite line is generally drawn as an immediate boundary of the volcanic arcs and the coastal districts that are known to belong to the andesite zone. But nobody knows whether the sialic crust ends exactly along that imaginary line. On the contrary it would seem much more probable that in many places the true extension of the sialic part is somewhat beyond the line, and that the true limit between sialic and basaltic sea-floor is presumably much more irregular than the nicely curved andesite line on Plates 4 and 7 would suggest. It will be seen in the following section that it is well to bear this point in mind when Lawson's speculation on the origin of the arcs is examined.

Petrographic provinces

In strong contrast to the uniformity of the basalts of the Pacific the island-arcs revealed a diversity of rock types belonging to two magmatic clans. Mostly the petrographic and petrochemical character of the rocks is of the calc-alcali or so-called pacific suite. Igneous rocks of the potash or so-called mediterranean type are also frequently met with. Their distribution in the East Indies may be seen in fig. 109 while fig. 110 shows the

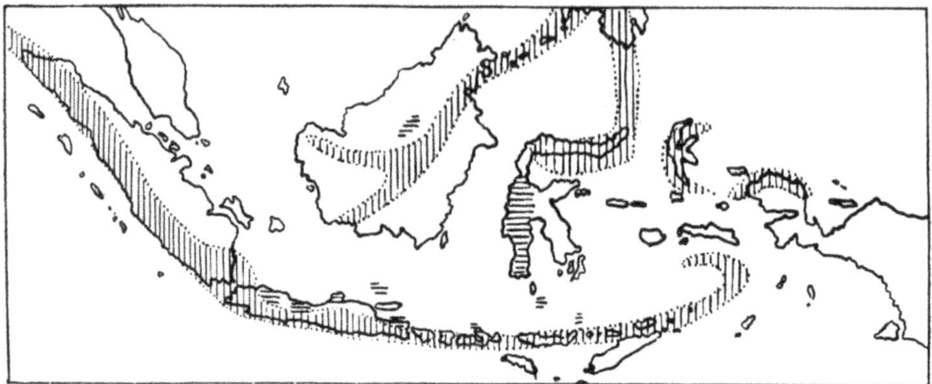

Fig. 109. Distribution of rock clans in the East Indies. Vertical shading: pacific clan; horizontal shading: mediterranean clan (From H. W. V. WILLEMS).

distribution of active volcanoes in the same region. In all 1220 rock-analyses have been reviewed by Willems in a paper especially devoted to the subject.

Van Bemmelen has discussed the origin and history of both the pacific and mediterranean magma-provinces in two papers. His conclusion is that the igneous rocks of the calc-alkali type, characterizing the geanticlinal

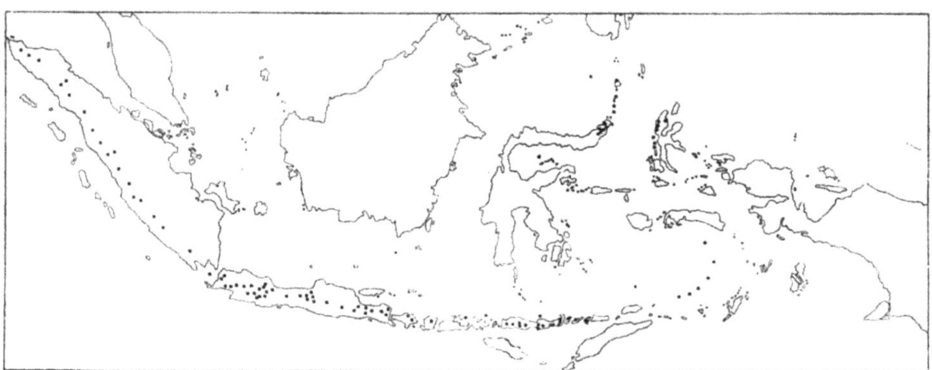

Fig. 110. Distribution of active volcanoes in the East Indies.

belts of the volcanic inner arc, are connected with certain orogenetic cycles. But they did not originate by differentiation of a primary or juvenile tholeiitic magma. Probably magmatization of pre-existing crustal rocks by ascending emanations from greater depths was a fundamental process. Besides, hybridisation or contamination of the ascending magma by assimilation and migmatization of sialic crustal rocks played a role of importance. If the plutonic rocks were entirely the products of juvenile origin and normal gravitational differentation they would be more basic in the deeper parts. Instead of such a sequence the rocks appear to be more acid towards the depth of the batholith (fig. 44). And this supports the view that the granitization was effected by migrating materials emanating from a rising migmatite front, thereby causing a progressive metamorphism of the existing crustal layers.

The alkaline rocks of the potash province are considered by Van Bemmelen as "pathological" products of which the origin is doubtful. Their formation may be the result of assimilation of comparatively large quantities of limestone and subsequent loss of soda in the hydrothermal phase.

A process of this kind in connection with the lavas of the Vesuvius has already been described at some length in Chapter IV (p.74), but according to other views, the alkaline rocks may be due to a more intensive action of magmatic emanations.

Igneous rocks in geanticlinal belts

It will be seen presently that the upward migration of magmatic emanations from the depth was especially active during epochs of increasing compression. During these epochs the geanticlinal areas tended to rise and apparently this process stimulated and activated the underlying earth-material.

Batholithic intrusions, mainly granodiorites and granites of Upper Neogene age, are known to occur on certain islands of the inner arc, viz. Sumatra, Java, Flores, Lirang and Wetar, and obviously the present volcanic action might be genetically related to such bodies. We owe to Van Bemmelen a clear insight into their mutual relations. It is worth while to quote a passage which summarizes the historical succession of igneous activity in the inner arc. Before doing so it should be mentioned that previous to the Upper Miocene a series of calc-alkaline volcanics called the Lower- and Mid-Miocene "old andesite formation" had accumulated in Sumatra.

"During the younger Miocene" — Van Bemmelen writes [1]) — "an orogenic phase occurred. The geanticlinal zone was lifted above sea-level and at the same time it has been block-faulted and intruded by acid magma. These intrusions nowadays are exposed as batholiths, stocks and bosses of coarse-medium grained granite and granodiorite with contemporaneous offshoots and dikes of dacitic and liparitic appearance. The whole old-andesite formation has been more or less altered by hydrothermal processes (such as propylitisation and locally gold-silver ores originated .."

"Thereafter, the geanticline sank down again and in some places this system of old-andesite formation with acid intrusions has been covered unconformably with younger Neogene (Pliocene) marine sediments."

"After this orogenic phase during the younger Miocene, with its concomitant acid intrusions and explosive paroxysms, in some places, the basic-intermediary volcanism temporarily came to rest, at other places however it seems to have continued uninterrupted. But soon it gathered again enough force to cause the general, formidable volcanic activity of the geanticlinal zone during the Plio-Pleistocene."

"These numerous younger strato-volcanoes of basic-intermediary,

[1]) Van Bemmelen 1938, p. IV 1—IV, 2.

calc-alkaline composition and Plio-Pleistocene age form an unconformable capping of the geanticlinal zone. This so-called younger andesitic series has not been altered (or only locally) by hydrothermal processes."

"In southern Sumatra this period of younger andesite eruptions was followed by a second uplift of the geanticlinal zone, which was accompanied by a second suite of acid (dacitic — liparitic) eruptions. These eruptions occurred chiefly along the remarkable rift-graben on the top of the geanticline, the so-called Semangko-graben...."

Why these details? Continually we are concerned in an attempt at understanding the challenging features of the remarkable island-festoons. It is clear, however, that an adequate theory of their formation should also account for the varied succession of phenomena presented by plutonic and volcanic rocks and the tectonic movements to which they seem to be related.

It appears from this short review that since the Upper Miocene the geanticlinal ridge of western Sumatra has been subjected more than once to a rising movement. During these epochs the arching geanticline became faulted and fractured. In its present state Sumatra shows longitudinal faults and a central rift-valley. Volcanism was very active along these faults and in many places its activity has continued along the tectonic fractures up to the present day.

The acid pacific magma is of the strongly explosive type. Ascending along the faults and graben of the updoming geanticline it gave origin to volcano-tectonic phenomena which are very impressive. A whole series of calderas and large volcano-tectonic depressions came into being on the crest of the geanticline.

Igneous rocks in geosynclinal basins

It is of importance to consider a different area of volcanic activity. Fig. 111 represents the historical succession of events in the Karangkobar region of Central Java. Apart from the rising migmatite front accompanied by an increasing magmatization of the crust, the figure clearly demonstrates the principal epochs of tectonic activity. The region is situated at the northern side of the geanticlinal belt of southern Java, where the latter is bounded by a geosynclinal trough. Volcanism started in Upper Miocene times with the formation of the submarine (geosynclinal) volcano Penjatan.

Its vents, cutting through the Miocene Merawa-shales (left part of fig. 111) are now exposed by erosion. They are composed of gabbro-dioritic rocks. Pliocene volcanism produced the volcanic products of the Ligung series (middle part of fig. 111) which is of andesitic composition. Finally the right part of fig. 111 shows the formation of the Pleistocene and more recent volcanoes. In the same region, but further to the north in the geosynclinal basin there occur intrusions which belong to the alkali or mediterranean series of igneous rocks.

Finally it should be noted that several epochs of tectonic activity are represented in the schematic fig. 111. After subsidence of the bottom during the Miocene, a warping movement took place in the Upper Miocene. Then after continued subsidence of the trough the region was subjected to a rising movement in the Pleistocene.

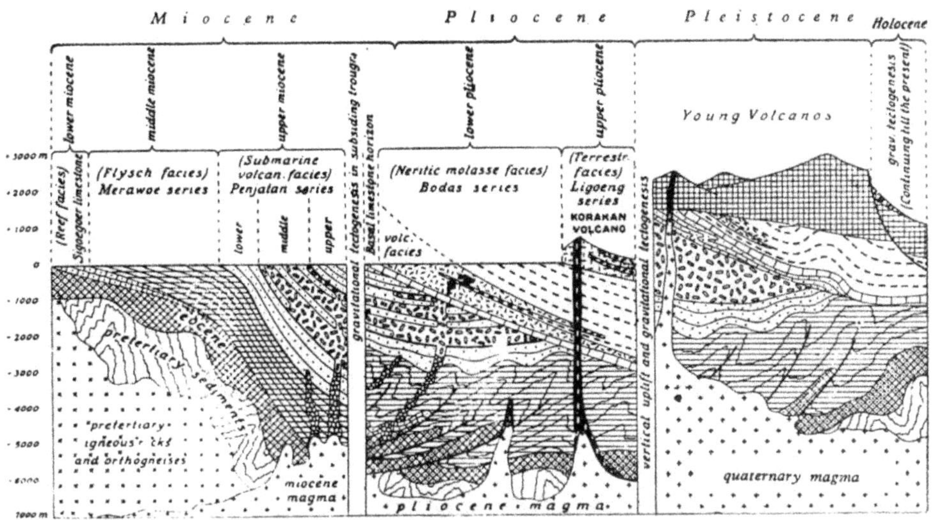

Fig. 111. Diagrammatic representation of epochs of plutonism and volcanism in the Karangkobar region, Java (From Van Bemmelen).

A theoretical synthesis of the formation of the East Indian island-arcs ought to account for these various expressions of plutonism, volcanism and tectonics!

Problems to be solved

The principal problems involved in the manifold manifestations of plutonism and volcanism during the origin and history of the island-arcs may now be shortly enumerated. A synthesis which aims at explaining the structural history and present condition of the island-arcs has to elucidate at least the following eight points.

(1) It has to be made clear why in a double festoon a volcanic arc is always the inner one, i.e. why it is always on the concave or inner side and never on the convex or outer side of a non-volcanic arc.

(2) We want to know why there were two phases of elevation and rejuvenated magmatic phenomena in the region of the East Indies inner arc, since Miocene times.

(3) The situation of the mediterranean rocks on the concave side of the inner arc is a further problematic point.

(4) It has to be explained, moreover, why active volcanoes are lacking on the islands of the outer arc.

(5) Then, why is active volcanism absent on the inner row north of the island Timor.

(6) Another feature which calls for explanation is the occurrence of a double row of volcanoes in the northern Moluccas, one on the northern

arm of Celebes and thence continuing into the Sangi islands, the other on Halmaheira and some neighboring islands.

(7) A further problematic point is the presence of volcanic rocks of Upper Tertiary age on Buru, Ceram, Timor and Sumba.

(8) Finally, the occurrence of serpentines and other ultrabasic rocks on many islands has to be explained.

We will consider other courses of data and then an attempt will be made to elucidate all these problematic features into a well-founded synthesis.

Deep-sea basins and troughs

A feature intimately related to island-arcs as defined in the introduction, is the presence of a deep and elongated furrow in front of the festoon. It belongs to the type called marginal deep in Chapter III. Among them are the deepest furrows known in the earth's crust and it seems obvious that their formation is due to some deep-reaching process of crustal deformation.

A further characteristic is the presence of deep troughs along the concave side of the outer arc, and finally a series of deep-sea basins which generally occur on the continental side of the inner arc.

In Chapter III the different types of troughs and basins associated with a double arc like the East Indian festoon were classified and compared with the continental basins. The comparatively recent time of their formation was demonstrated for the basins bordering the Asiatic continent in general, and the East Indies and the Bering basin more in particular. It will be seen, in the section under the heading "Synthesis", that converging evidence from various groups of data reveals some essential features of their origin and history. For the moment, however, we will stop to examine a hypothesis which was proposed by Lawson in the year 1923. For although Lawson's theory proves to be unsatisfactory the mechanism proposed by him has many interesting aspects.

Compensation for relief of load by denudation of the continental areas is held to be effected by a continentward inflow of heavier subcrustal matter from surrounding regions. At the boundary of continent and sea-floor the zone of flow would be lowered and the coast-region would become a shear-zone, the continent riding on the simatic floor which would thus creep landward. "On this rupture zone the crust of the sea floor would be thrust under the continental margin and elevate it as a ridge" [1]). The arcuate shape of the island-festoons would arise from the fact that a thrust plane slicing the earth's surface is necessarily arc-shaped. In addition the landward under-flow of the oceanic sima would result in the formation of open fissures which would be intruded by dunite from deeper parts of the sima. And because of the higher specific gravity of dunite the fissures would remain as hollows that are known to us as the marginal deeps in front of the island arcs (fig. 112).

---

[1]) Lawson, 1392, p. 367.

In order to explain the deep basins between the arcs and the present coast of the continents Lawson once more assumes a landward creep of the sima from beneath that region. He supposes the region to have been a broad coastal belt that had been approximately peneplaned, whereas back of this low continental margin a region with high relief would have existed, rising steadily in response to removal of load by erosion. Again the compensation for the erosional loss would have been effected "by an indraft of heavy rock in depth from the coastal region, causing the low lands and shallow epicontinental sea-floors to sink." Again the replacement of basalt by the heavier dunite would account for the submerged part of the continent lying deeper here.

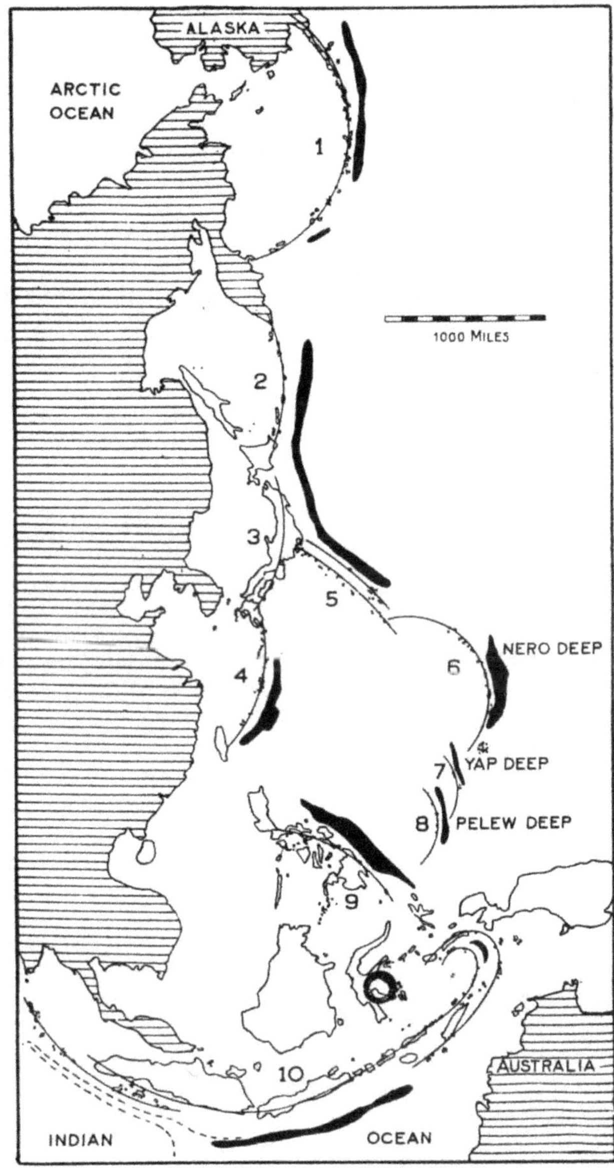

Fig. 112. Island-arcs bordering the eastern and south-eastern coast of Asia. (Adapted from Lawson). 1 Aleutian arc, 2 Kurile arc, 3 Japanese arc, 4 Riu-Kiu arc, 5 Bonin arc, 6 Mariana arc, 7 Yap arc, 8 Pelew arc, 9 Philippine arc, 10 Malayasian arc.

As regards the four arcs of the Bonin-, Mariana-, Yap- and Pelew islands, which are arranged *en échelon* each with a marginal deep in front, Lawson remarks: "it is not easy to regard this great compound arc as having been at one time the continental margin. To do so would be to claim that the sial extends out to east longitude 146 degrees in the latitude of Manila; and this would in some measure invalidate the hypothesis which has been advanced for the origin of the arcs nearer the mainland."

Hence Lawson tries to explain the origin of these arcs in two fundamentally different manners. Moreover, the sialic crust probably does extend so far out as these outer arcs. We need only recall the situation of the andesite line (p. 165), while arguments based on the gravity-field will be mentioned in the next section.

Apart from these difficulties it is assumed by Lawson that the edge of the former continent should have had the shape of the thrust planes. No explanation is given for the complicated double arc of the East Indies, bordered at its south-eastern side by Australia instead of by a simatic ocean floor, while the western part of the same arc is bounded by the sialic floor of the Indian Ocean.

Another striking feature of the arcs is the presence of volcanoes on the crest of the inner arc. Again Lawson's theory offers no explanation for this conspicuous phenomenon.

Finally one of the most serious objections against the theory is of a geophysical nature. Lawson assumed the marginal deep-sea furrows to be in isostatic equilibrium. On the other hand a defect of mass in their cross-sections would be rather what one might expect, having regard to their shapes as abnormal deep depressions of the crust. In order to reconcile these two opinions Lawson assumes intrusions of the heavier dunite from a deeper layer of the substratum into the sectors below the deep-sea furrows. The replacement of basalt which has a specific gravity of about 3.05 by dunite with a specific gravity of about 3.3 would account for the re-establishment of isostatic equilibrium of the furrow. However, since the announcement of the hypothesis, measurements of gravity have been carried out over some of the deeps. The results obtained revealed a situation which is not in accordance with Lawson's theory and compel us to accept a quite different view regarding the origin of the basins and troughs, as will be seen presently.

Isostatic anomalies

The interesting results of the gravity expeditions at sea carried out by Vening Meinesz from 1923 to 1932 in submarines belonging to the Royal Netherlands Navy are of much importance for more than one of the problematic points under discussion. One aspect concerns the boundary between continental and simatic areas. In addition, the results of the gravity expeditions at sea have largely contributed to a better understanding of several other phenomena. Among these are the deeper structure of the island-arcs including the origin of deep-sea troughs and volcanic belts accompanying them. These subjects will be considered subsequently.

The boundary of the Pacific basin

Speculations on the course of thermal convection currents under the earth's crust were advanced by Ampferer, Schwinner, Holmes, Pekeris, Escher, Griggs and Vening Meinesz. Entering into the various suggestions of their interesting hypotheses would take us too far out of the scope of the present subject. For the moment one asset of Vening Meinesz's notable results is of special interest. The question arises whether the boundary between areas of basaltic and sialic character coincides with the deeply dipping shear-zones that were deduced from physiographic and seismic data. Now indeed Vening Meinesz found that boundary "on the right place". On a crossing of the Pacific from San Francisco to the Philippines, via the Hawaian-islands, the gravity field shows positive anomalies until the island Guam is reached. Here rather strong negative anomalies reveal the presence of a crust of sialic character. It is here too that the Pacific basin proper is bordered by the Nero deep, and that the outcrop of a deep-reaching shear-zone was assumed on physiographic and seismologic grounds. It is the same boundary which from petrographic evidence is marked as the andesite-line.

It is worth while considering in some detail the gravity profiles at the border of the Pacific Basin proper.

Single Arcs

Single island-arcs at the Pacific border were crossed in two places. Both the gravimetric profiles show a similar phenomenon, viz. rather strong negative anomalies over the marginal deep-sea trough (fig. 113). The geo-

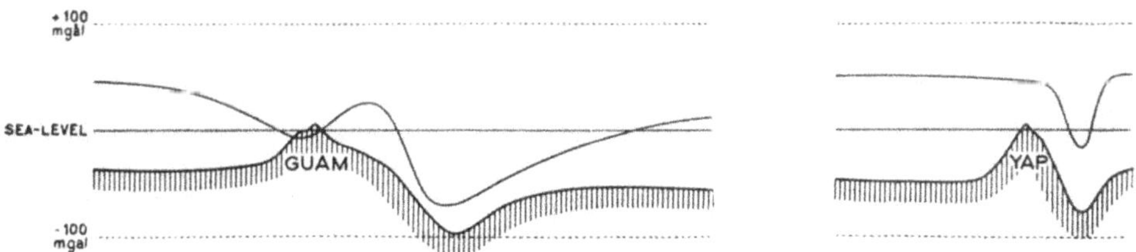

Fig. 113. Gravimetric profile across Guam and the Nero deep (left) and Yap and the Yap deep (right) (After Vening Meinesz).

logical interpretation of the profile across Guam and the Nero-deep suggests the existence of a root of light crustal matter which must have buckled downward so as to replace the pre-existing heavier matter beneath [1]). A geological interpretation is represented by the schematic block diagram fig. 114. The anomaly curve is asymmetrical and hence seems to be best explained by assuming that the eastern sector was thrust under the western

[1]) The most important events in the geological history of Guam, since Miocene times, are: (1) formation of a submarine volcano by extrusions of basaltic pillow lavas from fissures, now exposed as a complex of dikes, (2) an intermittent explosive phase giving origin to 300 meters of andesitic tuffs and shallow water agglomerates with interstratified shales and marls containing Miocene Foraminifera, (3) cessation of volcanism, (4) intense folding and overthrust faulting, (5) sedimentation of shallow-water limestones, (6) renewed faulting, (7) emergence and subaerial erosion, (8) submergence in the Pleistocene, (9) renewed emergences and submergences, partly due to changing ocean levels, five marine terraces being cut, (10) emergence of 5 feet. (Stearns, H. T. Geological History of Guam, Bull. Geol. Soc. Am. 51, 1940, p. 1948).

part. As a result the Nero-deep originated. The riding part of the crust bowed up and gave rise to the formation of faults dissecting the crust.

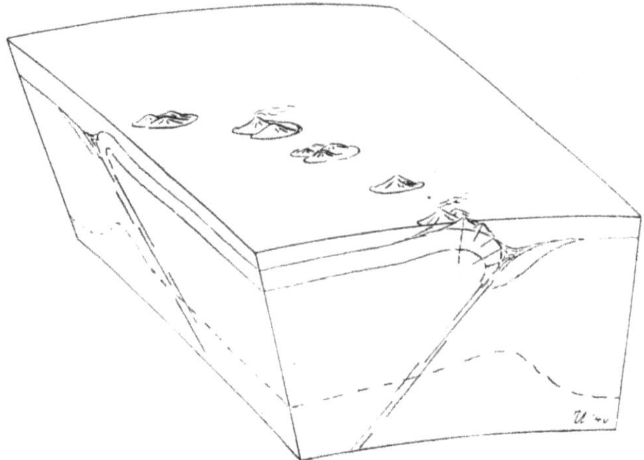

Fig. 114. Schematic blockdiagram of an island-arc of the Mariana type, showing the vulcanic arc, the deep-sea trough, the supposed under thrusting of the crust near the andesite-line, and the potential zone of shear.

Possibly a down-faulted part of the vault caused the second, though slighter downward bending of the anomaly-curve. Moreover the upward arching of this crustal part caused a relief of pressure. And accordingly emanations from the substratum ascended in these parts and caused volcanism at the surface (cf. p. 72). It is clear that the lavas will mainly belong to the pacific suite, since a sialic crust is present, and so, in fact, they are.

The place of the underthrusting was probably predestined by the boundary between two crustal areas of different composition. The location of the shear-zone, which is the site of numerous deep-focus earthquakes, is possibly due to the action of convection currents. And finally, these are probably also controlled by the different character of the ocean-floor on either side of the andesite line.

The second gravimetric profile across the Pacific border is also represented by fig. 113. It runs through the island Yap and the adjoining deep. Again the negative anomaly is over the deep-sea trough. Unlike the profile across Guam the anomaly-curve is symmetrical and no secondary root of sialic matter is indicated under the island Yap.

Accordingly a geological section must show a symmetrical root of the sial-crust. The site of the crustal buckle again appears to be accompanied by a deep-reaching shear-zone.

We mentioned already the symmetry of the gravimetric profile across the Yap-deep as contrasted with the asymmetrical character of the curve over the Nero-deep. This feature is explained by Vening Meinesz by the assumption that this part of the arc "encloses a greater angle with the direction of the compression".

Double arcs

Much stronger are the deviations of isostatic equilibrium in the complicated areas of the West- and East-Indies.

In the latter region Vening Meinesz made observations at 281 submarine stations. One of the most conspicuous results of his explorations is the establishing of a belt of strong negative anomalies of isostasy. The relation between submarine topography and isostatic anomalies in the East Indies is shown by the profiles of fig. 115, which should be compared to the profiles

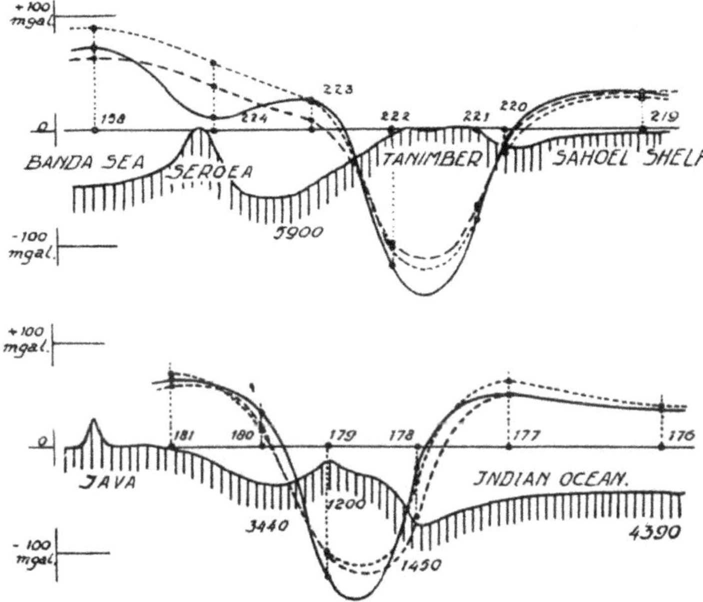

Fig. 115. Relation between submarine topography and negative anomalies of isostasy in the East Indies (After Vening Meinesz).

of fig. 113. It is clear that the situation of the negative belt is different from the Pacific border in one fundamental respect, viz. the presence of a sialic crust of much greater thickness. If the sialic part of the crust be comparatively thin and hence situated below sea-level — and moreover far from continental land — a deep-sea trough will come into being above the sialic root and a negative anomaly will be found over the deep (fig. 113). The result will, however, be radically different if the sialic crust protrudes above sea-level or at any rate if it is mainly of continental thickness. For large quantities of waste-products from the land will accumulate in the downward moving crustal wave. And when the crust buckles to form a downward root of sial, the contents of the furrow will necessarily become crumpled and squeezed out so as to form a ridge-shaped elevation above the zone of buckling (fig. 115). It goes without saying that in the first case considered a comparatively thin veneer of bottom-sediments was squeezed out, as well. But it did not form a factor of importance in the surface-topography. In the other case, however, a double arc is born.

These theoretical deductions seem to be confirmed by the situation of island-arcs and troughs in the East Indies. A comparison between topography and gravity-field clearly shows the belt of strong negative anomalies to follow the islands west of Sumatra and to continue over

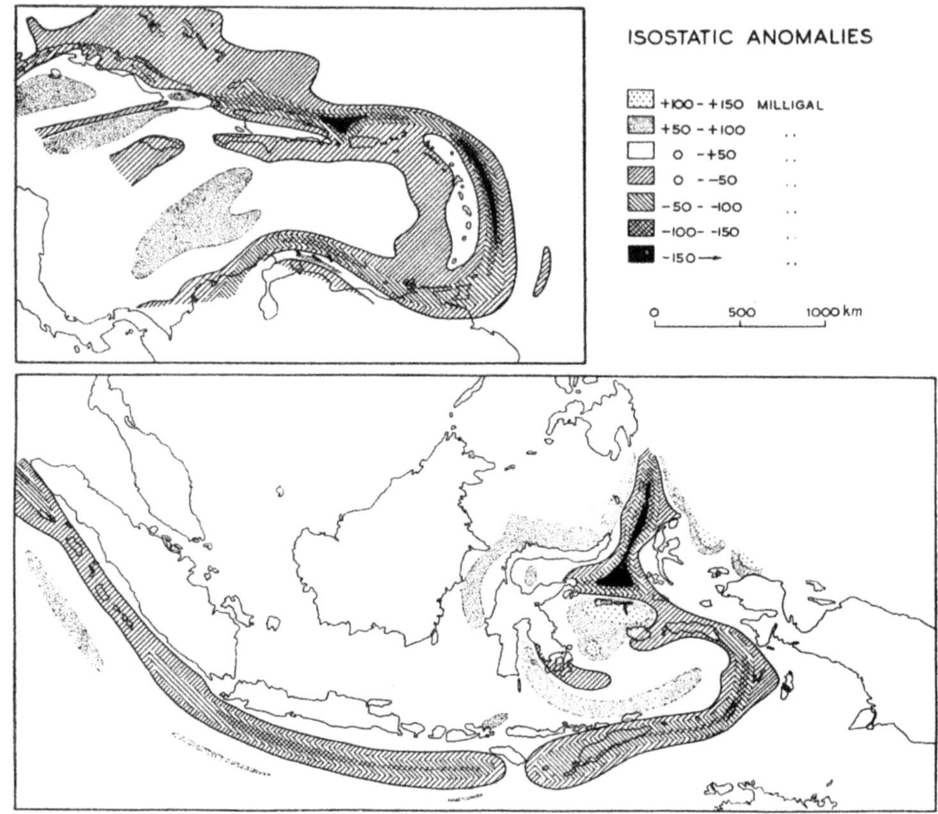

Fig. 116. Isostatic anomalies in the West Indies (after H. H. Hess) and the East Indies (after F. A. Vening Meinesz).

submarine ridges towards Ceram and Buru, via Timor and the Tanimber islands. And this forms a striking contrast to the location of the anomalies along the Pacific border in the vicinity of Guam and Yap where they coincide with the marginal deep.

A more or less similar pattern of anomalies was revealed in the West Indies by the explorations of Ewing, Hess and Vening Meinesz. An interpolation of the results may be seen in fig. 116. Remarkably enough it appears that the negative zone partly coincides with a deep-sea furrow, viz. the Brownson trough north of Puerto Rico. But to the east and southeast the axis of the negative zone lies over a ridge, which even partly projects above sea-level in the islands of Barbados, Tobago and Trinidad.

A proper understanding of these opposed relations will become possible only after a thorough study has been made of all available data bearing on the stratigraphy and structural geology, plutonism and volcanism, submarine topography, palaeogeography and geophysics of the West-Indies.

It is worth mentioning, however, that Hess has given the following preliminary explanation of the remarkable features of the negative belt. It will be seen that his interpretation agrees with the above given theory for the island-arcs of the Pacific and the East Indies. In 1939 Hess wrote: "During Mesozoic time, the continent of South America to the south may have afforded an abundant source of sediments. These sediments might be expected to thin and die out northward. Thus at the time of buckling presumably at the end of Middle Eocene, there may have been a thick layer of incompetent material on the crust in the area of the southern part of the present arc, but very little at the north end. When buckling occurred, the incompetent material might be expected to be squeezed up out of the core of the down-buckle, as in Kuenen's experiments, forming the ridge toward the south, but toward the north, where there was little such incompetent material, the axis of the down-buckle is actually indicated by a deep narrow trough."

The following section will be mainly devoted to the East Indian arcs and the isostatic anomalies will be amply taken into consideration together with other available sources of data.

Synthesis

General statement

Various sources of data have been discussed in the previous sections. Finally the structural history of the island-arcs requires to be examined. All such information will be considered in the present section and at the same time all other appropriate sources of data will come up for discussion, for the different phenomena discussed in previous sections are intimately correlated. Up to the present the highly interesting region of the East Indies has been more thoroughly examined in the fields of geology and geophysics than any other double island arc. Therefore it will be taken as a basis for an attempt at synthesis, and pages 178–200 will be practically entirely devoted to the East Indies.

At the outset I shall recall the following geophysical theory of Vening Meinesz. During epochs of increasing compression the earth's crust reacts by the development of large waves of two to four hundred kilometers in length [2]). Increasing compression will cause an increasing amplitude of the crustal waves till in one of them the strength of the crust was surpassed.

[1]) Hess, 1939, p. 46.

[2]) Cf. Vening Meinesz. Grav. Exp. II, p. 119 (one or two hundred km). According to a more recent paper by the same author (1940, p. 284) the amount is approximately 300—400 km. The wave-phenomenon also happened to occur in some experiments carried out by Kuenen in order to imitate and hence to elucidate the process of a down-buckling crust.

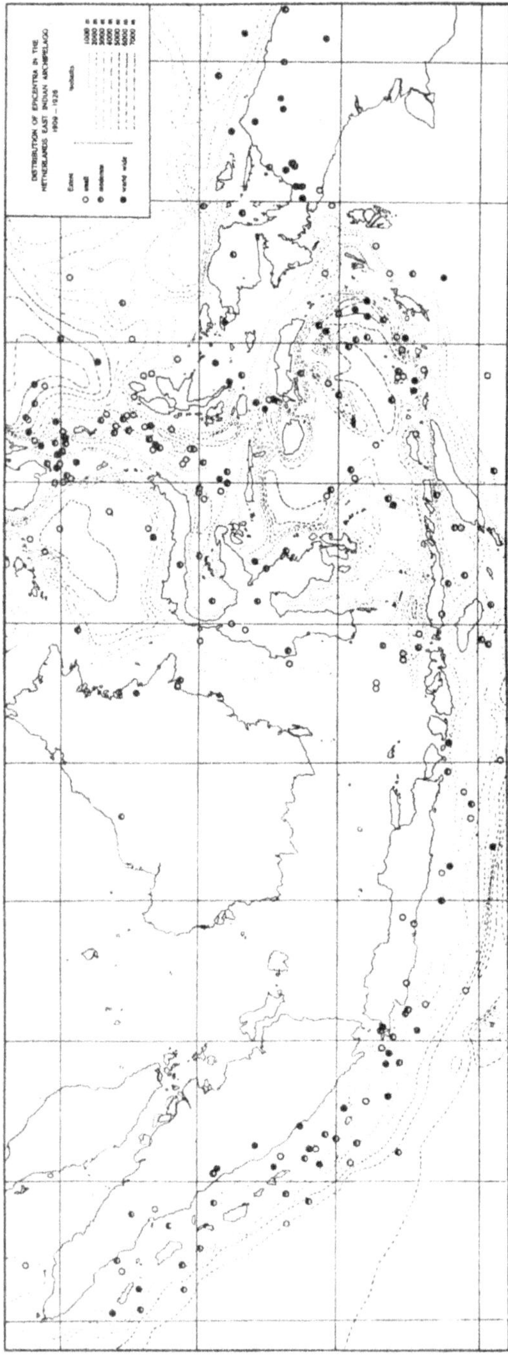
Fig. 117. Distribution of tectonic earthquake foci in the East Indies (From S. W. Visser).

The crust then broke at the weakest place of a downward wave and there buckled inwards so as to form a sialic root penetrating the simatic rocks beneath.

The relation between seismic and gravimetric phenomena will have to be considered in the first place. Then in consecutive order the corelatedness will be demonstrated of other characteristic (and problematic) features of the outer arc, the accompanying deep-sea furrows, the volcanic inner arc and the basins on its concave side. Displacements in the substratum as revealed by fields of positive anomalies of isostasy form the next subject. After that the question of evolutionary stages will be discussed. And, finally, some unsolved problems have to be outlined.

Site of the outer arc

The relation of the belt of negative anomalies of isostasy to the zone of normal earthquake-foci appears to be rather unmistakable in the western part of the East Indies, which is less complicated than the eastern part. It would seem that most of the strong earthquakes belonging to the normal class originate in the zone of negative anomalies and the accompanying deep-sea troughs (fig. 117). Other strong foci are known to be associated with some wellestablished fault-zones on the islands of the inner arc, for example in Java and Sumatra.

The most recent review of deep-focus earthquakes has been compiled by Gutenberg and Richter (cf. fig. 118). Apparently the distribution of the foci under Java and the Java Sea is not unfavourable to the assumption of their occurrence along a deep reaching potential zone of shear. This zone would seem to intersect the earth's surface in the area of the zone of negative anomalies and its adjoining troughs, which at the same time is a belt of seismic unrest at normal depths. At an earlier stage of our knowledge Berlage constructed a map showing seismic isobaths. On his construction the seismic surface dipping under Java and Borneo makes an angle of some 50 to 55° at the surface. A map differing in some details from Berlage's was published by Visser in 1943, but the angle of the seismic surface to the South of Java is also nearly 50°. Koning found about 55° for the area of the Banda Sea.

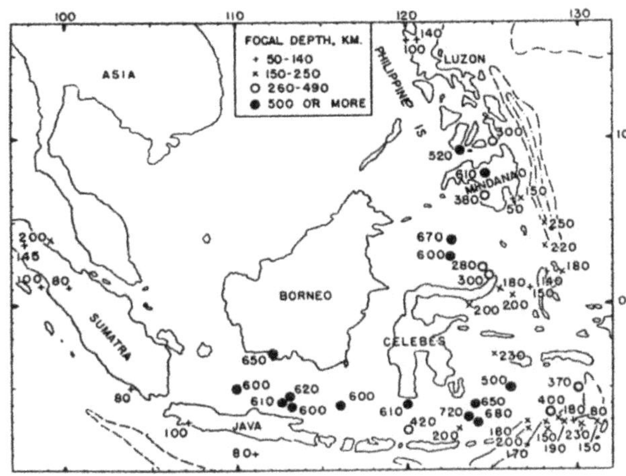

Fig. 118. Deep-focus earthquakes in the East Indies (From Gutenberg and Richter).

Apparently the crustal root — and hence the site of the islands arcs — was predestined to originate in a downward wave of the earth's crust which coincides with the outcrop of the deep-reaching potential zone of shear. Some theoretical suggestions on the origin of the shear-zone were discussed on p. 162. At any rate this zone and the zone of buckling appear to be correlated phenomena. We shall return to this remarkable relation on p. 185 and 204.

It will now be seen how, as a further consequence, the double island-arcs came into being.

Time of origin and epochs of rejuvenation.

The downward buckling of the sialic root had some far reaching geological consequences. For the islands situated in the belt of negative anomalies may be expected to reveal intensively folded and crumpled strata. Neighboring areas might show folding pertaining to the same time. A geological determination of the age of the epoch of compression would reveal the time of formation of the sialic root.

These considerations induced the present author to take up a critical examination of the available data in order to fix the stratigraphic sequences, including the epochs of compression, for the whole region of the East Indian Archipelago. At the outset of this coöperation with Vening Meinesz, we agreed that probably the effect had come into being in a recent geological past. Therefore the geological evidence was not sought further back in the past then Tertiary times. It is useful to mention these details for a thorough understanding of the obtained results. And for the same reason a few more historical notes may be of some interest. The results of Vening

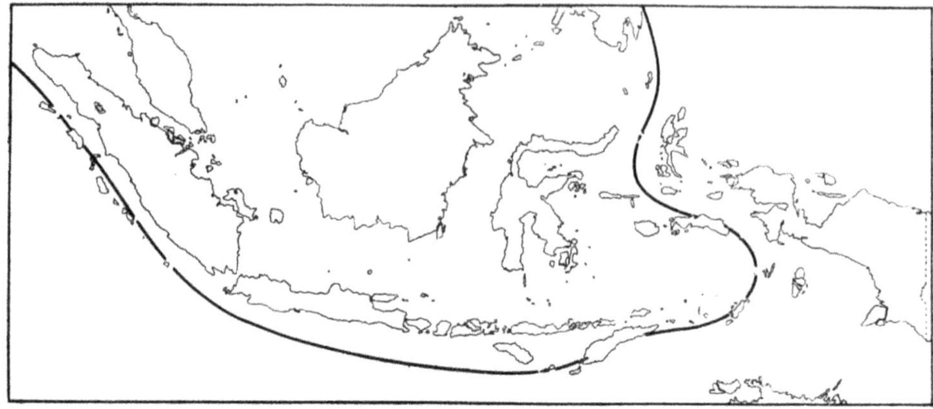

Fig. 119. Preliminary representation of the belt of strongly negative anomalies in the East Indies (After Vening Meinesz).

Meinesz provisional considerations were published in 1930, 1931 and 1932 [1]).

They show the presumed axis of the belt of negative anomalies as a

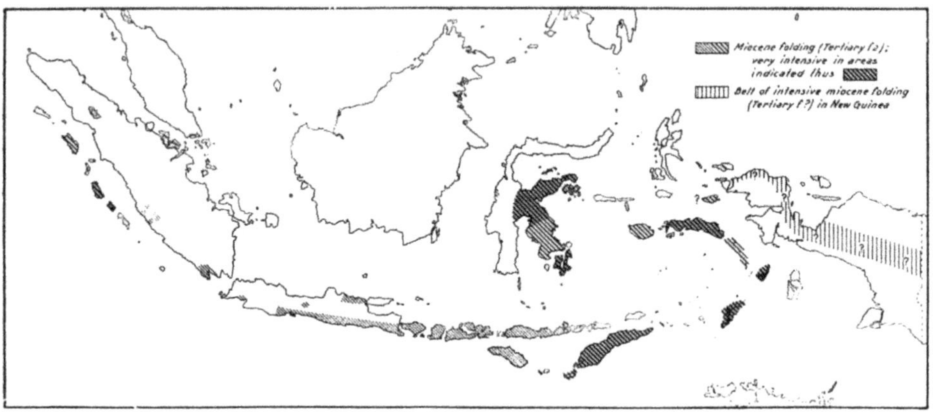

Fig. 120. Areas of Miocene folding in the East Indies.

double-bended curve (fig. 119). The results of the geological examination appeared in 1932 with some additions in 1933 and 1934 [2]). Part of it which is of direct importance here is reproduced in fig. 120. It shows the expected

[1]) F. A. Vening Meinesz. Maritime Gravity Survey in the Netherlands East-Indies. Tentative interpretation of the provisional results (Proc. K. Akad. Wet. Amsterdam, 33 1930; also in the Geogr. Journ. 77, 1931). Relevé gravimétrique maritime de l'Archipel Indien (Public. Comm. Géodésique Néerlandaise, Delft 1931) Gravity Expeditions at Sea 1923—1930, Vol. I, Delft, 1932).

[2]) J. H. F. Umbgrove. Het Neogeen in den Indischen Archipel (Tijdschr. Koninkl. Nederl. Aardrijks. Genootschap 49, 1932). Verschillende typen van Tertiaire geosynclinalen in den Indischen Archipel (Leidsche Geologische Mededeelingen 6, 1933) — Tijd en type der tertiaire plooiingszones in den Indischen Archipel (Tijdschr. Koninkl. Nederl. Aardrijksk. Genootschap 51, 1934).

congruence was not found between eastern Celebes and the axis of negative anomalies, for the latter leaves Celebes untouched and turns right on towards the Philippines, coming from Ceram and Buru. In the mean time the isostatic reduction of the gravimetric stations were made at the Bureau of the United States Coast and Geodetic Survey. And in such a complicated area like the Moluccas some surprising results — deviating from the provisional estimates — might be expected. Indeed, when the final results became available the congruence between geology and gravity-field appeared to be established in a convincing way, not the least so for eastern Celebes. The gist of the final report, published in 1934 [1]) may be seen in Plate 8, which should be consulted constantly in the course of the ensuing pages. It shows the results of the gravimetric survey combined with a synopsis of the structural history of the region [2]).

Moreover, fig. 121 summarizes the stratigraphic results and the different epochs of compression for the different zones. The figures I–III are the same as those appearing on Plate 8. Repeated reference will be made to them, for the following pages may be regarded as an elucidation of Plate 8.

Let us stop to consider first a few remarkable coincidences.

It appeared that the most recent epoch of compression in the islands belonging to the zone of negative anomalies occurred in a certain stage of the Upper Miocene called Tertiary f_2 in the stratigraphic taxonomy of the East Indies: notation I on Plate 8. In addition intense folding and overthrusting had been demonstrated from many of these islands.

On the other hand the same epoch of folding was demonstrated from the islands of the inner arc, e.g. Sumatra and Java, or more correctly: mainly from their western and southern parts respectively. Sumba too bears the same notation II on the map. But the folding was moderate in these regions. No crumpling and overthrusting took place comparable to that of zone I.

Among some other remarkable features the idio-geosynclinal basins should be mentioned (notation III). The moderate folding of their sedimentary contents took place at the end of the Tertiary. Again there is a striking congruence with the expectations of the geophysical theory. For their situation on Sumatra and Java clearly shows that they happen to occur in the downward wave that originated parallel to the wave which buckled downward. Some of the idio-geosynclines began to subside at the time of the Miocene epoch of compression, others already originated in the Early Eocene but these too reveal influence of the Miocene epoch of compression.

In short, it would seem that the crustal root had been formed in the Upper

[1]) F. A. Vening Meinesz, J. H. F. Umbgrove and Ph. H. Kuenen. Gravity Exped. at Sea 1923—1932, vol. II (Public. Netherl. Geodet. Comm., Delft, 1934).

[2]) A new map of anomalies (for regional isostatic reduction 1 = 60 km, t = 174,30 km) was published by Vening Meinesz in 1940 (Proc. K. Akad. Wetensch. Amsterdam 43, 1940). A nearly similar map may be seen in our fig. 116 on p. 176 (reproduced from Rutten, op. cit. 1900). It shows some slight differences if compared to Vening Meinesz' map of 1934, especially in the vicinity of Ceram and Buru. In Meinesz' map of 1940, however, a narrow negative strip connects the negative zone south of Java with the Timor zone, via Sumba. In my opinion such a connection seems improbable as will be pointed out on p. 190. A slightly different map of anomalies, published by Van Bemmelen (op. cit. 1938), shows also an interruption of the negative belt in the vicinity of Buru. The same holds good for the interpretation of gravity data reproduced by B. G. Escher in "Atlas van Tropisch Nederland" (1938).

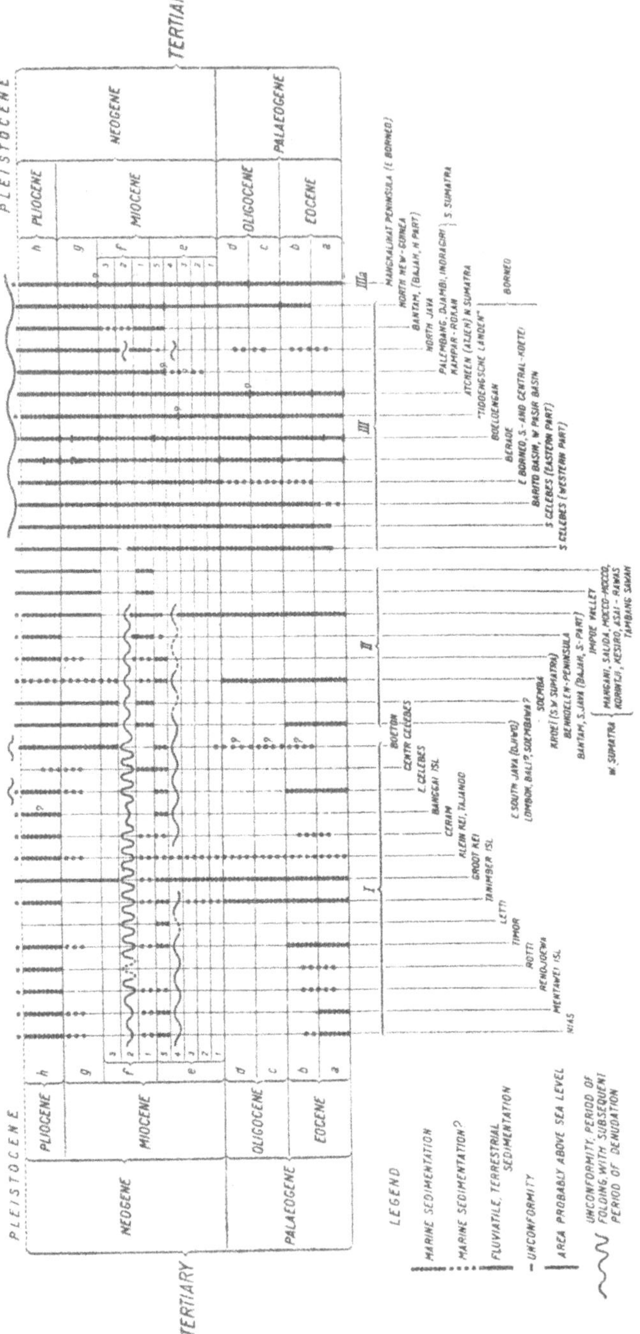

Fig. 121. Review of Tertiary stratigraphy in the East Indies.

Miocene. And hence the zone of negative anomalies would apparently date from the same time [1]).

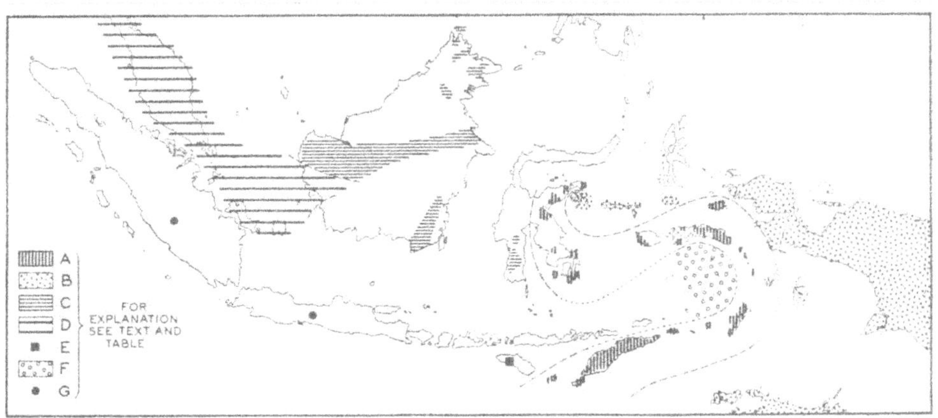

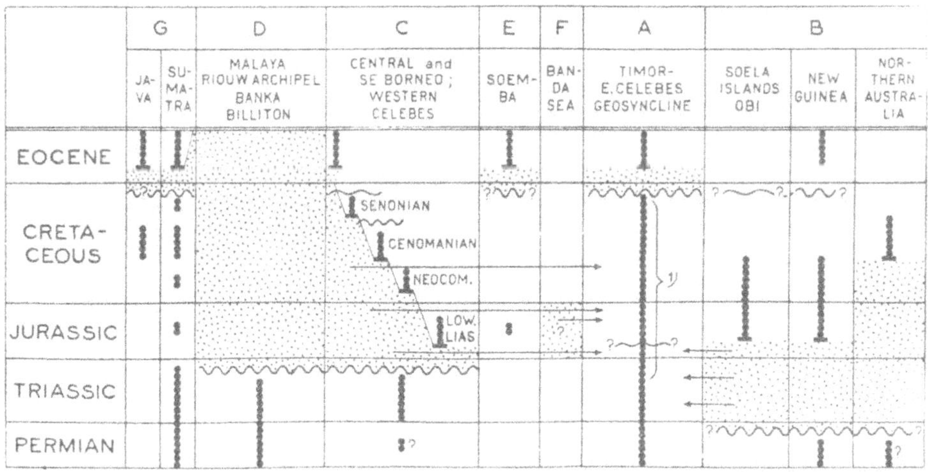

Fig. 122. Schematic representation of pre-Tertiary history of the East Indies.

A glance at fig. 121 and 122 will show the occurrence of earlier epochs of compression, one in the Lower Miocene, another in the Early Eocene, generally known as the Laramide epoch, etc. However, at the time when the Report was published nobody would have ventured to assume an origin

[1]) For a different opinion on the principal epoch of folding on Timor, as expressed by Brouwer c.s., see the Appendix, p. 337.

of the zone of buckling at an epoch more remote than the Miocene. The belief in a comparatively early re-establishment of the isostatic equilibrium seemed to oppose any assumption of a still greater antiquity for the phenomenon. To be sure an Upper Miocene origin seemed already a greater age than was expected at the outset. For it was generally admitted that, geologically speaking, the crust reacts swiftly to loading or unloading. And it is well known that the isostatic equilibrium is being comparatively quickly re-established in such areas as Scandinavia. In response to the melting of its ice-load that area regained half of its isostatic and elastic recoil during the last 20,000 years. On closer inspection, however, it appears that the experience gained from studying of the process of ice-loading is of no value for conclusions concerning isostatic anomalies that were produced by the formation of a mountain-root. For a quite different situation arises when a crustal down-buckle is formed. Not only does a sialic root protrude beneath the lower surface of the normal crust, but also a bulge of the lighter crust-material is forced down into the heavier layers of the crust. And even if the root disappears by melting and spreading the sialic bulge in the crust still remains and can be indirectly observed by the negative anomalies, as Bucher and Kuenen pointed out [1]).

Indeed, a year after the publication of the final Report I had taken up a similar critical study of the pre-Tertiary history of the East Indies [2]). And now the results, unexpectedly, opened a new aspect regarding the time of formation of the zone of buckling. A summary of results has been compiled in the accompanying table and map of fig. 122. Attention should be given here only to one interesting feature, viz. the zone indicated by letter A, the Banda geosyncline. It will be noticed at once that it coincides with part of the present belt of negative anomalies of isostasy. It now transpires that the structural history of zone A was fundamentally different from its immediate surroundings during the Mesozoic. It constituted a complicated geosyncline with an intricate structural history. During the Triassic waste-products were supplied to it from the present area of the Sulu islands, New Guinea and Australia. A second large area of denudation may be noticed in the western part of the archipelago. We will, however, not go into details here inasmuch as the only point of interest regarding the problem under consideration is the discovery of the Mesozoic Banda geosyncline, the site of which corresponds to the belt of negative anomalies of gravity of our own days!

Hence our opinion with regards to the time of origin of the zone of buckling has to be revised. It appears that, at several epochs the crust was buckled downwards along the same zone of weakness. The first down-buckling of the crust in this region may have happened in a very remote past.

Hess pointed out that serpentinized peridotites are always present in the most intensely deformed part of the strongly negative strip. They represent

[1]) Kuenen, 1936, Experim. pp. 201–202.

[2]) J. H. F. Umbgrove. De Pretertaire Historie van den Indischen Archipel (Leidsche Geologische Mededeelingen 7, 1935). An English summary was given in the 1938 paper cited in the Bibliography at the end of the chapter.

an ultramafic magma which was intruded during the first great crustal down buckling. Probably a peridotitic substratum is present below the basaltic layer. Assuming this, a very deeply penetrating root would reach the peridotitic magma, while at the same time, the latter would get an opportunity to invade the sialic root (cf. Ch. IV, p. 82). Hence a determination of the age of the serpentine-belt would fix the age of the first and greatest down buckling movement. In the West Indies this happened at the end of the Mid-Eocene [1]). In the East Indies, however, the serpentines and other ultra-basic rocks which were found on many islands of the outer arc, date from Triassic times in some places, whereas they appear to be more recent in other islands (e.g. Ceram). Hence, the first down-buckle may have occurred in a very remote past [2]).

Following upon the first buckling movement a new epoch of increasing compression caused a rejuvenation i.e. a further down-buckling of the sialic root [3]). And this process may have occurred repeatedly, at least three or four times since the beginning of the Tertiary.

As a further inevitable consequence we are led to the conclusion that the deep-reaching potential shear-zones also originated in a very remote past [4]) and were rejuvenated time and again during the later history of the belt.

Origin of the double arc

A further inspection of the structural history of the area will elucidate many other interesting features of the intricate pattern of the East Indian Archipelago. At the same time it will furnish some striking confirmations of the geophysical theory mentioned at the outset.

The schematic profile A of fig. 123 shows the crustal buckle of zone I on plate 8. To the left two parallel waves are drawn, one upward and one downward. Let us suppose that they respectively represent zones II and III of Plate 8 and, to begin with, let us confine our comparison between theory and facts to the western part of the archipelago. It may, moreover, be assumed that profile A represents a Tertiary epoch of compression and rejuvenation.

Sedimentation starts in the downward moving troughs III. Plutonic and volcanic activity begins in the upward moving geanticlinal belt II, as already sketched in a previous section (p. 167, 174) and as will be considered later in some detail (p. 191).

As soon as the compression in the crust decreases the process of restoration of isostatic equilibrium begins and thus the sialic root must have a strong tendency to rise. Consequently simatic material will flow towards

[1]) Hess, op. cit. 1938, p. 85.

[2]) Umbgrove, op. cit. 1935, p. 148; 1938, p. 68.

[3]) The rejuvenation of the physiographic aspect of continental mountain-chains is probably due to a similar effect (cf. Ch. II, p. 33).

[4]) In their summary of 1937, Gutenberg and Richter concluded (p. 298): "The true deep-focus shocks appear to be associated with events that took place very early in the history of the earth.... it is natural to suggest that there has been motion over a long period of geological time, by which the uppermost layers surrounding the Pacific basin have been displaced towards its center relative to the lower layers, and that no new zones of faulting or weakness have developed at great depth".

the belt of buckling and, as a further consequence, a furrow will form on either side of the folded zone I [1]). This stage is represented by profile B of fig. 123 (compare also fig. 15 and fig. 49, block III).

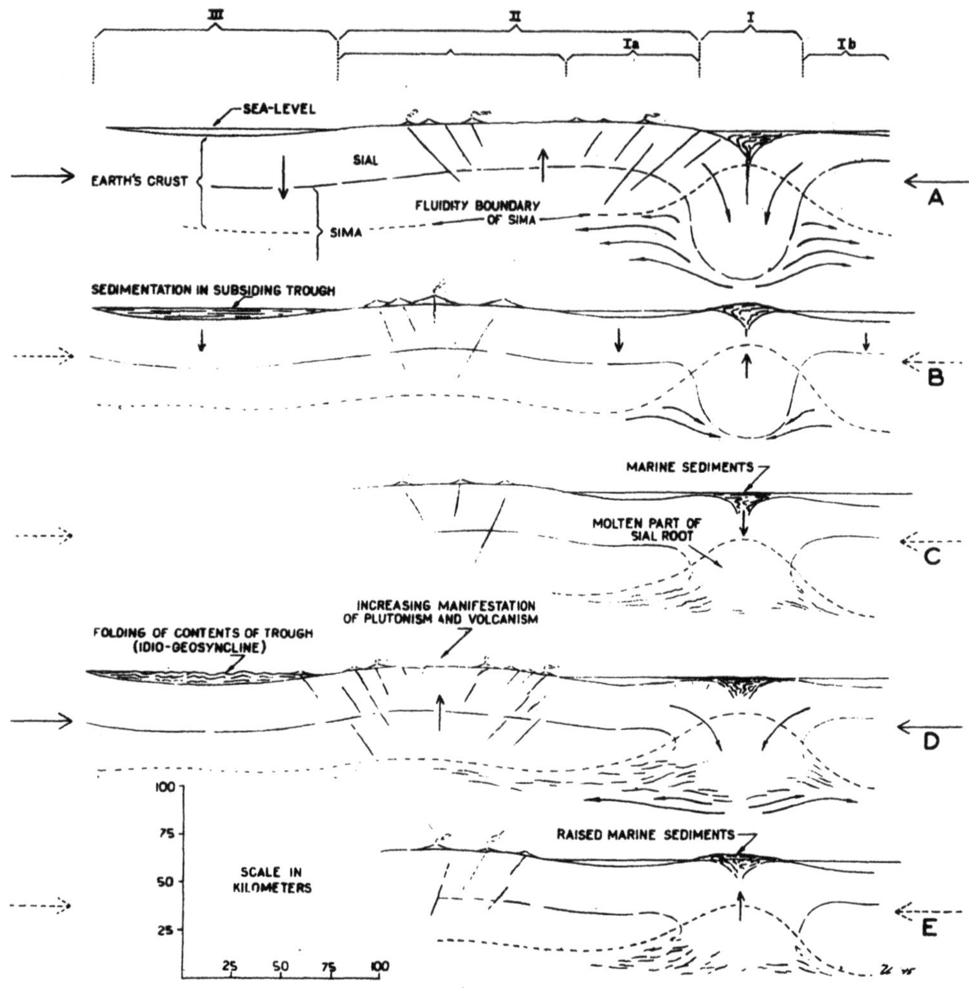

Fig. 123. Five schematic sections showing tentatively the development of a double island-arc. The zones I-III correspond with those of plate 8 and fig 121.

One of these furrows will come into being between the volcanic belt and the root (Ia), the other along its convex front (Ib). The first corresponds with the deep-sea basins between Sumatra and the row of islands to the

[1]) These furrows were attributed by Vening Meinesz (1940, p. 290) to a wave-formation of the crust due to increasing compression. The vertical movements in the Pleistocene and the belts of positive and negative anomalies were interpreted by him in the same way.

west of it. The second corresponds with the series of marginal deep-sea troughs running along their oceanic side. Hence the marginal deep of a double arc is genetically different from the marginal deep of a single arc! This question will be discussed at greater length on p. 188.

Whether the intervening belt of folded strata (zone I) appears as a submarine ridge or as an island-festoon, depends on the quantity of strata that was squeezed out. The second possibility is represented by profile B of fig. 123. In this case a double island-festoon has come into being, an inner volcanic arc and, parallel to it, an outer arc which at this stage is non-volcanic (cf. p. 192).

In all probability the downward folded root did not remain intact. It is reasonable to assume that the sialic matter of the root began to melt and spread. As a result the central part of zone I began to subside, forming a shallow depression, fig. 123, profile C. An actual depression of this kind is known as the central "graben" or "geosyncline" on Timor, containing a sequence of a few hundred meters of Pliocene sediments. Similar features are known from the Kei islands (fig. 124), Tanimbar islands (fig. 124), and Ceram. The ensuing epoch of compression at the end of the Pliocene, represented by profile D of Fig. 123, caused a moderate folding of the contents of zone III. The shallow depression on zone I was only slightly influenced at this stage, certain tectonic features at the margin of the shallow "geosyncline" being the only phenomena that have been noticed. The plutonic and volcanic processes of zone II became more active.

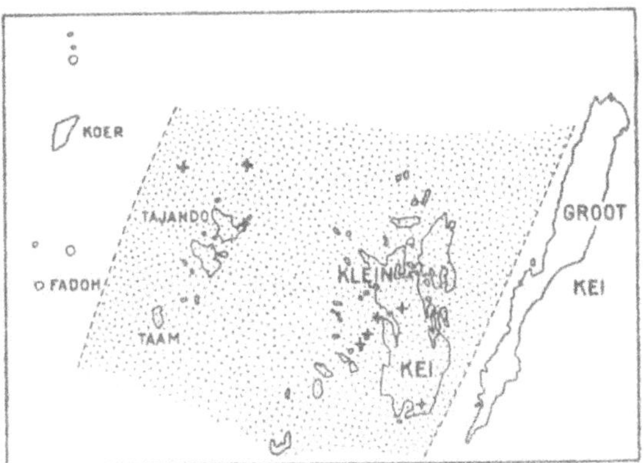

Fig. 124. Upper Tertiary depressions on the Kei- and Tanimber Islands. (After Fr. Weber).

The last stage, represented by profile E of fig. 123, shows the most recent period of decreasing compression. Again zone I rises isostatically. The contents of the central depression are now raised above sea-level, in places several hundreds of meters, and locally even more than a thousand [1]). Zone III also rises isostatically but to a slighter degree, and the plutonism and volcanism of the geanticlinal belt II decreases to its present though still active phase.

Again the root melts and spreads laterally in the substratum. A comparison of the gravimetric and bathymetric maps shows that the sialic root has spread, e.g., below the site of the Weber deep in the vicinity of the Kei islands.

According to the considerations just mentioned the topographic and bathymetric features of the East Indies should be of comparatively recent origin. They should all have come into being since at least the last epoch of strong compression in the Miocene. And this is what had been concluded from geological and morphological data [2]). It would appear from theory, moreover, that the deep-sea furrows Ia and Ib are of a still more recent origin than the depressions of zone III. In the same way the most recent elevation of zone I began later than the last rising movement of zone II. And the upward movement of zone I is of a fundamentally different nature from that of the geanticlinal updoming of zone II. Fig. 125shows the height of raised coral limestones and fluviatile terraces in the southern Moluccas. No data are available to fix their age exactly, nor would it be possible at the moment to discern such comparatively slight differences of age as are postulated by the theory. Mostly they are all designated as "Pleistocene" or "Plio-Pleistocene".

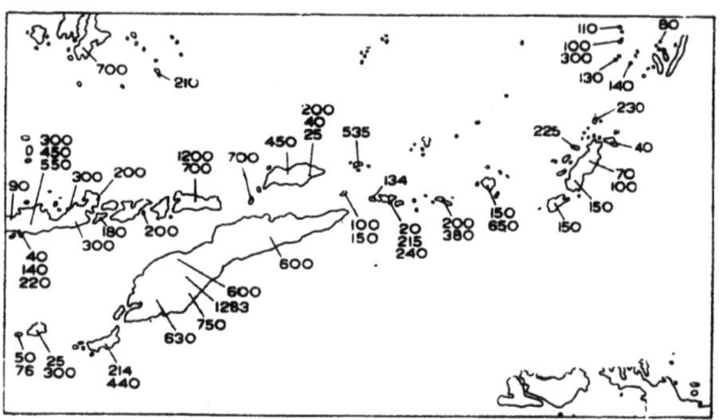

Fig. 125. Elevated reef limestones and terraces in the southern Moluccas. Height in meters.

The different structural elements will now be considered in more detail, proceeding from the external side of the arc to its concave inner side.

Deep-sea furrows

It has already been mentioned that the marginal deep of a double arc is genetically different from the marginal deep of a single arc. This statement is illustrated by fig. 126. Profiles 1 and 2 show the origin of a marginal deep of a single arc of the Mariana type, 1 over an asymmetrical sialroot, 2 over a symmetrical root.

---

[1]) Upper Tertiary sediments of the Masi-Bobot graben of Ceram are now at least 3000 meters above sea-level. Pleistocene sediments of Central Timor as much as 1280 meters.

[2]) See Umbgrove, op. cit. 1938, pp. 55–58.

Profile 3 is a diagrammatic representation of an arc of the East Indian type [1]) and corresponds with profile A of fig. 123. The axis of the marginal deep — indicated by the arrow — has shifted towards the right, i.e. the convex side of the arc, if compared with the marginal deep in profiles 1 and 2. Subsequent movements due to isostatic readjustments are much stronger in an arc of the East Indian type than in a single arc of the Mariana type. The processes shown by profiles *B–E* of fig. 123 cause a further oceanward displacement of the axis of the marginal deep, as illustrated by profile 3a of fig. 126.

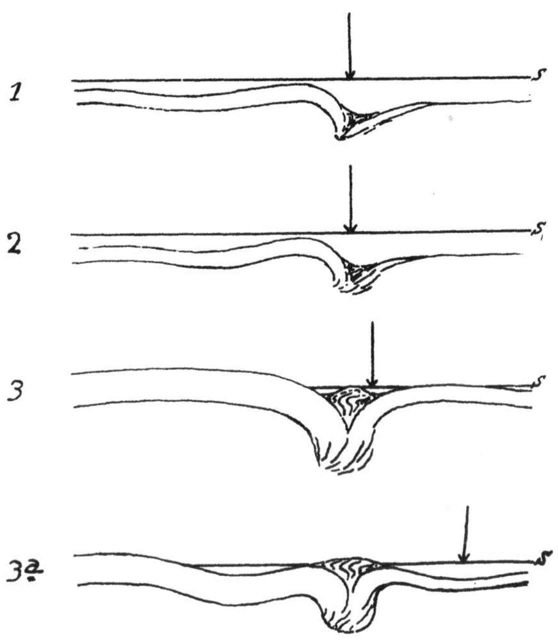

Fig. 126. Schematic representation of different types of marginal deeps. For explanation see text.

In order to check these theoretical deductions one must examine the pattern of a pseudo-single arc. The western and central parts of the Aleutians consist of a single row of islands but the eastern part of the festoon — near Alaska — is double. Now the bathymetric chart of the North Pacific Ocean clearly shows that the distance between the volcanic inner arc and the marginal deep increases [2]). This part of the arc shows the transition between the two types of marginal deeps. A similar phenomenon can be noticed with the Kurile and Japan trenches.

One of the consequences of the foregoing theoretical deductions concerns what might be called the problem of the island Sumba.

Obviously the island does not belong to the volcanic inner arc. Neither in stratigraphy nor in structural history is it intimately related to the islands of the outer arc, such as Timor and Roti.

Now compare the submarine relief with the gravimetric results (Plate 8). A submarine ridge partly emerging above sea-level in the shape of islands like Mentawei and Nias can be followed from off the coast of Sumatra to the neighborhood of Sumba. The axis of the belt of negative anomalies coincides with the ridge (fig. 115), which on either side is accompanied by a deep-sea trough. Along the external side is the marginal deep, while another series of troughs lies along the inner side, among which is the Java deep (fig.39).

[1]) Cf. Vening Meinesz. Gravity Expeditions at Sea II, 1934, p. 119 fig. 19.

[2]) Bathymetric Chart of the North Pacific Ocean compiled at the U.S. Hydrographic Office, preliminary edition, 1939.

The belt of negative anomalies terminates southwest of Sumba, only to reappear once more southeast of Sumba, continuing northeastward over the island Timor. And there a ridge is again accompanied on either side by deep troughs, viz. the Timor trough and the Sawu trough (fig. 39). So, the site of the island Sumba as well as the submarine topography in its vicinity seems to be intimately related to the interruption of the zone of negative anomalies [1]). Of course the question might be put in two different ways. One would be to ask whether the presence of the island Sumba caused the zone of buckling to become interrupted to the south of the island, and a further point would be to examine the cause-and-effect relation. The other attitude regarding the situation of Sumba might be formulated as follows. If the zone of buckling were continuous so as to unite the Timor ridge to the submarine ridge which is known to terminate S.W. of Sumba, the island Sumba would not exist, its site being occupied by a deep-sea trough!

According to our opinion Sumba stands as the exceptional evidence of a sort of terrain that elsewhere subsided so as to form the bottom of one of the series of deep-sea furrows between the outer and inner arcs!

The subcrustal migration of sialic material on either side of the root ought to cause a rising movement of the bottom of both the troughs Ia and Ib. The depth of the fluidity boundary of the sima (the Mohorovičić discontinuity) increases towards the regions of Ib, due to the thinning out of the sialic crust-layer towards the Indian Ocean. This situation causes the asymmetric spreading of subcrustal sial as shown by the profiles C, D and E of fig. 123. Therefore the rising movement of the bottom of trough Ia ought to surpass that of Ib. Now a glance at the bathymetric map shows the Ia trough between Sumatra and the Mentawei Islands to be much shallower than the marginal deep off the Mentawei Islands. The same holds good for the Ia and Ib troughs south of Java. At first sight these data seem in remarkable agreement with the theoretical deductions. However, between Sumba and New Guinea the bathymetric map shows exactly the reverse, the marginal deeps being shallower than the Ia troughs, although a symmetrical spreading might be expected due to the presence of the Australian continent. In this part of the East Indies the marginal deep is bounded by the large shelf between Australia and New Guinea, which was an extensive area of denudation during most of Tertiary and Pleistocene times. In the western part of the East Indies an extensive area of denudation existed along the opposite side of the arc, bordering the Ia troughs! It appears therefore that the bathymetric effect due to the sideways spreading of sub-

[1]) In a new map of regional isostatic anomalies, published by Vening Meinesz in his paper of the year 1940, a different view is expressed. He thinks that an interruption of the negative belt "would be difficult to understand because the question rises of the interruption without the block of Sumba taking part in the shortening". Therefore his map shows a narrow connection of some 0–50 milligals between the negative zone south of Java and across the island Timor, via the island Sumba. Gravity data are too scanty in this area to enable a final decision. The vicinity of Sumba undoubtedly deserves a narrower net of gravity-stations. According to my opinion data on structural history of the island are not in favour of Vening Meinesz' assumption (cf. note 2 on p. 181). Future investigations about the distribution of isostatic anomalies on Sumba and the neighboring seas, will prove which opinion is the right one. But we will present another argument in favour of our opinion presently when dealing with some volcanological aspects.

crustal sial is a factor of minor importance if compared with the effect due to sedimentation. Indeed, the amount of the movement may expected to be much smaller than e.g. the rising movement of zone I if the spreading of the sialic material of the root takes place in the form of thin films and filaments as suggested by profile C (fig. 123).

The volcanic inner arc

We started our disquisitions by assuming the formation of crustal waves some one for four hundred kilometers from crest to crest. The upward wave accompanying the zone of buckling on its continental side is more strongly developed than the corresponding wave on its convex side. It is moreover always a volcanic arc. The following explanation may be proposed for this state of affairs.

Subcrustal sima was pushed aside by the downward penetrating root (fig. 123 profile A) The curved shape of the arc involved a centripetal crowding of the sima on the concave side of the arc, whereas it could expand more freely on the convex side of the arc (fig. 127). This caused the arching of the geanticlinal belt to be sustained and augmented by the accumulating sima. Hence it became an ever more pronounced belt of active plutonism and volcanism, according to the mechanism described on pp. 72, 167, and 174. No such process happened to occur on the convex side of the buckled belt. And this may explain why a volcanic arc always developes at the concave side of the non-volcanic arc [1]).

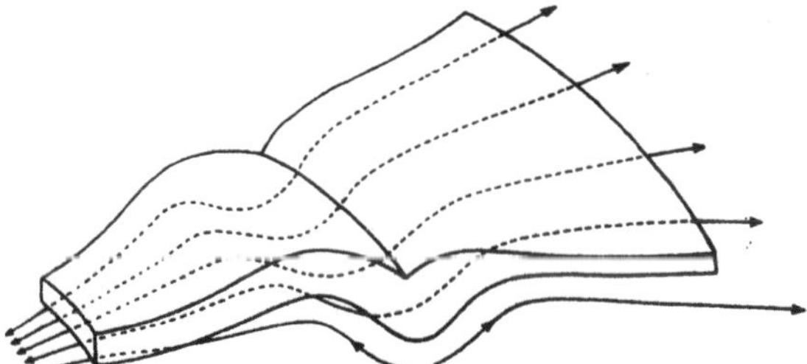

Fig. 127. Tentative interpretation of the subcrustal crowding of sima at the concave side of an island-arc and the origin of plutonism and volcanism in the inner arc.

According to our theoretical deductions plutonism and volcanism in the geanticlinal zone II would start during epochs of increasing compression. For arching of the zone induced relief of pressure at the underside of the crust and subsequent rising of a migmatite front as described on p. 72 and 174. Now the Miocene epoch of compression was followed by another one at the end of the Pliocene and accordingly two periods of increased volcanic activity might be expected in zone II, since the beginning of the Miocene. And in fact these have been found by field observations! (p. 167)

Quite different phenomena may be expected from the zone of buckling. For during an epoch of compression no geanticlinal arching or relief of pressure occurred. On the contrary a root of mobile sialic matter penetrated

[1]) Theoretically volcanism might originate at the convex side as well, though at a far less pronounced degree. Christmas Island was mentioned by Hess (1938, p. 73) as an example of an emergent outer geanticline. The submarine topography in that part of the Indian Ocean is, however, not examined in sufficient detail to substantiate his opinion. Undoubtedly two other examples cited by Hess are not to the point; viz. Aru Islands and part of New Guinea.

downwards and the belt was strongly compressed. No manifestation of volcanism is to be expected under such circumstances. During the ensuing period of decreasing crustal compression, the melting and expanding sialic root might give rise to ascending batholiths (cf. p. 79). Obviously, however, the mechanism of their formation was fundamentally different from that operating in the geanticlinal belt II. Seldom do the batholiths penetrate to a high level, so as to become revealed at the surface now exposed by the erosion of the arc. (The Adamello massif in the Alps may be cited as an example of a "high-level" batholith).

So, active volcanism is conspicuously absent in the outer arc of an island festoon. But volcanic rocks of Upper Tertiary age have been found on Timor, an island of the outer arc. It may be that these were derived from plutonism belonging to the buckled belt. Possibly, however, these Upper Tertiary volcanics originated in a different way. If we regarded once more the schematic illustration fig. 123 it will be noticed that originally the buckled zone was bordered immediately by rising crustal waves. In stage B of fig. 123, however, the geanticlinal belt was appreciably narrowed by the formation of the deep-sea trough Ib. Hence the volcanism of the remaining geanticlinal zone was also restricted. But we might expect to find volcanic rocks even as far as the present belt I. And we might add (1) that these igneous rocks should date from Upper Tertiary times ,(2), that their occurrence should be restricted to the concave part of zone I, (3) that they might possibly be of submarine origin, and (4) that they may be expected to occur on the bottom of trough Ib as well. Now, volcanic rocks and tuffs have indeed been found on the island Timor, and exactly where they might be expected to occur, viz. north of the central basin. Volcanic rocks of Upper Tertiary age have been found also on Sumba. But here they are widely distributed over the whole island. Once more this is in accordance with the view that Sumba has to be considered as a crustal part which, unlike the areas east and west of it, has not subsided so as to form the bottom of a deep-sea basin (fig. 128). The theory is supported by the fact that volcanic rocks on Buru and Ceram are restricted to the southern coastal district whereas they are widely distributed on Nusa Laut, Saparua Haruku, Amboina, and Amblau which are all situated to the South of Ceram and Buru.

Another remarkable feature is the double row of volcanoes in the northern Moluccas. The belt of negative anomalies is very strong between Celebes and Halmaheira. Anomalies of more than -200 milligals were found by Vening Meinesz. In addition the root shows a contour which is concave towards the west as well as to the east. Remarkably enough one row of volcanoes runs from northern Celebes to the Sangi islands: on the opposite side are the volcanoes of Halmaheira and the neighboring islands. The presence of a deep-sea furrow on either side of the zone of buckling would be in accordance with the theory. And they would both simultaneously represent the marginal deep type and the intramontane trough type. Matching the inference, deep-sea furrows are actually found to be present in the expected places. Kuenen marked them as marginal deeps on his map (reproduced in fig. 39). Moreover, we ought to assume two shear-zones in this area, one dipping westwards the other one eastwards, although it may be expected that the latter will be less pronounced. For the moment data on deep-focus earthquakes are

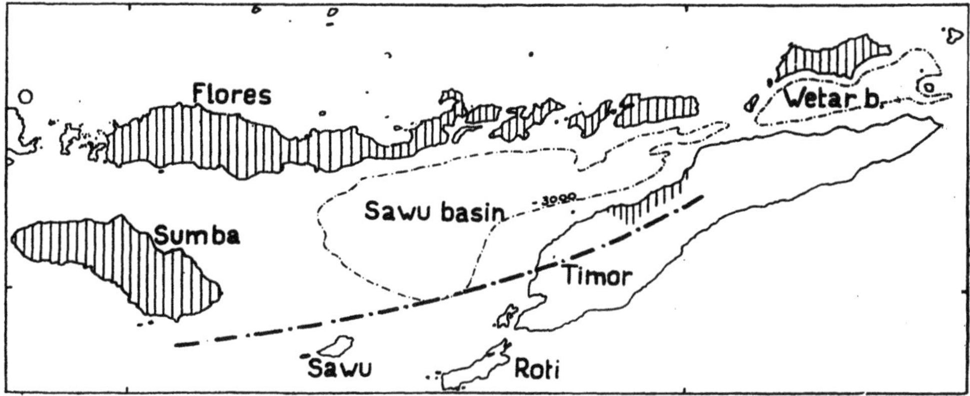

Fig. 128. Distribution of Upper Tertiary volcanic rocks (vertical shading) in the southern Moluccas. Their southern boundary is indicated by a dot-dash line. The Sawu- and Wetar basins are indicated by the isobath of 3,000 meters.

not numerous enough to allow a verification of the suggestion just given [1]). Many more data are desired, geological as well as gravimetric and seismic,

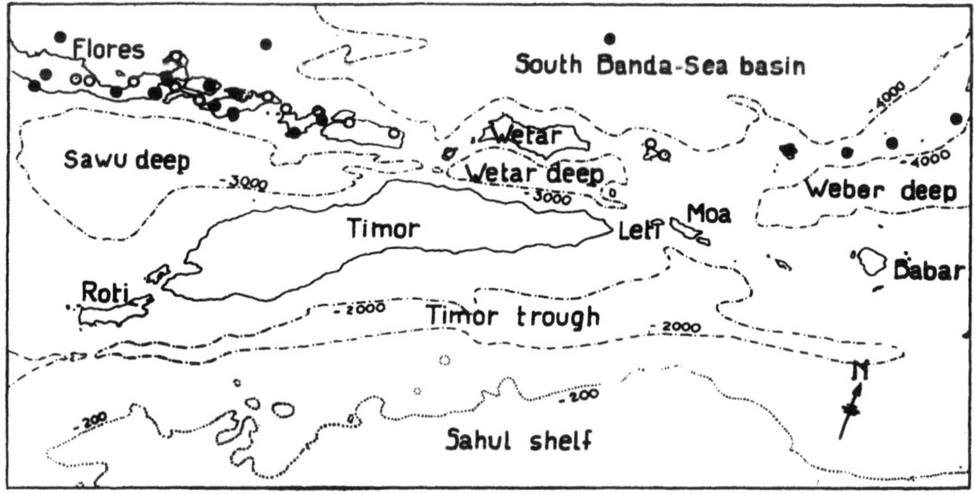

Fig. 129. Active volcanoes (black dots) and extinct volcanoes (circles) in the southern Moluccas (After H. A. Brouwer).

especially from the area of Halmaheira and New Guinea. Not until our knowledge is considerably augmented will geologists be able to build up a

[1]) It is true that G. L. Smit Sibinga published a map with seismic isobaths suggesting a seismic surface dipping in the direction of Australia. (Gerl. Beitr. z. Geoph. 51, 1937). Visser, however, pointed out that Smit Sibinga's map is of no value since it is based on too scanty and unreliable data.

well founded theory on the interesting problem of the double row of volcanoes and its associated phenomena, in place of tentative speculations of the type just given.

Finally, a feature of volcanological interest concerns the distribution of active and extinct volcanoes in the southern Moluccas, as shown by fig. 129. On many occasions Brouwer pointed out that volcanic action is extinct in those islands of the inner arc that are situated nearest to the island Timor, which belongs to the outer arc. No active volcanoes are found on Alor and Wetar. Proceeding from these islands towards the more western or eastern islands of the inner arc, the distance from the outer arc increases, and the number of still active volcanic vents increases as well. Probably this phenomenon is caused by the strong compression which the crust is undergoing here, as indicated by the formation of the marginal Timor trough and more especially by the subsiding Sawu trough [1]). For in this area the subsiding troughs had to become adapted to the limited space between the inner arc and the Australian Continent, as contrasted to their free development more to the west. This phenomenon is clearly revealed by the narrowing deep-sea relief as shown on the bathymetric chart (fig. 128, 129 and fig. 39).

Basins and troughs behind the inner arc

We may now proceed to consider the downward wave on the continental side of the volcanic geanticline or inner arc (zone III of Plate 8 and fig. 123).

Two possibilities should be considered separately in respect of a supply of waste-products from the surroundings.

One possibility is that the quantity of waste-products equals or even surpasses the rate of subsidence of the bottom. The trough will then gradually become filled up and appear as a geosynclinal basin. During an ensuing stage of increasing compression the contents of the furrow will become folded (fig. 33 and 123). It will be seen on Plate 8 and fig. 32 that regions of this type actually occur on Sumatra and Java. They are called idio-geosynclines and constitute oil-basins of high economic value. The strata underwent a moderate folding at the end of the Tertiary.

A comparison of fig. 109 (p. 166) and Plate 8, clearly reveals that most of the igneous rocks of the mediterranean suite are found in the idiogeosynclinal basins. Possibly the ascending magmatic emanations reacted with much lime-material from the marls and limestones that occur in comparatively great abundance in the geosynclinal basins (cf. pp. 74 and 168). On this view the location of the alkali-rock suites, mainly along the concave side of the inner volcanic arc, seems to be controlled by the occurrence of the limestone in the geosynclinal basins, the sites of which are in turn predestined by the formation of crustal waves accompanying the down-buckled belt.

The other possibility is that the rate of subsidence of the bottom surpasses the supply of sediments. In that case a deep-sea basin will originate and persist. The Flores trough might be cited as possible example (IX on fig. 39). In the eastern part of the East Indies the situation is more

[1]) Our explanation is different from Brouwer's in as much as the latter assumes an active influence of the Australian continent, more or less in the Wegenerian sense.

complicated on account of the double curvature of the arcs. However, the deep basins may be explained along the same lines of argument, their peculiar shapes being the result of a process of interference set up by the vortex-shaped arrangements of the buckled zone, as may readily be noticed from a comparison of Plate 8 and fig. 39 [1]).

Positive anomalies of isostasy

A field of positive anomalies — with an average of +20 milligals — covers the whole area of the East Indies outside the belts of negative anomalies. According to Vening Meinesz it is caused by lateral compression of the crust [2]).

Strips of stronger positive anomalies run parallel to the negative zone, south of Java. Vening Meinesz pointed out that the readjustment of equilibrium of the negative zone must have involved a tendency to rise. The adjoining zones would have a share in the same movement and would therefore also rise. The tendency to rise would explain the positive anomalies.

A different suggestion was offered by Vening Meinesz concerning the fields of positive anomalies in the Moluccas. These fields were ascribed to downward convection currents in the substratum.

So the fields of strong positive anomalies were explained along two different lines of speculation. One would feel more content if a single theory covered the whole of the problem, the more so since the distinguishing of the two separate types of positive fields seems rather artificial [3]).

In a more recent paper Vening Meinesz recognized the difficulties inherent to the dualistic character of his first attempt at explanation. He, therefore, tried to interpret the regular succession of positive and negative strips by a mechanical hypothesis [4]). The alternating belts would have been brought about by the formation of crustal waves. The waves would give origin to alternating deviations from the isostatic equilibrium of the crust, the upward wave causing positive anomalies and the downward wave negative anomalies. Continuing his considerations Vening Meinesz thinks that his theory finds support in the topographic features of the East-Indies. For he writes "..... the profiles of the ocean-bottom in many of the gravity profiles west of Sumatra and south of Java show a regular wave-like topography; the amplitude of the wave, i.e. the height of the topography, corresponds to the amplitude of the wave in the gravity anomalies when assuming no isostatic compensation, and this is of course as it ought to be according to our supposition because the wave is entirely a deviation from the isostatic equilibrium. We do not find this same relation to the topography as clearly elsewhere in the archipelago, but this may be explained by the

[1]) The northern Banda Sea Basin e.g. is, as it were, enclosed by the strongly curved zone of eastern and south-eastern Celebes. Instead of a continuous marginal deep three separate basins developed along the convex front of this zone, marked a, b and c on Kuenen's map, reproduced in our fig. 39.

[2]) Vening Meinesz, Grav. Exp. II, p. 134.

[3]) The same objections against Vening Meinesz' hypothesis were raised by Van Bemmelen (1938, pp. 61–67). His attempt at an explanation of the anomalies differs widely, however, from the theory that is presented here by the present author.

[4]) Vening Meinesz, 1940, p. 284.

complicated deformations of the surface layers and by the other surface effects of erosion sedimentation and volcanic activity."

As a matter of fact no such agreement between topography and positive anomalies is found in the eastern part of the archipelago. However, even the most complicated part of the Moluccas shows unmistakable evidence of a parallelism between belts of positive and negative anomalies. Here too their striking interrelation is a fundamental feature which calls for an explanation combining both negative and positive anomalies in a single synthesis.

The most important features may be summed up as follows. Comparing the gravimetric and bathymetric contours, one may notice some coincidences. A deep basin like the Celebes Sea shows positive anomalies of more than 50 milligals. The same holds good for the deep Makassar strait, the Gulf of Bone (between the southern and south-eastern arms of Celebes) and its continuation to the southern Banda Sea. From a further inspection, however, the following striking features may be noticed. (1) The relation to the submarine relief is not more than a rough approximation. (2) Neither the marginal deeps nor the troughs of the intramontane type (represented by 1, 2 and 3 in fig. 39) show marked positive anomalies. (3) In the western less complicated part of the archipelago zones of positive anomalies may be seen to run parallel to the strip of negative anomalies; nevertheless, the positive strips are not over the deep-sea furrow but roughly coincide with elevations of the bottom. It seems obvious that the local coincidence of a ridge and a belt of positive anomalies is purely fortuitous.

Hence the cause of the positive anomalies has to be sought not in crustal but in subcrustal phenomena! We shall therefore try to give another explanation. The following speculation may perhaps elucidate the problematic features just enumerated. The sial-root penetrated downward into the heavier substratum. Consequently the heavy masses of the substratum were pushed aside [1]). The consequences as regards the surface-features have already been discussed in previous pages. If it be assumed, however, that the deep-seated layer of heavy dunite was influenced in much the same way as the subcrustal layer of basaltic material, the result of the displacement of the heavy dunite would be revealed as a belt of positive anomalies on either side and at a comparatively short distance from the negative belt. The great depth of the process would explain why in one place the belt of positive anomalies corresponds to a ridge-shaped elevation of the crust, and elsewhere to a deep depression, as shown by the tentative block-diagram fig. 130.

The general principle of this speculation may account also for the distribution of gravity in the complicated area of the Moluccas. Here a kind of interference between the interwoven negative zones might be expected. And indeed, the highest positive values were found in the vicinity of the Sulu-Islands, i.e. in the immediate vicinity of the greatest negative values. The sialic root is exceptionally broad and deep here. The adjoining

[1]) Vening Meinesz found that a cross-section of 20 × 60 kilometers of the sialic root approximately corresponds to the curve of regional anomalies in the East Indies (Gravit. Exp. at Sea II, p. 121).

positive field is also broader and the positive values higher than anywhere else. And this is what might be expected according to the theory.

It will be remembered that the occurrence of the volcanic belt on the inner side of the non-volcanic arc was explained as an effect due to the

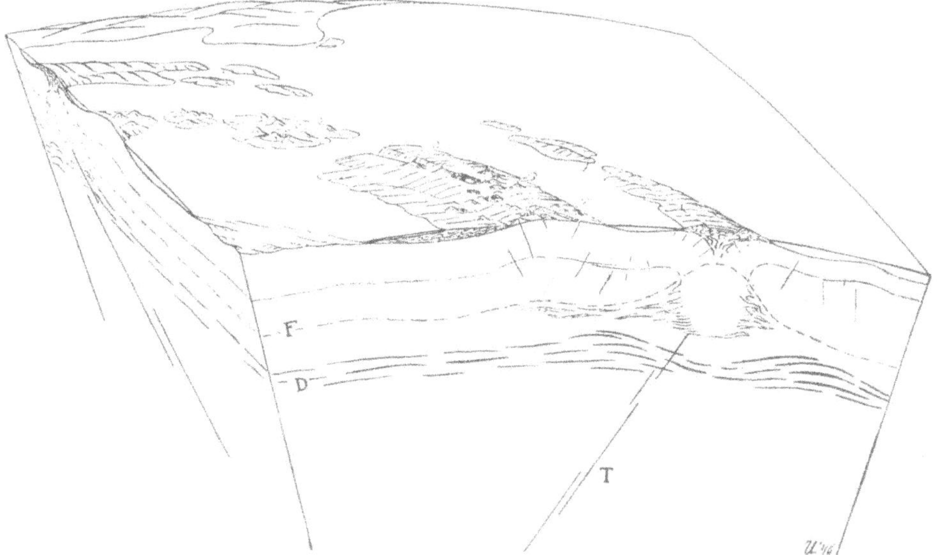

Fig. 130. Schematic block-diagram of a double island-arc of the East Indian type; T, potential zone of shear (deep-focus earthquakes); D, dunite; F, fluidity boundary.

curved shape of the negative zone (p. 191). Probably a similar process of centripetal crowding influenced the subcrustal ultrabasic sima. And therefore the positive anomalies are strong and appear in crowding and partly interfering belts along the concave inner side of the negative zone. But they appear only in much fainter degree at its convex outer side, e.g. west of Sumatra and south of Java.

The inverted arcs of Celebes

Doubtless the whirl-shaped pattern of the Moluccas is due to the presence of the continental block of Australia, including the Arafura sea and New Guinea. However, a mechanical interpretation of the puzzling island-arcs of the Moluccas is not yet possible. Our geological and geophysical knowledge of the adjacent regions is too scanty. This concerns especially New Guinea and Halmaheira on one side, and Borneo on the other side. From a geological point of view the island Celebes belongs to the Moluccas. Its remarkable four-armed morphology is due to a double arc which — unlike all the other arcs along the border of Asia — has its convex side turned towards the Asiatic continent. Obviously, the inverted position of the Celebes-arcs is caused by the sum of the forces interacting in the region between Asia and Australia. Accordingly, for the time being a mechanical interpretation of the inverted position of Celebes will be left out of discussion.

However, a few remarks may be made on some features which are intimately related to the inverted position of the arc and its position in the complicated pattern of the Moluccas.

The theoretical deduction formulated on p. 177 was substantiated by the structural history of the western part of the East Indies and a large part of the Moluccas. It will be seen that the main features of several other island-arcs are in accordance with the theory. At first sight, however, the theoretical scheme seems to break down if applied to Celebes. A belt of negative anomalies of isostasy coincides with the eastern and south-eastern arms of the island. These arms form an arc with its concavity towards the east. Hence — according to the theory — a volcanic "inner" arc ought to be present parallel to it somewhere in the Banda Sea. But instead of a volcanic arc a deep-sea basin has developed east of Celebes and a volcanic zone came into being on the opposite side. For the southern and northern arm of Celebes, as well as the uniting central part are characterized by numerous manifestations of volcanism dating from Upper Tertiary to sub-recent times. The same zone is moreover characterized by the abundance of granodioritic rocks and gneisses, zone 1 of fig. 131 [1]). Similar rocks are conspicuously absent from the eastern arc, where serpentines, lherzolites and other basic or ultrabasic rocks have a wide distribution (fig. 131 zone 2 and 3). A central zone of the island, running north-south, formed a depression with marine sedimentation, during the Upper Tertiary (fig. 131 zone 2). It is now characterized by a graben (the Posso graben) and separated from the western part of Celebes by a zone of mylonites. Finally three remarkable deep-sea basins have to be mentioned. On morphological grounds Kuenen united them in a special class (*a, b, c*, on fig. 39). Two of them, the Tomini and Bone basins, separate the eastern and western zones of Celebes; the other one — the basin of Makassar straits — lies in front of the "volcanic arc" (fig. 131).

In an attempt to explain the unusual arrangement of these morphologic and tectonic elements, attention should be focussed on the exceptional features of the Moluccas as a whole. We pointed out that the formation of the basin-shaped depression of the North Banda Sea was probably caused by interacting processes in the crust, due to the whirl-shaped arrangement of the zones of buckling (p. 195). The accumulation of displaced dunite masses was thought to be responsible for the large fields of positive anomalies in the same region. Probably these processes hampered the development of a volcanic "inner" arc east of the negative zone of Celebes. For the deeply subsiding crust under the Banda Sea basin acted as an antagonistic and dominant factor. And probably this effect was strengthened by the exceptionally strong accumulation of dunite in the same region (cf. pp. 196, 197). So, as a tentative explanation I suggest that these cooperating factors suppressed the formation of an "inner" arc in this area. Hence, the basaltic material of the substratum that was pushed aside by the downward movement of the root under the eastern part of Celebes had to flow towards

[1]) Cf. H. A. Brouwer. The major tectonic features of Celebes (Proc. K. Akad. v. Wet. Amsterdam 33, 1930) and H. A. Brouwer. Tektonik und Magma in der Insel Celebes und der Indionesische Gebirgstypus (Ibidem 44, 1941).

the opposite side, i.e. westward. Thus exceptionally the total volume of sima that was displaced by the root had to flow towards the convex side of the arc, where it set going the genetic processes responsible for the numerous granodioritic intrusions and volcanic manifestations found in the western part of the island.

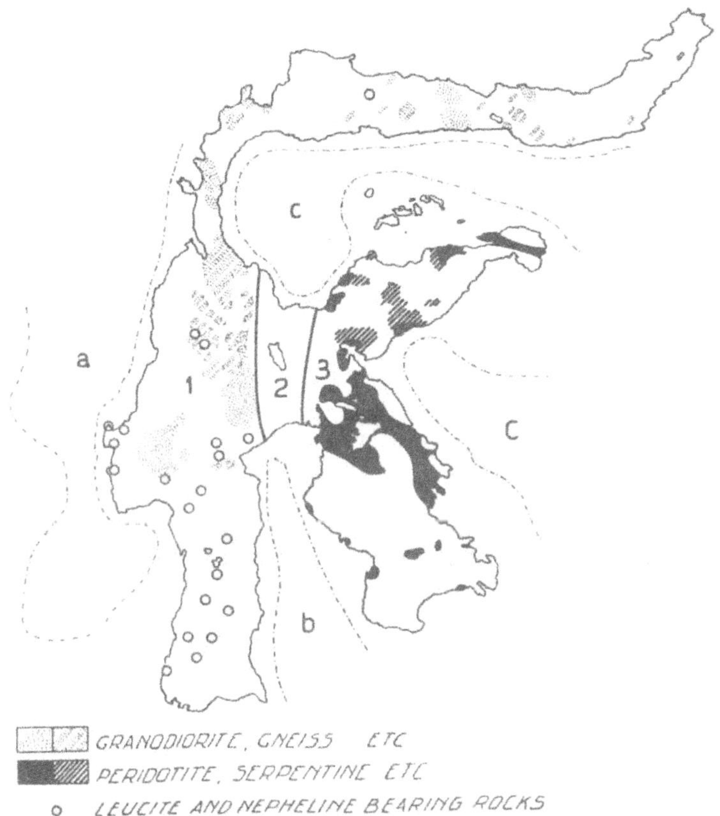

Fig. 131. Structural zones of Celebes.

The site of the dunite masses in the substratum is revealed by the fields and belts of strongly positive anomalies. In order to explain their distribution along the convex side of the negative zone the influence of all the major tectonic elements in the surroundings would have to be taken into account. Unmistakably the strongly negative zone between Celebes and Halmaheira had a paramount influence on the site and shape of the positive field in the Celebes Sea and its connection with the positive belt in the straits of Makassar.

However, in order to understand the total aspect of the positive fields as a whole we ought to take into consideration the gravimetric data on land, on Borneo, as well as on Celebes [1]). Accordingly a complete analysis of Celebes and its surroundings has to be postponed until more data become available. This applies also to the three deep-sea basins of fig. 131, the presence of the idiogeosynclinal basins in East-Borneo and S. Celebes, and finally to the different aspects of the zone of graben in central Celebes as compared with the deep-sea basins in its northern and southern continuation.

Evolution

Now that the origin and development of the East Indian arc has been sketched in its principal lines, the problem of some different types of island-arcs of the Pacific will be considered more closely. On p. 147 the question was raised whether different types of island-arcs represent different stages of evolution of a mountain-chain *in statu nascendi*. It will now be possible to give a well-founded answer to this question.

It was pointed out (p. 188) that the marginal deep of a double arc differs genetically from the marginal deep of a single arc. It was argued, moreover, that the quantity of waste-products from the land has to be considered as the principal factor causing the formation of either a single or a double arc which is composed of elements like those of the East Indian festoon, more especially the western part of the latter in which the complicated pattern of the Moluccas is lacking. The comparison is facilitated by Table VII, the figures I–IV corresponding to the same figures on fig. 123 and plate 8. According to the theory the Riu-Kiu islands and the East Indies should have passed through a paleogeographic stage of extensive land areas competent to produce great quantities of waste-products.

As to the East Indies the theory is supported by what is known of its paleogeographic evolution in Tertiary times. The region of the archipelago was probably an extensive land area at the end of the Mesozoic and the beginning of the Eocene. The increasing extent of several Paleogene and Neogene transgressions is illustrated by fig. 132.

A western land-area, including large parts of Borneo, Malaya, the present Java Sea and South China Sea, and Sumatra is still to be recognised in the Eocene. A second large land-area existed in the south-east including northern Australia and the present Arafura Sea. Is it mere fortuity that the zone of buckling appears as islands exactly where it runs along the areas that for a long time resisted the invasion of marine transgressions? In the Miocene, at last, the sea invaded these old land-nuclei over a larger extent, as shown by fig. 132.

Similar data are not available from the Riu-Kiu arc and its "hinterland". However, we might point out the following features. A deep basin with a maximum depth of 1248 fathoms is comparable to the idiogeosynclinal

[1]) It is e.g. uncertain whether the positive values in the Gulf of Bone and Makassar straits may be interpolated so as to form a zone across the southern arm of Celebes, or wether the positive zone of the Gulf of Bone has to be united with the positive values in the Gulf of Tomini. Compare the different interpretations by Van Bemmelen (1938) and Vening Meinesz (1934 and 1940).

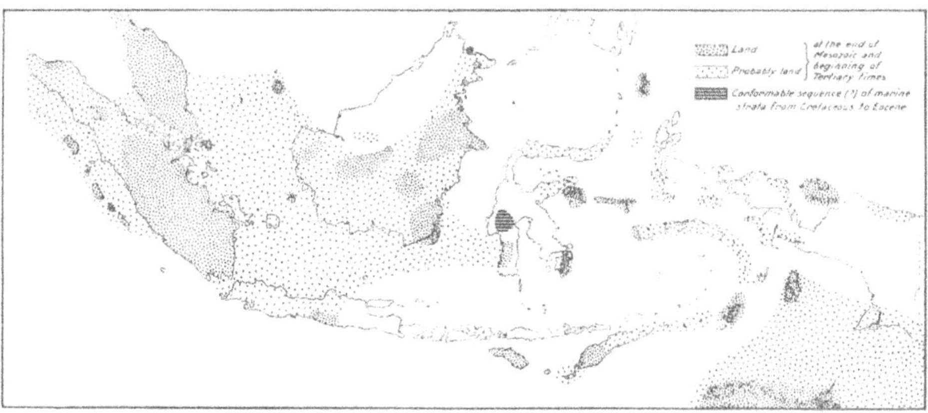

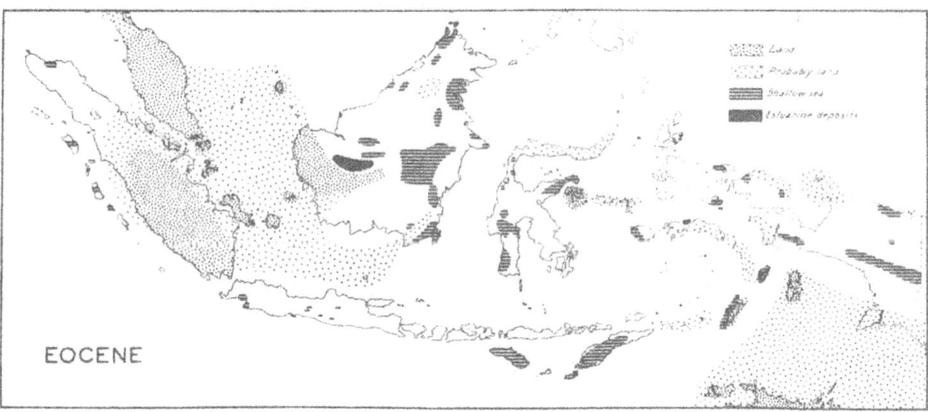

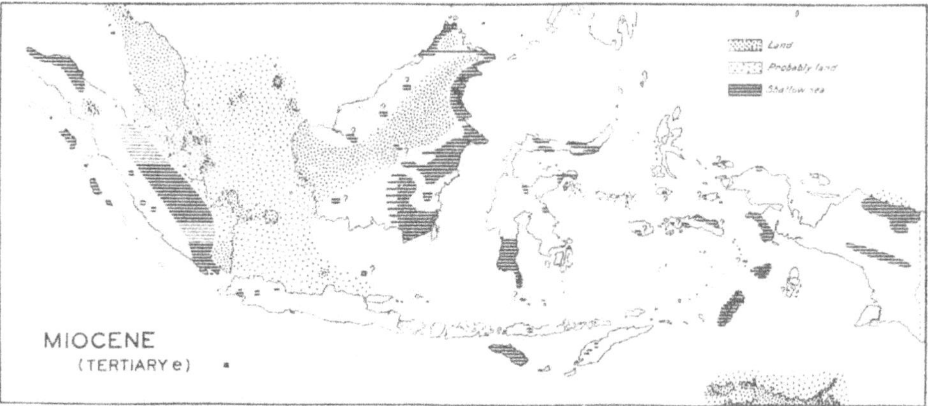

Fig. 132. Three tentative maps of paleogeographic conditions in the East Indies in late Mesozoic, Eocene and Miocene times. Land stippled, sea horizontally shaded.

TABLE VII. Comparison of East-Indian and Riu-Kiu Arcs. Depth in fathoms.

	Ia	I	Ib	II	III	IV
Average depth of ocean floor in front of the arc	Marginal deep	Zone of buckling (outer arc)	Intramontane troughs	First upward wave (volcanic inner arc)	First downward wave (basins behind the inner arc)	Area behind first downward wave
3000 f.	Java deep > 3300 down to 4000 f.	Nias-Mentawei-Timor islands	Mentawei trough > 550 f. down to 970 f. Java trough > 1650, down to 2900 f.	Barisan Mountains of West-Sumatra; S. Java; Lesser Sunda islands	Idio-geosyncl. basins of Sumatra and Java, Madura Straits, Flores trough (1650–2800f.)	Shallow Java Sea and S. China Sea, mainly 15–40 f.
3000 f. (Philippine basin)	Riu-Kiu deep > 3000 f. down to 3900	Nansei Shoto non-volcanic islands, from Kyushu to Formosa, (Taiwan).	Submarine trough between volcanic and non-volcanic arcs (no detailed bathymetric data available).	Volcanic Riukiu islands	Riukiu basin 1000 f. (max. depth 1248 f.)	Shallow East China Sea, mainly < 60 f.

TABLE VIII. Comparison of three single arcs with two pseudo-single arcs. Depth in fathoms.

		I + Ia + Ib	II	III	IV
Average depth of sea-floor in front of the arc	type of arc	Marginal **deep** (= zone of **buckling = outcrop** of shearing zone)	Volcanic Arc	First downward wave (Basin behind the arc)	Depth of sea-floor further continentward
	pseudo single arcs	Aleutian **trench** > 3000 f. (max. 4200 f.) outer non-volcanic arc, and trough behind it, near **Alaska**	Volcanic Aleutian Islands	Bering basin 2000–2090 f.	Northern and eastern Bering Sea mainly < 100 f.
	pseudo single arcs	Kurile **trench** 3000–4000 f. Tuscarora **deep** 4655 f. outer **non-volcanic** arc, and trough behind it, near Japan	Volcanic Kurile islands	Basin N.W. of Kurile Islands 1000–1500 f. max. 1836 f.	Sea of Okhotsk; N. part 50–100 f.; S. part 100–500 f.
3000	single arcs	Japan **trench** > 3500–4000 f. (**Rampo deep** 5771 f.)	Nanpo Shoto Isl. (From Oshima in the North to Torishima in the South)		
	single arcs	Bonin **trench** > 3000 f., **down to over** 4000 f. (Fleming **deep** 4730 f.)	Bonin islands	No detailed bathymetric data available	2000–2500 f. up to Philippine Basin (3000 f. and deeper)
	single arcs	Mariana **trench** > 3500 f.; Nero **deep** 5269 f.; **Mansyu** deep 5366 f.	Mariana Islands		

troughs on Sumatra and Java, or to the Flores deep. More continent-ward the sea is shallow, its depth seldom surpassing 50 fathoms. Hence a comparatively slight rising of the bottom or lowering of sea-level — or both combined — would provide extensive land areas behind the arc. The Riu-Kiu chain continues southward towards Formosa and northward into Kyushu which is one of the "continental" islands of Japan. Accordingly, the whole area between the arc and the present coast of Asia may be regarded as a "continental" region, which means that the sialic part of the crust is of continental thickness.

The outer arc is largely composed of strata ranging from pre-Carboniferous to Cenozoic. Parts of these sediments are composed of detritus, which could not possibly have been deposited under conditions of land- and sea-distribution such as prevail at present. Accordingly, the former existence of areas of denudation which have since sunk below sea-level has to be assumed. As a matter of fact Yabe pointed out that the area of the Riu-Kiu, Japanese and Kurile islands including the region from the 720 meter isobath to the present coast of the Asiatic continent, was dry land until sub-Recent times (late Pleistocene or later).

A similar conclusion holds good for the region behind the Kurile arc. The southern end of the arc is linked to Hokkaido, the northernmost island of Japan, whereas the arc continues northward into Kamchatka. Again there is a deep basin behind the arc with a maximal depth of 1836 fathoms, but the depth of the remaining part of the Okkotsh Sea does not surpass 50–100 fathoms.

The Aleutians represent another example of an arc with a "continental" sea-floor behind the arc. For, apart from the deep Bering basin, which again has to be regarded as a hollow comparable to those of zone III of the East Indies, the remaining northern and eastern part of the Bering Sea is mainly less than 100 fathoms deep. Remarkably enough the Kurile arc is a single arc, and the Aleutian arc is single over its largest extent. But its eastern end, where it continues into the American continent, is a double arc. The outer arc has developed over a comparatively short distance from Kodiak island towards the Kenai peninsula of Alaska, and it is separated from the volcanic inner arc by Shelikof straits and Cooks Inlet. The volcanic arc, too, continues into the continent by way of the Alaska Peninsula.

Both the Kurile and Aleutian arcs might be called pseudo-single arcs, for the Kurile-festoon, too, is double over a short distance, viz. near Japan.

For the Kurile and Aleutian arcs no paleogeographic data are available for making a comparison between the type of arc and the previous conditions of sedimentation, as was possible for the East-Indies [1]). But, the arcs so far considered have some striking features in common. For in a certain sense they might all be called "continental" arcs, which means that the sea-floor between the arc and the present coast-line of the continent lies comparatively high and corresponds to a sial-layer of continental thickness.

However, the situation of single arcs like the Marianas and the Bonin

[1]) The Japan-arc and the Philippines have been left out of consideration entirely because of lack of sufficient data.

islands is fundamentally different — see Table VIII —[1]). The average depth of the sea-floor behind the arc is of the order 2000-2500 fathoms or still more. This means that the sialic crust is thin, when compared with the areas considered so far. For, the deeply situated sea-bottom extends over an enormous distance, as much as 1500 miles from the central Marianas to the Philippines and the Riu-Kiu Islands. Hence sedimentation in this area is free of waste-products from the continent. Geographic conditions are very different from the regions considered so far because the sial-layer is much thinner. And it is not imaginable that the situation was fundamentally different even in earlier times. Under certain paleogeographic conditions a double arc of the Riu-Kiu type might have developed in the place of the Kurile arc or the Aleutians. But a double island-festoon of the Riu-Kiu type could not possibly come into being in the area of the Marianas, nor will it ever be able to develop in the future! The only possibility is the development of two or more volcanic arcs more or less parallel to each other owing to several shear-planes dissecting the crust near each other. This is what probably happened in the area of the Vulcan and Bonin islands (fig. 112).

Possibly the site of the shear-zones, as well as the location of the arcs, correspond with the boundaries of sial-layers of different thickness. So the Marianas and Bonin arcs developed near the boundary between the Pacific basin proper — which has no sial-layer — and the thin sial-layer extending from the Marianas as far as the Philippines and the Riu-Kiu islands. The latter developed on the transition of the thin sial-layer to one of continental thickness, whereas the Kurile and Aleutian arcs came into being on the boundary between a crust of the continental type and the basaltic floor of the Pacific.

For some distance the East Indian zones may be traced into the Asiatic continent, viz. in Burma[2]). For lack of sufficient detailed knowledge, it is not yet possible to follow them to the Himalayas, western Asia and Europe. The few facts that are known, however, justify the conjecture that this will probably be possible in the future. From the continuation of the East Indian zones into the Asiatic continent it appears, however, that the different physiographic aspects presented by an island-festoon and a continental mountain-chain do not represent different evolutionary stages. Whether an island-festoon or an arcuate structure on the continent, whether a deep-sea furrow or a "fore-deep" like the Siwalik trough, etc, merely depends on the geographic conditions that prevailed during the structural history of the region, more especially during the period preceding the last epochs of crustal compression. The thickness of the sialic part of the crust and the available quantity of waste-products from the land are the principal controlling factors. It was shown (p. 184) from the known history of the outer arc of the East Indies that the latter started as a geosynclinal belt in a remote past, at least as early as the Triassic.

Moreover, Mesozoic sediments of bathyal and abyssal facies types are

[1]) The bathymetric data are taken from the preliminary edition of the "Bathymetric chart of the North Pacific Ocean", compiled at the U.S. Hydrographic Office, mainly from sounding data by U.S. Navy Vessels.

[2]) See Umbgrove, 1938, pp. 65–69.

known from the island of Timor, demonstrating a very strong crustal relief in the vicinity of the present islands in a remote past. Hence, the region of this special belt went through a long and very complicated history which in itself shows some points of resemblance to the history of such mountain-chains as the Himalayas and the Alps. But the same facts make it clear that the East Indian festoons considered as a whole should not be interpreted as the embryonic stage of a mountain-belt which when fully grown, would be represented by the Alps and the Himalayas. For these regions are different in the same way as the Marianas differ from the East Indies: not because of their representing different evolutionary stages of one continuous process in the formation of mountain-chains.

Theoretically, i.e. in the ideal case of strictly unchanging conditions of crustal thickness and quantity of sediments, an island-arc of the Mariana type might come into view and vanish periodically, according to the alternating rhythm of epochs of increasing and decreasing compression in the earth's crust.

We have progressed far enough into the realm of speculation and refrain from picturing the future stages of a much more complicated area like the East-Indies. The scanty data known from its past history — e.g. the occurrence of deep-sea deposits in central Borneo, the presence of intense Mesozoic folding and overthrust strata in Sumatra and Java — clearly show that the past history was not a simple pulsating rhythm *in loco* and therefore we are not justified in expecting a simple repetition in the future of the Upper Tertiary rhythm that was pictured in the forgoing pages. But are we not equally unable to predict the future development of any other region, say southern Asia, Alpine Europe or western North America? We may once more point to the structural analysis of the continents (cf. p. 144). According to our present state of knowledge some regions seem to reveal a regular sequence of events, but in other areas the succession of tectonic belts seems capricious. Geologists have to admit frankly their inability to understand the cause- and effect-relations of the observed successions. Perhaps when more is known of Pre-Cambrian events the apparent capriciousness will be gradually replaced by a better understanding of past stages which will make possible a well-founded prediction of the future development of continental mountain-belts and arcuate island-festoons.

Unsolved Problems

In the introduction to this chapter the subject was limited to island-arcs associated with volcanism and a marginal deep-sea furrow. It appeared, moreover, that the Pacific festoons are invariably connected with a deep-reaching "seismic shearing-zone". We know that these features are characteristic for the western and northern boundaries of the Pacific Basin. The problem why similar phenomena are lacking along the southern and eastern borders of the Pacific Basin remained untouched.

The main problematic points may be indicated more precisely as follows.

(1) Melanesia is bordered by the Kermadec and Tonga deeps and a volcanic line runs from New Zealand to the Samoa islands. The problem is that

the trends are not arc-shaped but rectilinear. One might assume the existence of a shearing zone that dissects the crust almost perpendicularly. But deep-focus earthquakes are known in the region west of the Tonga deep as far as the Fiji islands (fig. 106) and their distribution is not associated with an arc-shaped zone of buckling; moreover deep-focus earthquakes have not been observed east of the belt. For the present, therefore, no adequate explanation can be suggested.

(2) South America is a region of considerable plutonism and volcanism. Most of the epicentres of deep-focus shocks have been noticed in a region eastward of the volcanic lines, whereas shocks of intermediate depth occur westward as far as the Pacific coast. Along the coast there is an interrupted series of deep-sea furrows (fig. 107). But remarkably enough no island-festoon has been developed.

Rittmann [1]) suggested that the western volcanic cordillera might be compared to an inner arc and the eastern cordillera to an outer arc. The position of the oceanic deeps and the arrangement of the earthquake foci oppose his hypothesis, which seems to me entirely unsatisfactory. The problem remains unsolved. The earthquake-zones and deep-sea furrows in S. America are interpreted by some authors as an arc which is concave towards the Pacific. Such a distribution of tectonic elements is exeptional. Now, two features deserve special attention. Firstly the marginal trough is of a more interrupted type than the marginal deeps along the western border of the Pacific. And, secondly, the distribution of the earthquake foci also displays some unusual features. Gutenberg and Richter [2]) summarized them very clearly in the following quotation:

"There are four groups of very deep shocks (600 to 660 km). The epicenters lie chiefly east of the Andes, with which they have no apparent connection. It is possible that these are four sections of a continuous active line, but it cannot be considered an established fact. In any case, there is no evidence of a contoured active surface, with depths decreasing westward toward the coast, as has been suggested by several authors. On the contrary, the writers now find no shocks in South America at depths between 290 and 600 kilometers; supposed instances of this kind have all been removed in the course of this investigation, having been assigned to greater or smaller depths. There is an equally evident gap in the geographical distribution".

When these features are explained it will perhaps be clear at the same time why no island festoon is present along the Pacific coast of South America. Meanwhile we have to wait for more data.

(3) For lack of sufficient data, geological as well as gravimetric, the entire region extending from New Guinea to Samoa has to be left out of consideration.

(4) No deep-focus earthquakes and no deep-sea furrow of the marginal deep type is known from the Pacific coast region of North America, though an island arc — partly crowned with volcanoes — is present along the coast of British Columbia. So far as I am aware only one attempt at explanation has been offered — by Schwinner — as illustrated by fig. 133. Accord-

[1]) A. Rittmann. Vulkane und ihre Tätigkeit 1936, p. 165.

[2]) Gutenberg en Richter, 1938, pp. 273–276.

ing to his hypothesis, the seismic districts of the Pacific border suggest a general overthrusting of the Asiatic, Australian and South American continents towards the Pacific Basin. The overthrusting would cause a system of shear-planes to develop in several blocks of the simatic bottom of the Pacific. Linear rows of volcanoes originated along the cracks. The movements of the blocks as surmised by Schwinner are indicated by the large arrows

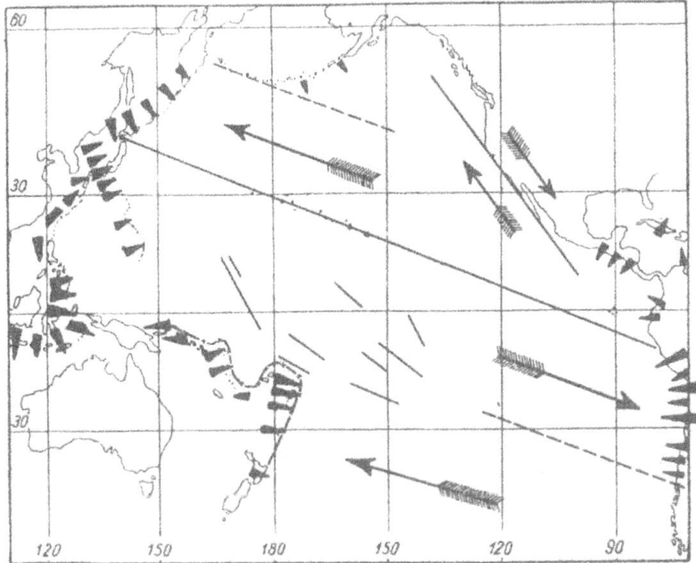

Fig. 133. **Major fault lines of the Pacific related to the sites of large thrust-planes with deep-focus earthquakes along the Pacific border. Thrust-planes are marked by rows of black triangles, their apex indicate the zone of normal shocks, their bases are directed towards the zone of deep foci. Supposed movements of blocks of the Pacific bottom are exhibited by arrows (From Schwinner).**

of fig. 133. It will be noticed that the direction of these movements is drawn parallel to the coast of North America, and parallel to the San Andreas and other great faults of California. This would explain the absence of a large continent-ward dipping thrust-plane as well as the absence of deep-focus earthquakes along the north-eastern margin of the Pacific.

However, his hypothesis does not explain the presence of an island-arc along the coast of British Columbia. Remarkably enough the arc is convex towards the Pacific. It is partly crowned by volcanoes and it seems to continue northward into the Aleutian arc. Apart from these deficiences of the theory, other conspicuous features remain unexplained viz: the Antillean island-festoons and the absence of island-arcs along the remainder of the coasts of the Atlantic and Indian Oceans.

Schwinner emphatically speaks of overthrusting continents instead of underthrusting of the Pacific bottom. He thinks a continent-ward movement of the Pacific improbable as it would cause the Pacific bottom to become stretched and enlarged towards all sides. Hence the fault lines of the Pacific are thought to be caused by compression due to the overthrusting continents that surround it, with the exception of North America.

It is clear, however, that movements along the supposed thrust-planes would induce relative movements of both the over-riding continental sector and the over-riden oceanic part. And as mentioned before (p. 150) the mechanism has to be described more accurately as underthrusting of the Pacific bottom. Moreover the so-called thrustplanes are potential zones of shear.

A different suggestion as regards the origin of the fault pattern of the Pacific is offered in Chapter XI (p. 319).

(5) Two regions along the eastern border of Asia have been left out of consideration, viz. the Japanese and the Philippine islands. The position of both is exceptional in so far as several arcs meet in these areas and form an intricate pattern. Fig. 104 shows the distribution of the volcanic zones of Japan and vicinity. One runs parallel to Sakhalin and, in Hokkaido, meets the volcanic arc of the Kuriles. Two parallel zones may be distinguished on the northern part of Honshiu or the main island of Japan, and again two zones may be seen on the southern part of Honshiu. In the south-western part they join the festoons of the Riu-Kiu arc. The festoons are double, the outer arc being composed of rocks ranging from pre-Carboniferous to Cenozoic formations. Finally the single Huzi volcanic zone comes in from the Pacific border and traverses the central part of the main island separating it into two distinct parts — north and south Japan. Obviously the intricate pattern of Japan depends on the joining and interacting of all these zones. The belt of negative anomalies follows the western slope of the marginal deep off Japan and it is accompanied by shallow earthquake foci. The foci of intermediate and deep shocks appear to fall on a surface dipping at about 30° degrees continentwards. This might be explained by asuming a sial root of the type shown by fig. 47 on p. 76. However, a detailed study of all the available data on the stratigraphy, paleogeography, structural history and geophysics of the region would be needed in order to elucidate the remarkable features of the Japanese islands.

A similar study would be needed for the Philippine islands which also appear to be a region where zones from various directions unite into an interlaced pattern. However, B. Willis' report of the year 1930 unmistakably testifies as to our still too scanty knowledge of the geography, geology and geophysics of the Philippine archipelago. It will have to be very considerably augmented before the configuration of these islands and the adjoining submarine topography can be understood.

Summary

(1) From a morphological analysis of the island-arcs bordering the Pacific side of the Asiatic continent the theory has been proposed by Lake that they are associated with large and deep-reaching shearing zones slicing the earth's crust. The form of an arc would then largely depend on the slopes of the shear surface dipping continentward from the Pacific. The radius of curvature of an island-arc would depend on the angle the shear-plane makes with the earth's surface.

Deep-reaching potential shear-zones dipping continentward from the island-arcs have been revealed by deep-focus earthquakes, but the angle a shear zone makes with the earth's crust greatly surpasses the angle that might be expected from the physiographic theory. So, if the curvature of the arc is due to a shear-zone, there must have existed a second zone which — unlike the zone found by seismologists — was restricted to the crust and eventually disturbed and obliterated by crustal processes. This, however, is a hypothesis *ad hoc* which will have to be abandoned as soon as a more plausible explanation of the shape of the festoon has been given. Recent theories on the origin of the deep reaching potential zone of shear were discussed on p. 162.

(2) Deep-focus earthquakes are apparently associated in some way with the border of the Pacific Basin. They occur in a large zone along its western side and in a smaller area in South America.

The character of the seismic disturbances clearly demonstrates that deep-focus shocks as well as normal tectonic earthquakes originate from the same causative mechanism, involving a shearing and faulting movement.

The zone of deep foci occurs at the continental side of a zone of intermediate focal depth and the latter is in turn succeeded by a zone of normal foci near the Pacific border, terminating in a deep-sea trough of the marginal deep type.

Probably all three classes of earthquake foci are situated in successive order along a surface — or a potential zone of shearing — which from a marginal deep-sea trough gradually slopes downward till under the continent it reaches a depth of approximately 700 kilometers. The slope or dip of the "seismic zone" attains 30°–55°.

From the various maps it appears that for the most part the zone of normal shocks "crops out" on the sea-floor within the marginal deep-sea troughs. Certainly this seems to be the case in the Japanese and South American districts, and probably it holds good for such regions as the western and south-western border of the Pacific Basin. In the double arc of the East Indies normal shocks are known to occur in a broader area extending from the marginal deep towards the inner row of islands. We will return to this phenomenon below (see 14).

Particularly noteworthy is the association of a marginal deep-sea trough with the outcrop of a seismic zone and the presence of a volcanic belt. This holds good for the island-festoons as well as for South America, where, however, an island-arc is conspicuously lacking.

(3) It appears from modern experiments that no discrepancy arises from the notion of deep-reaching shear planes in relation to the theory of thermal convection currents in the earth.

(4) The western boundary of the field of positive anomalies of gravity in the Pacific is found to coincide with the boundary of the island-arcs. It is at the same time a petrographic boundary (the andesite-line).

(5) Convection-currents transporting material with special magnetic properties towards more peripheral parts of the earth's interior from deeper realms would cause positive anomalies of terrestrial magnetism at the surface. There is, however, only a very rough agreement between the

fields of negative magnetic anomalies and areas where they might be expected to coincide with descending convection currents.

Conclusions regarding a relationship between deep-focus districts and anomalies of the vertical component of terrestrial magnetism stand in obvious need of confirmation by more detailed investigations.

(6) During epochs of increasing compression the earth's crust reacted by the development of large waves of two to four hundred kilometers in length. The amplitude of the crustal waves increased until in one of them the strength of the crust was surpassed. The crust then broke at the weakest place of a downward wave and buckled inwards so as to form a sialic root penetrating the simatic substratum.

The downward buckle of the sialic crust is revealed by a zone of strong negative anomalies of gravity.

(7) The development of a single or a double arc depends on the thickness of the sialic part of the crust and the quantity of sediment on the sea-floor above the crustal buckle.

One possibility arises where the sialic part of the crust is comparatively thin and hence situated deeply below sea-level. If the buckling zone be far from continental land, a deep-sea furrow will come into being above the sialic root, and the negative anomaly will be found above the deep. On the continent-ward side the deep-sea furrow will be accompanied by one volcanic island-arc which came into existence in the way described under (14). Examples are Guam with the Nero deep and Yap with the Yap deep. The anomaly curve across Guam and the Nero deep is asymmetrical. This seems best explained by assuming the eastern sector to be thrust under the western part. On the other hand the gravimetric profile across Yap and the Yap deep is more nearly symmetrical. Accordingly the existence of a symmetrical sial-root is more probable.

The other possibility arises where the sialic crust is of continental thickness and protrudes above sea-level for some time preceding the formation of the crustal down-buckle. Accordingly large quantities of waste-products from the land accumulated in the downward moving crustal wave. When the latter buckled down, its contents became crumpled and squeezed, so as to form a ridge-shaped elevation above the zone of buckling. A belt of negative anomalies is found over the ridge. Whether the ridge appears as a submarine rise or as an island-arc depends on the quantity of strata that were squeezed out and subsequently uplifted during the ensuing period of decreasing compression.

The Riu-Kiu arc and the East Indian festoon are types of arcs that were formed in this way. But only in the case of the East Indies are paleogeographic data available to substantiate the theory. The region of the archipelago was probably an extensive land area at the end of the Mesozoic and the beginning of the Eocene. During the ensuing Eocene and later transgressions a western area still remained above sea-level, including large parts of Borneo Malaya, the present Java Sea, the South China Sea, and Sumatra. A second large land-area existed in the southeast, including northern Australia and the present Arafura Sea. It is along these areas that the zone of buckling appears as an island-festoon, whereas the intervening strip developed only as a submarine ridge.

A shallow sea covers the area behind the Kurile and Aleutian arcs, apart from a deep basin that came into existence immediately behind the arc. These arcs are both continued into the continent; the Kurile arc into Kamchatka, the Aleutian arc into Alaska. In all probability the sial-crust behind the arc is of continental thickness. And we must ascribe it to paleogeographic conditions that the Kurile arc developed as a single arc instead of a double festoon. The same holds good for the Aleutian arc which is double only over a certain length of its eastern part.

The situation of the Marianas and Bonin arcs is fundamentally different. The sea floor behind the arc is some 2000-2500 fathoms deep or more. This means that the sialic part of the crust is comparatively thin over an area extending for more than 1500 miles westwards as far as the Riu-Kiu and Philippine islands. Sedimentation in this area is free from waste products from the land. Hence a double island-festoon of the Riu-Kiu or East Indian type could not possibly come into being in the area of the Marianas nor will it ever develop in the future.

(8) Possibly the "outcrop" of the potential shear-zones as well as the sites of the arcs, correspond with the boundaries of sial layers of different thickness. So the Marianas and Bonin arcs developed near the boundary between the sial-free Pacific Basin proper and the thin sial-layer extending from the Marianas to the Philippines and the Riu-Kiu islands. The latter developed where the thin sial-layer passes into one of continental thickness, whereas the Kurile and Aleutian arcs came into being on the boundary between a crust of the continental type and the basaltic floor of the Pacific.

(9) Given stable conditions of crustal thickness and such waste-products as are available, the Mariana-arc might appear and vanish periodically, according to the alternating rhythm of epochs of (a) increasing crustal compression, accompanied by arc formation and (b) decreasing compression, accompanied by decreasing volcanic activity and denudation of the festoon.

(10) Different physiographic aspects presented by such mountain-belts as the Alps, the Himalaya, the structural belts of Burma, and the island-festoon of the East Indies do not represent different evolutionary stages in a continuous process of mountain-making. The differences are mainly controlled by the thickness of the crust, the available quantity of waste-products from the land and the paleogeographic conditions that prevailed during the structural history of each region, more especially during the latest epochs of crustal compression.

(11) The time of origin of the crustal buckle as well as of the deep reaching shear-zone may lie in a very remote past. However, the features were rejuvenated time and again.

In the East Indies the first crustal down-buckling in the present zone of strongly negative anomalies of gravity occurred during the early Triassic at least — if not in more remote times. The most recent epoch of strong compression in the same belt dates from the Upper Miocene. Restoration of isostatic equilibrium is a rather slow process. For even if the root disappears by melting and spreading the sialic bulge in the crust still remains to give rise to a negative anomaly for many tens of million years afterwards.

(12) During a stage of decreasing compression in the crust the zone of buckling will rise in order to re-establish isostatic equilibrium. Consequently simatic material of the substratum will flow towards the buckled belt and as a further consequence, a furrow will form on the earth's surface on either side of the buckled belt. The external furrow is known as a marginal deep, while the furrow at the concave side of the root belongs to the intra-montane type. Hence the marginal deep of a double arc is genetically different from the marginal deep of a single arc (cf. under (7).

(13) The site of the island Sumba, as well as the submarine topography in its vicinity, turns out to be intimately related to the interruption of the belt of negative anomalies in that area. Sumba stands as the exceptional evidence of a sort of terrain that elsewhere subsided so as to form the bottom of a deep-sea furrow between the outer arc and the inner one.

Volcanic rocks of Upper Tertiary age on Sumba and northern Timor may possibly be explained as remnants of a situation prevailing before the formation of the deep-sea furrow between outer and inner arcs.

(14) The upward wave accompanying the zone of buckling on its continental side is more strongly developed than the wave on its convex side. It is moreover always a volcanic arc. These striking features may probably be explained in the following way. Subcrustal sima was pushed aside by the downward penetrating root. The curved shape of the arc involved centripetal crowding of the sima at the concave side of the arc, whereas it could expand more freely on its convex side. This caused the arching of the geanticlinal belt to be sustained and augmented by the accumulating sima and hence to become a belt of active plutonism and volcanism. For the arching of the geanticline induced relief of pressure at the underside of the crust. Accordingly emanations from the substratum ascended in these parts and by a process of migmatization of pre-existing crustal rocks formed acid batholiths. So, in a double festoon the inner arc always will be a volcanic arc. Plutonism and volcanism were rejuvenated during ensuing epochs of increasing crustal compression. During the same epochs longitudinal faults and rift-valleys in the crest of the geanticline developed. These are the loci of strong tectonic earthquakes of the normal class. The acid and highly volatile-rich character of the ascending migmatite front gave rise to strongly explosive vulcanism which resulted in the formation of volcano-tectonic depressions.

(15) Obviously the process responsible for the upward movement of the inner arc differs fundamentally from that which caused the rising of the outer arc. For the outer arc rises on account of the large sialic root, which tends to regain a state of isostatic equilibrium during a period of decreasing compression of the earth's crust. On the other hand, the inner arc rises as an upward wave of the crust during on epoch of increasing compression.

(16) In the outer arc, the downward penetrating root of sial prevented the development of plutonic and volcanic manifestations, such as those which characterize the inner arc during an epoch of increasing compres-

sion. During a period of decreasing compression acid batholiths growing upwards from the root seldom penetrate to a sufficiently high level to be revealed at the surface exposed by the erosion of the outer arc.

(17) The double row of volcanoes on North Celebes and Halmaheira seems to be related to the double-concave development of the zone of buckling, which can be inferred from the contours on the map of isostatic anomalies. Accordingly, a deep-sea furrow exists along each side of the root which may well be accompanied by two sets of deep-reaching shear-zones, one dipping westward, the other eastward.

(18) The absence of active vulcanism from the inner arc to the north of Timor, is probably controlled by the exceptionally strong compression to which the crust is subjected in that area, owing to the formation of the marginal Timor-deep, and more especially to that of the subsiding Sawu trough. For in this area the subsiding troughs on each side of the zone of buckling, had to become adapted to the limited space between the inner arc and the Australian continent.

(19) During an epoch of increasing compression a crustal downward wave formed along the continental side of the inner arc. If the quantity of available waste-products from the surrounding land-areas equals or surpasses the rate of subsidence of the bottom the furrow will gradually become filled up and will appear as a geosynclinal trough (idio-geosyncline).

If, on the other hand, the rate of subsidence surpasses the supply of sediments a deep-sea basin will originate (e.g. Flores deep).

The peculiar shapes of the Banda-Sea basins are due to a process of interference between neighboring crustal waves, resulting from the vortex-shaped pattern of buckled zones.

(20) The site of the alkali-rock provinces, mainly along the concave side of the inner volcanic arc of the East Indies, may be due to the effect of magmatic emanations on the marls and limestones which occur in comparatively great abundance in the idiogeosynclinal troughs mentioned under (19), the site of these troughs being, in turn, predestined by the formation of the crustal waves accompanying the downward-buckled belt.

(21) Fields of stronger positive anomalies of isostasy show a clear parallelism and relationship to the belt of strongly negative anomalies. They partly coincide with ridge-shaped elevations of the crust, partly with deep depressions. Therefore the cause of the anomalies has to be sought not in crustal but in sub-crustal phenomena. When the sial-root penetrated downward the heavier masses of the substratum were displaced. Probably a deep-seated layer of heavy material, such as dunite, was influenced in much the same way as the sub-crustal basalt-layer (cf. 14). Such displacements would be revealed as a belt of positive anomalies on either side of the negative belt and at a comparatively short distance from it. The more complicated pattern of negative zones in the Moluccas suggest phenomena of interference, but the situation of the positive fields may be explained according to the same causative principle.

(22) The four-armed morphology of Celebes is an expression of a double arc, which — unlike the other arcs along the border of Asia — has its convex side turned towards the Asiatic continent. The inverted position of the Celebes

arcs is caused by the sum of forces interacting in the region between Asia and Australia. It forms part of the whirl-shaped pattern of the Moluccas, which is due to the presence of the continental block of Australia, including the Arafura Sea and New Guinea. However, a mechanical interpretation of the puzzling island-arcs of the Moluccas, including Celebes, is not yet possible, as our geological and geophysical knowledge of the adjacent regions is too scanty. This especially concerns New Guinea and Halmaheira on the one hand, and Borneo on the other.

(23) Among the exceptional features related to the inverted position of the zone of buckling in East Celebes are the absence of a volcanic "inner" arc along its concave side and the numerous manifestations of Upper Tertiary to sub-recent volcanism and plutonism on its convex side.

As a tentative explanation it is suggested that two cooperating factors suppressed the formation of an "inner" arc; one being the interacting processes in the crust that caused the floor of the Banda Sea to subside deeply in the shape of a wide basin; the other being the accumulation of dunite masses that were laterally displaced from the zones of downward-buckling. Both factors worked in opposition to the formation of an "inner" arc. Hence, in this exceptional case, the basic material of the substratum that was pushed aside by the root under the eastern part of Celebes, had to flow towards the convex side of the zone of buckling, where it set going the genetic processes responsible for the numerous granodioritic intrusions and manifestations of Upper Neogene volcanism in the western division of the island.

References

ARGAND, E., *Sur l'Arc des Alpes occidentales* (Eclog. Geolog. Helvetiae, 14, 1916, pp. 179–182).

BARRABÉ, L., *La signification structural de l'arc des Petites Antilles* (Bull. Soc. Geol. de France 12, 1942).

BELTZ, E. W., *Principal sedimentary basins in East Indies* (Bull. Americ. Assoc. Petrol. Geolog. 28, 1944).

BEMMELEN, R. W. VAN, *Igneous Rocks of the Karangkobar region (Central Java) and the origin of the Malayan potash provinces* (De Ingen. Nederl. Indië, 4, 1937).

BEMMELEN, R. W. VAN, *On the origin of the Pacific Magma types in the volcanic Inner Arc of the Soenda Mountain System* (De Ingen. Nederl. Indië, 5, 1938).

BEMMELEN, R. W. VAN, *The distribution of the regional isostatic anomalies in the Malayan Archipelago* (De Ingenieur Nederl. Indië, 5, 1938).

BERLAGE, H. P., *A provisional catalogue of deep-focus earthquakes in the Netherlands East Indies* (Gerl. Beitr. Geoph. 51, 1937).

BRIDGMAN, P. W., *Shearing phenomena at high pressure of possible importance for geology* (Journ. of Geology, 44, 1936).

BRIDGMAN, P. W., *Reflections on rupture* (Journ. Appl. Phys. 9, 1938).

BROUWER, H. A., *Exploration in the Lesser Sunda Islands* (The Georg. Journ. 94, 1939).

BROUWER, H. A., *Geological Expedition to the Lesser Sunda Islands* (IV, 1942).

BRYAN, W. H., *The Relationship of the Australian Continent to the Pacific Ocean, now and in the past* (Journ. and Proceed. of the R. Soc. of New S. Wales, 78 1944).

BUCHER, W. H., *The pattern of the Earth's mobile belts* (Journ. of Geology 32, 1924).

BULL, A. S., *Note on the origin of island arcs* (Proc. Geolog. Assoc. 55, 1944).

BULLEN, E. K., *Composition of the Earth at a depth of 500–700 Km* (Nature 142, 1938).

BUWALDA, J. P., *Recent horizontal shearing in the coastal mountains of California* (Proc. Geol. Soc. Am. 1936).

CLOOS, H., *Künstliche Gebirge* (Natur und Museum 1929).

ESCHER, B. G., *On the relation between the volcanic activity in the Netherlands East Indies and the belt of negative gravity anomalies* (Proc. K. Akad. Wet. Amsterdam 36, 1933).

GREGORY, J. W., *The Banda Arc* (The Geogr. Journ. 62, 1923)

GRIGGS, D. T., *Deformation of rocks under high pressure* (Journ. of Geology 44, 1936).

GUTENBERG, B. and RICHTER, C. F., *Depth and geographical distribution of deep-focus earthquakes* (Bull. Geolog. Soc. America 49, 1938).

GUTENBERG, B., and RICHTER, C. F., *Evidence from deep-focus earthquakes* (in Physics of the Earth 7, 1939).

GUTENBERG, B., *Viscosity, strength and internal friction in the interior of the earth* (in: Physics of the Earth, 7, 1939).

GUTENBERG, B. and RICHTER, C. F., *Seismicity of the Earth* (Geol. of America Soc. Spec. Paper 34, 1941, and Bull. 56, 1945).

HANZAWA, S., *Topography and Geology of the Riu-kiu Islands* (Sci. Reports Tohoku Imp. Univ. 17, 1933).

HANZAWA, S., *Geological history of the Riu-kiu Islands* (Proc. Imp. University, 11, 1935).

HASKULL, N. A., *The motion of a viscous fluid under a surface load* (Physics 16, 1935 and 7, 1936).

HEISKANEN, W., *Über die Struktur und Figur der Erde* (Geol. Beitr. Geoph. 57, 1941).

HESS, H. H., *Island Arcs, gravity anomalies and serpentine intrusions* (Proc. 17th Intern. Geolog. Congr. Moscow 1937).

HESS, H. H., *Gravity anomalies and Island arc structure with particular reference to the West Indies* (Proc. Americ. Philos. Soc. 79, 1938).

HESS, H. H., *Recent Advances in interpretation of Gravity-Anomalies and Island-Arc Structure* (Adv. Rep. Commiss. on continental and Oceanic Structure 1939).

HEISKANEN, W., *The gravity anomalies on the Japanese Islands and in the waters East of them* (Public. Isost. Inst. Intern. Ass. Geodesy. 13, 1945).

HOBBS, W. H., *Mechanics in the formation of arcuate mountains* (Journ. of Geology 22, 1914).

HOBBS, W. H., *The Asiatic Arcs* (Bull. Geolog. Soc. America 34, 1923).

HOBBS, W. H., *The unstable middle section of the Island Arcs* (Verh. Geolog. Mijnbouwk. Genootschap 8, 1925).

HOBBS, W. H., *Mountain Growth. A study of the S. W.-Pacific Region* (Proceed. of the American Philosophical Soc. 88. 1944).

HOLMES, A., *The thermal history of the Earth* (Journ. Washington Acad. Sci. 23, 1933).

HONDA, H., *On the types of the seismograms and the mechanism of deep earthquakes* (Geoph. Magaz. Tokyo 5, 1932).

KONING, L. P. G., *Over het mechanism in den haard van diepe aardbevingen* (Acad. Thesis, Amsterdam 1941).

KOTÔ, B., *Morphological Summary of Japan and Korea* (Journ. Geol. Soc. Tokyo 22, 1915).

KOTÔ, B., *The Rocky Mountain Arcs in Eastern Asia* (Journ. Fac. Sci. Univ. Tokyo, sec. ser. vol. 3, pt. 3, 1931).

KUENEN, PH. H., *Geological Results, part I* (The Snellius Exped. 5, 1935).

KUENEN, PH. H., *The negative isostatic anomalies in the East Indies, with Experiments* (Leidsche Geolog. Mededeelingen 8, 1936).

LAKE, PH., *Island Arcs and Mountain building* (Geogr. Journal 78, 1931).

LAWSON, A. C., *Insular Arcs, Foredeeps and geosynclinal seas of the Asiatic Coast* (Bull. Geolog. Soc. America 43, 1932).

LEITH, A. and SHARPE, J. A., *Deep-focus earthquakes and their geological significance* (Journ. of Geology 44, 1936).

LINK, T. A., *An echelon structure of the Japanese archipelagoes* (Japan Journ. Geolog. Geogr. 5, 1927).

MACELWANE, J. B., *Evidence on the Interior of the Earth devised from seismic sources* (in Physics of the Earth 7, 1939).

MURRAY, H. W., *Profiles of the Aleutian Trench* (Bull. Geol. Soc. of America, 56. 1945).

PEKERIS, C. L., *Thermal Convection in the Interior of the Earth* (Monthly Not. R. Astron. Soc. Geoph. Suppl. 3, 1936).

RUTTEN, L., *Het parelsnoer der Antillen en de gordel van smaragd* (Tijdschr. Kon. Nederl. Aardrijksk. Genootschap 1940).

SCHUPPLI, H. M., *Geology of oil basins of East Indian Archipelago* (Bull. Americ. Assoc. Petrol. Geolog. 30, 1946).

SCHWINNER, R., *Vulkanismus und Gebirgsbildung* (Zeitschr. f. Vulkanologie 5, H. 4, 1920).

SCHWINNER, R., *Seismik und Tektonische Geologie der Jetztzeit* (Zeitschr. f. Geoph. 17, 1941).

SCHWINNER, R., *Der Begriff der Konvektions-Strömung in der Mechanik der Erde* (Gerl. Beiträge Geoph. 57, 1941).

SITTER, L. U. DE, *Deep-focus Earthquakes* (Geologie en Mijnbouw 1, 1939).
SOLLAS, W. J., *The figure of the Earth* (Quart. Journ. Geolog. Soc. London 59, 1903).
STILLE, H., *Die Angebliche junge Vorwärtsbewegung im Timor-Ceram Bogen* (Nach. K. Ges. d. Wissensch. Göttingen 1920).
SUESS, Ed., *The face of the Earth* (3. 1908).
TEICHERT, C., *Upper Paleozoic of Western Australia, correlation and paleogeography* (Bull. Americ Ass. Petrol. Geol. 25. 1941).
THOM, W. L., *The relation of deep-seated faults to the surface structure features of Central Montana* (Bull. Am. Assoc. Petrol. Geolog. 7, 1923).
TOKUDA, S., *On the echelon structure of the Japanese archipelago* (Japan. Journ. Geolog. Geogr. 5, 1927).
UMBGROVE, J. H. F., *Geological History of the East Indies* (Bull. Americ. Assoc. Petrol. Geol. 22, 1938).
UMBGROVE, J. H. F., *Different types of Island Arcs in the Pacific* (The Geograph. Journ. 106, 1945).
VENING MEINESZ, F. A., *Gravity and the hypothesis of convection currents in the Earth* (Proc. K. Akad. Wetensch. Amsterdam 37, 1934).
VENING MEINESZ, F. A., UMBGROVE, J. H. F. and KUENEN, PH. H., *Gravity Expeditions at Sea 1923–1932*, vol II (Netherlands Geodetic Commission, Delft 1934).
VENING MEINESZ, F. A., *The Earth's crust deformation in the East Indies* (Proc. K. Nederl. Akad. Wet. Amsterdam 43, 1940).
VENING MEINESZ, F. A., *Deep-focus and intermediaire earthquakes* (Proc. K. Nederl. Akad. Wet. Amsterdam, 49, 1946).
VISSER, S. W., *On the distribution of earthquakes in the Netherlands East Indian Archipelago* (Verh. K. Magnet. en Meteorolog. Observat. Batavia 22, 1930).
VISSER, S. W., *On anomalies of terrestrial magnetism* (Proc. 5th Pacific Sci. Congr. Canada 1933, vol. 3, 1934).
VISSER, S. W., *Magnetic anomalies in the Netherlands East Indies* (Proceed. 5th Pacific Sci. Congr. Canada 1933, vol. 3, 1934).
VISSER, S. W., *Some remarks on the deep-focus earthquakes in the international seismological summary* (Geolog. Beitr. zur Geophysik 48, 1936).
VISSER, S. W., *Aardbevingen met zeer diepen haard.* (Tijdschr. Kon. Nederl. Aardrijks. Genootsch. 54, 1937).
VISSER, S. W., *A connection between deep-focus earthquakes and anomalies of terrestrial magnetism and gravity* (Terrestr. Magn. and Atmospheric Electricity 1937).
VISSER, S. W., *Seismic isobaths in the East Indian Archipelago* (Gerl. Beitr. z. Geoph. 53, 1938).
VISSER, S. W., *Seismologie* (Noorduyn, Groningen 1943).
WADATI, K., *On shallow and deep earthquakes* (Geoph. Magaz. 1, 1928).
WADATI, K., *On the activity of deep-focus earthquakes in the Japan islands* (Geoph. Magaz. 8, 1934).
WILLEMS, H. W. V., *On the magmatic provinces in the Netherlands East Indies* (Verh. Geolog. Mijnb. Genootsch. Geolog. Ser. 12, 1940).
WILLIS, B., *Geologic observations in the Philippine Archipelago* (Nat. Research Council of the Philipp. Isl., Bull. 13, 1937).
YABE, H., *The latest land connection of the Japanese Islands to the Asiatic Continent* (Proc. Imp. Acad. 5, 1929).
YABE, H. and TAYAMA, R., *Bottom relief of the seas bordering the Japanese Islands and Korea Peninsula.* (Bull. of the Earthquake Research Institute. Tokyo Imp. Univ. 12, pt. 3, 1934).

CHAPTER VIII

THE FLOOR OF THE OCEANS

"....it appears that the higher continental masses and the depressed oceanic basins came into being very early in the history of the earth" (CH. SCHUCHERT)

Introduction

The antipodal distribution of continental blocks and oceanic receptacles ranges among the most peculiar features of the earth. The deep-sea area of the North Pole is situated directly opposite the Antarctic Continent. The six non-polar continents are grouped in pairs, with a roughly meridional orientation, and form an elongated triangle with its apex pointing southwards. Conversely, the oceanic sectors extend in an opposite direction. The broad bases of the continents, arranged in a nearly continuous ring around the North Polar Basin, have their antitype in the broad, circum-Antarctic ring of water. The counterpart of the Indian Ocean is found in North America and that of Australia in the Atlantic north of the equator, etc. (fig. 134). Of course, there are exceptions to this antipodal distribution of land and water and there is no question of a rigorous crystallographic symmetry but the fact remains that only 1/20 of the whole land-surface is antipodal to land [1]). ***The distribution of the continents is not hap-hazard but displays a certain regularity***. This exceptionally prominent feature of the globe, in which land masses are opposed to oceanic areas, ***reminds*** one of a

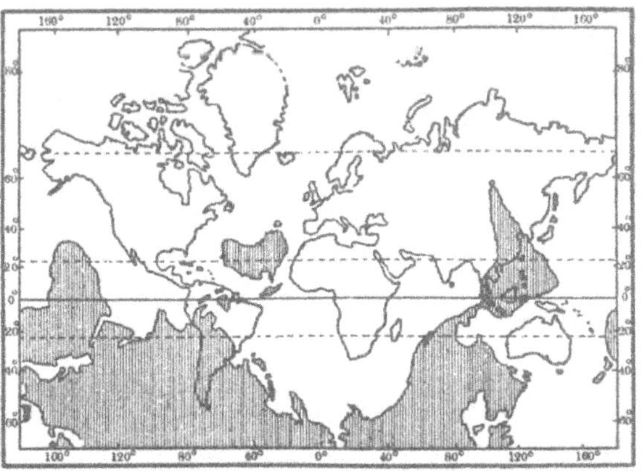

Fig. 134. Antipodal distribution of major geographic units (From J. W. Gregory).

[1]) Greenland and the Arctic Islands of North America are situated opposite Victoria and Wilkesland. New Zealand is antipodal to the Iberian Peninsula, Patagonia to part of North China and Grahamland to the Taimyr Peninsula.

tetrahedron, i.e. a crystallographic body in which a rib is always situated opposite a plane. It has given rise to a great deal of speculation, but it need hardly be said that this characteristic is such a remarkable one that any valid theory as regards the origin of continents and oceanic receptacles ought to be able to furnish an explanation of their tetrahedral arrangement [1]. It is clear, however, that this question involves the whole intricate problem of the origin of continents and ocean-floors, and there are many conflicting views on this subject. Some are of the opinion that the continents and oceans represent permanent features. Another hypothesis, diametrically opposed to the preceding one, is that the oceanic receptacles originated as a result of the submergence of land-masses. Others associate their formation with a drifting-apart of continents, while a fourth group claims that at least two of the ocean-floors were formed by stretching of continental blocks. All these possibilities will be discussed below, but it appears advisable to begin with a brief review of all that is known at present of the bottom relief of the oceans, while a comparison with the results of geological and geophysical research is obviously warranted at this stage.

The major characteristics of the bottom-relief

As more and more data become available, it becomes increasingly evident [2] that the relief of the ocean-floors presents some very curious characteristics.

A comparison between the various oceanic sectors will make it immediately clear that the relief of the Atlantic and Indian Oceans — while differing to a considerable degree from that of the Pacific, especially in the case of the North Pacific basin proper, i.e. the area limited by the andesite line and a southern demarcation running approximately from the Fiji to the Galapagos Islands — concur in many respects. The morphology of these areas is matched by their geological structure, for seismic and petrographic data both clearly show that the bottoms of the Atlantic and Indian Oceans are very different from that of the Pacific. The generally accepted view at present is that no sialic layer exists beneath this Pacific sector, though such a layer is in fact assumed to extend under the Atlantic and the Indian Ocean.

These broad outlines make it possible to divide the morphological characteristics into four major groups:

[1]) The geometric pattern of the surface of the earth gave rise to the well-known tetrahedral theory of Lowthian Green. (See J. W. Gregory, the plan of the Earth and its causes — The Geogr. Journ. 13, 1899, pp. 236—245, with Green's tetrahedral map of the world opposite p. 336). Among the later exponents of similar theories should also be mentioned Gregory, de Lapparent, Arldt and Kober. For a discussion of these hypotheses we refer the reader to "The unstable Earth", by J. A. Steers (Methuen, London 1932) and to Bucher's "The deformation of the earth's crust" (1933).

[2]) The following considerations are based on the deep-sea chart of the Atlantic by Stocks and Wüst (1935), Schott's chart of the Indian Ocean (1935), the chart of the John Murray expedition for more recent data on this area, a chart by Leahy (1938) and another fine chart of the Pacific by the U.S. Hydrographic Office (1939). A clear picture of diverse morphological questions will furthermore be found in publications by such authors as Cloos, Wüst, Schott and Mecking.

(1) The Atlantic Ocean, and the western part of the Indian Ocean, with their numerous and comparatively flat-bottomed basins, separated by relatively narrow and steep ridges.

(2) The large North Pacific basin, with its frequent linear and at times intersecting ridges and troughs. No mention will be made of the manifold basins in eastern and southeastern Asia, since these formations are situated inside the andesite line. Nor will we discuss Melanesia, which comprises the East Indies from a geological and morphological point of view. Most authors assert that the andesite line represent the true boundary of the

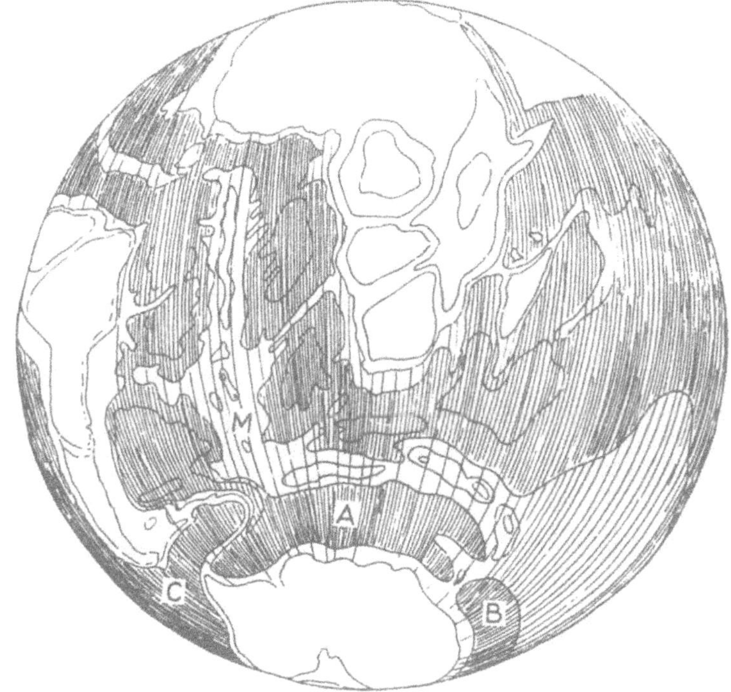

Fig. 135. Sketch of the submarine relief of the Atlantic and Indian Ocean floors surrounding Antarctica (After H. Cloos) A, Atlantic Antarctic Basin; B, Indian Antarctic Basin; C, Pacific Antarctic Basin; M, Mid-Atlantic Rise.

Pacific basin proper, and the real problem may thus be stated to begin in this area. Thus too, when dealing with the problem of the Atlantic, no mention will be made of the Mediterranean and West Indian basins. Moreover, these regions have already been discussed in Chapter III and the Appendix.

(3) The eastern part of the Indian Ocean, the south-western part of the Pacific and the three Antarctic basins.

These basins are in so far similar that a plain bottom-relief with hardly any subdivision is found in these areas. The three elongated deep-sea basins around the Antarctic continent (fig. 135) may be referred to briefly as the Atlantic, the Indian and the Pacific Antarctic basins [1]).

The above conception of the morphology of the basins may possibly prove to be more complicated than we suppose it to be, as none of them has been examined in detail.

The eastern part of the Indian Ocean and the south-western part of the Pacific form vast receptacles with a symmetrical arrangement in respect to Australia and Melanesia, the concave sides being turned towards one another.

(4) The North-Polar basin. Little is known of the details of this receptacle (cf. the basins of group (3), which it would appear to resemble in some respects), and no further reference will consequently be made to this basin [2]).

On the other hand special attention will be paid to the oceanic sectors of groups (1) and (2).

The Atlantic Ocean and the western part of the Indian Ocean

We will begin with a summary of a series of features which illustrate the significant congruence of these oceanic sectors and the surrounding continents, showing that in many respects the topographic limits between continental and oceanic areas do not correspond with structural limits.

The regions appearing under group (1) are characterized by a great number of deep-sea basins with comparatively flat bottoms surrounded by relatively steep ridges. The whole structure is in many respects very similar to that of the intervening African continent (fig. 136). This similarity is further accentuated by the fact that Africa, besides containing many more basins than South America, is also flanked by far more submarine basins than the American sector west of the Mid-Atlantic Rise. The morphology of the southern

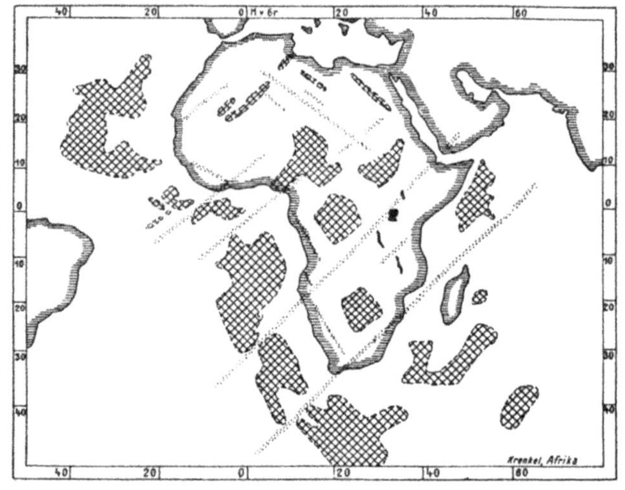

Fig. 136. Basins and ridges on the African continent and in the adjacent deep-sea (From E. Krenkel).

[1]) The Atlantic Antarctic basin extends eastwards from Grahamland to the southern point of Africa. It is separated from the Indian Antarctic basin by the Kerguelen Rise, and reaches as far as the ridge connecting Tasmania with Antarctica. It is succeeded by the Pacific Antarctic basin, which — extending eastwards to South America and Grahamland — is bounded on its northern extremity by the long, so-called East Pacific Rise.

[2]) The latest bathymetrical chart of this area can be found in a publication by Stocks (1939). An earlier map by Holtedahl shows the geology of the surrounding continents.

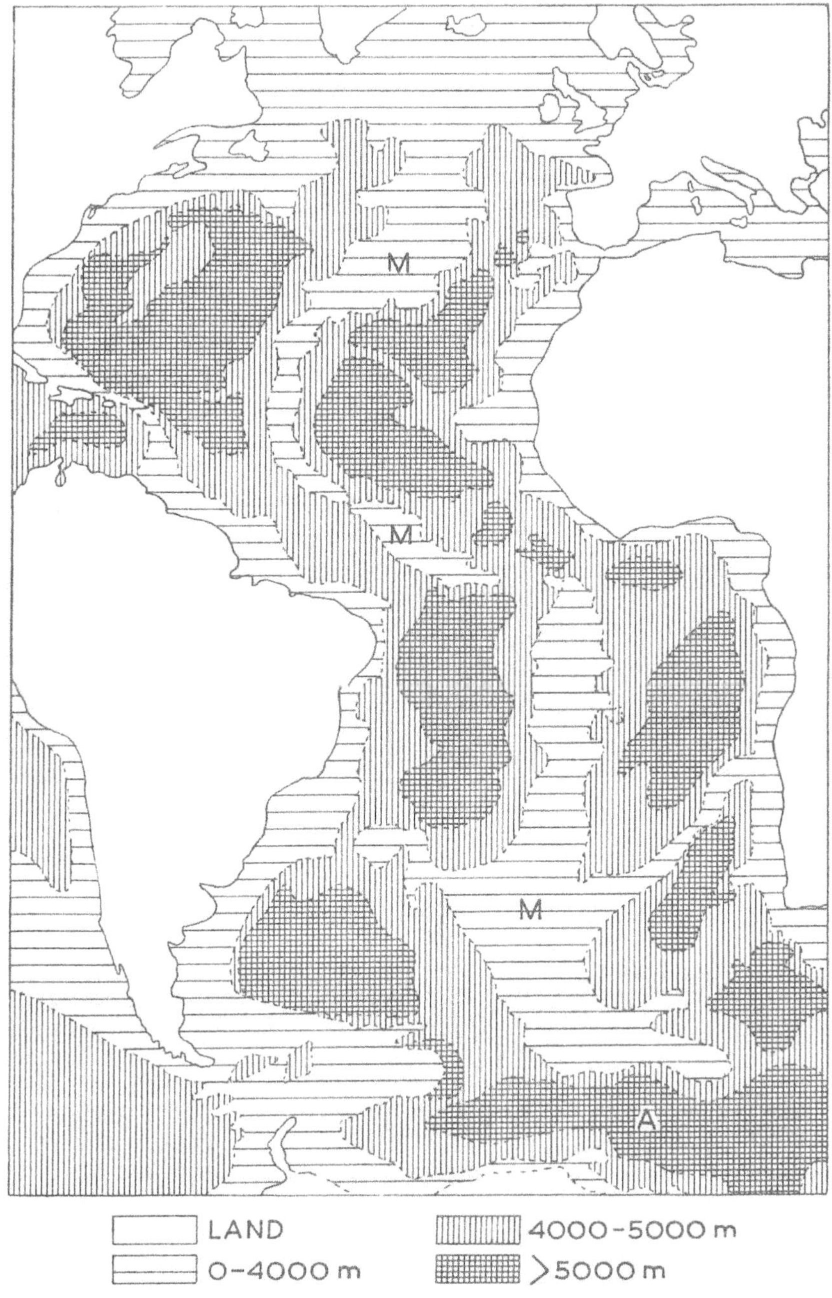

Fig. 137. Bathymetric chart of the Atlantic Ocean. M, Mid-Atlantic Rise; A, Atlantic Antarctic basin.

Atlantic blends, as it were, with that of the Atlantic Antarctic basin (fig. 135 and 137). The Mid-Atlantic Rise bends eastwards in the same way as the loop of the Southern Antilles and ultimately forms the boundary of the Antarctic basin, joining the Kerguelen Rise. In the Indian Ocean an analogous ridge bends eastwards around the Indian Antarctic basin.

A series of ridges with a roughly NS. trend, separating the continental and submarine basins in NS. alignements [1]), can be observed on the South American and African continents, parallel to the Mid-Atlantic and the Mid-Indian Rise. The thresholds between the continental basins consist mostly of Pre-Cambrian rocks. Two main strikes dominate in Africa since the early Pre-Cambrian: a NE.–SW., or so-called Somali strike, and a NW.–SE., or so-called Erythraean strike. Both played an important part in the whole later history of the continent. The same strike is again apparent in the narrow submarine ridges separating the deep-sea basins (fig. 137). The same holds good for the northern part of the Atlantic. Bucher [2]), for instance, observes: "The map of the North Atlantic shows that the rises which separate the individual basins consist of lines of swells comparable to a certain extent to that which in eastern North America runs from southwestern Ontario through the Cincinnati and Nashville domes, separating the Appalachian from the east-central sedimentation basin. The whole pattern of the ocean-floor seems, in fact, comparable to that of the basins and swells of the continental areas outside the great orogenic belts, although the scale is larger both horizontally and vertically on the oceanic surfaces".

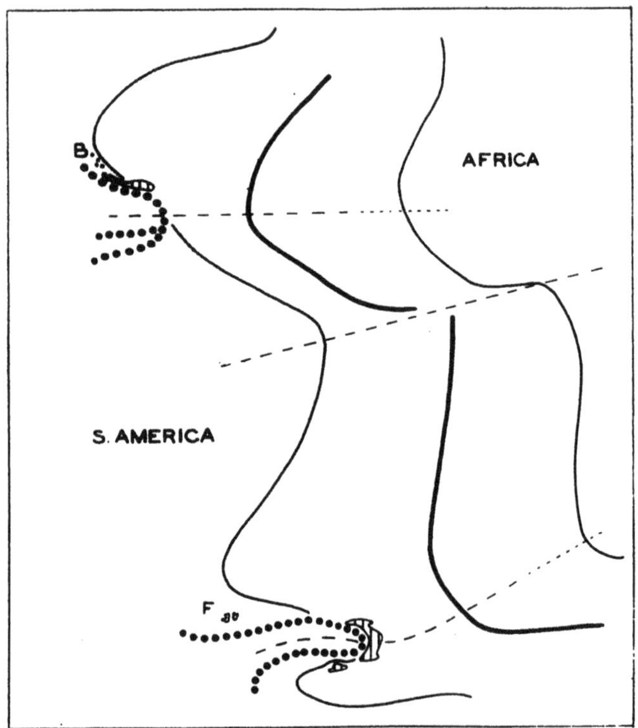

Fig. 138. Symmetry of the northern and southern Atlantic. (After H. Stille).

[1]) Krenkel's configuration has been followed. [2]) Bucher, 1940, p. 492.

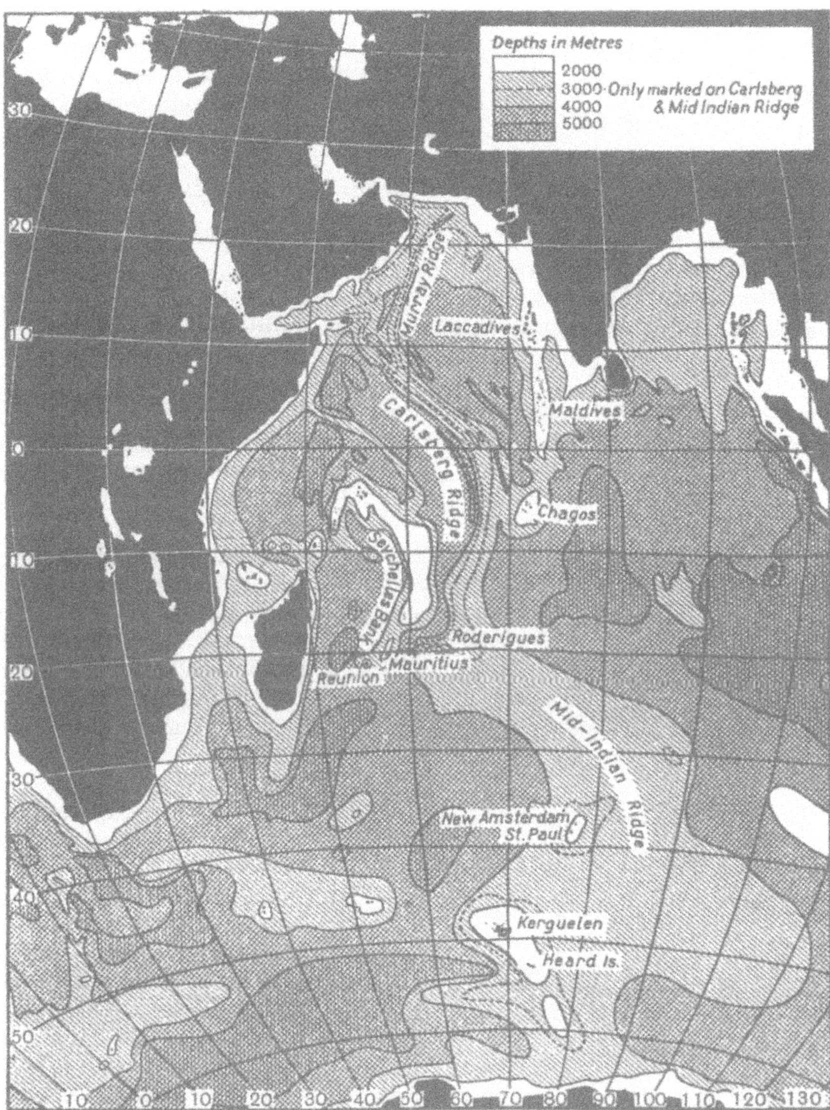

Fig. 139. Bathymetric chart of the Indian Ocean (from J. D. H. Wiseman and R. B. Seymour Sewell).

Mention should be made here of the opinion of Cloos [1]), who asserts that a domeshaped elevation with a fault-system, accompanied by volcanism and similar to that observed on the European and African continents, exists in the Azores, and this leads him to conclude that there can be no fundamental difference between this part of the Atlantic and a continent.

[1]) Cloos, Geol. Rundschau, 30, 4a, 1939.

Another striking feature, as pointed out by Stille (who illustrated it most suggestively in one of his recent papers) is the symmetry of the northern and southern Atlantic sectors (fig. 138). These may be described as forming one another's counterpart. This applies to the submarine characteristics (e.g. the convex arc of the Mid-Atlantic Rise bends eastwards twice), as well as to the features above sea-level (e.g. the arc of the northern Antilles with the shelf of the Bahamas and the arc of the southern Antilles with the Falkland Islands shelf, etc.).

Symmetrical features are also found in the Indian Ocean, and deserve special attention, as half of the symmetrical structure appears on the African continent. The new chart of the John Murray expedition shows the existence of a series of widely-curving ridges in the north-western part of the Indian Ocean. The best-known elements are the Seychelles Bank, the double Carlsberg Ridge and the Murray Ridge (fig. 139). Wiseman and Sewell report that the ridges form the almost exact mirror-reflection of the famous tectonic rift-valleys in East Africa (fig. 140). This analogy is emphasized by the fact that the Carlsberg ridge is also a belt of tectonic earthquakes and volcanism (fig. 141). This consequently constitutes another point of similarity between sub-oceanic and continental sectors. Wiseman and Sewell assume that an Upper-Tertiary age may probably be attributed to the Carlsberg Ridge. The Murray Ridge probably fuses with the Carlsberg System, and this probably connects in the south with the ridges upon which the Lacadive-, Maldive- and Chagos-Islands rest. The latter appear to unite as the Mid-Indian Rise, which can be traced to Antarctica via New Amsterdam, the Kerguelen, and the Heard Islands.

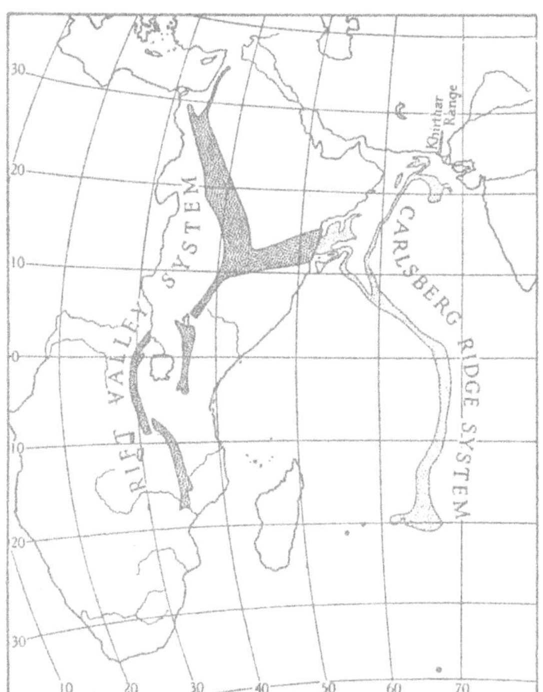

Fig. 140. The systems of the Carlsberg ridges and the Great Rift-Valley. (From J. H. Wiseman and R. B. Seymour Sewell).

It will be clear that, in case the basins of the north-western area correspond with old structural elements, their frames should at any rate have been subjected to a "rejuvenation" (a phenomenon also observed on the continents) in comparatively recent times. The same applies *mutatis mutandis* to the relief of the Atlantic. Young fault-systems, graben, Upper Tertiary strata and Pleistocene vulcanism are found in Iceland, the Azores and St. Paul. Unfolded Upper Tertiary sediments occur in the first two areas. Other young volcanic regions are Ascension, Tristan da Cunha, St. Helena, Gough Island and Bouvet Island. Moreover, the seismic unrest of the Mid-Atlantic Rise proves that the movements are still being continued [1]).

Petrographical data obtained in the oceanic areas stress the results of geological and seismic observations and point to the sialic composition of the floor [2]).

So far attention has chiefly been paid to the pattern of the submarine relief. Yet it is obvious that valuable information can also be obtained from the geology of the surrounding continents. We will consequently give a brief summary [3]) of a few of the most important points that call for special attention when dealing with the problem of the ocean-floors.

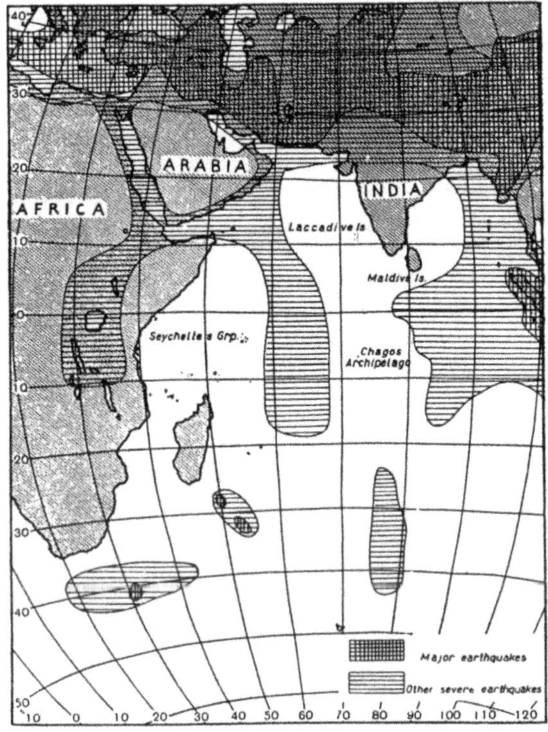

Fig. 141. The earthquake belts of the Indian Ocean and its surroundings (After Heck).

(*a*) Scandia. This area lay above sea-level till the Lower-Tertiary according to Holtedahl and Frebold. Its former extent is purely hypothetical.

(*b*) Appalachia extended at least 200 miles beyond the eastern coast of North America (cf. Barrell, Schuchert and Ewing), and now lies submerged to a depth of at least 3,000 to 4,000 meters. Geophysical data seem to imply

[1]) cf. Bucher, 1940, p. 494—495.

[2]) Plutonic rocks are known to occur in some small islands of the Indian Ocean, e.g. in Juan de Nova (between Africa and Madagascar). The Seychelles, Socotra and the western sector of the Kuria Muria Islands (granite,syenite), while dubious finds (which were not actually discovered as bedrock) have been reported to be present in the Comores. Clay- and chlorite-schists have been observed in Mauritius, and dolomitic limestone, Tertiary petrified wood, mica-diorite, felsite-porphyry, labradorporphyrite, schists and siliceous slate in the Kerguelen. Miocene and Eocene limestone rest on basalt and trachyte in Christmas-Island. Only a few indications of these rocks are found in the Atlantic. Granitic xenoliths occur in the volcanic lavas of Ascension and probably in Tristan da Cunha. Plutonic rocks, sediments of varying age and metamorphic strata are found in a few of the Cape Verde Islands and the Canaries.

[3]) The geology of the coastal areas was discussed in Chapter II and VI, and is also dealt with in the Appendix.

that this area began to founder in the Triassic and that this process continued until at any rate the Tertiary (fig. 25 and 26).

(*c*) The seismic method of investigation was also applied by Bullard and Gaskell to the submarine geology of the eastern part of the Atlantic i.e. along a line extending 170 miles WSW. from the Lizard. These authors revealed the presence, in this area, of a submerged block probably consisting of igneous rocks covered by Triassic strata.

(*d*) In his monograph on the Congo basin Veatch writes as follows: "A consideration of the source of the sediments forming the Lubilash of the western Congo, likewise, points to a more elevated, though not necessarily mountainous, land mass occupying the present depressed coastal plain region and extending into what is now the Atlantic Ocean".

(*e*) Faulting of presumably Pliocene or post-Pliocene age occurred along the coasts of Africa, Arabia and India around the Arabian Sea, and movements of a similar age caused part of the Oman-Indus arc to break down into the Arabian Sea; according to Wiseman and Sewell.

(*f*) The southern part of the South African Variscides and the probable connection with the Cedar Mountains disappeared into the sea. Nothing is known of the original extent of this land mass.

In none of these cases need one think of blocks of enormous continental extent, though many investigators are indeed known to have jumped at this conclusion, while others arrived at a similar result on the plea that the folded chains (the Pre-Cambrian, Caledonian, Variscian and Alpine belts) terminate abruptly along the coasts of the Atlantic. Still others base their conclusions on the analogy of the submarine morphology and that of the continents [1]).

Many coasts are indeed formed by faulting [2]). Although these areas have been discussed previously (Ch. II and VI, pp. 38 and 111) the possible consequences of the idea of tremendous subsided blocks will be dealt with subsequently. Biogeographical data will play no part in this discussion, as these should never be allowed to influence the solution of any geological problem *a priori*, even though such data may prove to be valuable if compared with geological results.

Recapitulating, a comparison between the submarine relief of the oceans and the geology of the surrounding continents may therefore be said to lead to the following conclusions:

1. The bottom of the Atlantic and the floor of the western part of the Indian Ocean are built up of similar (sialic) material as the continents.
2. The same events (i.e. the formation of basins, ridges, and rift-valleys) occurred in the oceanic sectors as well as the bordering continents.
3. The structural lines determining the location of the continental and submarine basins date from very early Pre-Cambrian times.

---

[1]) Stille, for instance, speaks of a "Destruktionsfeld, ein Feld der Einsenkung und des Einbruchs, herausgeschnitten aus dem uralten Südkontinent der in Resten heute noch östlich und westlich von ihm auftragt".

[2]) Born writes: "Es ist das Lissaboner becken nicht anders als eine Teilerscheinung der in Abbröckeln begriffenen West Europa". (Born 1932, p. 698).

4. The submarine relief, however, indicates that it underwent the same process of "rejuvenation" in comparatively recent times as the continents.

5. The submarine basins need not have originated simultaneously, any more than the continental basins, which are known to have formed in very different periods.

6. The pattern resulting from the arrangement of the three oblong basins around Antarctica, the blending of the submarine relief of the adjoining areas and the symmetry of the bordering oceanic sectors must surely have some special significance. Cloos attributed this constellation to the influence of the rotation of the earth. It would in any case appear to reflect the activity of subcrustal forces, which would then have produced this pattern in the early Pre-Cambrian.

7. The arrangement of the submarine basins and their increasing depth and width towards the south until confronting the Atlantic and Indian Antarctic basins on a broad line, reveal two large wedge-shaped oceanic sectors with their apexes pointing northwards, while the South American, African and Australian continents point in the opposite direction. There is no doubt in my mind that this situation can be said to constitute a fundamental feature, resulting from primordial and deep-rooted forces. The same applies to the tetrahedral arrangement of the continents and their antipodal relations to the oceanic depressions, as mentioned in the introduction.

8. Several blocks of unknown size have foundered very deeply. Many coasts are bounded by faults. Nevertheless, the characteristics mentioned under 1-7 should not induce us to suppose that the whole floor of either the Atlantic or the Indian Oceans constitute foundered continental blocks. However, this possibility will be examined below.

9. The oceanic and continental sectors differ chiefly as regards the thickness of the sialic layer. This layer is thinner under the former sectors, and situated at a great depth. It can be stated to constitute one of the major problems of our globe.

The Pacific Ocean

The large North Pacific basin, situated north of an imaginary line running from the Fijis to the Galapagos Islands is characterized by many linear ridges and troughs which occasionally intersect one another, thus giving rise to a pattern which is met with in no other oceanic area. This linear arrangement, which is such a prominent feature of many Polynesian Islands (fig. 142), is believed to have been caused by the rising of magma along faults and fissures in the floor of the ocean, which in this case appears to be composed of basic rocks (crystalline sima). The lineaments have relatively straight courses and are extremely long. In the opinion of Betz, Hess and others the fissures have to be considered as transcurrent faults. The possible cause of their origin will be discussed in Chapter XI (p. 318).

Many volcanic peaks in the central part are crowned with atolls, and nothing is consequently known of the nature of the eruptive rocks in their foundation. A quantity of data could nevertheless be obtained outside this area, upon which many investigators — especially Lacroix [1]) — based conclusions of a general kind. There seems no doubt that the bottom of the Pacific Basin proper, i.e. the area inside the petrographic boundary known as the andesite line consists of olivine basalt (cf. p. 165, and Plates 4 and 7).

In spite of the above, there would seem to be no doubt that sialic blocks have foundered in some parts of the Pacific. A few of these masses were

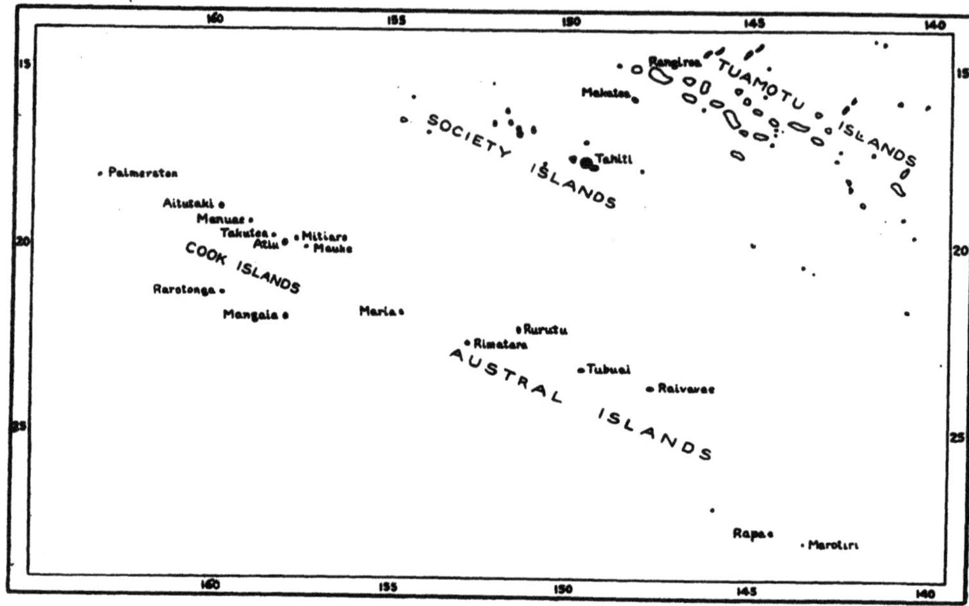

Fig. 142. Linear arrangement of island groups in the Pacific (From L. J. Chubb).

discussed in Chapter II. If we exclude such areas as Choco and the much-debated region of Burckhardtland, there would appear to be ample justification for the assumption that Cascadia represents a "border-land" of unknown extent, which probably acted as an area of denudation to the geosyncline of the Sierra Nevada up to Jurassic time, supplying it the while with detritus. It has probably sunk to a depth of 3,000 to 4,000 m below sea-level [2]).

In addition it should be noted that a land-mass of unknown dimensions has undoubtedly sunk into the Pacific off the coast of South America [3]).

The assumption that the eastern margin of Melanesia is a fault zone, on top of which formed the series of islands extending from New Zealand to Samoa, would logically imply that a block must have foundered east of this area.

[1]) See also Chubb (1934) and Smit Sibinga (1943).

[2]) A summary of various data on Burckhardtland and a number of other regions appeared in my publication of the year 1937; see also the Appendix.

[3]) Gerth reported that a large stretch of mountain-chains foundered near Payta, along the west coast of South America (Reg. Geol. d. Erde, 1939, p. 2), and also writes that "ein Zweig des paläozoischen Gebirges sich nach N.W. in das Gebiet des heutigen pazifischen Ozeans fortsetzte" (Ibid. p. 6).

In fine, blocks of unknown extent may undoubtedly be said to have sunk in the marginal zones of the Pacific, but there can be no question that these constituted large sialic blocks, as enormous stretches of the floor of the Pacific appear to be built up of rocks of basaltic composition.

The distribution of isobathic areas

The distribution of isohypses on the continents and isobathic contours in the oceans has led to interesting speculations. A graphic representation for the whole world, i.e. a so-called hypsographic curve of the earth's surface shows two frequency maxima, corresponding with two favoured levels, one occurring at 100 m above sea-level, a second at a depth of 4,700 m below it. In other words the continental surfaces are situated at an average altitude of 5,CC0 meters above the ocean-floors. Wegener argued that the lower level should be regarded as the surface of the sima, whereas the upper level was that of the sial. These two levels of equilibrium would give rise to two frequency maxima, which would simply be controlled by Gauss' law of errors [1]). I wish to draw attention, however, to a few salient features of the submarine relief as known to us at present. The following table shows the distribution of iscbathic areas in the Atlantic and the North Pacific Basin, according to Kossinna [2]). The areas are given in millions of square kilometers, the depth in km.

Depth	0.0–0.2	0.2–1.0	1–2	2–3	3–4	4–5	5–6	>6	Total area	Average depth in m
N. Atlantic Ocean	2.6	1.8	1.8	3.0	6.7	**11.5**	8.8	0.6	36.8	3,788
S. Atlantic Ocean	2.0	1.5	1.2	3.2	9.2	**16.2**	13.1	0.2	45.6	4,036
N. Pacific Ocean	1.1	0.8	0.8	1.9	8.8	21.2	**33.2**	3.0	70.8	4,753

The most frequent depth of the Atlantic appears to occur at the 4–5 km level, whereas that of the North Pacific is generally one km deeper! [3]).

It seems to me that the explanation is obvious, viz. the deep level of the Pacific is controlled by the simatic nature of its bottom, the 1,000 m higher level of the Atlantic floor is due to the presence of a sial-sheet and the still higher level of the continents is the result of the fact that the latter consist

[1]) It has to be admitted that a hypsographic curve for e.g. a Jurassic or Upper Paleozoic earth-surface would probably deviate in many, though not in fundamental respects from the present one. But I cannot endorse Bucher's criticism (1933, p. 52–59) that Wegener's interpretation of the two frequency maxima would be valueless.

[2]) Kossinna, 1933, p. 884.

[3]) In the words of Kossinna (1933, p. 885): "Nur ein grosses Meeresgebiet, der allerdings noch wenig erkundete Nordpazifische Ozean, weicht erheblich von den übrigen ab. Die 5000–6000 m Stufe nimmt hier allein 46,8 v. H. des Areals ein. Auf mehr als der Hälfte seiner Fläche (51 v. H.) ist der Nordpazifische Ozean über 5000 m tief. Die Flachsee und die Stufen von 200 m bis 3000 m haben dagegen eine geringere Ausdehnung als in allen anderen Ozeanen. Dieser Verteilung der Tiefenstufen entspricht die enorme Tiefe des Nordpazifschen Ozeans von 4753 m, welche die des Nordatlantische um rund 1000 m übertrifft".

of sial-slabs of a considerably greater thickness than the sial-sheet under the Atlantic.

We pointed out that the morphological features of the Atlantic and the Pacific are fundamentally different. If we glance at the new bathymetric chart of the Atlantic Ocean by Stocks and Wüst, it appears at once that the system of the Mid-Atlantic Rise and its adjoining ridges is situated at an average depth of 4 km or less (from 2–4.5 km), but the average level of the basins is 5 km (from 4.5–6 km)! Which one of these two is the original level of the Atlantic floor, the higher or the deeper one? If the basins are to be considered as secondary subsided formations, the primary surface of the Atlantic floor would be the higher level! Moreover, it seems very probable that the major positive relief-features of the Pacific Basin proper, as compared to those of the Atlantic, are largely due to later accumulations of lava on the primary surface, which seems to lie at an average depth of 5.5 km below sea-level. These considerations stress the fundamental differences between the average depths of the Atlantic and Pacific ocean-floors. On the other hand, the rejuvenation of the relief is a factor which should be taken into account. Thus, the result which we arrive at here is that at least three distinct levels of primary importance may be observed. These are represented by the surface of the continents, the floor of the Atlantic and that of the Pacific. We will return to these features at the end of this chapter.

The problem

Now that the major characteristics of the sub-oceanic relief have been considered and before entering into a discussion of the various hypotheses which try to explain these features, the main points of the problem can be summarized as follows. (1) The "tetrahedral arrangement" of the continents, their antipodal relations to the oceanic receptacles and their peculiar shapes. (2) The occurrence of at least three favoured levels on the earth's surface and their respective vertical distances. (3) The occurrence of a thin sial-sheet in the Atlantic sector and part of the Indian ocean. (4) The presence within these areas of a remarkable structural pattern dating from the early Pre-Cambrian and the occurrence of certain tectonic features which have their counterpart on the adjacent continents. (5) The congruence of orogenic belts on either side of the Atlantic ocean, terminating abruptly on the coast. (6) The absence of a sial-sheet over the floor of the Pacific Basin proper as well, probably, as in a few other oceanic areas mentioned above under groups 3 and 4.

Continental drift

Alfred Wegener's much-debated hypothesis of continental drift seemed to give an elegant and simple solution of many problems of the earth's crust, but it is clear that this original conception (fig. 156) cannot be made to agree with our present knowledge of the floors of the Atlantic and Indian Oceans, for these are now generally believed to be sialic layers. Moreover, the structural pattern of these ocean-floors and their striking similarity to that of the surrounding continents cannot be explained by the drift-hypothesis. However, it might on the other hand be assumed that our continents occupied their present site as a result of sliding, which occurred in such a way that a sialic block of continental thickness was stretched. The stretched portion would then simultaneously have sunk and thus formed the bottom of the Atlantic and Indian Oceans (fig. 156). This deduction characterizes du Toit's hypothesis of drift. In 1937 this same author wrote the following in a publication on the submarine relief of the Atlantic: "A pattern like this over so gigantic a region asks for a single controlling cause, and finds its proper answer only in continental sliding". To this he added a year later: "The relief of the Atlantic bottom shows all the characters of a stretch basin — particularly in the symmetrically-set, though crooked, Mid-Atlantic Rise with its lateral branches that reach out to either shore following north-easterly and north-westerly trends."

Yet I agree with Cloos that the striking similarity of the structures of the basins and surrounding ridges as compared to those of the adjacent continents is incompatible with the idea that these originated as a result of drift. Nor do I find it possible to reconcile any of the above conclusions as regards the Atlantic and Indian Oceans with the hypothesis of sliding continents, since the features of the bottom relief at all events make it clear that if the floors of said oceans originated as a result of stretching *this process must have occurred during the Pre-Cambrian.*

The same holds good regarding several other types of drift-hypotheses, including those of Gutenberg, Grabau, and Chelikowsky [1]).

The above would be quite sufficient to reject the hypothesis of du Toit and others, but I would like to examine another argument, which has recently appeared in several publications. Biological and paleontological data may be able to furnish more or less convincing evidence in favour of trans-oceanic land-bridges, but not — generally speaking — of the nature of such land-connections. We must make an exception for those data which seem to indicate that the distance between the continents had originally been a shorter one. Gerth, for example, argues that the similarity of the marine Devonian strata of South America and South Africa pleads for a shorter distance between Africa and America during the Devonian. 50% of the Lower Devonian species of the Falkland Islands and 30% of the Lower Devonian species of Brasil and Bolivia correspond with those of the Souter African Bokkeveld beds. Gerth claims that this can only be explained if we assume that South Africa lay nearer South America in former times [2]). Still, these facts cannot be said to contribute any arguments in favour of the idea of continental drift. The same applies to the assertions of Gerth, du Toit

[1]) The fact that the bottom of the Atlantic and the Indian Oceans, and the main part of Melanesia, are covered by sial is the principal reason why Gutenberg suggested a modification by supposing that the continents did not break and drift like ice-bergs but that they flowed apart, so that a connection of sialic material remains between them across the floor of the Indian and Atlantic Oceans. So he "assumes that the continents have changed their distances and positions by plastic flow but that otherwise, the history of the continents and oceans corresponds roughly to the sketch given by Wegener (1936, p. 1609). As to Wade's modification of the drift theory, see p. 243 note [1]).

[2]) Gerth 1934, p. 66.

and others as regards a stratigraphic and structural similarity of the Gondwanides and Cape Mountains. It should be noted that the existence of analogous structures and stratigraphical sequences were reported by Holtedahl in such widely-separated regions as Spitsbergen and Scotland! (fig. 143). Yet no one would think of continental drift in this last case! Moreover, I wonder whether this hypothesis would be used to explain the following: the Maastrichtian fauna of Western Australia is closely allied to that of the Arigalus beds of the Trichinopoli district [1]) of India; and Spath [2]) recently

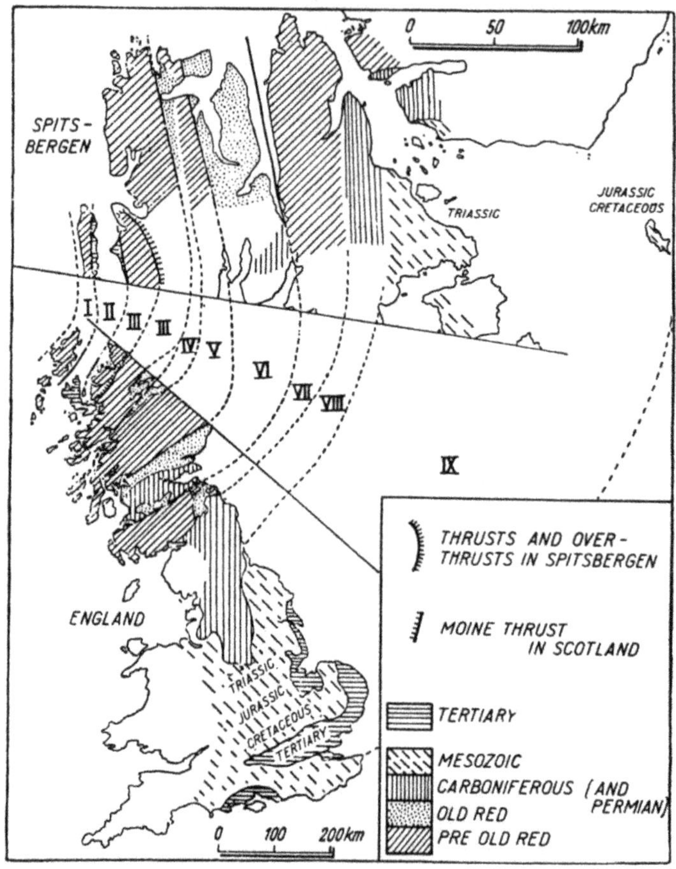

Fig. 143. A comparison between the geological structures of Scotland and those of northern Spitsbergen (After Holtedahl).

described the Ammonites of Geraldton, Western Australia, and revealed that they closely resembled those of the Bajocian in England!

The distribution of the Gondwana-flora, including plant remains like *Glossopteris* and *Gangamopteris* (cf. p. 289 and fig. 178), has often been con-

[1]) Teichert. Australian Journ. Sci. 2, 1939, p. 86. See also below (New Zealand). p. 165.

[2]) Journ. R. Soc. W. Australia, 25, 1939, p. 123.

sidered as in favour of the theory of continental drift. However, Longwell has given an example of similar problems presented by the distribution of modern plants belonging to the primitive dogwoods of the genus *Cornus*. The distribution of the genus, containing about 40 species and its near allies, (fig. 144) "poses greater problems than that of the *Glossopteries* flora, since it involves greater oceanic distances and more formidable climatic barriers. Moreover, the hypothesis of continental drift does not help explain the dispersal of dogwoods — rather it increases the difficulties. Obscure ways in which plants and also some animals must have been distributed during long geologic time intervals are suggested by other examples from the vast Pacific region. When these numerous problems are viewed together, emphasis that has been placed on the *Glossopteris* flora as "compelling evidence" of once-continuous lands seems dangerously near the unscientific procedure of selecting evidence to support a favored theory."

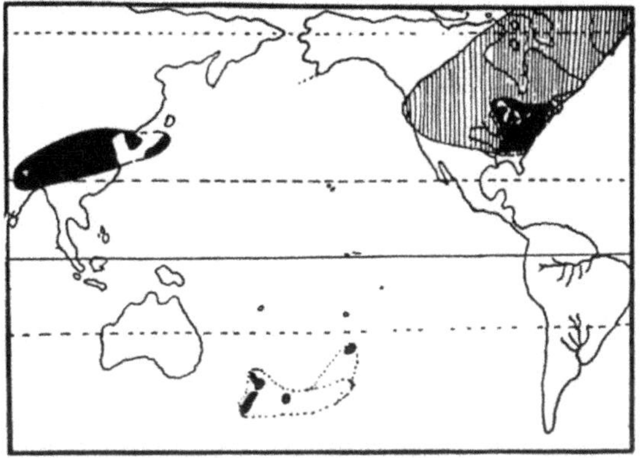

Fig. 144. Distribution of the genus *Cornus*. Modern forms are found in the black areas. Vertical lining shows known Cretaceous distribution (From Ch. Longwell).

No other examples will be mentioned, for it would take us too far to discuss all the difficulties inherent to the hypothesis of drifting continents [1]) (we will deal more fully with some points of this hypothesis in Chapter IX). Besides, the conclusions arrived at in the preceding and following chapters cannot be said to agree with the idea of an enormously extensive drift of continents *since the Paleozoic*. The rejection of the theory that the bottom of the Atlantic and Indian Ocean originated owing to drift during the Paleozoic, or in even more recent times leaves us with two rival hypotheses. The first is the idea of submerged continents, the second the hypothesis of permanence.

Submerged continents

Many geologists, following the example of Haug and Suess, have attached great importance to the reconstruction of large continents (fig. 145) which presumably sank to the bottom of the oceans.

Termier [2]), for instance, wrote: ". . . . beaucoup de nos abimes océaniques sont des gouffres relativement récents où se cachent des portions de l'ancien domaine continental; et si l'on pouvait vider ces gouffres de l'eau qu'ils

[1]) For a more detailed examination of these difficulties see a publication of mine of 1937. A few recent papers *pro* and *contra* are mentioned in the bibliography at the end of the Chapter (Grabau, du Toit, Holland, Longwell, Simpson and Stetson).

[2]) Termier, 1926, p. 359.

contiennent, on verrait, au fond, des fragments de vieilles montagnes ou de vieux plateaux, qui se sont jadis étendus à la surface, sous la bienfaisante caresse du soleil".

Yet many geologists have refused to accept the idea of enormous submerged continents as, apart from objections of a geophysical nature which will be discussed later, such a hypothesis would complicate the water-economy of the oceans in a hopeless manner. The continents would (assuming that the volume of water of the oceans has remained constant) need to have been permanently flooded prior to the submersion of these huge blocks but this conflicts with all available data. The alternative would be that the volume of water would have increased steadily and to an enormous extent since the foundering of continental blocks. Yet this seems highly improbable. This point was clearly illustrated by Kuenen in the

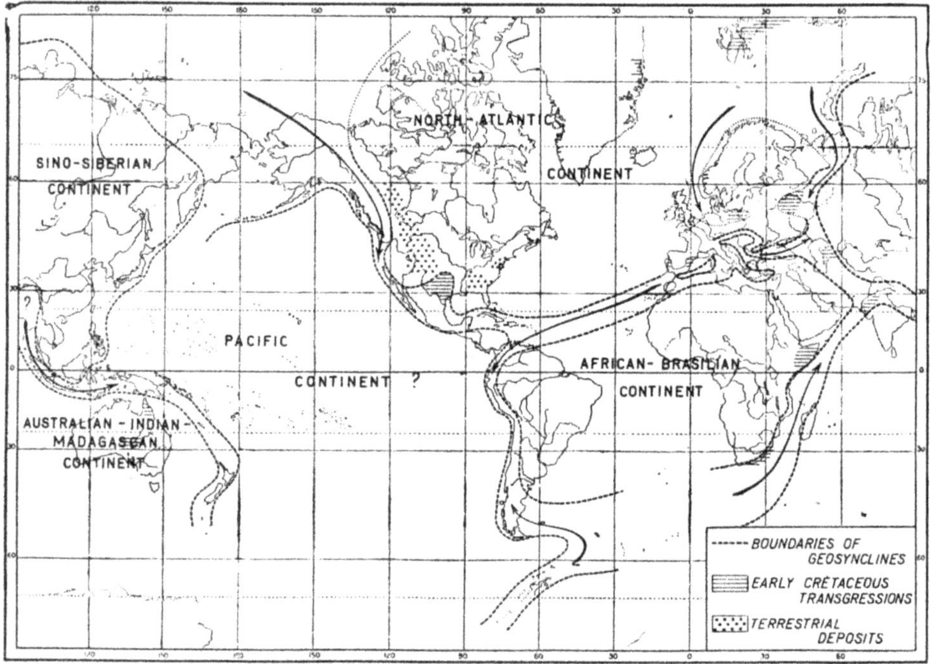

Fig. 145. Hypothetical reconstruction of trans-oceanic continents in the Lower Cretaceous (After E. Haug).

accompanying graph (fig. 146), which includes Walther's and Twenhofel's opinion, as well as his own. Kuenen writes in this connection [1]: "Walther held that there was probably no deep-sea until the end of the Paleozoic because no Paleozoic faunal elements have been found in the present deep-sea faunas. But this assumed absence of old deep-seas is only one of several possible causes that might explain what is after all a rather vague conception. It necessarily implies that the earth had an unimportant hydrosphere

[1]) Kuenen, 1937, p. 460–462.

for some 1500 million years and then acquired its present volume of sea water at the rate of a few km³ per year or a few times the speed at which volcanic rocks are produced. Add to this that salts had to be furnished on the same gigantic scale and the impossibility of Walther's conclusion becomes manifest.

"Twenhofel's assumption, though less extreme is still unacceptable. He says:

'In this inquiry it is assumed that Pre-Cambrian and Paleozoic deep-seas existed, and that if the deep-seas of all time are considered they would be equivalent to deep-sea, with area like that of the present, extending as far back as the beginning of the Devonian'.

"In our graph, the volume of the oceans is plotted against geologic time, showing the views of Walther, Twenhofel and the writer. In Twenhofel's view the area abc must be equal to cde.

"There are many arguments for and against permanency of the oceans. In the opinion of the writer the most forcible are: the absence of fossil deep-sea sediments from the continents and the impossibility of sudden production of water and salts since the juvenile stage of the earth. For

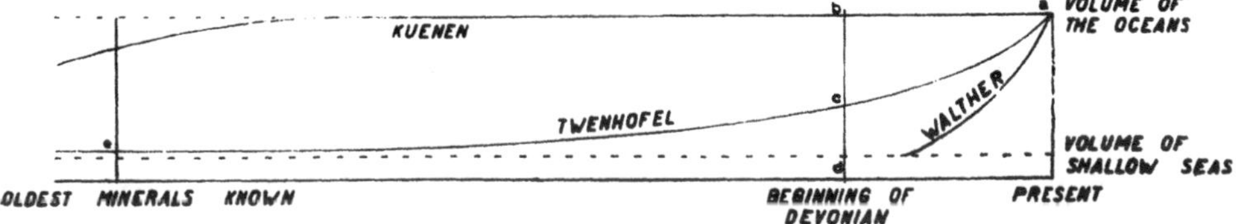

Fig. 146. Total amount of oceanic waters in geological time according to the views of Walther, Twenhofel and Kuenen (After Ph. H. Kuenen).

post-Algonkian times doubt as to the existence of permanent ocean basins is quite unreasonable; although their position and shapes may have varied through continental drift and although isthmian links may have arisen and disappeared".

This question is closely related to that of the total amount of sedimentation in the deep-sea and to the whole complex of geochemical processes of the continents and oceans. So far, however, there seems to be no evidence of an annual increase of a few km³ in the amount of the oceanic waters, and it is quite as improbable that the salinity increased at the same rapid rate. Walther's curve would therefore appear to be too extreme, and — unlike du Toit, whose point of view is expressed in a publication of 1940 on the subject of the Pacific Ocean — I am of the opinion that Kuenen's graphic representation is probably more in accordance with the facts than Twenhofel's curve.

Two conflicting deductions are found side by side if we adhere to the hypothesis of submersion. The first asserts that vast continental blocks foundered in the course of time. The second claims not only that the total amount of the oceanic waters has not increased proportionately since the Cambrian, but that it has even remained almost constant. Should one nevertheless cling to the theory of submerged continents, the only alternative would be to assume that while vast blocks were being submerged in one area, parts of the ocean-floor of an almost identical size were being elevated in others, i.e. — to cite one example — that the bottom of the Pacific which was originally situated at a far greater depth, rose while blocks sank into the Atlantic and Indian Oceans, and that the total amount of water and its surface level consequently remained almost wholly unaltered. These considerations are indicated by the line B–C in the diagram in fig. 147. The line A–C on the other hand, illustrates the idea of a relatively stable floor of the Pacific. In addition to this, both graphs show the periodic cadence of the floor of the Pacific, as deduced from the rhythm of transgressions and regressions.

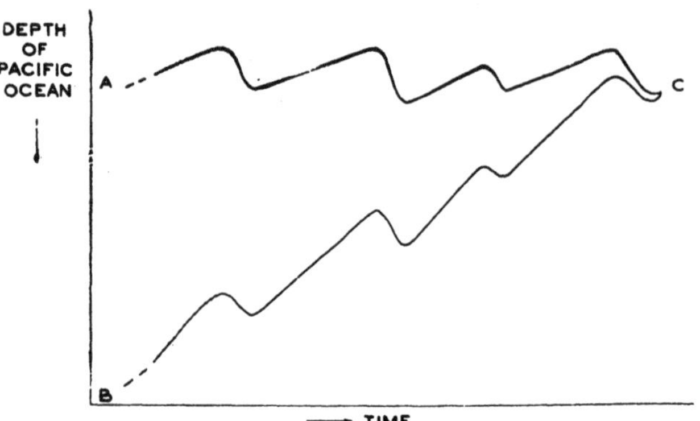

Fig. 147. Schematic diagram illustrating the ideas of a relatively stable (A—C) and a rising floor (B—C) of the Pacific Ocean.

It is not quite clear, however, why such opposed movements should have occurred in areas of almost equal extent. Nor is it clear why these movements should have occurred in such a way that the sea-level remained comparatively stable. Besides, the above assumption gives rise to further complications, for if the phenomena can be attributed to this process, it might be asked what happened in the substratum. To begin with, it remains entirely problematical (in spite of Barrell's speculative though interesting reflections on this subject [1]) why a given area such as, for instance, the present floor of the Atlantic should have been submerged, while the neighbouring blocks remained standing as continents. Moreover, this process involves more than just submersion, and cannot merely be compared to a block of wood which is pressed down more deeply into the water. Seismic data show that the blocks must have grown much thinner during the act of submersion (fig. 156). This might be attributed either to stretching, or else to the melting of sialic material from the lower part of the crust, which subsequently vanished. The first alternative was discussed in the preceding paragraph. This stretching-process might possibly have occurred during

[1]) Barrell made some interesting reflections on the possible cause of the subsiding movement, suggesting that the foundering might have been the result of "the weight of magmas of high specific gravity rising widely and in enormous volume from a deep core of greater density into these portions of an originally lighter crust". Barrell claims that this regional foundering occurred especially in primordial times (though it did not cease altogether then, owing to the periodic accumulation of heat from radioactive processes deep beneath the crust), and draws a comparison with the lava plains of the moon. Yet this hypothesis, too, fails to explain the presence of an upper sialic layer of the Atlantic type, or the absence of such a layer in the Pacific basin proper, and the same applies to the comparatively important thickness of the continental bucklers.

an early stage of the earth's history, but it contributes nothing towards the solution of the problems which have arisen as regards Cambrian and later times, as the features of the sea-bottom (which were dealt with extensively above) appear to contradict the idea of a more recent process of stretching. The second contingency would only force us to accept some additional *hypothesis ad hoc*, with no sound basis whatever.

The above clearly shows that the hypothesis of submerged continents leaves many points unsolved, and we consequently proceed to the third hypothesis with the unpleasant sensation of having been twice disappointed.

The hypothesis of permanence

The aspect of the continents and oceans has not escaped frequent alterations. We need only think in this connection of the East Indian basins, Melanesia and Appalachia. Some parts have been submerged, even though such regions as Cascadia, Appalachia and Scandia were no larger than

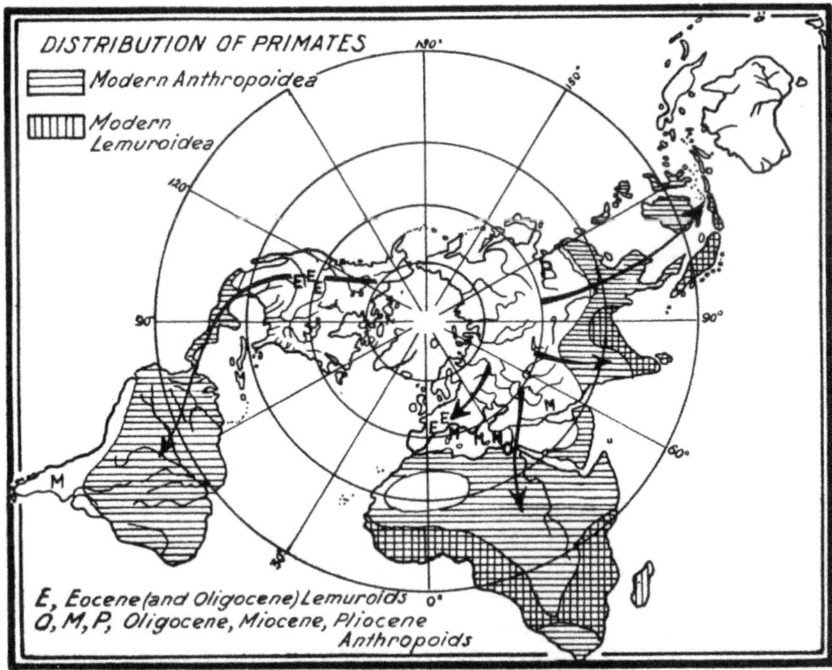

Fig. 148. The distribution of Primates according to the hypothesis of permanence. (After W. D. Matthew).

"border-lands", as Schuchert called them. The idea of permanence is consequently untenable in its extreme sense. But just as Schuchert and Willis proclaimed themselves advocates of the theory of permanence, in spite of the border-lands and isthmian links which they themselves reconstructed, so, too, we can speak *cum grano salis* of permanence as regards the oceans as long as we refrain from reconstructing huge submerged continents (fig. 148).

This hypothesis will have to be based on three premisses: (1) that comparatively thin sialic layers existed in the floors of the Atlantic and the Indian Ocean since at least Cambrian time; (2) that in spite of this, relief-features such as basins and ridges resembling those of the intervening African continent originated within these blocks, and that these were brought about by the same subcrustal processes; and (3) that the amount of oceanic waters has not changed to any important degree since the Cambrian.

Among the features which are liable to change in one or perhaps several

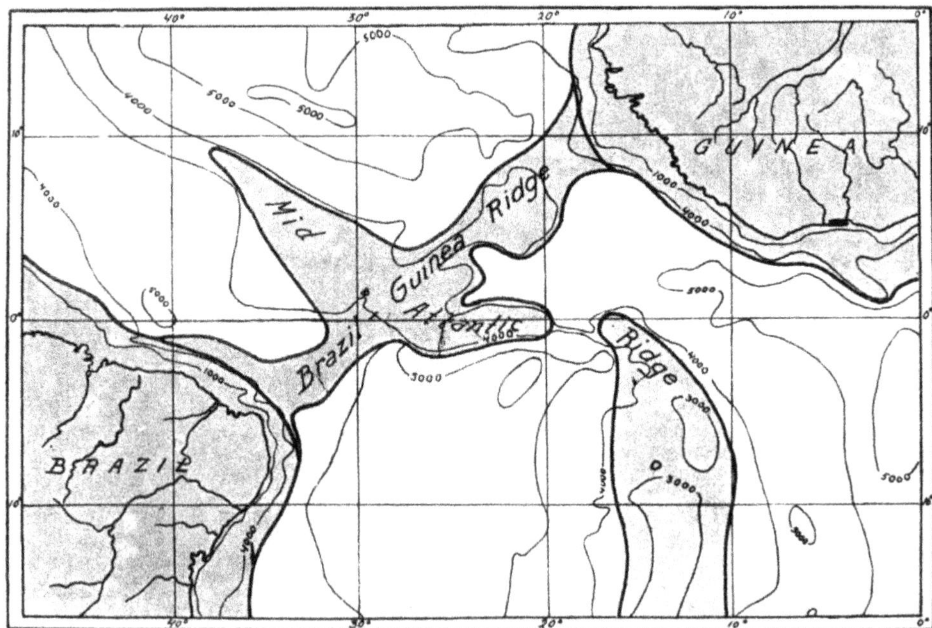

Fig. 149. A transatlantic isthmian link (After B. Willis).

respects are the details of the submarine relief. For example, here too we need only think of such foundered blocks as Appalachia, the possibility of widely diverging periods of origin of the basins of the Atlantic and Indian Oceans with their possibly later movements, and the "rejuvenation" of the relief, which is so strikingly illustrated by the Mid-Atlantic Rise and the Carlsberg System. These movements remind one at once of the "isthmian links" of Schuchert and Willis (fig. 149). Reflections of a highly interesting though speculative nature were made by Willis and Nölke on the emersion and submersion of narrow trans-oceanic isthmi. Their origin and submersion will probably remain a mystery for some time to come, but the inference that such features had in fact been formed and then vanished would not seem to be an improbable one. Cloos [1]), too, when dealing with the analogy

[1]) Cloos, 1937, p. 344.

of the characteristics of the continental and oceanic sectors (Atlantic and Indian) drew attention to the fact that the narrow zones surrounding the continental basins can look back upon a particularly agitated history, with important vertical upward and downward oscillations. I consequently incline towards the view that no essential objection can prevent us from accepting the idea of isthmian links as postulated by Schuchert and Willis. Of course, it cannot be denied that the existence of land-connections during certain given periods cannot be proved geologically. Each so-called reconstruction of a trans-oceanic land-bridge msut necessarily retain a hypothetical character. Yet such land-bridges need not be discarded *a priori* as mere products of the imagination.

Another changeable factor is the periodical variation of the distance between the floor of the sea and the surface of the continents. This question was discussed in Chapter V.

A few fundamental questions still remain to be solved, however, viz. how to explain (1) the congruence of the abruptly ending folded belts on both sides of the Atlantic; (2) the way in which the relatively thin and deeply situated floors of the Atlantic and Indian Oceans came to be formed; (3) what caused the Pacific floor to be sial-free; and (4) how the antipodal arrangement of continents and ocean floors came into existence. These problems will be treated in two separate sections.

Trans-oceanic mountain-chains

I would first like to draw attention to a difficulty which may be said to characterize every kind of hypothesis in which a certain degree of permanence is assumed in oceanic floors of the Atlantic type. For how are we to explain the abruptly ending Caledonian and younger folded mountains on either side of the ocean? Stratigraphy and tectonic data show that a strip of unknown extent is lacking. Yet the morphology of the floor of the ocean contains no indication of submerged Caledonian or other folded chains. Moreover, Haug remarked quite rightly that the theory of trans-oceanic mountain-chains requires, amongst other things that a transoceanic geosyncline, including an area of erosion on at least one of its sides, should have existed prior to the said mountain-chains. It was this that induced Haug to reconstruct those continents, the existence

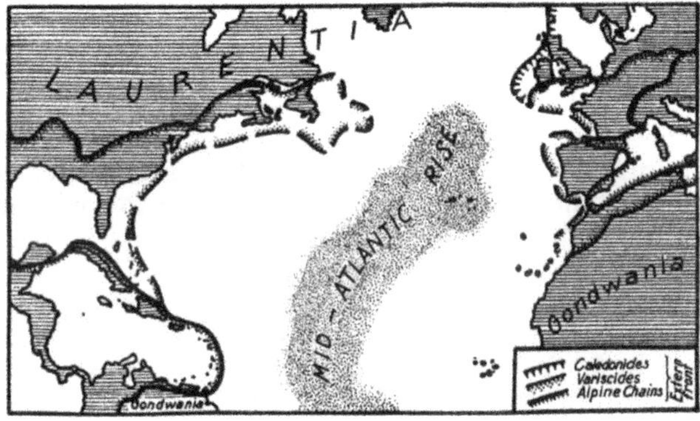

Fig. 150. A hypothetical reconstruction of non-transatlantic mountain-chains (After H. Stille).

of which we rejected on the grounds mentioned above. We can therefore easily understand that Stille doubted the former existence of transatlantic mountains [1]), and fig. 150 shows how Stille imagined the Caledonian and Variscian Mountains to have bent back originally towards the continents (see also Plates 1 and 2). Yet even so, another fact remains, viz. that the abruptly ending belts are situated approximately opposite one another on either side of the ocean. The following suggestions may help to overcome these difficulties. Let us, for example, suppose that, owing to the influence of subcrustal processes, the earth's crust had reached a stage where it became possible for a geosyncline to form within a certain zone. Let us also assume that this zone continued under an area such as that of the Atlantic. In that case a geosyncline and folded chain would potentially be able to originate within a certain zone of the floor of the Atlantic. Yet, such a thing would never happen, for insufficient erosion products would be available for the purpose of filling the furrow. Should a subsiding trough originate within this zone and be buckled, the result would be a sialic root of considerably smaller dimensions, for the sialic layer of the Atlantic is a much thinner one than that of a continent. Should this nevertheless happen, the isostatic anomaly would be much smaller than that in the contemporaneous extension of this zone on the continents. This would help to show that a trans-oceanic connection need never have existed above sea-level, and that the stratigraphy, tectonic structure and epochs of folding of mountain-belts are more or less analogous on either side of the ocean in spite of the absence of trans-oceanic mountain-belts. It would then no longer be necessary to look for the formation and subsequent disappearance of a missing link which had never been visible above sea-level — nay, which had never even existed.

Moreover, the above considerations show a way in which to test the hypothesis of permanence. For if transoceanic chains foundered in the ocean, gravimetrical observations might be expected to reveal them as a strip of distinct deviation of the isostatic equilibrium, occurring in the prolongation of the abruptly ending zones of folding on the continents and presenting anomalies of the same intensity as those in this particular continental belt, On the other hand, it would be more in accordance with the hypothesis if these anomalies were non-existent or much fainter than those in the adjoining continental strips. So far as I know, gravimetrical observations at sea are not numerous enough and far too widespread at present to enable us to test this surmise. The same applies to a possible testing of sunken border-lands and isthmian links.

To return to the working-hypothesis — we might add the following. It was shown above that terrestrial dynamics caused a similar pattern to form in both the continental and the oceanic sial-sheets during the consolidation of the surface. It should therefore not be wondered at that both areas responded in an identical fashion to the subcrustal forces of later times. Hence phenomena such as the formation of basins and furrows, graben and dome-shaped elevations and the rejuvenation and foundering of blocks obviously occurred in both these sectors. It should be noted, however, that

[1]) Stille, 1939, p. 343, 345.

there is one phenomenon which would never manifest itself in oceanic areas viz. the formation of trans-oceanic mountain-chains. The one factor to prevent the formation of the latter is the absence of detritus resulting from atmospheric denudation.

It might be objected that no attempt has been made to explain the foundering of border-lands or the sinking of certain oceanic basins. These phenomena, however, occur in quite the same way on the continents, and the problems arising from their origin and history are not confined to oceanic sectors. The submergence of border-lands, for instance, cannot in itself be regarded as an unusual phenomenon. For it should be noted that blocks which were first submerged, then elevated, and then once more submerged and elevated, are also met with on the continents. The sub-oceanic features and the similar continental characteristics cannot be explained at present, for our knowledge of Pre-Cambrian history and terrestrial dynamics is not yet extensive enough. Hence the fact that no adequate explanation can be given of these features does not imply that the above working-hypothesis is a deficient one.

The origin of continents and oceanic receptacles

The concept of permanency of the oceanic receptacles implies the conviction that these major depressions date from some very early period of the earth's history and that their position has remained essentially unchanged since then. But now the question arises when and how the "permanent" distribution of continental and oceanic sectors came into being in such a way as to display an antipodal arrangement and a "tetrahedral" plan. What were the factors that caused the concentration of a northern belt of triangular continents alternating with a southern belt of oceanic units distributed "like a pair of cog-wheels with interlocking teeth"? [1]). When, on page 230, the problem of the ocean-floors was outlined in six different points, this question was mentioned under number one. Thus far, however this question remained unsolved, even untouched. As a matter of fact thinking of the terrestrial processes that must have operated in the earliest days of the infancy of our globe, means to enter inevitably a domain which is characterized by very scanty available data and is therefore wide open to unbridled speculation. But why should we not enter it if every one who wants to join us on our geopoetic expedition into the unknown realm of the earth's early infancy is warned at the beginning that probably not a single step can be placed on solid ground? [2])

It is believed by some geologists, especially by petrologists like Lawson and Rittmann, that the Earth's surface in its initial stage was composed of a basaltic melt. It will be interesting to see how they deduced sialic continents from a primary homogeneous surface-layer of basic composition.

A second group of theories lays stress on a happening of cosmic dimensions: the birth of the moon out of the body of the earth. In this group will have

[1]) J. W. Gregory, 1899, p. 227.

[2]) An interesting review of older theories of Prof. Lapworth, Sir John Lubbock, Lord Kelvin a.o., was given by Gregory in his paper of the year 1899, cited in the bibliography at the end of this chapter.

to be considered the theories of Osmond Fisher, Pickering, Escher a.o. They start with a globe surrounded by an outermost layer of sialic rocks of continental thickness, and they try to deduce the origin of continental bucklers and oceanic depressions from the cosmic catastrophe which gave rise to the genesis of the moon in a way suggested by G. H. Darwin.

Then, in the third place, we will start anew with the idea of a world-encircling layer of sial and try to deduce the desired distribution of continents and oceans without the aid, however, of the hypothetical events of the lunar catastrophe. Some efforts in this direction have been presented by the present author and by Vening Meinesz.

In 1932 Lawson formulated a theory to account for the formation of continents from a crust which originally consisted of sima only. In his opinion an upper layer of basaltic rocks covered a deeper substratum of dunite. Basaltic continents were produced by orogenic stress. Erosion products from these basaltic continents were carried to the primeval oceans, the basic parts went in a state of solution, while the sialic elements became concentrated, and when these waste-products sank sufficiently they were fused together so as to form acid, granitic rocks. By this process the continents became continually more sialic. Then, the sialic material differentiated into a double layer of granite above and diorite beneath.

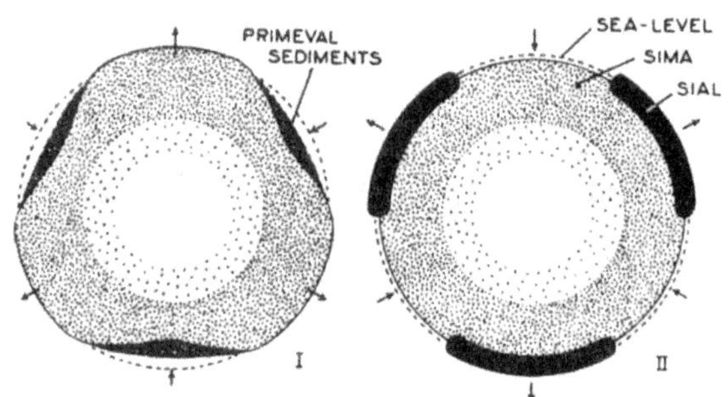

Fig. 151. Schematic representation of the origin of sialic continents on a primeval earth-surface of alkali-basalt, according to the hypothesis of Rittmann.

Seven years later the Swiss petrologist Rittmann introduced a theory which in its main elements is closely akin to that of Lawson. It differs from it, however, in one respect. According to Rittmann the sialic waste-products which assembled in the primeval oceans — see fig. 151 — were migmatized by gases emanating from the substratum. By this process they grew specifically lighter. To restore isostatic equilibrium the sialic floors of the primary oceanic receptacles emerged to form sialic continents. On the other hand the original bucklers of alkali-basalt subsided to form the bottom of the present oceans!

A weak point in both these theories is the formation of the primary basaltic continents. For, in the first place, it is not clear how these protuberances could have formed under the preservation of isostatic equilibrium. Nor is it clear how these units could have risen isostatically during the long continued rise above sea-level necessary to provide the primeval oceanic depressions with thousands of meters of erosion products [1]). Moreover, the primary disturbance causing the genesis of continental domes out of the basaltic surface layer comes like a *deus ex machina*. Without accepting this initial stage, the further deductions of the theory would become impossible. Finally, the theories leave the question of the tetrahedral or antipodal plan of terrestrial elements untouched. Nor does it explain the existence of three hypsometric levels, nor the remarkable structural pattern of the bottom of the Atlantic, etc. etc.

Disappointed we return, and at once we depart again to share another group of excursionists in the realm of the earth's infancy. It was G. H. Darwin who announced the hypothesis of the moon's origin not as a little sister but as a daughter of the earth. The main points of his luminous theory were set forth in Chapter I (p. 6). Several geologists have tried to imagine the consequences of such a hypothetical birth upon the structural history of the earth's surface-layers.

If we accept the resonance theory of the moon's origin it seems highly probable that a great part of the earth's outer shells entered the moon's formation. But what happened to the remaining sialic matter of the earth? There are two possibilities, viz. either the terrestrial remnants of sial were able to flow out and unite again as one continuous shell, enveloping the entire world, or they were unable to do so, being too rigid. The last possibility is basic to the theories of many authors. In their opinion the earth was originally enveloped by a sialic shell of continental thickness. After the moon's disruption the remaining sial-fragments of the earth would have formed our continents, floating on the heavier simatic substratum that forms the bottom of the oceanic receptacles (fig. 156). Long ago the question was raised whether the Pacific might not be considered as a scar brought about by the separation of the moon. Our continents should be compared to the acid slags floating on the surface of the heavier melt in a blast furnace.

Practically the same idea can be found in the theories of Osmond Fisher, Taylor, Wegener, Schwinner, Mohorovičić and Escher. The two last named authors even tried to calculate the thickness of the lunar sial by computing the mass of terrestrial sial that originally filled the gaps of the oceanic sectors, fig. 152. (Both assumed an original crust of continental thickness and no account was taken of the presence of the sial-flakes of the Atlantic and Indian oceans).

It is at all events clear, however, that neither the continents nor even their so-called nuclei (representing the innermost structure of the continents) should be regarded as undisturbed remnants of that distant and turbulent period in our planet's infancy. The preceding chapters showed that these

[1]) In this respect Wade's theory of continental spreading (cf. p. 231) which also starts from a basaltic surface has some slight advantage. His primeval polar continents were thought to become bare land by the hypothetical process of a retraction of the oceans from the polar areas towards the equatorial belt. The water was supposed to pile up in that region as a result of (1) a greater speed at which the earth rotated on its axis and (2) a greater gravitational pull by the moon which then was nearer to the earth. According to Wade's theory the polar continents became covered by a sheet of sial by the "agencies at work which would effect oxidation, hydration, erosion, and sedimentation". And he continues (p. 1808): "As soon as these sial-sheets became thick enough and sufficiently consolidated to upset the hydrostatic equilibrium at the poles, both sheets would begin to move and the problem resolves into one concerning two fluids — a lighter one resting on one more dense".

"acid slags" had from the early beginning a very complicated history and that the continents (parts of which have been breaking down until quite recently) formerly occupied a more extensive area than they do at present. We know, moreover, that the history of the submarine relief of the oceans

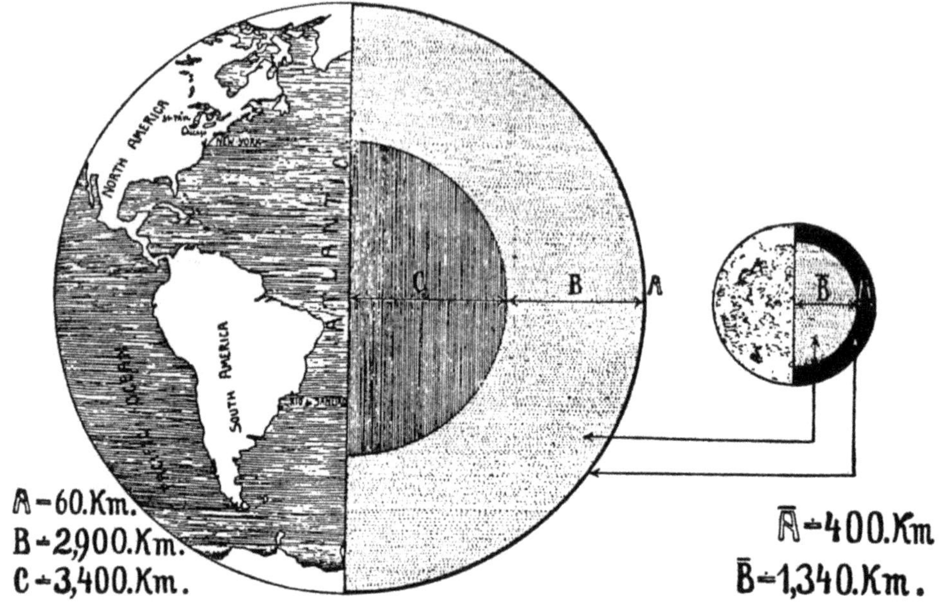

Fig. 152. Internal constitution of the Moon and the Earth (From Mohorovičić).

is a far from simple one. Another point is that the bottom of the Atlantic and Indian oceans differ considerably from the floor of the Pacific.

This last important point is, however, taken into account by Pickering's hypothesis. For, firstly, his supposition that the scar left on our globe's surface when the moon was torn from the earth's body must be represented by the Pacific, explains why that ocean-bottom is free of sial. Secondly, he supposes the scar originally to have been much larger than the present Pacific. ". . . . three-quarters of this crust was carried away" — he wrote — "and it is suggested that the remainder was torn in two to form the eastern and western continents." So we could very well imagine that *at this stage* some continental masses drifted apart and that between them the floor of the Atlantic and Indian oceans formed as a result of stretching. Here Pickering's theory is much to be preferred above Wegener's and du Toit's theories of continental drift. For in a previous section (see p. 231) much stress was laid on the convincing arguments which show that provided the floor of the said oceans originated as a result of stretching this process must have occurred during the early Pre-Cambrian and decidedly not since Cambrian or still less in post-Carboniferous times, as Wegener *cum suis* wished us to believe.

But again the question arises, why these sial-flakes drifting apart like large ice-floes, came to be settled down in their peculiar tetrahedral arrangement.

Another question of primary importance, however, is: *why should the sialic material have flowed out to form a continuous shell on the moon and not on earth?* The following points should not be forgotten in this connection. It is questionable, although quite possible that prior to the postulated disruption, the sialic material had differentiated into the outer parts of our globe, but was not yet in a solid state. Every one will admit (see fig. 4) that the outer shells of the earth were of such a fluid composition that they readily responded to the tidal forces which deformed the earth's geoid before as well as after the moon's separation. It will also be admitted that part of these outer shells shifted towards the moon and that the scar of the separation was closed smoothly. Moreover, there does not seem to be the slightest doubt that when the earth, recovering from its amputation, was moulded into its new shape, the simatic layers — even the uppermost — formed a fluid mass uniting into a continuous shell. It seems, therefore, quite unreasonable to suppose that the thin upper layer of sial (see fig. 7) would have been the only one incapable of reacting in the same way [1]. And further if the Pacific has to be considered as the scar that originated by the birth of the moon what can be said about the North Polar Basin which probably also has no cover of sial?

Finally, the preceding considerations are based on the resonance theory of the moon's origin and are valueless if we accept the opposite view, viz. that the moon and the earth formed simultaneously as two separate bodies (see Chapter I).

For the third time we start on a trip in the realm of speculation. Our considerations will in this case have to be based on the supposition that the earth was initially enveloped by a continuous sialic layer gradually solidifying (fig. 154, A, B) and floating on a denser basic substratum [2].

It may be that this layer formed after the moon's disruption, but one might similarly assume that the moon was not severed from the earth.

According to the present author, the only plausible way in which continental blocks might have originated from a world-encircling sialic layer, growing gradually cooler and solidifying, is that a process of thickening occurred, resulting from folding and drifting owing to the action of convection currents [3]: This process contracted the sialic layer, and sial-free parts (such

[1] Even the lighter simatic material would have been lacking in the Pacific sector according to Schwinner (1935, p. 316) and this would have prevented the sialic layer from covering this area. The fact, however, that the volcanoes of the Pacific are built up of basaltic material clearly shows that this opinion can not be upheld.

[2] A similar idea about an originally world-encircling layer of sial may be found in Bowie, Geszti, Daly (1938, p. 176) and Wegener (4e Aufl. 1929, p. 207).

[3] This is the same idea, after all, as that met with previously when retracing the earth's history from its more recent stages to the remote ages of its very beginning (see Chapter IV). Since the oldest continental shields are known to consist of intensely folded belts, the question naturally arises whether the continents were not the result of a process of thickening, i.e. of a periodical process of buckling of an originally thinner and world-encompassing sialic layer. A tentative and questioning reference was made to these considerations in Chapter IV.

as large portions of the present Pacific area, with the possible inclusion of the oceanic bottoms mentioned in groups 3 and 4, or parts thereof, see page 219) were therefore formed.

Fig. 153 shows diagrammatically how a continent may possibly have

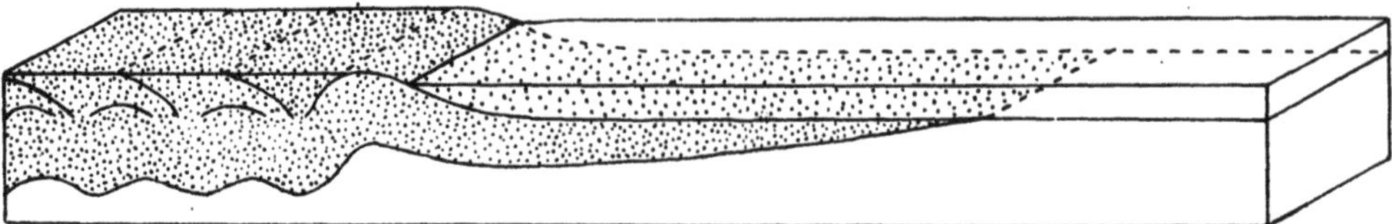

Fig. 153. Diagrammatic and tentative illustration showing the hypothetical formation of a continent and a sial-free ocean-floor.

originated through periodical buckling and drifting of an originally thin sialic layer enveloping the whole earth, owing to the action of convection currents. In several of the preceding paragraphs it was pointed out that the Atlantic floor might have formed as a result of stretching in the sense Bucher and du Toit attached to this process, but special emphasis was laid on the fact that in that case it would have occurred at any rate in the Pre-Cambrian, as shown above by morphological and geological arguments. It may be that during the process of drifting of these primordial continents a sial-sheet was torn into two or more parts, each moving in a different direction and producing one or more thinly stretched sial-flakes between them. In this way (and only during these primordial times) a thin sial-sheet like that which probably exists under the Atlantic may possibly have originated as a result of stretching of either a primordial continent or an intervening primary sial-layer (fig. 154C). If the sial of our present continents were imagined to spread out as thickly as the sial-layer of the Atlantic, it would probably cover a larger area than that of the present sial-free ocean bottoms. A floor such as that of the Atlantic cannot therefore be regarded as a remnant of the primary world-encircling sialic shell. On the contrary, it would seem to have grown thinner as a result of stretching during the early Pre-Cambrian. It would be impossible to say just now whether a fragment of the primary sialic layer of original thickness exists in some part of the world to-day.

It is obvious that this process must have attained its most important effect in the early Pre-Cambrian, as the whole later process of buckling and folding of the continents is known to have evolved only in these existing Pre-Cambrian bucklers [1]).

This is indeed remarkable and will have to be examined before any other features are discussed.

In the beginning the continents grew steadily larger, but this has not been repeated since at least the Cambrian or probably even earlier times, as the geosynclines and folded chains that are known to have formed since then

---

[1]) Even on those rare occasions in which abyssal sediments have been observed, it seems unlikely that these areas should be regardul. as exceptions. This is at the most doubtfed

originated in a basement which had already been folded previously [1]). Hence we cannot escape the conclusion that some event of fundamental importance must have occurred during the early part of the earth's history causing the surface history of primeval times to differ from that of subsequent periods. By this we mean that from a certain critical moment onward continental growth in a "horizontal" direction ceased.

The most plausible explanation seems to be that the world-encircling solid crust, had become so thick that subsequent growth was no longer possible. Such youthful caprices as are illustrated in fig. 153 and 154 C, D, may be expected to have ended as soon as the earth had attained a physical state more or less resembling that which it has at present. From that time onward the terrestrial forces were imprisoned and became operative as *subcrustal* processes (fig. 154, E).

At that time the crust consisted of basic rocks in the sial-free parts of the surface, just as it does now, and it may have contained both sialic and simatic rocks in those regions of the world where sial-flakes were present. It must have been of varying thickness, though it need not of course have equalled its present-day thickness. We will call this state the "consolidated surface of the earth".

Prior to the formation of a world-encircling crust, the terrestrial dynamics probably evolved with the same periodicity as later on. Yet the effect produced thereby on a non-consolidated surface was apparently a very different one than that produced on a solid crust by the "deeply imprisoned titans". Before the formation of a consolidated surface, the sialic upper layer was folded into continental thickenings. This was the period of continental growth. With the solidification of the earth's surface this process came to an end.

Complete homogenity must, if ever it existed, necessarily have been disturbed by several factors.

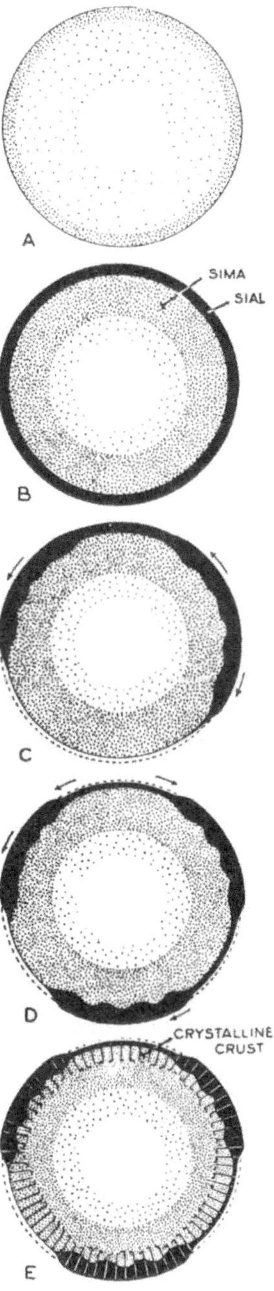

Fig. 154. Schematic representation of the origin of continents and oceanic receptacles in the early Pre-Cambrian by a process of buckling and drifting of an originally sialic layer enveloping the whole earth.

[1]) A fuller illustration of the preceding consideration will be found in Chapter II and the notes on Plates 1 to 5 in the Appendix.

One needs to think only of the earth's flattening at the poles, the excentricity of its orbit, the inclination of the earth's axis, and the action of the tides.

A few years ago Bowie formulated his opinion regarding a primeval process of crustal buckling as follow (1935, p. 447): "It is impossible on a molten earth gradually cooling to have the light material which must have been present around the whole earth, pushed together into great masses to form continents. There is no force available inside or outside the earth which could have caused such a segregation of surface matter".

As mentioned before, however, the process of folding and drifting of the sialic surface layer is ascribed by us to the action of convection currents. Their periodical occurrence must at one time have found the sial-sheet cooled and solidified to such a degree that the result was the formation of continental thickenings.

If the above hypothesis may be said to contain a germ of truth, the problem of the continents and ocean-floors would consequently be associated with periodic phenomena of crustal folding which had, however, already produced its main effect in the early Pre-Cambrian. A few points will now be examined more closely.

Let us consider the moment that these primeval surface-features became petrified, as it were, by the formation of a world-encircling solid crust of sufficient thickness (fig. 154D). Among the first topics to be mentioned in this chapter was the noteworthy structural pattern of the floor of the Atlantic and the surrounding continents. It was shown that this pattern had probably originated in the early Pre-Cambrian and attention was drawn to the part which these old structural lines repeatedly played in later times (pp. 219–224) We arrived at the same conclusion when dealing with the continental basins (p. 54). It might therefore be asked whether it is not obvious that this ancient pattern dates from the period of consolidation of a solid crust.

In the preceding paragraphs we came to the conclusion that three favoured levels occur on earth, represented by (1) the continental surface, (2) the floor of the Atlantic, and (3) the bottom of the Pacific Ocean. It need hardly be said that these results fit in well with our working-hypothesis. The difference in the level of the primeval floors of the Pacific and Atlantic is seemingly due to the difference of the materials composing each floor. On the other hand, buckling of the primordial sialic crust was responsible for the continental thickenings. Their remarkable thickness and high level, as compared to that of the bottom of the Atlantic, is controlled by five factors, viz. (1) the thickness of the original world-wide sialic layer, (2) the compressive forces during the periodic phases of early Pre-Cambrian buckling, (3) the leveling effect of subaeral erosion, (4) the Pre-Cambrian stretching of sial-sheets of the Atlantic type, and (5) the local thickening of the continental basement-complex as a result of the formation of later geosynclines and mountain-belts.

The above considerations indicate the main outlines of the new working-hypothesis. I do not wish to suggest that this hypothesis solves the intricate problem of the ocean-floors. I merely followed a line of thought which had hitherto been neglected, and as all the previous hypotheses are difficient in many respects, the new "working hypothesis" may stimulate others in their efforts to unravel the mystery of the ocean-floors. I fully realize how many difficulties are inherent in this new aspect of the old problem.

It will be seen that some of the most attractive aspects of the theory were explained as well by Pickering's trend of thought. And the weak points in both the theories are again the same. For no adequate explanation was

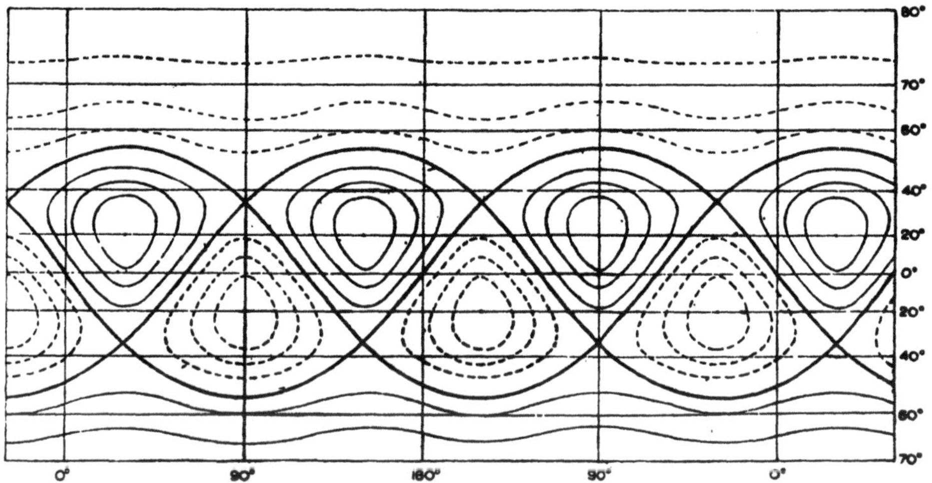

Fig. 155. Distribution of primeval convection-currents in the surface layers of the earth. Full drawn lines represent rising currents; dotted lines descending currents, each for 0, 1/4, 1/2 and 3/4 of the maximal intensity (From Vening Meinesz).

given for the antipodal distribution of the major units of the earth's surface, nor for the triangular shape of these elements [1]).

Quite unexpectedly, however, a mathematical and geophysical treatise by Vening Meinesz seems to throw some light on this side of the problem.

1) I would like to elucidate a few remaining points. The hypothesis to which we referred above is based on the assumption that a continuous sialic layer, which was orginally enveloped by a thin layer of water, extended over the entire world. Oceanic basins with a steadily increasing volume of water originated *ipso facto* as the continental blocks were created. It would consequently be wrong to presume that the continents emerged from the deep-sea, and that we might therefore expect to find the continents covered by a veneer of deep-sea sediments. On the contrary, deep-sea deposits are known to be almost non-existent in the present continents. Yet this cannot be said to conflict with our working-hypothesis, the more so as the volume of water has undoubtedly grown since primeval times (see fig. 164, where we find this opinion expressed even in the top curve!).

As pointed out in Chapter II, the whole history of geosynclines and folded chains since the Cambrian took place on a basement which had already been folded during the Pre-Cambrian. It is impossible to describe the sub-crustal processes which were responsible for the distribution and sequence of folding on the continents (as shown on plates 1 to 5). It is evident, however, that a repeated buckling of continental areas — a process which may possibly also have taken place in the Pre-Cambrian — must have caused the lower side of the continents to become extremely irregular. This feature is in fact corroborated by both seismic data and the different altitudes of the mountain-belts, which are more or less in a state of isostatic equilibrium.

In the year 1944 he published a preliminary paper on the distribution of continents and oceans. The main line of his argumentation runs as follows. It is supposed that a system of convection-currents must have come into existence during the primordial phase of cooling of the earth. One possibility is that the rising currents carried sialic components to the surface where these acid constituents solidified *in loco*, no further displacements taking place on the surface. Another possibility is based on our above-mentioned theory of an originally continuous layer of sial envelopping the whole world. The sialic layer then was thought to have been thrust together by the action of subcrustal convection-currents. Now, Vening Meinesz tried to prove that the system of convection-currents had to become arranged in a very peculiar manner, viz. in such a way that the breadth of a current is double its height. Admitting that the terrestrial system, at least in a primordial stage of cooling, reached down to a depth of 2900 kilometers, Vening Meinesz found as their most probable distribution: eight currents, four rising ones alternating with four descending currents. Hence, each current ought to occupy an octant of the outer parts of the globe and it seems plausible to suppose that the axis of two of these octants coincided with the rotation-axis of the earth. The resulting distribution is shown by fig. 155 in Mercator-projection. This remarkable result suddenly seems to make clear how the antipodal distribution and the triangular shape of the major terrestrial elements possibly came into being. For the figure clearly shows how these features are graphically represented in the scheme as a necessary result of the mathematical formulae.

Facts, many more facts are wanted. Future investigations undoubtedly will bring more unexpected data — and theoretical points of view. For the moment we may say that undoubtedly some remarkable results have been obtained, but we may safely say also that the nuclear point, i.e. the question of the origin of continents and oceanic depressions is and remains a baffling problem. Science progresses and every one will be convinced that one day scientists will be able to dissipate more and ever more the fog that at present makes us stumble when we try to penetrate — along various paths — into the sacrosanct of the unknown.

Summary

Four major morphological units can be observed in the relief of the Ocean-floors. They are: (1) the Atlantic and the western part of the Indian Ocean; these areas are characterized by numerous basins with comparatively flat bottoms, separated by relatively narrow and steep ridges; (2) the large North Pacific basin. The most prominent characteristics of the latter are the diagonally-arranged ridges; (3) the eastern part of the Indian Ocean, the south-western part of the Pacific, and the three Antarctic basins, all of which are marked by but a slight diversity of relief; (4) the North Polar basin, which would appear in many respects to resemble the basins of the third group. We will not consider the areas mentioned under (3) and (4), for data on these regions are still very scarce.

An examination of the Atlantic and Indian Oceans, and a comparison with the available geological and seismic data, shows that the bottom probably consists of sialic material. Internal forces caused similar events to occur in this sialic layer and the surrounding continents (e.g. basins and ridges were formed, and eventually rift-valleys). The basins and ridges have a structural pattern and symmetrical arrangement which seem to date from the early Pre-Cambrian, but the relief of the ocean-floor underwent the same rejuvenation as the continents in comparatively recent times (typical examples are the Mid-Atlantic Rise and the Carlsberg Ridge). There can hardly be any doubt that some blocks of as yet unknown extent (such as Scandia, Appalachia and the Arabian Sea) have become submerged.

The arrangement of the linear ridges in the Pacific, crowned with many volcanoes, is attributed to the magma finding its way upward along faults and fissures in the ocean-floor. This floor of the Pacific appears to be built up by a rigid crystalline layer of basic rocks (crystalline sima). Sialic blocks seem to have foundered in the marginal areas around the Pacific (e.g. Cascadia), but the fact that the major part of the floor of the Pacific is composed of crystalline sima conflicts with the idea of a submerged Pacific continent.

The most frequent depth of the Atlantic appears to occur at a level of 4–5 km, whereas the North Pacific is generally one km deeper. The explanation seems obvious, viz. the deep level of the Pacific is controlled by the simatic nature of its bottom, the 1,000 m higher level of the Atlantic floor is brought about by the presence of a sial-sheet and the still higher level of the continents is due to their being composed of sial-sheets, which are considerably thicker than the sial-flake under the Atlantic.

The available morphological, geological and geophysical data cannot be said to agree with either the hypothesis of a large continental drift in comparatively recent times (fig. 156B), or with the hypothetical assumption that the floors of the Atlantic and the Indian Ocean originated in the Paleozoic or at an even more recent date as a result of stretching of continental blocks (fig. 156C).

The hypothesis of large submerged blocks is inconsistent with the water-economy of the oceans (fig. 156A). For if the volume of water of the oceans is assumed to have remained relatively stable, the continents would have had to be permanently flooded prior to the foundering of the continental blocks, and this conflicts with all that is known of its geological history.

The alternative, viz. that the volume of the oceanic waters has increased by a few km^3 per year is equally improbable. The only way out of the impasse would be to suppose that, while large blocks were sinking in one area, other equally extensive blocks were being raised elsewhere to the same extent. Such a phenomenon would in itself be unacceptable, but apart from this objection it would be difficult to understand why certain continental blocks should have submerged to the exclusion of others. The last point to be noted is not only that submersion itself gives rise to insurmountable objections, but that the thinning of the gigantic blocks forms an additional difficulty. Many fundamental difficulties will therefore have to be solved if we are to explain the problem along the lines of the hypothesis of submerged continents.

The hypothesis of permanence should be taken *cum grano salis*, for sialic blocks of restricted size are known to have foundered as "border-lands". If the bottom of the Atlantic and Indian Oceans may be assumed to have existed as thin sialic layers since at least the Paleozoic, and the total amount of oceanic waters to have remained almost constant since then, two changeable factors remain. The first is represented by the details of the

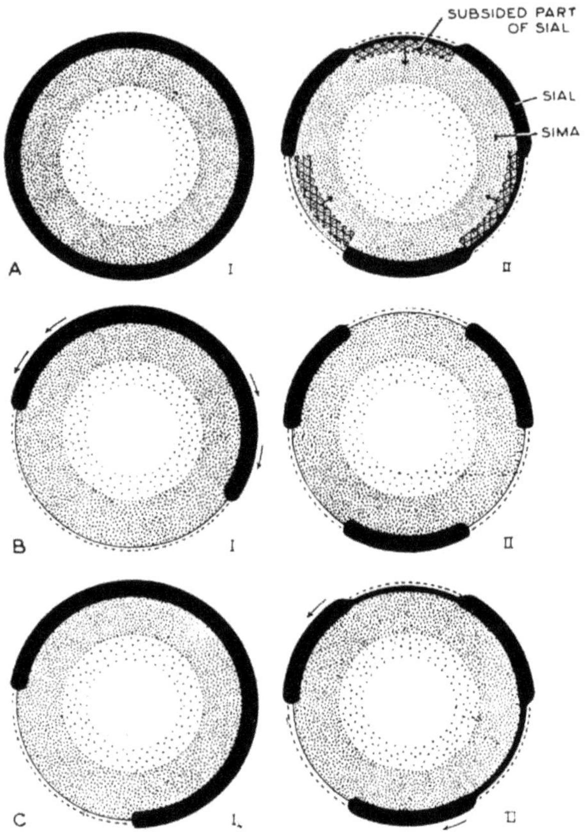

Fig. 156. Schematic representation of three groups of hypotheses on the formation of continents and ocean floors. A, by subsidence (Suess, Haug); B, by drifting (Wegener); C, by stretching (Du Toit); I, Upper Paleozoic, II, later stage.

submarine relief, the second by the periodic changes in the vertical distance between the ocean-floor and the surface of the continents. The first includes different periods of formation of (and later movements within) the basins of the Indian and Atlantic Oceans, as well as a rejuvenation of the relief (e.g. the Mid-Atlantic Rise and the Carlsberg Ridge), and the formation and submersion of isthmian links. The origin of these movements must necessarily remain problematical, but the similarity of the features of the oceanic and continental areas hardly leaves any doubt that they might have occurred. This brings us to the fundamental potentiality of isthmus-shaped trans-oceanic land-communications. The second changeable factor is the pulse of the oceanic floor as derived from the world-wide trans- and regressions described in Chapter V.

The trans-Atlantic similarities may possibly be explained by the suggestion that if a geosynclinal zone continued under an area like the Atlantic, no mountain-belt could have originated in the oceanic sector, since no area of denudation and therefore no detritus was available. Hence mountain-belts on either side of the Atlantic may match as regards their respective stratigraphy, epochs of folding and tectonic structure in spite of there never having been a trans-oceanic connection above sea-level. We will be able to test this idea as soon as gravimetric observations at sea have become more numerous.

However, the following major problems remain to be solved, viz. the comparatively thin sialic layer covering the floor of the Atlantic and Indian Oceans, the absence of such a layer in the Pacific, the antipodal arrangement of continents and oceanic depressions. One group of theories surmising primary continental masses of alkali-basalt comes into conflict with the principle of isostatic equilibrium.

Many authors assert that the earth originally was enveloped by a sialic shell of continental thickness. If we accept the resonance theory of the moon's origin, it seems highly probable that a great deal of the outer shells of the earth entered the formation of the moon, and the remaining terrestrial sial-sheets would have formed our continents, floating on the heavier substratum that constituted the bottom of the oceanic depressions (fig. 156B). But why should the sialic material have formed a continuous shell on the moon and not on earth? Besides, several objections, showing that this point of view is open to serious doubt, were raised against this hypothesis in the preceding paragraphs.

The only alternative — leading to a new working-hypothesis — appears to be that the earth had initially been enveloped by a continuous sialic layer which gradually solidified and floated on a denser basic substratum. It may be that this layer originated after the moon's disruption, but it might similarly be supposed that the moon was not severed from the earth.

The only possible way in which continental blocks might have originated from a world-encircling sialic layer is that the latter was thickened by folding and drifting owing to the action of convection currents. The sialic layer is contracted by this process, and sial-free parts, such as the present North Pacific Basin, were thus formed (fig. 154D). During the process of drifting, a thin sial-sheet — such as may probably be said to exist beneath the Atlantic —, may have originated as a result of stretching of either a primordial continent or an intervening primary sial-layer. The occurrence of three favoured crustal levels agrees with this surmise. An event of fundamental importance was the completion of a world-encircling solid crust, which had gradually become so thick that subsequent continental growth was no longer possible.

Moreover, it would seem plausible to assume that the curious structural pattern which characterizes both the floor of the Atlantic and the surrounding continents and which dates from early Pre-Cambrian times came into existence during the above process of crustal consolidation.

In recent years some interesting geophysical considerations have been proposed regarding the question of the antipodal arrangement of the major geographical units and their triangular shapes. It may be that their peculiar distribution was caused by a certain system of convection currents occurring in the early days of our globe's infancy. As a whole, however, the question of the origin of continents and oceanic depressions still remains an unsolved problem.

References

BANDY, M. C., *Geology and petrology of Easter Island* (Bull. Geol. Soc. America 48, 1937).
BARRELL, J., *The origin of ocean basins* (pp. 39–43 of Chapter I, in "The Evolution of the earth" 1919).
BARRELL, J. *On continental fragmention, and the geologic bearing of the moon's surficial features* (Americ. Journal of Sci. 13, 1927).
BETZ, F. and HESS, H. H. *The floor of the North Pacific Ocean* (The Geogr. Review, 32, 1942).
BOWIE, W. *The origin of continents and oceans* (Sci. Monthly 41, 1935).
BUCHER, W. H. *The deformation of the earth's crust* (1933).
BUCHER, W. H. *Submarine valleys and related geologic problems of the North Atlantic* (Bull. Geol. Soc. America 51, 1940).
BULLARD, E. C. and GASKELL T. F. *Seismic methods in submarine geology* (Nature 142, 1938).
CHELIKOWSKY, J. R. *A new idea on continental drift* (Americ. Journ. Sci. 242, 1944).
CHUBB, L. J. *The structure of the Pacific basin* (Geolog. Magaz. 71, 1934).
CLOOS, H. *Zur Gross-tektonik Hochafikas und seiner Umgebung* (Geol. Rundschau, 28, 1937).
CONWAY, E. J. *Mean Geochemical data in relation to Oceanic evolution* (Proc. R. Irish Acad. 68B no. 8, 1942, and no. 9, 1943).
DALY, R. A. *Architecture of the Earth* (1938).
DALY, R. A. *Strength and structure of the Earth* (Prentice Hall New York, 1940).
DU TOIT, A. L. *Our wandering continents* (1937).
DU TOIT, A. L. *The origin of the Atlantic-Artic* Ocean (Geol. Rundschau 30, 1938).
DU TOIT, A. L. *Observations on the evolution of the Pacific Ocean* (Proceed. Sixth Pacific Sci. Congr. California 1939, vol. I, 1940).
DU TOIT, A. L. *Further remarks on continental drift* (Americ. Journ. Sci. 243, 1945).
ESCHER, B. G. *Moon and Earth* (Proceed. K. Akad. Wetensch. Amsterdam 47, 1939).
ESCHER, B. G. *De asymmetrische gedaante der aarde en haar oorzaak* (IJdo, Leiden 1946).
EWING, M., MILLER, B. L. a.o. *Geophysical investigations in the emerged and submerged Atlantic coastal plain* (Bull. Geol. Soc. America 48, 1937).
FARQUHARSON, W. I. *Topography* (John Murray Exped. Rep. I, Nr. 2, 1936).
FIELD, R. M. *Structure of continents and ocean basins* (Journ. Washington Acad. Sci. 27, 1937).
FIELD, R. M. and JONES, D. M. *Geophysical explorations of ocean-basins* (Quart. Journ. Geol. Soc. London 1938).
FISHER, O. *Physics of the Earth's Crust* (1889).
FURON, R. *La Paléogéographie* (1941).
GERMAIN, L. *Le problème de l'Atlantic et la zoologie* (Ann. Geogr. 22, 1913).
GERMAIN, L., JOUBIN, L. et LE DANOIS, E. *Une esquisse du passé de l'Atlantique Nord* (La Géographie 40, 1923).
GRABAU, A. W. *The Rhythm of the Ages* (Peking, Hermi Vetch 1940).
GREGORY, J. W. *The plan of the Earth and its Causes* (The Geogr. Journ. 13, 1899).
GREGORY, J. W. *The geological history of the Atlantic Ocean* (Quart. Journ. Geol. Soc. London 85, 1929).
GREGORY, J. W. *The geological history of the Pacific Ocean* (Quart. Journ. Geol. Soc. London 86, 1930).
GUTENBERG, B. *Structure of the Earth's Crust and the spreading of the continents* (Bull. Geol. Soc. America 47, 1936).
GUTENBERG, B. and RICHTER, C. F. *Structure of the crust. Continents and oceans* (Ch. XII in Physics of the Earth VII, 1939).
GUTENBERG, B. *The structure of the Pacific Basin as indicated by earthquakes* (Science, 90, 1939).

HAUG, E. *Les géosynclinaux et les aires continentales* (Bull. Soc. Geol. de France, 3e sér. 28, 1900).

HÄGBOM, A. G. *Die Atlantislitteratur unserer Zeit* (Bull. of the Geol. Inst. of the University of Upsala, vol. 28, 1941).

HOLLAND, TH. H. *The permanence of Oceanic Basins and Continental Masses* (London 1937).

HOLLAND, TH. H. *The evolution of continents. A possible Reconciliation of conflicting evidence* (Proc. R. Soc. Edingburgh Sect. B. 61, 1941).

JARDETZKY, W. *Über die Ursachen der Spaltung und Verschiebung der Kontinente* (Gerl. Beitr. z. Geoph. 26, 1930).

JONES, O. TH. *Explorations in ocean basins* (Quart Journ. Geol. Soc. 1937).

KENNEDY, W. G., and ANDERSON, G. M. *Crustal layers and the origin of magmas* (Bull. Volcanol. Ser. II, T. 3. 1938).

KOSSINNA, E. *Die Erdoberfläche* (Handb. der Geophysik II, 3, 1933).

LACROIX, A. *La constitution lithologique des îles volcaniques de la Polynésie australe* (Mém. Acad. Sci. France 59, 1927).

LACROIX, A. *Clipperton, îles de Paques et Pitcairn* (Ann. de l'Institut Océanogr. N. Sér. 18, 1939).

LAWSON, A. C. *Insular arcs, fore deeps and geosynclinal seas of the Asiatic Coast* (Bull. Geol. Soc. America 43, 1932).

LEAHY, L. TH. *The configuration of the bottom of the South Pacific* (C. R. Congr. Intern. de Géographie Amsterdam vol. 2, sect. IIA, 1938).

LITTLEHALES, G. W. *Configuration of the ocean basins* (Bull. Nat. Research Council 85, 1932).

LONGWELL, CH. R. *Some thoughts on the evidence of a continental drift* (Americ. Journ. Sci. 242. 1944).

LONGWELL, CH. R. *The mobility of Greenland* (Americ. Journ. Sci. 242, 1944).

LONGWELL, CH. R. *Determinations of geographic coordinates in Greenland* (Science, 100, 1944).

MECKING, L. *Ozeanische Bodenformen und ihre Beziehungen zum Bau der Erde* (Peterm. Geogr. Mitt. 1940).

MOHOROVIČIĆ, S. *Das Erdinnere* (Zeitschr. f. Angew. Geophysik 1, 1925).

MOHOROVIČIĆ, S. *Über Nahbeben und über die Konstitution der Erd- und Mondinnern* (Gerl. Beiträge zur Geophys. 17, 1927).

NANSEN, F. *Bathymetric map of the Arctic basin* (Americ. Geogr. Soc. Spec. Public. 7, 1927).

NÖLKE, F. *Zur Tektonik des Atlantischen Beckens* (Geol. Rundschau 30, 1939).

PICKERING, W. H. *The Place of Origin of the Moon. The Volcanic Problem* (Journ. of Geology 15, 1907, Geolog. Magaz. 61, 1924).

RITTMANN, A. *Über die Herkunft der Vulkanischen Energie und die Entstehung des Sials* (Geol. en Mijnb. I, 1939 and Geolog. Rundschau 30, 1939).

SCHOTT, G. *Geographie des Atlantischen Ozeans* (1926).

SCHOTT, G. *Geographie des Indischen und Stillen Ozeans* (1935).

SMIT SIBINGA, S. *On the petrological and structural character of the Pacific* (Verh. Geol. Mijnb. Genootsch. 13, 1943).

SONDER, R. A. *Zur Tektonik des Atlantischen Ozeans* (Geol. Rundschau 30, 1939).

SCHUCHERT, Ch. *The problem of continental fracturing and diastrophism in Oceanica* (Americ. Journ. of Sci. 42, 1916).

SCHUCHERT, CH. *Gondwana landbridges* (Bull. Geol. Soc. America 43, 1932).

SCHWINNER, R. *Lehrbuch der physikalischen Geologie* (I, 1936).

SIMPSON, G. G. *Mammals and the nature of continents* (Americ. Journ. Sci. 241, 1943).

STETSON, H. T. *Modern evidences for differential movement of certain points on the earth's surface* (Science, 100, 1944).

SVERDRUP, H. U., JOHNSON, M. W. and FLEMING, R. H. *The Oceans* (New York, Prentice Hall, 1942).

STILLE, H. *Zur Frage der transatlantischen Faltenverbindungen* (Sitzungsber. Preuss. Akad. Wissensch. 11, 1934).

STILLE, H. *Geotektonische Probleme im Atlantischen Raume* (Kais. Leopoldinische Carol. Deutsche Akad. d. Naturf. Halle 1937).

STILLE, H. *Queratlantische Faltenverbindungen* (Geol. Rundschau 30, 1939).

STILLE, H. *Kordillerisch-atlantische Wechselbeziehungen* (Geol. Rundschau 30, 1939).

STOCKS, TH. *Neues zur Morphometrie des Atlantischen Ozeans* (Ann. d. Hydrographie 67, 1939).

STOCKS, TH. und WÜST, G. *Die Tiefenverhältnisse des offenen Atlantischen Ozeans* (Meteor. Exp. III, 1, 1935).

Symposium on the geophysical exploration of the ocean bottom (Proceed. Americ. Philos. Soc. 1938).

TERMIER, P. *Les Océans à travers les Ages* (Bull. Inst. Océanogr. Paris 1920); Idem in "*A la gloire de la Terre*" (1926).

THOM, W. T. *Position, extent and structural make up of Appalachia* (Bull. Geol. Soc. of America 48, 1937).
UMBGROVE, J. H. F. *Palaeogeographie der Oceanen* (Tijdschr. Kon. Nederl. Aardr. Genootschap 54, 1937).
UMBGROVE, J. H. F. *On the origin of continents and Ocean Floors* (Journ. of Geology, 244, 1946).
VAUGHAN, T. W. *International aspects of Oceanography* (Nat. Acad. Sci. Washington 1937).
VENING MEINESZ, F. A. *Gravity over the Hawaian archipelago and over the Madeira area, conclusions about the Earth's crust* (Proc. Nederl. Acad. Wetensch. 44, 1941).
VENING MEINESZ, F. A. *De verdeeling der continenten en oceanen over het aardoppervlak* (Versl. Nederl. Akad. v. Wet. 53, 1944).
WADE, A. *A new theory of continental spreading* (Bull. of the Americ. Ass. of Petrol. Geol. 19, 1935).
WEGENER, A. *The origin of continents and oceans* (New-York. Dutton 1924).
WILLIS, B. *Isthmian links* (Bull. Geol. Soc. America, vol. 43, 1932).
WILLIS, B. and WASHINGTON, H. S. *San Felix and San Ambrosio, their geology and petrology* (Bull. Geol. Soc. America 35, 1924).
WISEMAN, J. D. H. and SEYMOUR SEWELL, R. B. *The floor of the Arabian Sea* (Geol. Magaz. 74, 1937).
WÜST, G. *Die Gliederung der Weltmeere* (Peterm. Geogr. Mitteil. 1936).
QUENSEL, P. D. *Die Geologie der Juan Fernandez Inseln* (Bull. Geol. Instit. Upsala, 11, 1912).

CHAPTER IX

ICE-AGES

"We are apt to think that our present conditions are normal, but that is far from being the case". (A. P. COLEMAN).

Introduction

The earth's aspect is a very recent one from a geological point of view. Everyone knows that Greenland and Antarctia are at present covered by extensive ice-caps (fig. 157). Yet only one tick back on the geological clock large parts of America and Europe were covered by analogous ice-sheets of enormous dimensions. In Europe the ice-front extended from the British Isles to the Netherlands, and from there to Germany and North Russia, covering the whole of Scandinavia. Another ice-sheet covered the Great Lakes district of North America and practically the whole of Canada (fig. 158). Indisputable traces of large glaciations in South Africa, South America, Australia and a few other regions date from a still more distant past — the Upper Paleozoic (fig. 159) — or even remoter times, the Pre-Cambrian, when plant-life had not yet imposed itself upon the land, and the seas were only populated by Invertebrates (fig. 160). Rocks whose origin cannot be explained unless we assume that they were formed under the same conditions of heat and drought as are now found in the Sahara (fig. 165) are met with in many places in Europe and America. On the other hand, the fossil remains of a luxuriant vegetation are still observed in such bitterly cold regions as Spitsbergen and Greenland.

The most absorbing topic of changing climates has not only attracted geologists, but also investigators in other branches of science such as astronomers, meteorologists, paleontologists and biologists, and the hypotheses that attempt to find a solution for the earth's climatological problems are so numerous and so varied that Daly was induced to remark in this connection: "Geologists do not know the causes", and somewhat further: "At present the cause of excessive ice-making on the lands remains a baffling mystery".

However, the veil of mystery around these climates has been lifted to some extent by the increasing amount of data, and the minute studies and lucid suggestions of a large number of investigators. An attempt will now be made to unravel the main outlines of the chaotic mass of data and opinions and special attention will be paid to those which I consider to be the most

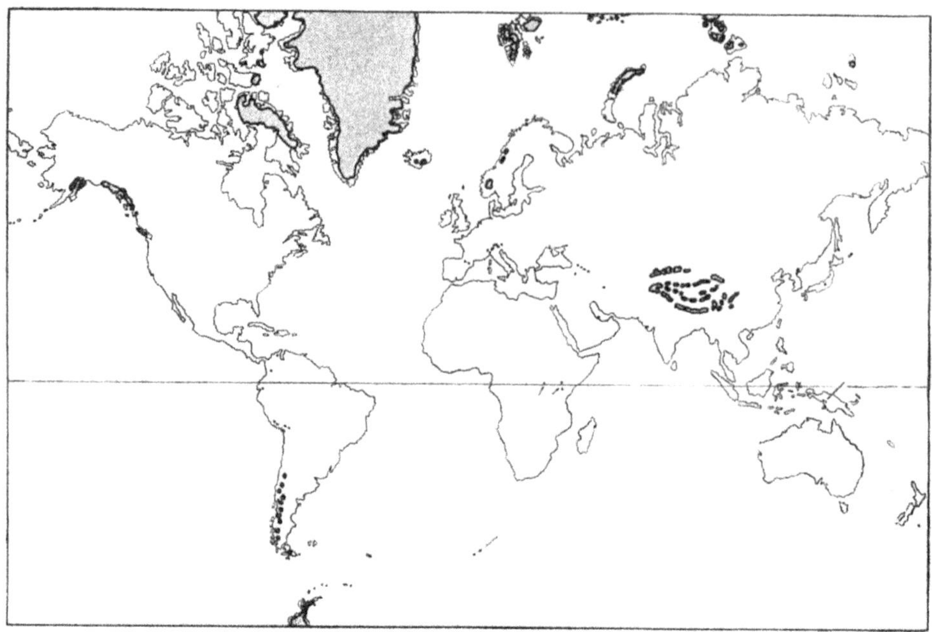

Fig. 157. Ice-sheets and glaciers of the present time.

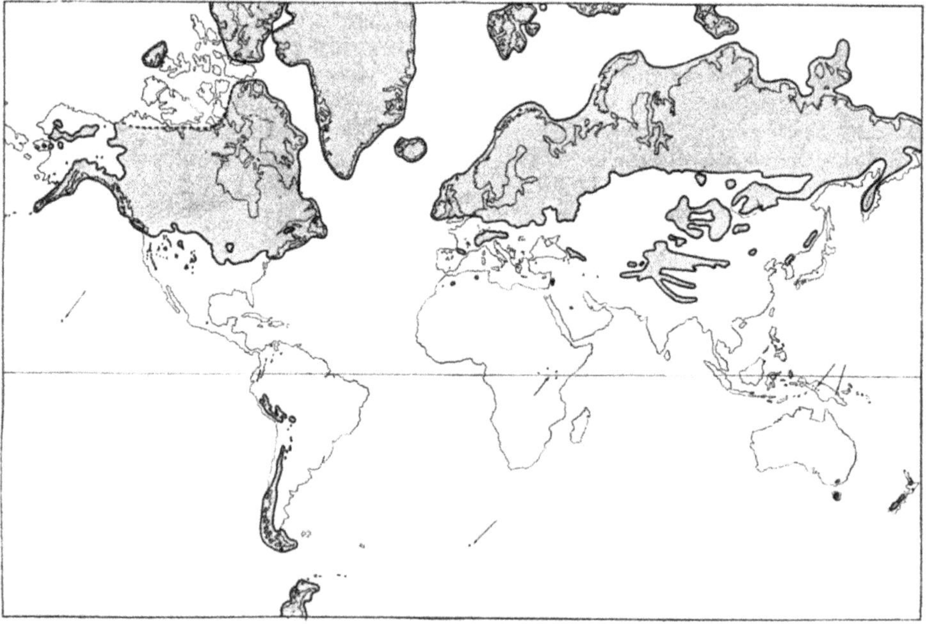

Fig. 158. Ice-sheets and glaciers during maximum glaciation in the Pleistocene.

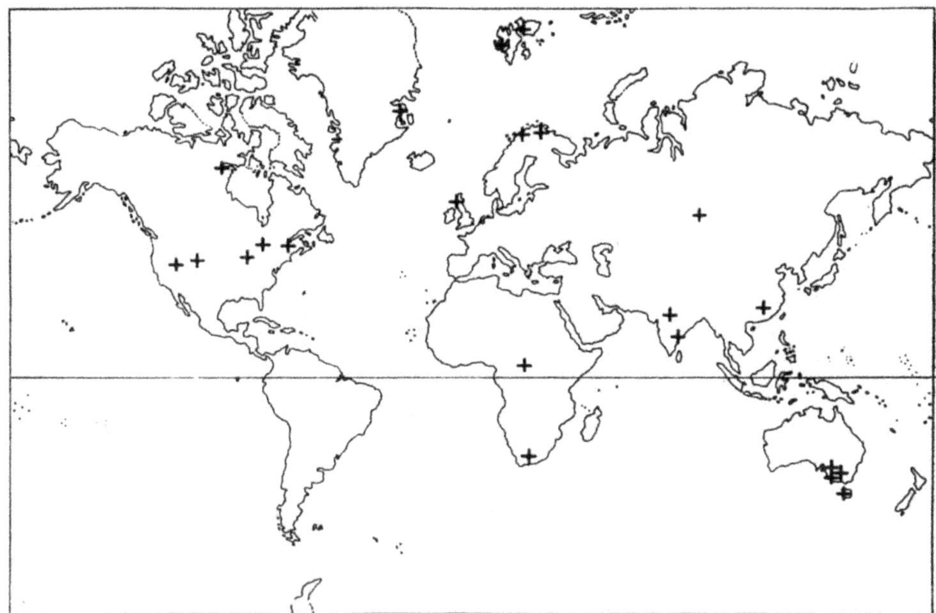

Fig. 159. Upper Paleozoic tillites.

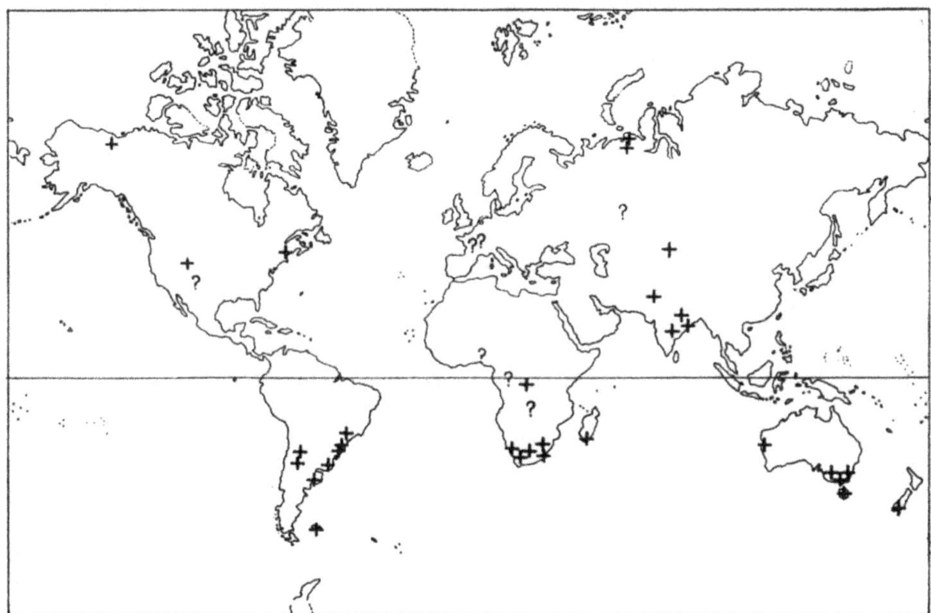

Fig. 160. Pre-Cambrian tillites.

important. I hope to be able to show the approximate state of present knowledge on the subject of the climates, and will also emphasize those aspects of the problem which are still wrapped in much uncertainty.

The periodicity of climates

If noted in a proper order on the geological time-scale, periods of glaciation and periods that were characterized by a mild climate up to high latitudes will be seen to give rise to a curious phenomenon, viz. a large climatic periodicity. Some 250 million years elapsed between the beginning of the Upper Paleozoic glaciation and the Pleistocene ice-ages. Evidence of another extensive glaciation i.e. towards the close of the Pre-Cambrian, is found if we reach back an additional 250 million years. Holmes showed that this periodicity can probably be retraced several times in still remoter times, though there are as yet too few data to enable us to speak of a phenomenon recurring with perfect regularity [1]). A periodicity of 250 million years would appear to be a rough, but fairly accurate estimate.

The duration of a glaciation can only be indicated in one case, viz. in that of the Upper Paleozoic, which continued for about 25 to 30 million years. Should the series of ice-ages, beginning with that of the Pleistocene (ca. one million years ago), be attributable to a phenomenon with a cause similar to that of the Upper Paleozoic glaciation, the present period can consequently be said to constitute an intervening stage of partial deglaciation. Many glacial and interglacial stages may follow the present situation before a new period is reached with conditions comparable to those which existed from the Upper Permian until the Pleistocene.

Speculation on what might happen if these ice-caps were to grow again excessively is one that cannot but fill one with dread. Prosperous agricultural and economic centers, thriving cities and important ore deposits would all disappear under the irresistible march of the ice, and an alarming decrease in the areas fit for population would ensue. Yet let us suppose that the reverse were to happen, i.e that the ice were to melt entirely. The resulting state of affairs would not be a much happier one. It can easily be calculated that the level of the sea would be raised some 40 to 50 [2]) meters if the ice were to melt over the whole world, and 12% of the land would consequently be flooded. Proud cities such as Amsterdam and Batavia would sink down in a muddy sea-bottom, and the lower parts of the sky-scrapers of New-York would be turned into shelters for lobsters. However, we are not concerned

[1]) The figures for the Pre-Cambrian glaciations are 500, approx. 600, approx. 800, approx. 1,000, probably > 1,000 and probably > 1,500 million years (A. Holmes, The age of the earth, 1937, p. 217).

[2]) If the ice were to melt over the whole world, the volume of water would amount to about 18 million cubic kilometers (Daly, p. 11), assuming a mean density of 0,9 for the land-ice. The total area of the oceans occupies 361 million square kilometers. This melting process would consequently cause the level of the ocean to rise 50 meters. The added weight of this water, however, would press down the ocean-floor to some extent, and the continents would be slightly elevated as a result of the displacement of subcrustal matter from the oceanic sectors. The change in the ocean-level consequently ultimately amounts to less than 50 meters.

with what might happen in the future, but with what happened in the past, and it is these last problems which we now intend to examine.

An intervening period between two glaciations (e.g. from the Cambrian to the Upper Carboniferous, and from the Upper Permian to the Pleistocene) is marked by the absence of strongly differentiated climatic zones. The shape of the terrestrial globe has, of course, always been responsible for the fact that the polar regions constitute the colder areas of the earth, and the equatorial zones the warmer ones. The inclined rotation-axis of the earth has similarly always been responsible for more or less pronounced seasons. On the other hand, the seasonal changes during some periods were so small that several writers prefer to speak of a uniform climate for the whole world. The climate would appear to have been much milder up to a short distance from the poles in former times. This is shown (1) by the luxuriant Tertiary vegetation in such high northern latitudes as Greenland, where the only existing plant life at present is the tundra, and (2) by the reef corals that flourished up to much more northern and southern latitudes. Though the latter are only found in warm seas, they are known to have existed as extensive coral reefs in the Tertiary seas of North America, and were distributed as far as Maastricht (and even Denmark) in the Senonian.

Special phenomena that occurred during certain periods show that the climate was less differentiated than during the major periods of glaciation. This is how some authors interpret the tillites, which may be observed in the Eocene beds of North America and Antarctica, the Jurassic of California (fig. 161), the Upper Silurian and Devonian of South Africa and Alaska, the Ordovician of Europe and North America (fig. 162), and possibly a few other localities and periods (the Triassic of Central Africa?) [1]. No one regards these tillites as the products of former ice sheets. They are thought to constitute marks left by more or less local mountain glaciers and piedmont glaciers. It may still be questioned whether the formation of the glaciers in different regions did not in each case result from a combination of local circumstances. There is reason to believe, however, that these old moraines of Eocene, Jurassic, Devonian, Upper Silurian and Ordovician time justify the assumption that minor periods of cold prevailed during the long interval of time that separated two major periods of glaciation. This enables us to draw a *schematic* curve of the climate such as that in Table II.

A mild and equable climate, affecting almost the whole world and extending even as far as polar latitudes, prevailed during such aeons of our earth's history that this condition may be considered to have represented the earth's normal climate. The major glaciations, on the other hand, and the minor periods of cold which we have just mentioned may be assumed to form "abnormal" and relatively short deviations.

The earth's normal climate

What was the earth's aspect during normal climatic conditions? The remains of a Mesozoic vegatation are found in Greenland, and as far as

[1] Fig. 157–162 have been adapted from publications by Antevs, Backlund, Brooks, Coleman, Norin, Teichert and others.

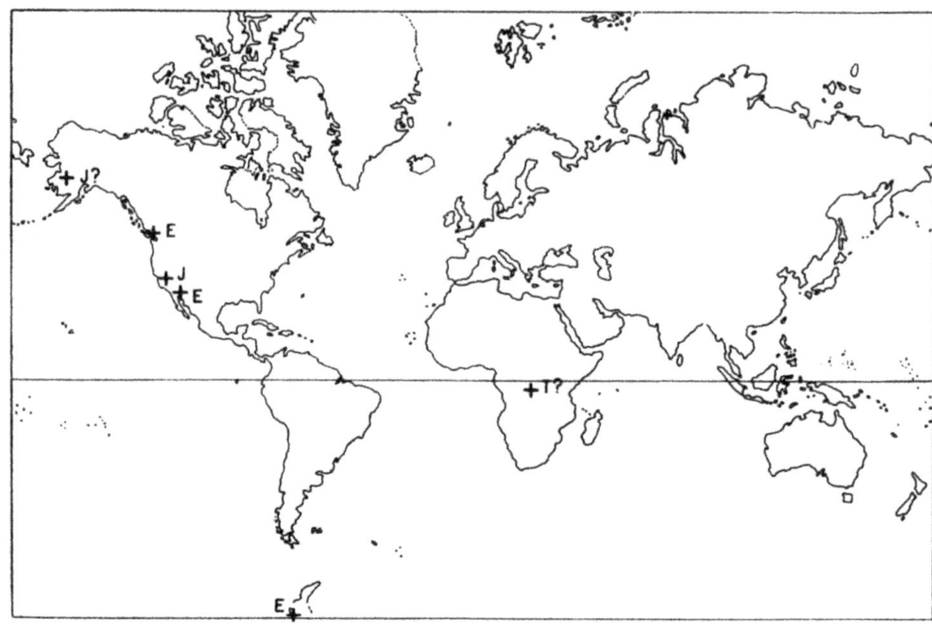

Fig. 161. Mesozoic and Eocene tillites. T Triassic, J Jurassic, E Eocene.

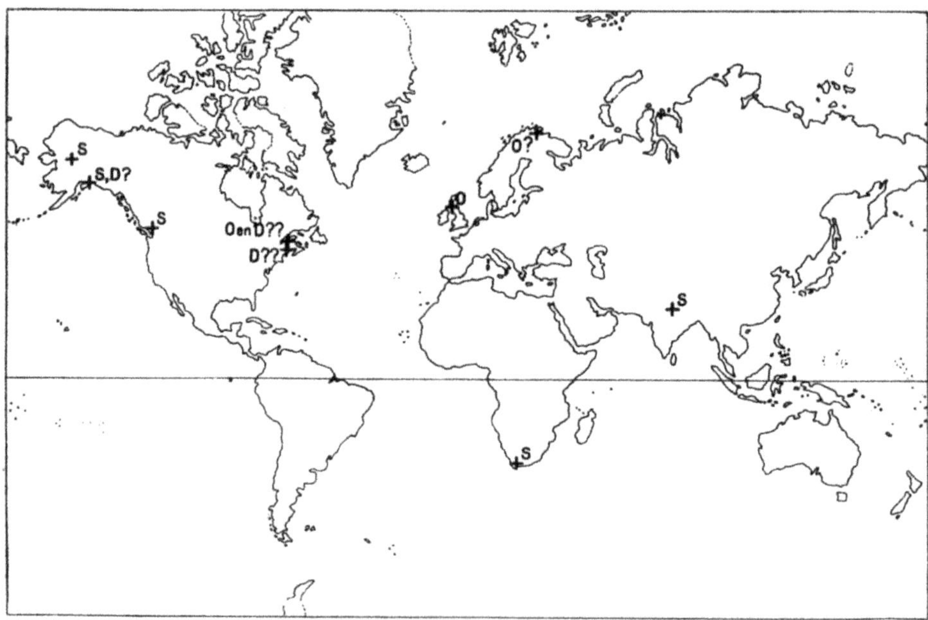

Fig. 162. Lower Paleozoic tillites. O Ordovician, S Silurian, D Upper Devonian.

Grahamland, also between (and including) New Zealand and North America. Darrah [1]) elucidates this question as follows: "The Jurassic floras of such widely separated regions as the Antarctic continent, England, India and North America have so very many plants in common that the essential uniformity of the vegetation all over the earth at that period is well known. The succeeding flora of the Cretaceous displays almost as great a uniformity in the composition. Not only were these floras homogeneous, but they appear to have flourished under more equable conditions than those now prevailing over most of the earth". A few pages further, Darrah states: "The rapid radiation of the diversifying flowering plants (of the Upper Cretaceous) certainly is indicative of an equable climate and an absence of barriers of migration".

Recent discoveries have made it clear that the climate need only change a little to bring about a recurrence of a luxuriant growth up to such high northern latitudes as Greenland. Since the retreat of the ice towards the beginning of this century, the remains of an autochthonous vegetation and of settlements dating from the eleventh and twelfth centuries have been brought to light in this region, and indicate that at that time the transitory melting of the ice-caps took place on a much larger scale than during the last forty years.

Under the mild climatic conditions of the Mesozoic, the land-fauna, including organisms such as the giant reptiles, were able to migrate from Central Asia to America (via the Straits of Bering) and Africa without meeting any unusual barriers of temperature (fig. 163).

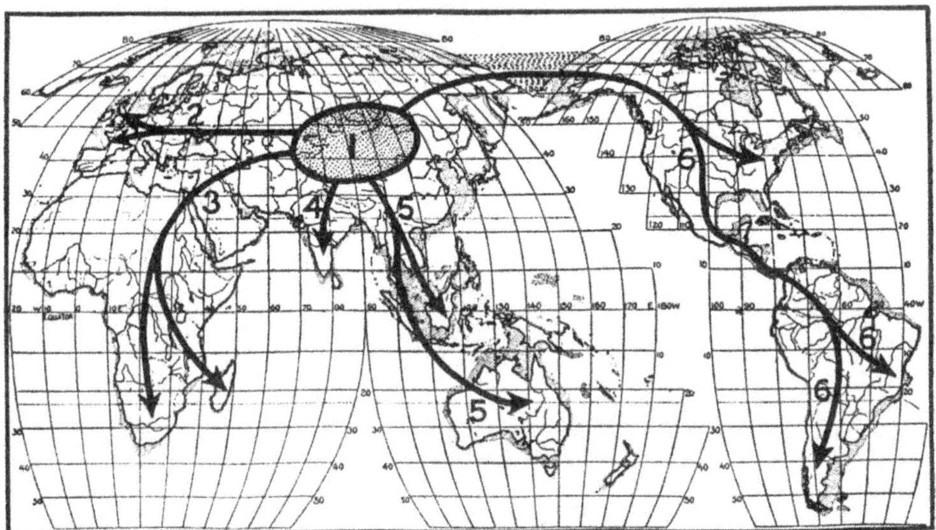

Fig. 163. Upper Jurassic — Lower Cretaceous routes of migration of the giant Dinosaurs (After H. F. Osborn).

Such an evenly distributed climate also influences the atmospheric and

[1]) Darrah, 1939, p. 171, 180.

oceanic circulation, producing less intensive horizontal and vertical movements. One of its of consequences was that the deep-sea received a smaller supply of oxygen. It may be assumed that marine animals that had adapted themselves to colder water could only have thrived in the northern and southern polar seas. These have indeed been described from Jurassic and Lower Cretaceous deposits, and are known as the boreal fauna. A specific geographical position may cause this boreal fauna to extend locally up to a shorter distance from the equator than elsewhere. The remaining marine fauna is almost entirely uniform. Everyone will agree that though Neumayer distinguishes between four types of Mesozoic marine faunas the Mediterrane-Caucasian, Himalayan, Japanese and South-Andine "provinces", his classification merely divides the biocoenoses of the warm seas into separate geographical provinces. The Cambrian Trilobites, too, for instance, could be subdivided into an Atlantic and Pacific province, but this does not mean that they lived under different climatic conditions.

The reef-building Rudists of the Upper Cretaceous (fig. 164) have often been

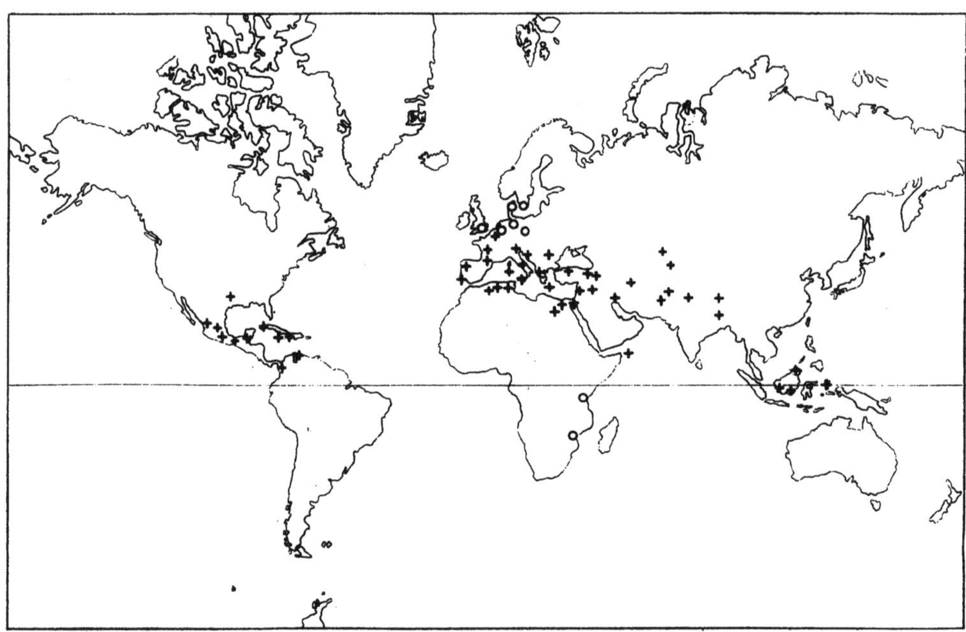

Fig. 164. The distribution of Upper Cretaceous Rudists, + normal types, O dwarf types (After Dacqué).

cited as examples of warm-water organisms. It is noteworthy that dwarf-forms have been found in addition to the larger normal types in their most northern occurrences. Some attribute this to the influence of the boreal zone. A like phenomenon is observed in their southern extension, in East Africa, where they occur from lat. 5° to 20°S (lat.50° to 55° N. on the other side of the equator). It is clear that the existence of a southern cold sea-current coming from Antarctic regions must be taken into account if their distribution is to be explained.

It would not surprise us if those zones of the earth which may be said to be predestined to become belts of arid degradation had had a greater extent during a "warm-period" than they have at present. There can hardly be anything unusual in the distribution of the deserts during a period of normal geological climates if we remember that the present distribution of land and sea still exerts a strong influence on the disposition of arid areas. Fig. 165 indicates the location of former Mesozoic desert formations, together

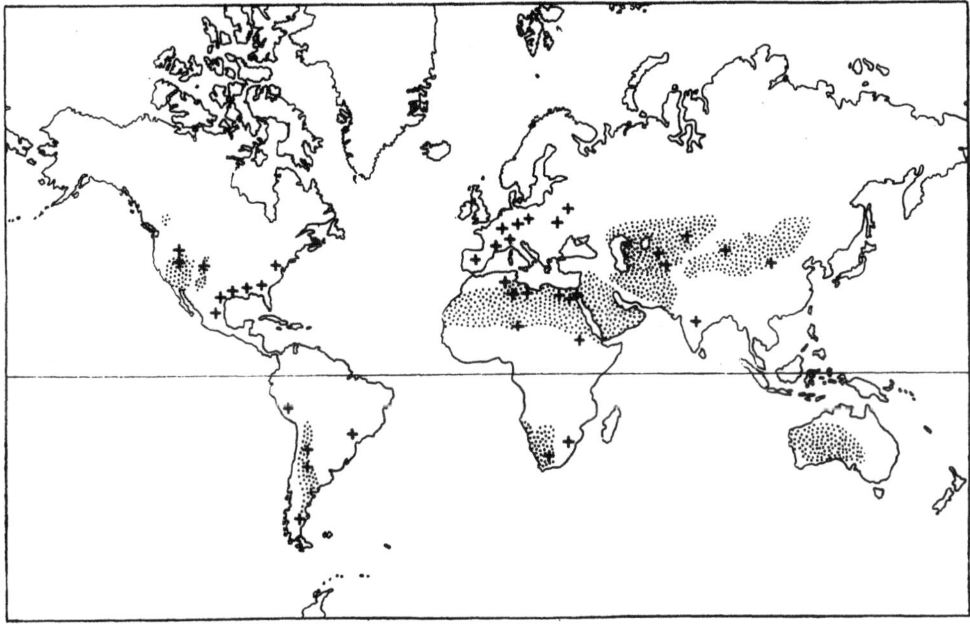

Fig. 165. Desert deposits of Mesozoic times (+), after A. Wegener, and desert belts of the present time (stippled).

with those areas which are now occupied by deserts. Brooks pointed out at great length that the earth's normal climate need not be the result of an important change in the position of the continents and poles. Berry [1]) illustrated this statement as follows as regards the distribution of the flora: "Fossil plants have been found at six general horizons in the Arctic, namely: Late Devonian, Mississippian, Late Triassic, Late Jurassic and early Lower Cretaceous, Upper Cretaceous, and Eo-Oligocene. They are frequently associated with coal seams, and this, together with roots in place in the underlying clays, the presence of fresh-water diatoms, mollusks and aquatic beetles, proves that the plants grew near where they are found fossil and were not drifted from lower latitudes".... "The distribution of the localities

[1]) Berry, 1929, p. 236.

around the poles conclusively discredits the hypothesis, now fashionable in certain quarters, that there has been a change in position of the pole. A botanical analysis of the floras shows clearly that the term tropical, so frequently applied to them, is highly improper, and that they are, for the most part, distinctly cold temperate. The associated petrified woods, with the single exception of the *Dadoxylon* from the Mississippian of Spitsbergen, show well marked seasonal rings.

"Detailed comparisons of these Arctic floras with contemporaneous floras from lower latitudes, especially in the case of the younger, post-Paleozoic floras, show unmistakable evidence for the existence of climatic zones. It was further shown that these northward extensions of temperate floras in the past were all comtemporaneous with or immediately followed times of unusually widespread continental submergence and sea extension, and that they lived under the domain of oceanic as opposed to continental climatic conditions.

"It was concluded that low lands and expanded seas were amply sufficient to account for the necessary ten to fifteen degrees northward swing of isotherms demanded by the botanical character of the floras. This, with the concomitant alterations in wind and current circulation, combined with Brooks' meteorological results, are considered to account for all of the observed facts".

Nor need we revert to the idea of continental drift to explain the occurrence of periods of abnormal climate or glaciations, to which we will confine ourselves in the following paragraphs.

The abnormal character of present conditions

It should be remembered that we are living in an abnormal climatological age, albeit in an intervening stage of milder glaciation or partial deglaciation. The actual distribution of fauna and flora in distinct biogeographical units determined by the climatic zones is a phenomenon of very recent origin, and merely dates from the close of the Tertiary (cf. Chapter X). The extreme specialization and differentiation of the flora and fauna form another unusual aspect of our time (cf. Chapter X). For example, flowering plants (Angiospermae), which are first observed in the Upper Mesozoic, dominate the plant-world, while the adaptive radiation of the mammalian fauna only began to develop in the Tertiary. What do we know of the ecological and biological conditions necessary for the maintenance of a biocoenosis during the Mesozoic and Paleozoic? It has frequently been shown that an attempt to base conclusions as regards the climatological significance of fossil finds on present conditions is bound to meet with failure. The rhinoceros, and a type of elephant —the mammouth—populated north-western Europe during the bitterly cold Pleistocene. Yet the near relatives of these animals are now found exclusively in the tropics! This in itself is quite enough to show that the greatest care should be taken by investigators before basing conclusions on the present situation. It was shown above that a very different atmospheric and oceanic circulation must have prevailed formerly, during periods of a normal climate. It will also be obvious that far more intensive denudation must have influenced the land during a period like the Lower Paleozoic, for no vegetation protected the earth at that time. These examples may help one to obtain some idea of the many and occasionally very difficult problems which confront the paleontologist and climatologist in their attempts to interpret the geological past.

This point needs emphasizing when an attempt is made to find a solution for the climatological changes in the earth's history. For many authors put the problem in such a way that two wholly different questions have to be solved simultaneously. The first deals with the cold ages, the second with the warm climates. These scientists begin by using the present climatic conditions as a basis upon which to build further conclusions, and then try to explain both the glaciations and the cause of the climatic change which transformed the climate into an almost mild and equable one all over the world. The problem should be put quite differently, however, for in following the above method we assume the present situation to be the normal one and to characterize the whole geological past, and thus take it for granted that any other situation must be an abnormal one. Investigators such as Köppen and Wegener undoubtedly lost sight of the abnormal character of actual conditions when determining the climatic zones of former geological periods, for they sought to reconstruct climatic zones equivalent to our own for all these formations, and any conflicting evidence in the past was simply attributed to hypothetical changes in the position of the poles and continents. They thus construed a continuous, non-periodical process, instead of the periodicity of abnormal climates.

The problem

As already stated, the whole question should be approached from a different angle. We should break with the habit of considering our present conditions to be normal ones. Once this is done, the main problem left is: what caused the abnormal or cold deviations of the earth's climatological history? In other words, to what impulses should we attribute the glaciations? The problem is fourfold, for the following points have to be explained: (1) the major periodicity of approximately 250 million years, including the intervening minor cold spells; (2) the relatively rapid growth of an ice-cap and its subsequent equally rapid decay; (3) the occurrence of interglacial stages; and (4) the site of an ice-cap.

The relation between periodicity and physico-geographical factors

Many authors have tried to connect climatic changes with other events within the earth itself, or with processes upon its surface. The Pleistocene ice-ages succeeded the folding of such belts as those extending between the Alps and the Himalayas. The Upper Paleozoic glaciations appeared as the Variscian Mountains began to emerge, and the glacial phenomena of the Pre-Cambrian coincide with the so-called Huronian period of mountain-building. The level of the continents was a comparatively high one compared to that of the sea. Glaciations can thus be said to be observed during periods in which we find much land and few epicontinental seas. The intervening ages, however, are characterized by a relative tectonic quiet and low continents, large parts of which were flooded by the seas. These periods were marked by a mild climate up to high latitudes. The relation of epochs of mountain-building to glaciations was revealed by Leconte in 1889, and in the same way Wright, Upham, and especially Ramsay and Ruedemann attempted to explain the phenomenon of glaciation. These authors assert that the ice-caps originated as a result of the extension and particularly the altitude of land masses. Before going any further, however, a few other phenomena, which are likewise related to periods of mountain-building and oscillations of the sea-level, will have to be mentioned, since these enable us not only to form a particularly clear picture of the peculiar intricacies of the problem, but also of the various lines along which a solution has so far been sought for the occurrence of glaciations.

Several writers stress the possible influence of a changing amount of carbon-dioxide in the atmosphere. This gas, which is transparent to light, though opaque to heat, would act in the atmosphere like a glass pane in a hot-house and retain the heat, and the temperature would then increase proportionally to the quantity of CO_2 present in the air (Arrhenius). Chamberlin claims that a considerable amount of CO_2 was locked up in the formation of limestones and dolomites during periods of continental emersion (regression), and that the climate was consequently transformed into a cold one. On the other hand, periods of submersion, or peneplanation and transgression would have set this CO_2 free, and this would then have produced a milder climate.

The course of the sea-currents has undoubtedly changed (and the salinity of the oceans oscillated to some extent) in the course of time. Changes have doubtlessly also affected the atmospheric circulation and the cloudiness of the sky. All these variations were due to the repeated alterations in the distribution of land and sea.

Other investigators (especially Humphreys) argue that a great quantity of fine dust is flung into the atmosphere — and even the stratosphere — during violent volcanic eruptions, and that the dust continues to linger in these higher regions during a considerable lapse of time, scattering and reflecting the solar radiation. Since these eruptions are known to have occurred with exceptional frequency during certain periods, it follows that the atmosphere may have been cooled during these stages.

Many factors had an undeniable influence on the climates during the changes which were constantly affecting the face of the earth. It is almost impossible, however, to describe each one's influence separately, or to estimate their combined total effect quantitatively, one of the chief difficulties being that our paleogeographical maps are incomplete in several respects, and also very schematic. Yet it cannot be denied that mountain-building and the major periods of glaciation are in fact correlated in one way or another.

Still, intensive folding and subsequent mountain-building are known to have occurred during several other periods — e.g. the Caledonian, the

Early and Late Cimmerian (Nevadian), and the Laramide. As stated above, a few of the less important glacial deposits may possibly be related to these periods. The tillites of Ordovician time and the Silurian and Devonian have been assumed to be related as such to Caledonian mountain-building, and those of the Jurassic and Eocene to the Late Cimmerian and the Laramide revolution respectively. The Variscian and Alpine Mountains (as shown in Chapter II) were far more extensive than the Caledonian, Cimmerian and Laramide chains. Though it is extremely difficult — and at times almost impossible — to reconstruct a satisfactory comprehensive map of the above periods, the following conclusions may safely be drawn: (1) that the periodicity of 250 million years as indicated by the major periods of glaciation is also manifested in the epochs of mountain-building (2) that a rhythm of minor amplitude, as observed in the less accentuated deviations from normal climatological conditions in the Ordovician, Silurian, Devonian, (Triassic?), Jurassic and Eocene can probably be said to coincide with epochs of mountain-building of far lesser geographical importance than those which accompanied the major rhythm of 250 million years referred to above. This is illustrated by the schematic graph in Table II [1]).

The deeper causes of periodicity

A comparison between the various processes of folding and movements of the sea-level shows that these heterogeneous phenomena are distinctly correlated. There also appears to be some relation to magmatic cycles; and the alternating increase and decrease of compression of the earth's crust seems to be directly connected with the above events, as explained in the preceding chapters. In short, everything shows that a whole series of different phenomena reflect the influence of a deeper and widespread impulse. The term "deeper" should be taken in its figurative and literal sense for the phenomena are attributed to deep-seated forces. It would take us too far to examine this difficult and intricate problem at this point, but it should be noted that several diverging hypotheses have been put forward on this subject by Joly, Holmes, Bucher and — recently — Griggs, which though differing in many respects, agree on one point, for each attributes the origin of the phenomena to subcrustal forces. This vague designation will have to suffice for the moment. However, this much can be said now: the periodical processes in the earth's interior may be thermal cycles, i.e. they may be accompanied by periods of an increased accumulation of heat alternating with periods of decreasing temperature. Folding would then have occurred in the last phase. This hypothesis implies that it cannot be a mere accident that these phenomena preceded periods of mountain-building and the abnormal "cold" climates.

It should not be assumed, however, that a fall in the temperature of the atmosphere is due exclusively to these subcrustal conditions. For there still remain the changes in the earth's physiographical aspect (mountain-building, regression, a decrease in the amount of CO_2 in the atmosphere, a less intensive

[1]) Similar illustrations are found in Schuchert, Dacqué and Brooks.

circulation, etc. all of which have a considerable influence on atmospheric conditions). To these should be added a further factor, which is known to intensify the fall of the annual mean temperature once the latter has begun to decrease. This factor will now be examined.

The rapid growth and subsequent retreat of an ice-cap

Brooks argues that the ice-cap has a cooling effect on the atmosphere above and along its edge and that this will become an important factor once the ice-cap has begun to grow. The additional cooling influence will cause the ice-cap to extend beyond the boundaries which would have existed without such an effect, i.e. it will cause it to expand beyond the polar regions where an initial fall in the temperature would result in the formation of an ice-cap. On the other hand, the temperature is known to rise steadily in the areas between the poles and the equator, and a state of equilibrium will consequently be reached at a given moment between both counteracting factors. An exhaustive review of this subject will be found in Brooks, but I would like to draw attention to a few results of this author.

The ice-cap extends at an increasing rate once it acquires a given size, for its area increases by the square of its radius. The opposite effect, however, — i.e. the rising temperature from the poles to the equator — is a gradual one and proportional to the distance from the pole (normal horizontal gradient). Observations in polar regions seem to bear out this statement, revealing that the cooling effect of an extending ice-cap will in fact increase almost proportionately to its surface till the growing circumference of the ice-cap has attained a radius of approximately ten degrees. The more distant regions cease to exert their full effect in case of a larger area. For though the cooling effect will continue to increase, the average effect — expressed in km^2 — will begin to diminish (see the following table and fig. 166).

Radius of ice-cap, in °	Total cooling on edge of ice-cap, °F	Rise of temperature due to normal horizontal gradient, °F
0	0.6	0.0
1	1.05	0.9
2	2.44	1.8
3	4.6	2.7
4	7.8	3.6
5	11.8	4.5
6	14.5	5.4
8	17.2	7.2
10	18.2	9.0
15	20.4	13.5
20	21.3	18.0
25	22.2	22.5

Brooks concludes that if the initial winter cooling of an open polar ocean drops to a mere 0.6°F. (0.35°C) below the freezing point of the water, the result will be the formation of an ice-cap, which at first extends rapidly to a latitude of approximately 78° and then more slowly to about 65° (i.e.

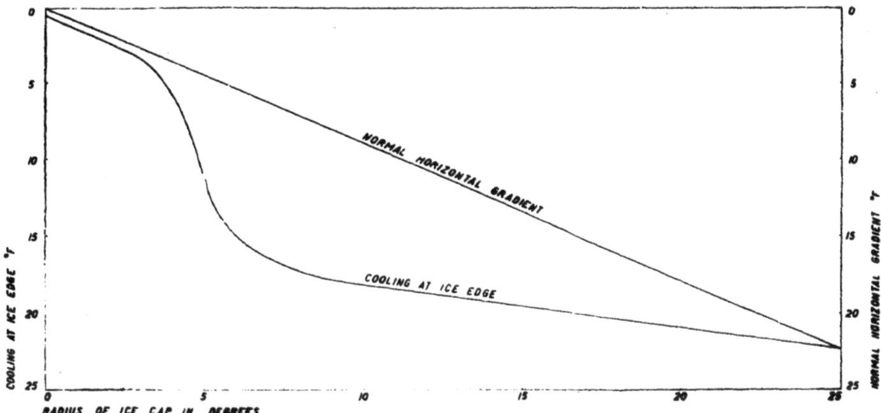

Fig. 166. Cooling power of an ice-cap (After Brooks).

assuming that no land influences or warm sea currents arrest its growth), and the ultimate lowering of the winter temperature, owing to the initial fall of 0.6°F., will then amount to 45°F. (25°C). This applies to ice-sheets on sea as well as to those on land. Brooks attached such importance to the cooling power of an ice-cap that he devoted the first chapter of his book to this subject.

This much seems clear, however: the annual mean temperature will, once it has begun to fall as a result of some cause, be lowered still further the moment an ice-sheet is created.

The same so-called cooling power of an ice-cap may probably also explain the extremely rapid growth and subsequent equally swift decay of the Pleistocene ice-sheets. The retreat of the last Scandinavian ice-cap has been depicted in fig. 167 and 168. The effect in this case is seen to consist of a slow and subsequently faster retreat of the border of the ice-cap. It should be noted, however, that another factor helped to accelerate the retreating movement. For the ice-front began to yield at an increasing rate as the ice-cap melted and grew thinner. The extreme right of the curve in fig. 168 finds its answer in one more factor. The land began to rise as it was freed of the load of the ice-sheet, and the highly elevated remnants of the ice-cap consequently melted at a much slower rate.

The following constitutes another important point, in my opinion. An extensive retreat of the sea is one of many factors that initiate glaciation. Once the ice-caps begin to extend, however, they will cause the sea-level to drop. The maximum lowering of the sea-level by glacial control must have attained approximately 100 meters during the Pleistocene. We know now that a great many shallow seas were reclaimed by the continents during

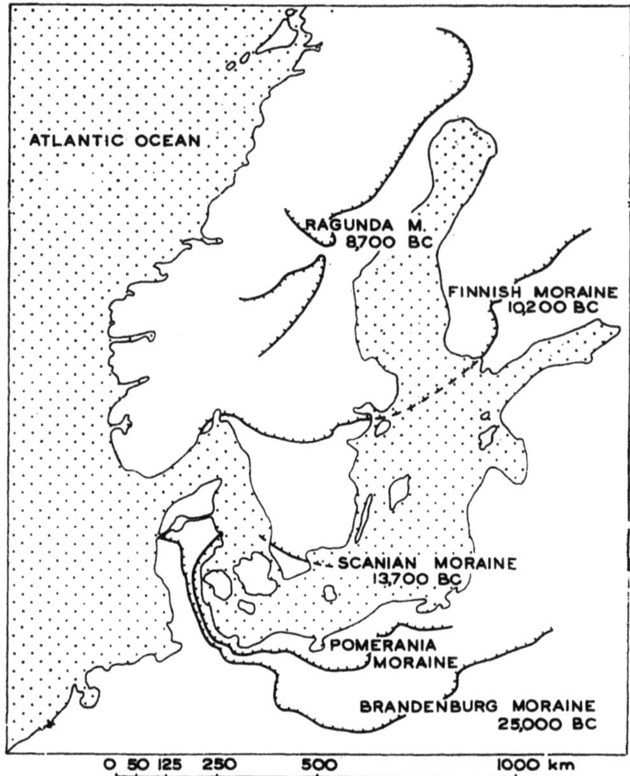

Fig. 167. Retreat of Fennoscandian ice-cap since the beginning of "post-glacial" times. (After R. A. Daly).

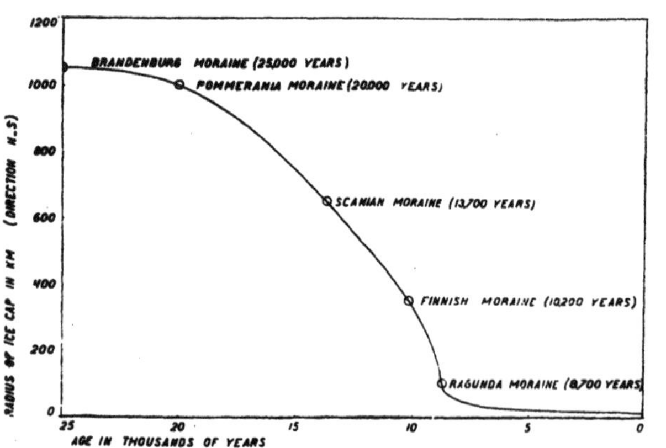

Fig. 168. Graph of the retreat of the Fennoscandian ice-cap, based on fig. 167.

this period, and that these same areas were drowned once more in post-glacial times as the ice-caps began to melt. An example of this is found in the drowned rivers of Pleistocene Sunda-land (fig. 60). This extension of land-areas as a result of glacial lowering of the sea-level favours glaciation and therefore this has to be considered as another autocatalytic factor.

Interglacial periods

The third aspect of the problem — viz. the interglacial periods that divide one long period of glaciation into a series of ice-ages — will now have to be examined. It is possible to distinguish between four separate ice-ages, — including three interglacial stages and our own "post-glacial" period (this may appear to be an euphemism!) — from the Pleistocene up to the present day.

The investigations by Milankovitch and Spitaler are of particular importance for this part of the climatic problem. Milankovitch calculated the radiation of the sun during the last 600,000 years. The heat which different zones of the globe obtain from the sun depends on three astronomical variables — viz. the obliquity of the ecliptic, the eccentricity of the orbit of the earth, and the cyclic variation of the perihelion. The obliquity of the ecliptic is responsible for the seasons. The greater the obliquity, the more contrast we find in the seasons. The limits of variation vary between 22° and 24½° in 40,400 years according to Milankovitch.

The orbit described by the earth around the sun is an ellipse, with the sun at one of the foci. The distance between the focus and the centre of the ellips, is called the earth's eccentricity and, expressed in terms of the major axis, varies between zero and approximately 0.07 in about 100,000 years. The third variable is the cyclic variation of the perihelion with a period of 20,700 years, i.e. the so-called precession of the equinoxes. Should the maximum effects of these three variables coincide the result would be a minimum radiation. It was on this basis that Milankovitch calculated and constructed his radiation-curve for the sun. This curve contains a number of small peaks and several higher ones. Some authors — particularly Köppen, Zeuner and Soergel — claim that a remarkable similarity exists between this curve and the conclusions regarding glacial and interglacial stages arrived at previously by geologists.

Nevertheless, it should not be concluded that the problem can be considered to be definitely solved, for several circumstances will dampen our enthusiasm in this respect. To begin with, the curve is far from simple, and can be interpreted in several ways. It makes no reference to the way in which the combined atmospheric factors (such as circulation, cloudiness, evaporation and absorption) react to a greater or lesser amount of radiation. Köppen, for instance, supposes that periods with a less than normal summer temperature and short, mild winters favour the development of ice-sheets, but Croll advocated the exact opposite — i.e. that periods with exceptionally cold winters and short, warm summers provide the most favourable conditions for the evolution of such sheets. Besides, Spitaler's curve deviates to such an extent from that of Milankovitch's, that both may be said to agree only superficially [1]). These authors thus arrive at very different results as regards the chronology of the ice-ages. The following table shows the astronomical time-scales of the four glacial stages as reconstructed by Köppen (Zeuner) and Spitaler, and includes a list based on geological evidence, after Brooks. The figures are in thousands of years B.C.

	Spitaler	Köppen	Brooks
Fourth Ice-Age	250–150	181– 66	50– 28
Third Ice-Age	600–450	236–183	140–110
Second Ice-Age	1,000–800	478–429	430–380
First Ice-Age	1,350-1,150	592–543	—

The problem of radiation was likewise examined by Brooks. The relation of solar radiation to the appearance of sunspots is undoubtedly a complicated one. Moreover, it is difficult to predict the effect which a change in the

[1]) Extensive consideration of the chronology of the Pleistocene based on Milankovitch's hypothesis of solar radiation was published by Zeuner in 1945 and 1946.

radiation of the sun will produce on a climate. Notwithstanding these difficulties, Brooks may safely be said to have shown a distinct connection between solar variation and general atmospheric changes. A great sunspot maximum marked the year 1870, and since then (apart from certain fluctuations, such as an eleven-year cycle) there has been a gradual decrease of sunspots. During this time the average annual temperature and pressure rose in North Asia, Europe, Iceland and part of North America. Moreover, the glaciers and ice-caps have, on the whole, been retreating — especially since the beginning of the twentieth century.

In fine, all that can be said at present is that glacial and interglacial stages may possibly, and even probably, be attributed to changes in the amount of radiation dependent upon several cosmic variables, and that a correct interpretation of the climatic effect of these variable factors, and an accurate reconstruction and interpretation of the radiation-curve, must remain reserved for the future.

Of course, these astronomical factors were already active before the beginning of the Pleistocene. Yet, the fact that they only produced ice-ages during certain distinct periods proves that glaciation must have been initiated by one or more additional causes. The latter were discussed above, and it was pointed out that the major periods of abnormal climates are controlled by geographical changes and deep-seated terrestrial processes.

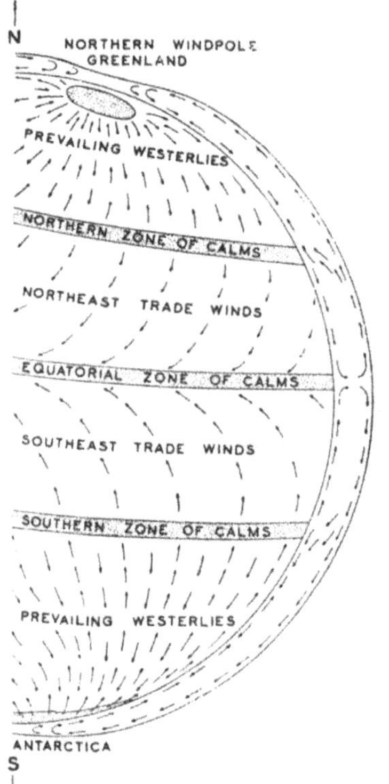

Fig. 169. Diagram of the atmospheric circulation (After W. H. Hobbs).

The geographical location of an ice-cap

As mentioned above, the last part of the problem deals with the site of an ice-cap once glaciation has become possible.

A few examples were given of normal climates, and it was seen that arid regions, for instance, will change with corresponding alterations in the distribution of land and sea.

If the earth's surface were homogeneous, with the exception of an ice-cap stationed in the present areas of Greenland and the South Pole, the atmospheric circulation would be approximately equivalent to that shown by Hobbs in the accompanying diagram (fig. 169). The actual situation deviates from this theoretical conception because of the special distribution of land and sea.

In this schematic diagram two zones of high pressure will be seen to run parallel to the equator, and these are predestined to become belts where the climate is subjected to an arid degradation. The present geograph-

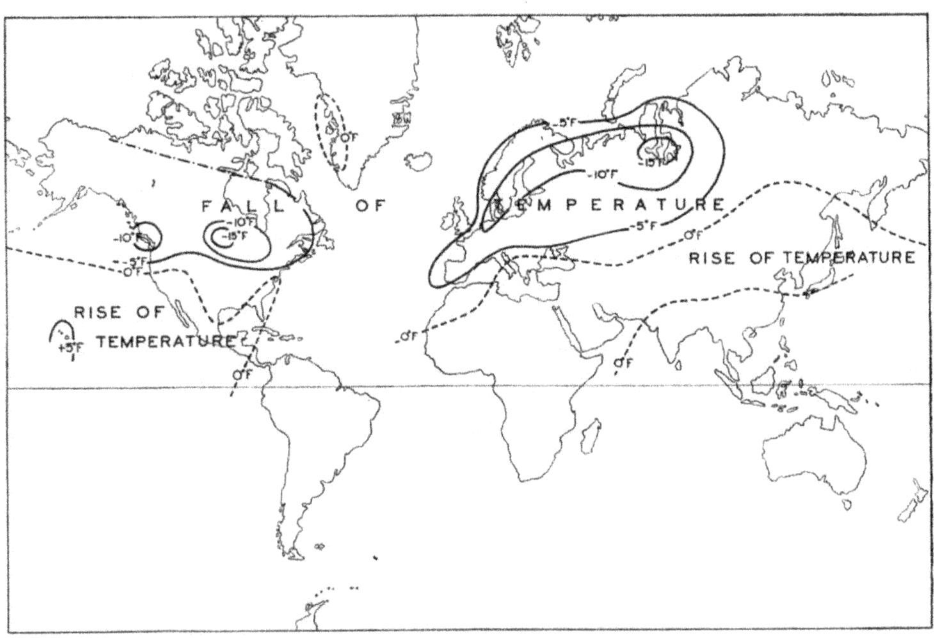

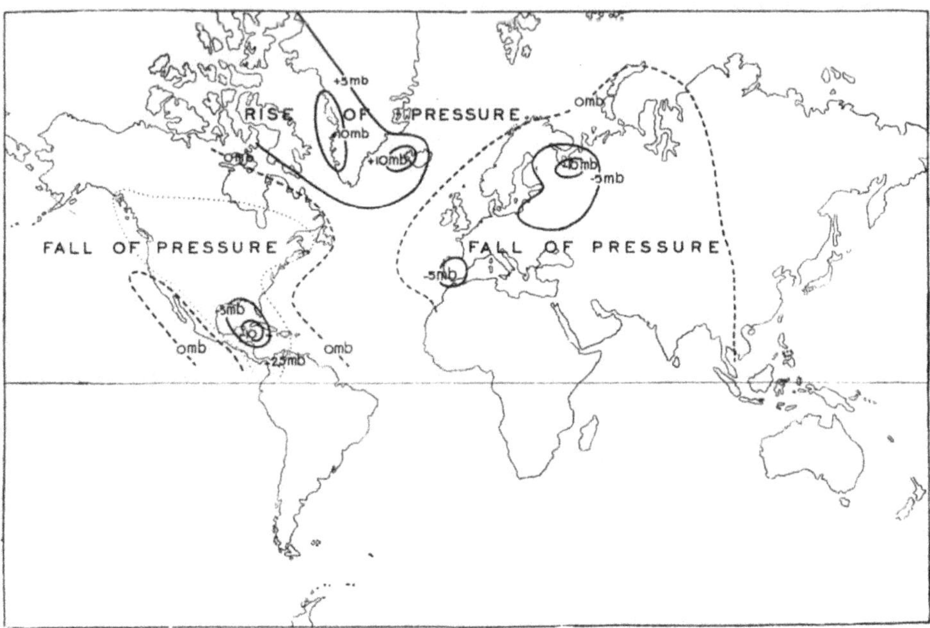

Fig. 170 and 171. Changes of mean temperature and mean pressure with increase of the relative sunspot number to 200 (After C. E. P. Brooks) mb = 1 millibar = 1.0197 g/cm².

ical conditions, however, are responsible for the fact that this is only partly true. To mention two examples: they caused the arid zone to shift northwards in Central Asia, and the same conditions were responsible for the absence of an ice-cap in Siberia, though this last area is known to constitute the coldest region on earth. The distribution not only of the temperature, but also of pressure, water vapor and the whole complex of oceanic and atmospheric circulation has a most important bearing on these deviations from the schematic diagram.

This will also help to make it clear why the large Pleistocene ice-sheets are found in the northern hemisphere, and Brooks even tried to show why they originated in those localities in which they are actually known to have occurred. To understand his arguments, it will be necessary to return once more to the influence of solar radiation. As mentioned above, Brooks revealed that a decrease in the frequency of sunspots since 1870 has clearly been accompanied by a corresponding increase in the mean annual temperature and pressure over specific areas in North America. Iceland, Europe and northern Asia. With the aid of observations obtained from these regions and the logical premiss that a growing number of sunspots would have a contrary effect, Brooks constructed two maps (fig. 170 and 171), showing the distribution of the temperature and pressure in accordance with a hypothetical increase in spottedness up to an average of 200. This "relative sunspot number" means that approximately one five-hundredth part of the sun's visible disc is covered by spots. The figure for the sunspot maximum in 1870 was 139, and 32 for the eleven-year period from 1904 to 1914. Fig. 170 indicates a fall in temperature of more than 5° F. (approx. 3°C.), and of even as much as 15° F.(approx. 8°C.) over Canada, Northern Europe and the western part of Siberia, together with a rise in the temperature above the regions situated south of these areas. Fig. 171 depicts a decrease in the pressure over the continents, and an area of increased pressure over the oceans. These reconstructions present conditions which may very well have prevailed towards the beginning of the Pleistocene ice ages!

It should not be forgotten that these results are based on the present distribution of land, sea and mountain belts. There is indeed good reason to suppose that the position of the continents during the Pleistocene did not differ from their present situation. However, let us reach back somewhat further into the past. The plant-bearing deposits of Eocene time causes Chaney to conclude that the position of the European continent in respect to America has remained unaltered since as far back as the Eocene. The accompanying maps make it easy to follow his argumentation. Those Eocene plant units that bear evidence of the formerly existing climates have been grouped in fig. 172, and "similar fossil floras, which are considered to represent essentially similar climatic conditions" have been connected by lines which Chaney referred to as isoflors. These are also felt to represent Eocene isotherms. Fig. 172 should be compared with fig. 173, which includes present-day isotherms for the month of January and the modern flora-types. The similarity between our isotherms and those of Eocene time is apparent,

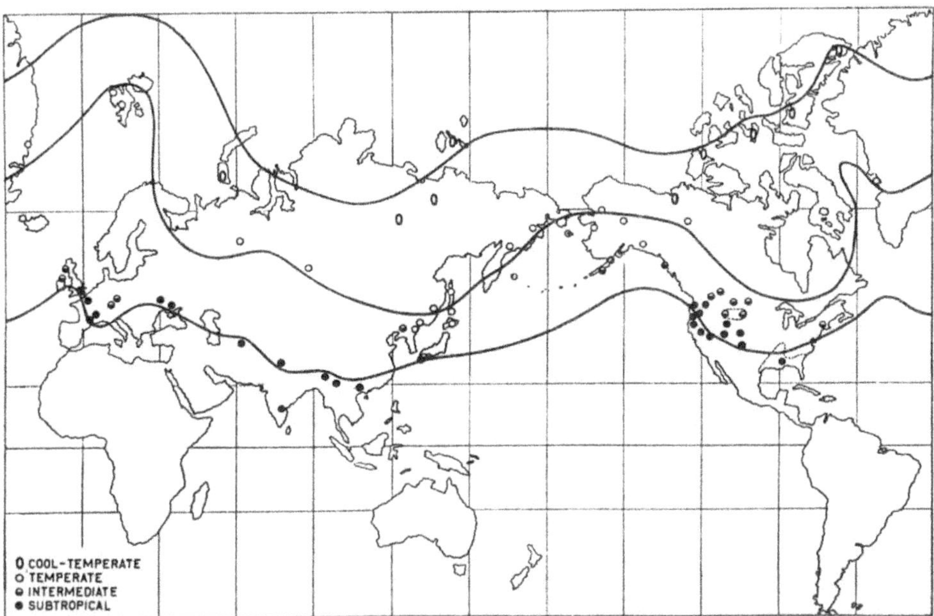

Fig. 172. Eocene "isoflors" in the Northern Hemisphere (From R. W. Chaney).

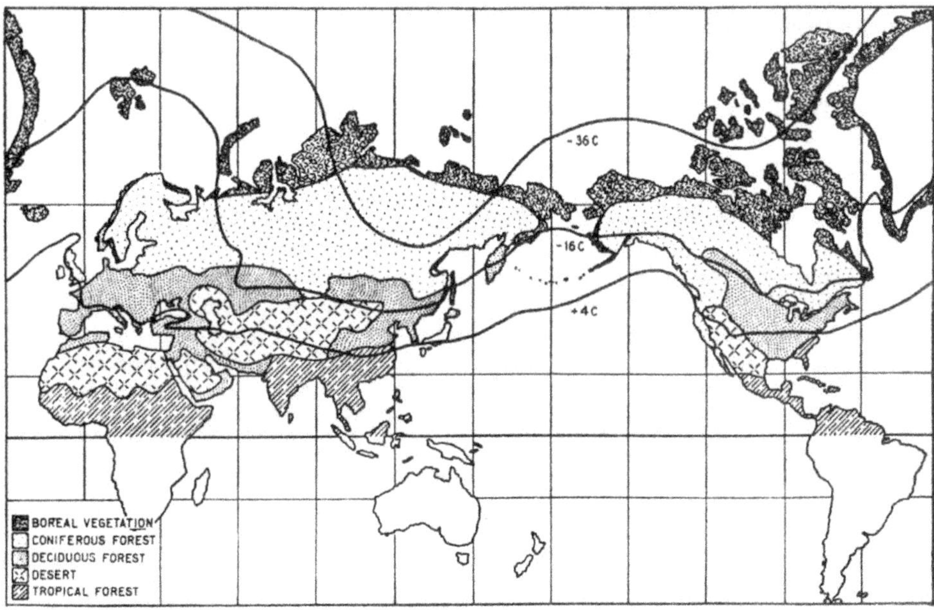

Fig. 173. The distribution of modern vegetation and January isotherms in the Northern Hemisphere (From R. W. Chaney).

and Chaney consequently wrote: "Present departures of isotherms from coincidence with parallels of latitude are interpreted from the tempering effects of ocean currents on the western sides of continents in middle and high latitudes, and from the increasing continentality toward the east. It seems entirely reasonable to assume that a similar distribution of vegetation and temperature during the Eocene resulted from the same controlling factors, and to conclude that land and sea relations, as well as planetary circulation, were essentially like those of to-day. Brooks shows a map of part of Europe, after F. von Kerner, with a similar trend of Tertiary isotherms, based on wholly different evidence".

Like Berry, whose view was mentioned above, Chaney [1] concludes that all paleobotanic data plead against the idea of a shifting of continents and poles since the Eocene, and his final verdict is summed up as follows: "The pattern of land vegetation around the Northern hemisphere indicates that the factors controlling air and water circulation have been essentially the same since the Eocene — that North America and Eurasia have stood in their present positions since the dawn of the Cenozoic. Evidence of land plants of earlier times, less completely known and interpreted, seems also to refute the hypothesis of continental drift. Forrests under compulsion of climatic change, rather than the continents on which they live, appear to have been the wanderers during the history of life upon the earth."

This brings us to the problem of the position of the Upper Paleozoic ice-sheets. Chaney's results, together with Brook's noteworthy investigations, make it clear that the unusual distribution of Upper Paleozoic and even more ancient glaciations could in all probability be explained without continental drift if only accurate paleogeographic maps could be reconstructed for those distant ages.

However, some authors regard the distribution of the tillites in the Upper Paleozoic as one of the most important arguments in favour of the hypothesis that vast changes affected the position of the continents and poles. It was explained above that the reconstruction of climatological zones, based on the present situation, as put forward by Köppen and Wegener, was a fundamentally wrong one for several reasons. For, to begin with, these investigators assume that the present conditions are the normal ones; but — as Coleman remarked — "that is far from being the case". In the second place, they do not explain — nay, even obscure — one of the most prominent climatic features, viz. the periodicity of abnormal climates. There can be no doubt about the occurrence of this last phenomenon, for, as shown above, it takes place in conjuntion with other periodical processes in the earth's crust.

As mentioned previously, the abnormal climates may be explained by the combined influence of widely diverging factors, among which, however, the shifting of poles and continents has played no part.

I would now like to examine a few points, of which no account has been taken so far. Astronomical data do not support the view that the position

[1] Chaney, op. cit. p. 486.

of the earth's rotation-axis has undergone any significant change, nor does the astronomer find it possible to suggest a cause for such a hypothetical occurrence. Moreover, the comparatively rapid genesis of glaciation would require equally rapid hypothetical changes, but since the earth is a gyroscope and thus has a powerful tendency to keep the axis of its rotation constant, any sudden alteration in its direction would only bring about a world-wide catastrophe [1]. Historical geological data prove that nothing of the kind has ever happened.

On the other hand, it might be assumed that the position of the earth's crust has undergone some changes in respect to the core. This would result in a different distribution of the continents in respect to the axis of rotation, and the position of the latter would then undergo a slight alteration owing to this process. It was this type of "shifting of poles" that was used by Köppen and Wegener when constructing their paleogeographic maps. Milankovitch published a mathematical treatise on this subject, showing that the North Pole lay in the vicinity of the Hawaiian Islands in the Paleozoic, and that since then it had migrated to its present position *via* Alaska.

Mathematical calculations should always be welcomed when dealing with geological problems, for they frequently sift faulty arguments and incorrect suppositions. Yet an entirely accurate and unimpeachable mathemathical calculation may sometimes lead to wrong conclusions. As Huxley observed [2]: "Mathematics may be compared to a mill of exquisite workmanship, which grinds your stuff to any degree of fineness; but nevertheless what you get out depends upon what you put in; and as the finest mill in the world will not extract wheat-flour from peascods, so pages of formulae will not get a definite result out of loose data". The premisses of Milankovitch's calculations are such that its results can be said to be the reverse of what might be termed definite proof. For one of the assumptions is that no change has affected the sialic shell, and another that an activity, referred to by Milankovitch as a "Polfluchtkraft", was in existence, etc. Yet these conjectures are so questionable that his conclusions may be said to lose all value, as was shown at great length by Schwinner.

However, there is more than that. It will be seen, in Chapter XI, that the premisses of Milankovitch's calculations are entirely wrong and that his conclusions are of no value at all (p. 302).

I would like to mention a few additional points at this stage. The fusion of the continental blocks into a single, large primordial continent, which is known as Pangea, would provide an adequate explanation, in the opinion of the drift hypothesis, of the present widely-scattered remnants of the Upper Paleozoic glaciations, for they would have united as one extensive ice-cap in the South Pole. Yet several authors, including Coleman, doubt whether such an extensive land mass would have acted favourably on the formation of an ice-cap. The reconstruction of a continent such as Pangaea was discussed exhaustively by Brooks, who examined to what extent climatic evidence might be said to favour the idea of continental drift. One conclusion of this investigator was that the relative ages of the glacial

---

[1] Coleman, op. cit. 1926, p. 263.

[2] *Fide* Knowlton. p. 565.

deposits in the southern hemisphere are incompatible with Wegener's view of moving poles. Another point is that tillites occurred in the northern hemisphere, particularly in North America, north-western Siberia, (where the ice probably radiated from the existing Kara Sea Massif). It should be remembered that the existence of an enormous ice-cap in Asia during the Pleistocene only became known quite recently. The Paleozoic of Asia may consequently hold more surprises in store for us in this respect! The presence of tillites has been reported in numerous other areas. Some may have been brought thither by local mountain glaciers or floating ice-masses. Others, such as those existing in Texas, are probably tectonic breccias, but the so-called Squantum tillite near Boston is so thick and widespread that geologists have not hesitated to assume that these fossil moraines very probably constitute remnants of a vast ice-sheet. In Wegener's reconstruction of Pangaea, the latter is situated in his equatorial rain-zones of Carboniferous time and in the centre of his Permian desert belt. These tillites consequently conflict with the theory of drift, and Wegener therefore attempts to "explain them away tacitly" [1]. Brooks expressed the opinion that the Squantum-tillites might be explained without a shifting of poles and continents, and assumes that they were formed under the influence of a cold sea, extending from the Arctic region as far as the eastern part of America [2]. Brooks devoted a whole chapter to the causes of the abnormal climate of Upper Paleozoic time (these are dealt with at great length in his Chapter XV), and also showed that a possible westward drift of both the American continents in respect to Europe could not affect his solution in any way, though the change in the position of other continents would only augment the difficulties of the problem. As a matter of fact, Brooks not only pointed to the defects of the hypothesis of drift, but also attempted to furnish a constructive and extensively argumented explanation of the question of geological climates, in which "epeirophorese" played no part whatever.

We are living in an age of ice-caps. The distribution of land, sea and mountain chains, etc. are responsible for the position of the ice-caps as seen above. It was likewise shown that the site of the Pleistocene ice-caps can be understood without a hypothetical shifting of continents. The emplacement of the Upper Paleozoic ice-caps, too, could probably be explained if we had all available data as regards the distribution of physiographic factors in the distant past at our disposal.

My own conclusion may be said to concur with that of Schuchert, which the latter expressed in 1923. When asked what had brought about the Upper Paleozoic glaciations, this author replied [3]: "Certainly never the

[1] Brooks, p. 266.

[2] It is far from certain, however (see fig. 174), whether a cold sea existed in North America, which would thus explain the presence of the above Squantum tillites. Some geologists have drawn a sea-arm in this region, but others Tdicate land in this area instead of water. inhis may be enough to show that Brook's explanation, too, should be taken *cum grano salis*. Still, no other alternative is left us, for it would be impossible to construct a paleogeographic map which would not be subject to criticism, or which would not at times even give rise to severe controversy.

[3] Schuchert, 1923, p. 1081.

impossible Wegenerian hitching together in close embrace of the present masses of South America, Africa, India, and Australia. In this hypothesis I see only a drunken sialic upper crust hopelessly floundering upon the sober sima. Let us return to a tangible geology. We all now know that the earth underwent one of its greatest crustal disturbances during the Carboniferous, beginning late in the Lower Carboniferous (toward the close of Viséan and Chesterian time), attaining a first climax in middle Upper Carboniferous time, and a second maximum early in the Permian. The supercrust was undergoing one of its periodic revolutions, and the world was then as scenically grand as it is to-day. It is in the youthful topography, the enlarged continents and the peculiar connections of the continents that seemingly are to be sought the reasons for the Permian ice-age." To this he added, some nine pages further on: "If it can be shown that Australia was connected in Permian times with Antarctica, then this holding in of so much of the cold waters of the Antarctic Ocean, combined with moist climates in the southern hemisphere and the general highland condition of so much of the world in early Permian time, will be the explanation for the peculiar position of the continental ice-masses of the southern hemisphere".

Schuchert and Willis later illustrated the supposed distribution of land and sea, as well as that of cold and warm ocean currents, and tried to explain the Upper Paleozoic glaciation without the occurrence of alterations in the present position of continental blocks. These authors, like Brooks, are of the opinion that a solution must take into account the existence of former land connections between Africa and Australia, and between Africa and South America. Many investigators have based similar conclusions on other grounds (i.e. on various geological arguments, and the distribution of land plants, animals and marine faunas).

The preceding chapter explained that though each reconstruction of transoceanic land-bridges retains a purely hypothetical character, the idea that land-bridges existed in former times should not be wholly discarded. A "not impossible" reconstruction will be found in fig. 174.

At some time in the future, when science will have progressed still further, it will perhaps be possible for geologists to draw a complete and accurate paleogeographic and paleoclimatologic map of the Upper Paleozoic. A more profound knowledge of the geological history of the oceans will perhaps have been acquired by that time. Until then, however, a "not impossible" reconstruction such as that appearing in fig. 174 will have to replace a more detailed and satisfactory explanation of the remarkable glacial remnants of Upper Paleozoic time.

The restricted influence of cosmic factors

The influence of cosmic phenomena was met with twice in the foregoing paragraphs. Huntington and Visher, in their "solar cyclonic hypothesis", went so far astoregard a great increase in sunspots as the only primary factor capable of bringing about ice-ages. Yet there are two objections to this theory: (1) that a periodicity of the fluctuation of solar radiation, such as is sought when attempting to explain the earth's abnormal climates, is, of course quite unknown, and that its existence has never been made acceptable; (2) that abnormal climates, as previously stated, are connected with internal terrestrial processes (mountain-building, magmatic cycles, etc.). The influence of sunspots on subcrustal processes and its effect on epochs of folding, mountain-building, regressions, etc. would therefore have to be shown first. Huntington and Visher furnish tables purporting to show the apparent relation of an increase in spottedness to an accompanying increase

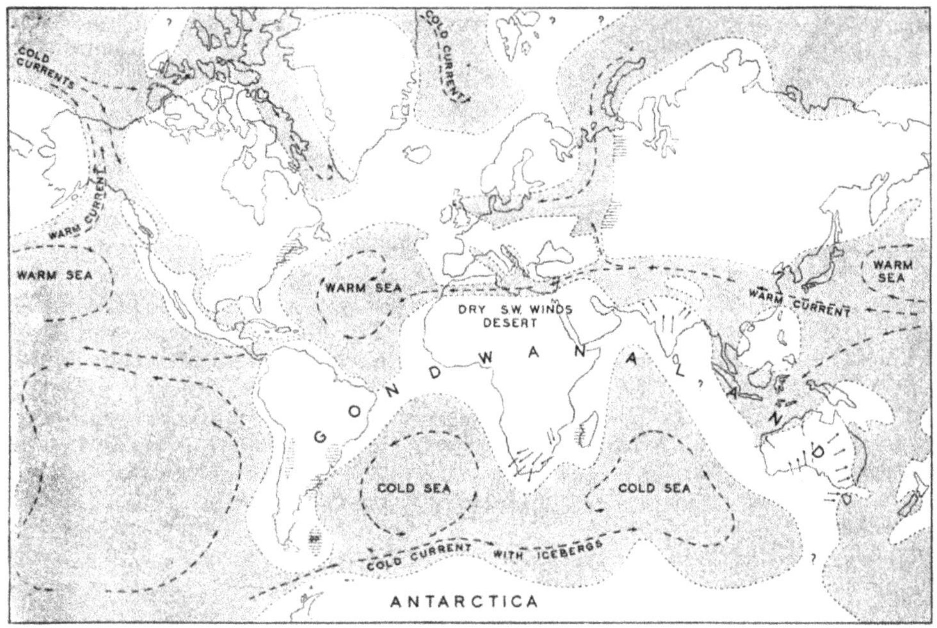

Fig. 174. Hypothetical reconstruction of geographic conditions during the Upper Paleozoic ice ages. Glaciated regions are indicated by horizontal shading and small arrows; the latter show the direction of ice-flow in South Africa, India, and Australia.

in tectonic earthquakes, and retrace this relation *via* changes in the atmospheric pressure. They derive the hypothetical and major periodicity of sunspots from another theoretical process, in which the sun is supposedly affected by changes in the position of other heavenly bodies. Their combined views thus attribute a direct influence on the climate, and an indirect one on subcrustal events to cosmic phenomena. An illustration of their argumentation appears in fig. 175. However, this figure also includes the effect of geographical factors, since a hypothesis which attempts to explain everything solely by means of cosmic phenomena seems to me to be hopelessly incompatible with present geological data.

Nevertheless, it may be that the influence of cosmic phenomena, as represented in this scheme, is not sufficiently restricted. Joly had, indeed, already observed that it was not improbable that the periodicity of subcrustal processes was stimulated by a cosmic effect and Schwinner subsequently argued that after the supposed disruption of the moon from the earth a resonance effect would have occurred as many as eleven times (i.e. the period of the tides would have coincided eleven times with the natural free period of vibration of the earth), and he assumes that this would have caused the periodic phenomena of mountain building (fig.176). Yet, it makes a considerable difference, from a fundamental point of view, whether we restrict the cosmic influence in this manner, or attribute the dominating role to astronomic factors, as in the "solar cyclonic hypothesis".

Besides, even when restricted thus, the above astronomical influence can be stated to constitute no more than a hypothetical factor. Prof. Oort recently assured me that the only exceptionally long periodicity established so far in astronomy is the rotation of our galactic system, which revolves in approximately 200 million years. This figure agrees fairly well with that of the major geological rhythm of 250 million years. This congruence is indeed remarkable, but for the moment it is hard to see how the astronomical cycle could be connected with atmospheric and subcrustal processes.

Summary

The Canadian geologist Coleman, who gleaned a considerable knowledge as regards glacial phenomena from direct observation over the entire world, remarked towards the end of his book on Ice-ages that he did not consider himself competent to propose a hypothesis which would prove to be more satisfactory than the existing ones (it should be noted, however, that he did not hesitate to draw attention to the defects in all the previous theories). Coleman ended, however, with these words: "In my opinion no single cause, as expressed in one of the theories proposed, can accomplish this; and any final solution of the complicated problems involved must come from a conjunction of general and local causes. Some combination of astronomic, geologic and atmospheric conditions seems to be necessary to produce such catastrophic events in the worlds' history".

I regard this modest view as the only correct one. The problem is undoubtedly a very intricate one. I have attempted to emphasize the necessity of combining the various cooperating factors, and my motives may be summarised as follows (see fig. 175 and 176): an almost world-wide and temperate climate prevailed during such a considerable lapse of time that we may consider it to have constituted the normal climate. Deviations from this normal state of affairs, which produced large glaciations, occur with a periodicity of approximately 250 million years. These are related to the principal Pre-Cambrian, Upper Paleozoic and Pleistocene periods of mountain-building that influenced large parts of the globe. Less intensive climatic variations with a shorter periodicity occur during the intervening periods and are probably related to mountain-building of minor geographical significance. Mountain-building is closely related to various other groups of phenomena (regressions, changes in the pressure of the earth's crust, magmatic processes, etc.), which point to a common and widespread cause: i.e. to the activity of periodical subcrustal processes. The latter might also be thermal cycles, and the fall of temperature might then possibly

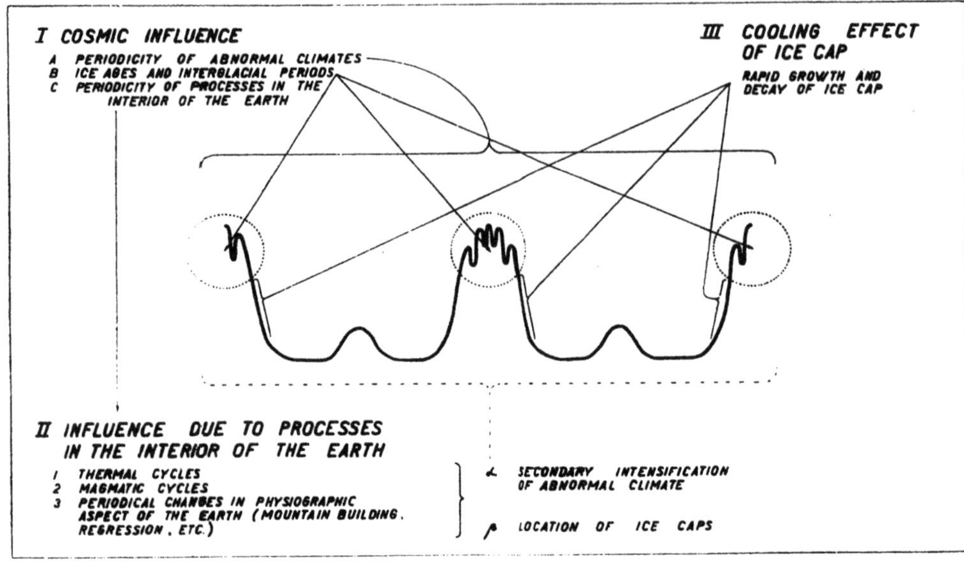

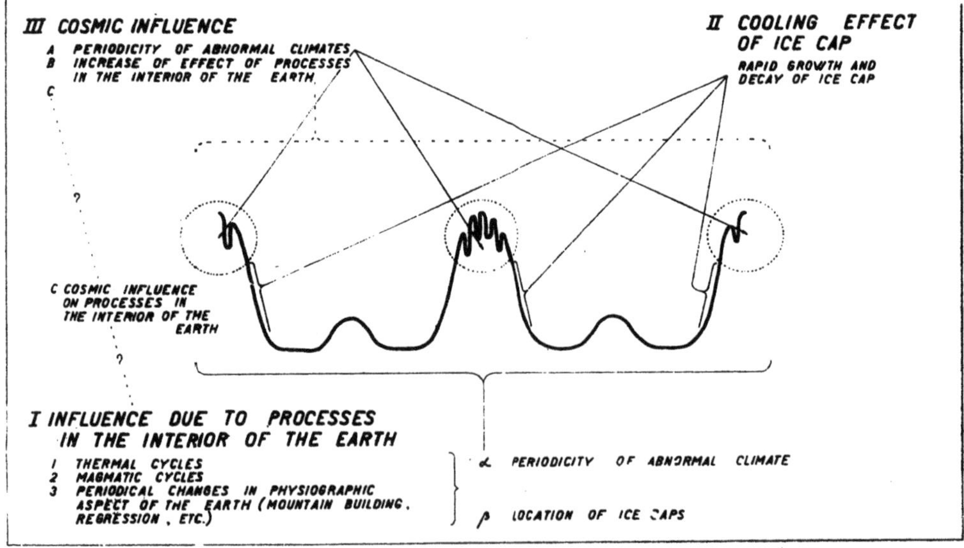

Fig. 175 and 176. Schematic representations of two possible explanations of glaciation and ice-ages.

concur with periods of mountain-building and glaciation. When the alterations in the earth's physiographic aspect (with the possible inclusion of subcrustal cooling) produce local ice-caps, they suddenly begin to expand and assume unusual proportions as a result of the cooling effect exerted by an ice-sheet on the temperature in the surroundings. The special position of ice-caps shows that they are controlled by physico-geographical factors: the distribution of land and sea, mountains, cyclone routes, etc. Nevertheless, the influence of astronomical factors cannot be denied. To begin with, it may be possible that an increase in the amount of sunspots finds expression in the atmosphere, i.e. in a decrease in the temperature and pressure above certain continental areas. And in the second place, the division of glaciations into ice-ages and interglacial stages may probably be ascribed to cosmic influences, and may probably be derived from changes in the obliquity of the ecliptic, the variable eccentricity of the earth's orbit and the precession of equinoxes.

All these effects might be combined in two ways. One would be to assume (fig. 176) that the periodicity of the abnormal climates is activated primarily by internal terrestrial processes. Their effect would suddenly increase to a considerable degree owing to the cooling influence of an ice-sheet. The influence of cosmic phenomena, moreover, would increase the effect still further, and these phenomena would explain the appearance of interglacial stages. The subcrustal processes may possibly also be stimulated by cosmic phenomena.

The alternative (fig. 175) would be that the abnormal climates and the initiation of events in the interior of the globe are the direct result of a hypothetical cosmic phenomenon. The subcrustal processes would therefore merely be related in point of time to abnormal climates, and would exert a mere secondary effect on the latter, at the most. One cosmic phenomenon is in fact known to possess a periodicity of the same order as that found in the major periodicity of 250 million years on earth, viz. the rotation of the Galactic System. It is not yet clear, however, how this astronomical period might be connected with subcrustal events.

In my opinion the solution of this tangled problem must be sought in one of these contingencies. The question, at present, is whether the primary cause must be attributed into internal terrestrial processes or to cosmic factors Some may incline to the first and others to the second alternative when looking for a probable explanation of the periodical occurrence of abnormal climates, but it cannot be denied that it is impossible — or expressed more optimistically, not yet possible — to determine the primary cause of this phenomenon. Daly's negative view, which was mentioned at the beginning of this chapter, can thus be said to be correct up to this point. I hope, however, that I have succeeded in showing some of the features of the periodical occurrence of the remarkable phenomena of glaciation and ice-ages.

References

ANTEVS, E. *Maps of pleistocene glaciations* (Bull. Geol. Soc. America, 40, 1929).
BACKLUND, H. G. *On late palaeozoic glaciations in the northern hemisphere* (Uppsala Universitets Arsskrift 1929).
BERRY, E. E. *Climatic significance of arctic fossil floras* (Bull. Geol. Soc. America, 40, 1929).
BROOKS, C. E. P. *Climate through the ages* (London, E. Blenor 1926).
BUCHER, W. H. *Deformation of the Earth's Crust* (Bull. Geol. Soc. America 1939).
CHANEY, R. W. *Tertiary forests and continental history* (Bull. Geol. Soc. America, 51, 1940).
COLEMAN, A. P. *Ice ages* (Macmillan Co 1926).
COLEMAN, A. P. *Ice ages in the geological column* (Bull. Geol. Soc. of America, vol. 50, 1939).
COLEMAN, A. P. *The last million years. A history of the Pleistocene of North America* (Univ. of Toronto Press, 1941).
DAQUÉ, E. *Grundlagen und Methoden der Paläogeographie* (p. 376–476 Die Paläoklimatologie 1915).
DACQUÉ, E. und WEGENER, A. *Paläogeographie* (in Enzyklopädie der Erdkunde, 1926).
DALY, R. A. *The changing world of the ice age* (1934).
DARRAH, W. C. *Principles of Paleobotany* (1939).
DU TOIT, A. L. *Our wandering continents* (Oliver and Boyd, Edinburgh 1937).
EVERDINGEN, E. VAN, *Verklaring van geologische klimaten* (Tijdschr. Kon. Nederl. Aardrijkskundig Genootschap 57, 1940).
FORREST, H. E. *The Atlantean Continent. Its bearing on the great ice age and the distribution of species* (Whitherby, London 1933).
FURON, R. *La palaeogeographie* (1941).
GERTH, H. *Die Tertiärfloren des südlichen Südamerika und die angebliche Verlagerung des Südpols während dieser Periode* (Geolog. Rundschau. 32, 1941).
GRIGGS, D. A. *Theory of mountain building* (Americ. Journ of Science 1939).
HOBBS, W. H. *The glacial anticyclones* (1926).
HOBBS, W. H. *The glacial Anticyclone and the continental glacier of North America* (Proc. Americ. Philosoph. Soc. 86, 1943).
HOBBS, W. H. *Nourishment of the Greenland continental glacier* (Journ. of Geology 52, 1944).
HOLMES, A. *A contribution to the theory of magmatic cycles* (Geol. Magazine 63, 1926).
HOLMES, A. *The age of the earth* (1937).
HUMMEL, K. *Welteislehre und Geologie* (Natur und Volk, 69, p. 129, 1939).
HUNTINGTON, E. and VISHER, S. S. *Climatic changes* (Yale Univ. Press 1932).
JOLY, J. *The surface of the Earth* (1925).
JOLY, J. *The theory of thermal cycles* (Gerlands Beiträge zur Geophysik 1928).
KERNER-MARILAUN, F. *Paläoklimatologie* (Berlin, Borntraeger, 1930).
KNOWLTON, F. H. *Evolution of geologic climates* (Bull. of the Geol. Soc. of America, vol. 30, 1919).
KOPPEN, W. und WEGENER, A. *Die Klimate der geologischen Vorzeit* (Berlin, Borntraeger 1924).
MILANKOVITCH, M. *Théorie mathématique des phénomènes thermiques produits par la radiation solaire* (Paris. Gauthier-Villars 1920).
MILANKOVITCH, M. *Säkulare Polverlagerungen* (Handb. d. Geophysik I, Lief. 2, 1933).
MILANKOVITCH, M. *Astronomische Mittel zur Erforschung der Erdgeschichtlichen Klimate* (Handb. d. Geophysik IX, 1938).
NORIN, E. *An occurrence of late palaeozoic tillite in the Kuruk-tagh mountains, C. Asia* (Adress G. F. Stockholm Meeting 1929).
OESTREICH, K. *IJstijd en actualisme. Opmerkingen naar aanleiding van de Verhandelingen van Dr. C. G. S. Sandberg: Ist die Annahme von Eiszeiten berechtigt?* (Tijdschr. Kon. Nederl. Aardr. Genootschap 58, 1941).
OSBORN, H. F. *Ancient Vertebrate life of Central Asia* (Livre Jubilaire. Soc. Geol. de France II, 1930).
RAMSAY, W. *The probable solution of the climatic problem in geology* (Geol. Magazine 61, 1924).
RUEDEMANN, R. *Climates of the past* (Geologie der Erde, North America 1939).
SALOMON-CALVI, W. *Die Permokarbonischen Eiszeiten* (Leipzig, Akad. Verlag 1933).
SANDBERG, C. G. S. *Ist die Annahme von Eiszeiten berechtigt?* (Ed. Sijthoff, Leiden 1937).
SAURAMO, M. *The Quarternary geology of Finland* (Bull. Comm. Geol. Finlande 89, 1929).
SCHUCHERT, CH. *The Earth's changing surface and climate* (in R. S. Lull. The Evolution of the Earth, Yale Univ. Press 1919).
SCHUCHERT, CH. *The paleogeography of Permian time in relation to the geography of earlier and later periods* (Proceed. Second Pan Pacific Sci. Congres, vol. 2, 1926).
SCHUCHERT, CH. *Gondwana landbridges* (Bull. Geol. Soc. of America, vol. 43, 1932).

SCHWINNER, R. *Sind grosse Polverschiebungen möglich?* (Gerlands Beiträge zur Geophysik 43, 1934).
SCHWINNER, R. *Lehrbuch der physikalischen Geologie* (I, 1936).
SIMPSON, G. C. *World climate during the Quarternary period* (Quart. Journal Royal Meteorol. Soc. 60, 1934).
SOERGEL, W. *Die Vereisungskurve* (1937).
SPITALER, R. *Die sommerliche und winterliche Bestrahlung während der quartären Eiszeit* (Meteorolog. Zeitschrift 1939).
TEICHERT, C. *A tillite occurrence on the Canadian Shield* (Report of the fifth Thule Expedition, Vol. I, Nt. 6, 1937).
UMBGROVE, J. H. F. *Oorzaken van IJstijden* (Natuurkundige Voordrachten, Mij. Diligentia, 's-Gravenhage 1941).
VENING MEINESZ, F. A. *The determination of the Earth's plasticity from the postglacial uplift of Scandinavia* (Proc. Kon. Acad. v. Wetensch. Amsterdam, vol. 40, 1937).
WAGNER, A. *Klimaänderungen und Klimaschwankungen* (1940).
WANLESS, H. R. and SHEPARD, F. P. *Sea level and climatic changes related to late Paleozoic Cycles* (Bull. Geol. Soc. of America, vol. 47, 1936).
WATERSCHOOT VAN DER GRACHT, W. A. J. M. VAN, *The Permocarboniferous exotic boulders in the so-called "caney shale" in the northwestern front of the Ouachita mountains of Oklahoma* (Journal of Geology, vol. 39, 1931).
WILLIS, B. *Isthmian links* (Bull. Geol. Soc. of America, vol. 43, 1932).
ZEUNER, F. E. *The Pleistocene Period. Its climate, chronology and faunal successions* (The Ray Society London, Monogr. 130, 1945).
ZEUNER, F. E. *Dating the past* (Methuen, London, 1946).

CHAPTER X

THE RHYTHM OF LIFE

"... . changing environmental conditions stimulate the sluggish evolutionary stream to quickened movement".

(R. S. LULL)

Introduction

Environmental factors have had a decided influence on the evolution of Life. It can hardly be an accident, for instance, that geological phenomena such as regression and the Laramide mountain-building towards the close of the Mesozoic, should have corresponded with the equally important adaptive radiation of placental mammals. Then, too, can it be mere chance that the Psychozoic — which received its name from the fact that Man, the specialist of spiritual and intellectual differentiation, began to extend his supremacy over the world at that time — coincides with an exceptionally intensive phase of mountain-building and glaciation?

Many more examples would be given, but we will confine ourselves to the above two. In one of his well-known treatises, Matthew emphasized the influence of the climate on organic evolution, regarding it as the most important environmental factor to effect biological changes. This significant primary role was attributed by Szalai to the various manifestations of mountain-building, and by Wilser to the varying intensity of solar radiation. Other factors, too, have undoubtedly played a more or less important part, but it will be clear that it is scarcely possible to estimate each influence separately, since all are related to one another to a greater or lesser degree. It would lead us too far to examine the absorbing biological problems inherent in the changes to which the geneplasm and hereditary constitution were obviously subjected. For the moment we merely wish to point to the frequent concurrence of changes in the physical and organic world. As Lull remarked in the closing lines of his chapter on "The Pulse of Life":

"Thus time has wrought great changes in earth and sea, and these changes acting directly or through the climate have always found somewhere in the unending chain of living beings certain groups whose plasticity permitted their adaption to the newly arising conditions.

"The great heart of nature beats, its throbbing stimulates the pulse of life".

The evolution of the fauna was not only affected by many changing terrestrial factors but also adapted itself to the evolution of the flora. The boundary between the Paleozoic and the Mesozoic lies on a higher stratigraphic level than that between the Paleophytic and Mesophytic, and the Cenophytic, too, began at an earlier date than the Cenozoic! The effect of external changes on the evolutionary stream of life might therefore *a priori* be expected to reflect itself more distinctly in the flora than in the fauna. That this influence on the flora is in fact very evident will be shown in the following paragraphs.

The periodic differentiation of the flora

The flora appears to have been entirely uniform during the Devonian, judging by what is known of its distribution during that period. An almost equable land-flora also characterizes the Lower Carboniferous (fig. 177). This uniformity became far less pronounced in the Upper Carboniferous, and separated into four more or less well-defined provinces (fig. 178). One of these flora-types is called the Atlantic-Chinese by Seward, the Arcto-Carboniferous by Darrah and the Eurameric by Jongmans. It includes such well known genera as *Calamites, Cordaites, Lepidodendron, Sigillaria,* as well as many *Pteridospermae* and ferns, and covers parts of eastern North America, Greenland, Europe and North Africa. This flora-type extended eastwards as far as Korea and Sumatra during the middle division of the Upper Carboniferous, mingling with plants of the Gigantopteris flora [1]). The latter is also known as the Cathaysia flora and developed in China, but it has also been found — remarkably enough — in Texas and Oklahoma! The third botanical province comprises the Glossopteris flora with, as its most outstanding genera: *Gangamopteris, Glossopteris (Vertebraria). Phyllotheca, Gondwanidium* and *Dadoxylon,* to which should also be added such types as *Lepidodendron* and *Psaronius.* The latter are frequently found among the Eurameric flora. No sharp line can be drawn between the different provinces, and a particularly strong mixture of Eurameric and *Glossopteris* plants seems to exist in Rhodesia [2]). Jongmans recently reported that he had come across such elements as the Cathaysia and *Glossopteris* flora among Upper Paleozoic fossils in New Guinea [3]).

The same author restricted the name Gondwana flora to the *Glossopteris* plants of the southern continents, and speaks of the more or less analogous plant-finds around Angaraland (Russia, North Central-Asia and East Asia) as the Angara flora. Seward, who refers to the last flora-type as the Kusnezk flora, draws attention to the fact that fossil plants very similar to the Kusnezk flora were discovered in Permian strata (in the uppermost part of the Lower Permian) in Arizona!

Thus four distinct botanical provinces can be observed among the flora of the Upper Carboniferous and the Permian. This differentiation, which corresponded with tremendous mountain-building and important changes in the distribution of land and sea, together with exceptionally widespread glaciation (which, according to investigations by du Toit, David and

[1]) EC in fig. 178.
[2]) Seward, p. 246. EG. in fig. 178.
[3]) CG in fig. 178.

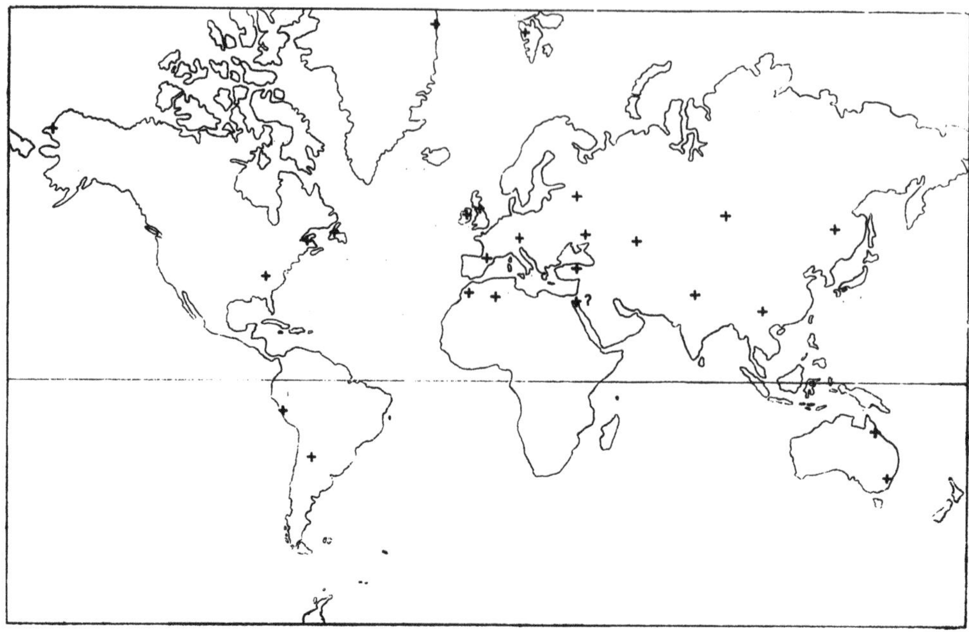

Fig. 177. The distribution of a selected number of localities where the cosmopolitan flora of the Lower Carboniferous has been found (After A. C. Seward).

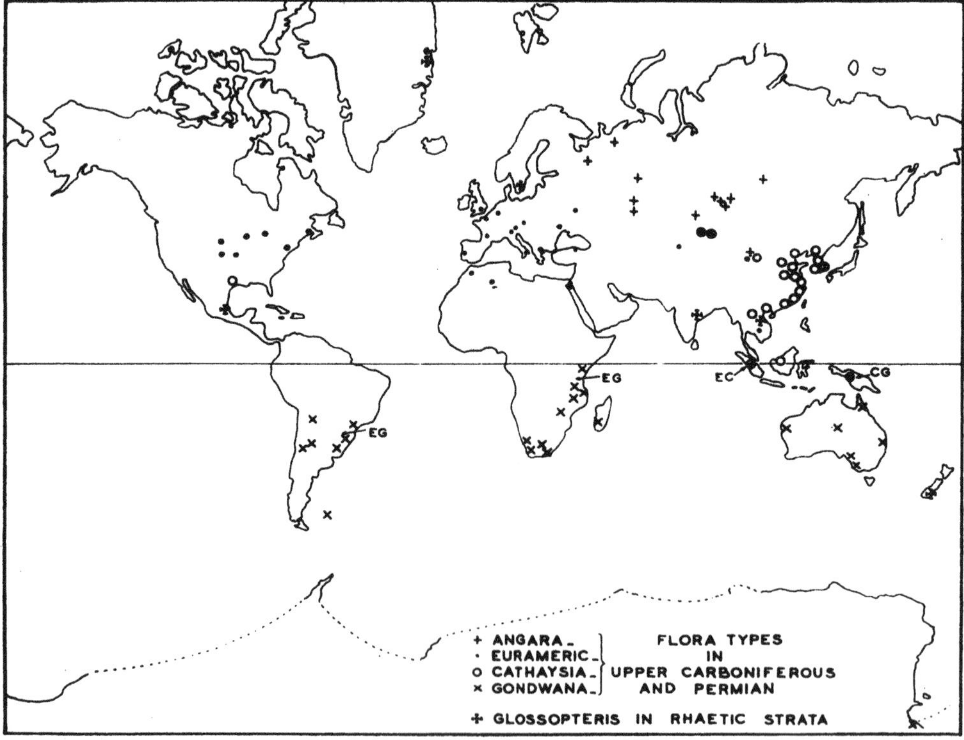

Fig. 178. Upper Paleozoic flora types.

Süssmilch, began in the Lower Carboniferous and continued with a few phases up to the Lower Permian) can hardly be said to be due to mere chance. On the contrary, the important alterations in the earth's physiographic aspect during the Upper Paleozoic probably stimulated the differentiation of the flora.

It is therefore especially noteworthy that the "levelling" process to which the climate was subjected during the Mesozoic (see Chapter IX) should have been accompanied by a cessation of the extreme differentiation of botanical provinces. For it is well-known that after the Upper Paleozoic glaciation the land flora developed into what might almost be called a world-wide uniformity during the Triassic, Jurassic, Cretaceous and even up to the Tertiary, though this uniformity was distinctly less accentuated during the last period. This uniformity was described by Seward [1]), and Darrah [2]), too, observes: "Perhaps the most remarkable feature of the early Mesozoic flora is its homogenity, both in space and in time. The flora does not vary greatly from the Middle Triassic to almost the end of the Lower Cretaceous. Of course, the specific content of the floral succession changes, but the generic content is strikingly constant".

A sudden and as it were explosive evolution of the Angiospermae may indeed be observed in the Upper Cretaceous. In this case, too, it could hardly be claimed that the fact that this intensive evolution coincided with a phenomenon of world-wide importance, — i.e. one of the largest transgressions that have ever occurred in geological history — is merely a question of change. Seward [3]) consequently assumes that these vast changes in the physiographic aspect of our globe had an undeniable influence on the evolution of plant-life on the continents.

The widespread uniformity of the climates had ceased to exist in the Lower Tertiary (i.e. after the period of Laramide mountain-building) and was immediately followed by a zonal differentiation of the flora, which — like that observed in the Miocene [4]) — was roughly arranged according to latitudes. This differentiation was far less pronounced than it is at present, for the latter has resulted from the strong climatic changes since the Pleistocene, and these were responsible for the distribution of our plant-life into the actual botanical provinces. The steadily expanding ice-sheets subsequently drove the widely-scattered, circumpolar Miocene-Pliocene flora of Europe, Asia and North America in a southerly direction, separating them into smaller units as the ice pushed forward. The plants of the northern hemisphere survived where they were able to migrate southwards, and thus found it possible to return north and repopulate the glaciated areas as soon as the ice began to retreat. Darrah [5]) writes in this connection: "The Miocene and early Pliocene floras were practically circumpolar in distribution and generally spread over the northern hemisphere with the zonations already indicated. The oncoming of colder conditions probably forced the plants southward. Presumably there were three principal avenues for

---

[1]) Seward 1931, p. 332–433, 368, 371, 404, 406, 454–455.
[2]) Darrah, p. 154.
[3]) Seward, p. 404.
[4]) Darrah, p. 187, 192, 193.
[5]) Darrah, p. 195, 196.

migration or escape, and these were determined by the prevailing direction of the mountain systems. One avenue of migration was across the lowlands of eastern Asia, along the great valley systems and the coastal plains. It has been suggested that the richness of the existing flora of China is due to the intermingling or persistence of northerly species with those already established.

"In North America the north-south trend of the mountain ranges permitted fairly free migration, and many species of plants which came by this route and those which are relict in place, enrich the American flora. This is the explanation for the marked similarity between the floras of eastern North America and eastern Asia, first observed by Asa Gray, and since demonstrated by many others.

"This resemblance extends not only to many genera with closely related species, but to a considerable number of identical species. A study of the flora of the Island of Yezo, which lies to the north of the main island of Japan, discloses that more than twenty-six per cent of its plants are found also in North America.

"The third migration route for the floras of the north, as they were forced southward by the increasing cold, was down the Scandinavian Peninsula and adjacent areas into central and western Europe, but here the physiography is very different from that of North America and eastern Asia. The great mountain systems, i.e. the Alps, Pyrenees, Carpathians, Balkans and Caucasias all trend east and west, and thus prevent further southward migration. Many of the northerly Miocene and Pliocene plants actually reached western Europe, as is shown by their presence in the Reuverian flora, but, as the cold increased, they were forced against the mountain barriers and perished.

"The Teglian flora of the Late Pliocene is very different from the Reuverian flora, and apparently had two centres of origin, one from Scandinavia, and the other from the mountains of central Asia. It was a cold temperate assemblage, and hence when it was pushed southward by the advancing cold, much of it was able to survive".

Should this prove to give a satisfactory explanation of the distribution of the Pleistocene and actual flora types we might enquire wether the distribution of flora types in the Upper Paleozoic were not, ***mutatis mutandis*** due to fundamentally analogous circumstances. It would take us beyond the scope of the present chapter, however, to examine this interesting biogeographical question.

The most important point of all, as far as we are concerned, is that the two majors periods of strong differentiation of plant life correspond with two major periods of mountain-building and glaciation of the Upper Paleozoic and Pleistocene.

References

BARRELL, J. *Influence of Silurian-Devonian climates on the rise of air-breathing vertebrates* (Bull. Geol. Soc. of America, vol. 27, 1916).

BARRELL, J. *Probable relations of climatic change to the origin of the Tertiary ape-man* (Scientific Monthly 1917).

BARRELL, J. *Relation between lapse of time and organic evolution* (Bull. Geol. Soc. America, vol. 28, 1917).
CAMP, CH. L. *Pretertiary environment and the Vertebrate Record* (Proceed. Sixth Pacific Sci. Congr. 3,1940).
DARRAH, W. *Principles of Paleobotany* (Ed. Chronica Botan. Cy. Leiden 1939).
JONGMANS W. J. *Die Kohlenbecken des Karbons und Perm im U.S.S.R. und Ost-Asien* (Jaarversl. Geol. Stichting Heerlen over 1934–1937, 1939).
JONGMANS, W. J. *Beiträge zur Kenntniss der Karbonflora von Niederländisch Neu-Guinea* (Ibid. 1940).
JONGMANS, W. J. *Das Alter der Karbon- und Permflora von Ost-Europa bis Ost-Asien* (Palaeontographica 87B, 1942).
LULL, R. S. *The pulse of Life* in "*The evolution of the Earth*" (Yale Univ. Press 1919).
MATTHEW, W. D. *Climate and evolution* (N. Y. Acad. Science vol. 24, 1915).
NICHOLS, G. E.; WOODRUFF, L. L.; PETRUNKEVITCH, A.; COE, W. R.; WIELAND, G. R.; DUNBAR, C. O.; LULL, R. S. and HUNTINGTON E. *Organic adaptation to environment* (Yale Univ. Press 1924).
SEWARD, A. C. *Plant life through the ages* (Cambridge Univ. Press 1931).
STEBBINS, G. L. *Additional evidence for a holarctic dispersal of flowering plants in the Mesozoic era* (Proc. Sixth Pacific Sci. Congr. 3, 1940).
SZALAI, T. *Der Einfluss der Gebirgsbildung auf die Evolution des Lebens* (Palaeontologische Zeitschrift Bd. 18, 1936).
TEICHERT, C. *Gangamopteris in the marine Permian of Western Australia* (Geolog. Magaz. 79, 1942).
UMBGROVE, J. H. F. *Leven en Materie* (Nijhoff, Den Haag, 3e edit. 1946).
WILSER, J. L. *Lichtreaktionen in der fossilen Tierwelt* (1931).
ZIMMERMANN, W. *Die Phylogenie der Pflanzen* (Ed. Fischer, Jena 1930).

Chapter XI

LINEAR PATTERNS

"Partout, même dans les pays où les couches ont conservé a peu près leur horizontalité, les formes du sol offrent le reflet d'innombrables cassures internes qui s'y répercutent en dessins significatifs". (A. Daubree)

Introduction

A rocky surface of appreciable extent, be it examined in a mine, a quarry or a natural exposure, generally reveals two or more systems of dislocations. Sometines these are mere joints (diaclases) along which the rock opens up under the hammer of the geologist. Elsewhere fissures may be seen, seldom open, sometimes filled up by minerals or ore-deposits that condensed from exhalations or hot solutions. Apart from these features, the geological explorer frequently comes across localities showing unmistakable evidence of displacement along prevailing fracture-planes. The origin of these tectonic elements may be very different and may vary according to several local circumstances. The orientation of the joints and fissures may be a question of cleavage due to special localised regional circumstances of pressure or tension. It is a well-known fact that granite batholiths and other plutonic bodies have their characteristic joint-systems, the prevailing directions of joints depending on the stress caused by the shape of the surrounding walls and the load of the strata that originally covered it. A special kind of jointing originates in a cooling lava flow. Striking dislocations are associated with folded mountain-chains and their foreland. And it is obvious that older high-situated rock masses or "massifs" exert a marked influence on the direction of the faults that formed in their surroundings. In short, a geologist paying attention to the joints, fissures and faults, in general to the rectilinear patterns of a certain district, will always look for an explanation of their origin in the distribution of the structural elements in their immediate vicinity. Regardless of the question whether his endeavors are rewarded or not he will often notice that the orientation of the joint and fracture patterns reveal an apparent constancy throughout wide areas. This impression is strengthened by the evidence of morphological features of the earth's surface which to a large extent seem to depend on the structural pattern of the underlying formations. Frequently the relief-plan of a landscape shows significant lines, be it in its drainage system or shore-lines, escarpments or arrangement of volcanic vents, or some other physiographic features. "However expressed, these approximately right lines in the plan of the earth's surface have been designated *lineaments* — significant lines in the earth's face". [1])

Among the first students of these linear features were some English pioneers (John Phillips — whose treatise was published as early as the year 1828 — Samuel Haughton and Robert Harkness), two Norwegians (Kjerulf and Brögger) and the Frenchman Daubrée —. The notable and voluminous publications of the last-named geologist attracted general attention because he was one of the first who attacked a series of geological problems from the experimental side. One of his illustrations is reproduced in fig. 179. It represents a drainage network in France, showing two bisecting rectangular sets of lineaments, one set running north-south and east-west the other northeast-southwest and northwest-southeast.

Fig. 179. Lineaments as a repeatative pattern in a drainage network in northern France (After Daubrée).

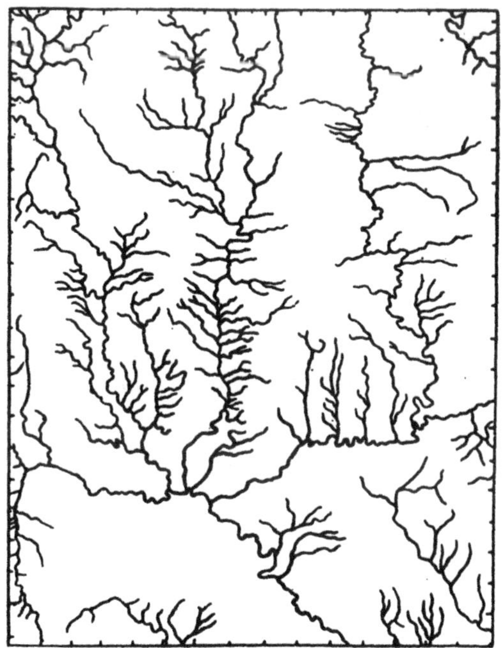

Fig. 180. Pattern of rivers in the area about Rockland, Michigan (From Hobbs).

Several authors are convinced that these two sets of lineaments are encountered not only in different parts of the European continent but in the other continental areas as well. An American example is given in fig. 180. Not always are both of them found and it appears that the NW–SE and NE–SW system is the dominating one. Again fig. 181 gives an example. And in one set a special direction may widely overshadow its counterpart (see fig. 142 and fig. 144).

Lapworth was among the first to recognise the value of these repetative lineamentsets for morphological and structural problems of the earth's crust as a whole . He attributed the shape of the continents and oceanic depressions to a double intercrossing series of folds or waves — "On the surface of the globe" — Lapworth wrote as early as

[1]) Hobbs, 1911, p. 144.

1892 — this double set of longitudinal and transverse waves is everywhere apparent. They account for the detailed disposition of our lands and our waters, for our present coastal forms, for the direction, length and disposition of our mountain-ranges, our seas, our plains and lakes".

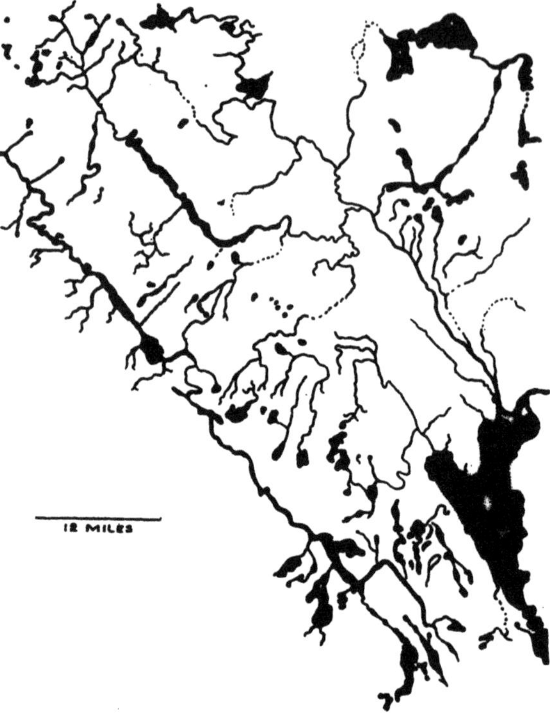

Fig. 181. Drainage map of a part of the district of Nipissing (From Hobbs).

Local circumstances may cause a special pattern of dislocations of no planetary importance. Some areas may reveal complicated arrangements of joints and fractures which could hardly, if ever, be analysed into a few orderly systems of lineaments. But wherever some orderly sets are discovered they may be classified according to either one or two of the diagonal directions or to the other system which runs meridionally and perpendicular to it. This anyhow, was also the strong conviction of Hobbs[1]) who, summarizing, came to the following conclusion [2]):

"The recognition within the fracture complex of the earth's outer shell of an unique and relatively simple fracture pattern, common to at least a large portion of the surface obscured though it may be in local districts through the superimposition of more or less disorderly fracture complexes, must be regarded as of the most fundamental importance. It points inevitably to the conclusion that more or less uniform conditions of stress and strain have been common to probably the earth's entire outer shell".

Planetary systems

The African continent is a good example of the predominance of lineaments in the structural pattern of the earth's surface. Again the meridional and east-west set is not so strikingly developed as the set which bisects it diagonally. It will be remembered that these structural lines of Africa were mentioned in Chapter VIII (p. 222) and that the same trends may also be

[1]) Hobbs, 1911 p. 158.

[2]) Hobbs 1911, p. 163.

recognised in the surrounding ridges separating the deep-sea basins of the adjoining part of the Atlantic (fig. 136 and 137). Now these lineaments are not purely morphological features. They are at the same time substantiated as geological units. A further point of importance is that their significance as structural lines or zones, dates from early Pre-Cambrian times. Hence it is concluded that the planetary significance of these lineaments dates from the earth's primeval days. Their role as lineaments was variable. They are known as linear ridges separating basin-shaped depression. Locally, and in certain periods, real fault-structures may appear, following the same direction. At present, e.g., a line of volcanoes may be traced from the Cameroon fissure on the African continent oceanwards along the Guinea-ridge (fig. 182, compare fig. 136 and 137).

Other examples of lineaments may be seen in fig. 194 which represents

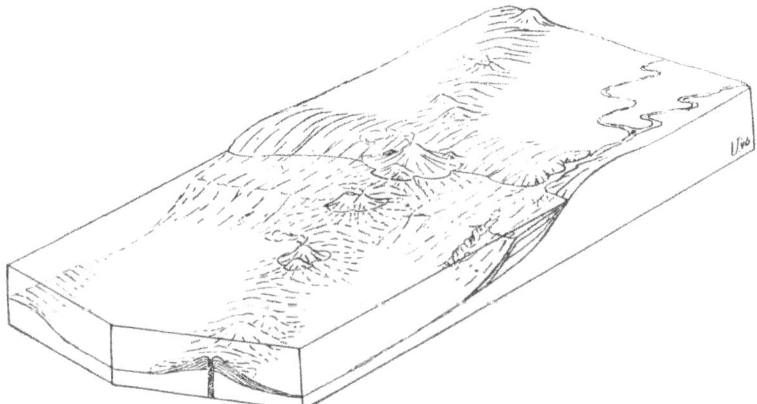

Fig. 182. Schematic block-diagram of the Guinea-line of volcanoes continuing from the Africa continent into the Atlantic.

a schematic map of the surroundings of the North Sea and Paris basins. Some of the lines represent long-enduring ridges or axes of elevation: The Pennines in England (a_5) and the axis of Erkelenz (a_3) forming the western and eastern boundaries of the North Sea basin since such remote times as the Paleozoic.

Most of the lineaments marked c_1 to c_4 are known in France as "anticlinals" but on some of them true faults have come into existence. On the other hand the lines marked a_1, a_2, b_3 and d_3 are large faults bordering the graben of the Rhine, the fault-blocks of the Limburgian coal district and the central graben of the Netherlands. In many places a correlation between lineaments and seismological data seems apparent. An illustration is reproduced in fig. 183 which shows the "tectonic" lines in southern Europe as deduced by Sieberg from the interpretation of seismological facts. According to Sonder most specialists of tectonics of that area would agree with the general trend of these lineaments (although perhaps some allowance must be made for the possibility that the "seismic lineaments" are not wholly uninfluenced by certain prevailing tectonic conceptions).

As has been already mentioned there are good reasons to believe that the planetary lineaments date from a very early period of the earth's history. If this assumption be correct, it follows that these features, when buried under younger strata or tectonic structures, become revealed again in the overlying cover. In this way some of the oldest features of our globe have been repeatedly rejuvenated. And they have to be considered as active elements even in the youngest tectonic zones of our globe. Perhaps we may understand the otherwise incomprehensible knot of Alpine chains in the surroundings of the Mediterranean as partly due to the influence of some very old lineaments. Though deeply buried, they still actively exercise their modelling power on the Tertiary mountain chains (cf. fig. 183 and Plate 4).

It is a rather widespread belief that the origin of faults with a certain well defined strike dates from a special period, whereas faults with a markedly different strike would date from another well defined period. In certain areas this conviction is founded on sound arguments and sometimes it can be shown that one set of faults not only originated in a special period but that its formation was associated with the deposition of valuable ores or the intrusion of a special magma-type. If, however, the set of dislocations is not merely of local interest but belongs to a system of planetary significance, then — according to the considerations just mentioned — we are only justified in saying that at that special moment a certain set of dislocations with a special direction of strike was inherited by the rocks under consideration from some older tectonic structure hidden in the structural units underlying them.

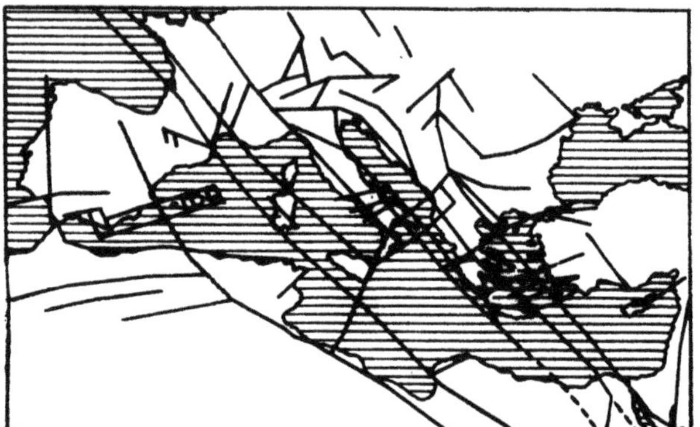

Fig. 183. Seismo-tectonic lineaments in the surroundings of the Mediterranean (After Sieberg; from Sonder).

Some authors, however, have doubtless overrated the relation between direction and time of origin of a fault system. As a typical example I may mention Philipp who once advanced the opinion that the directions of the principal fault-lines of north-western Europe changed from W. and W.N.W. in the Upper Jurassic towards N or N.N.E. in the Oligocene and thence E.N.E. in the Upper Tertiary and Pleistocene. He added the hypothesis that their rotation could have been caused by a large displacement of the poles! In the meantime it has been shown that some faults with a meridional strike date from much older periods. Moreover large and well defined faults with N.N.W. direction dating at least from the Upper Paleozoic appear to have been of paramount influence in the structural history of the Netherlands (see fig. 194). Therefore Philipp's hypothesis has to be abandoned in as much as it is inconsistent with well-established facts.

Finally mention should be made here of the occurrence of large shear-planes which are especially characteristics of the surroundings of the Pacific. They appear in association with deep-focus earthquakes and their outcrops at the surface are not rectilinear but more in the shape of loxodromes. This is one of several reasons why they were discussed in Chapter VII, which was especially devoted to arcuate structures.

The Problem

For the sake of clearness this interesting problem may be divided into three parts. The first part concerns the period of origin of the planetary system of lineaments. We saw just now that it must doubtlessly date from a very remote past. Its primeval antiquity is unanimously postulated in the theories — to be summarized in the next section — that in the course of time have been advanced by various authors. The second part, then, concerns the origin of the prevailing directions of the lineaments. Under the present circumstances special attention must be given, in a separate section, to the possible influence of displacement of the poles. The third part of the problem concerns the phenomenon of repeated rejuvenation of the lineament-sets. It will be treated under the heading "Periodical Rejuvenation", for it will be seen that the process of rejuvenation can be clearly correlated with several of the periodical processes of the earth's crust.

Origin of the lineaments

Older theories

Most authors who have formulated their opinion on the question of the origin of planetary lineament-sets agree that, whatever the cause may be, the process of their formation occurred in the earth's early infancy. One of the oldest views [1]) regarding the cause of the planetary lineaments was offered by G.H. Darwin in connection with his well known theory on the origin of the moon from the earth. He suggested that after the moon solidified and revolved round the earth at a short distance a new equatorial protuberance formed. This time, however, the moon pulled backward the terrestrial matter which then formed the major structural lineaments of the globe. Owing to the moon's strong influence on the equatorial girdle of the earth, this zone would revolve more slowly than the non-equatorial parts. As a result the dominant strike of the terrestrial lineaments would have become northeast-southwest in the northern hemisphere and northwest-southeast south of the equator. Some morphological and structural trends fit rather well into his scheme but other lines of equal prominence do not agree at all.

A fantastic hypothesis, surmising that in primeval times the northern part of the earth had a lower angular velocity than its other regions, was advanced by Prinz, a Frenchman. He attempted to make clear that — as a result of differences of velocity — the land-masses in the southern

---

[1]) See the publication of J. E. Gregory, op cit. 1899.

hemisphere were gradually sailing eastwards at a swifter rate than those of the northern hemisphere. Between the two parts a major zone of weakness would form running from the Malay archipelago along the Alpine belt toward the Mediterranean and further towards the Caribbean region. Moreover certain lines of torsion would have resulted which we would now recognise as the principal lineaments of the earth. As a document of historical interest his map is reproduced in fig. 184.

Another theory was tentatively put forward by Lapworth in the year 1892. He was inclined to believe that the primeval origin of the lineaments was mainly the result of a shortening of the earth's diameter, due to the loss of original heat into outer space [1]). It is clear that his opinion is rather akin to our suggestion given *en passant* in Chapter VIII (p. 248) viz., that the epoch of lineament formation was contemporaneous with the completion and consolidation of a solid crust around the globe.

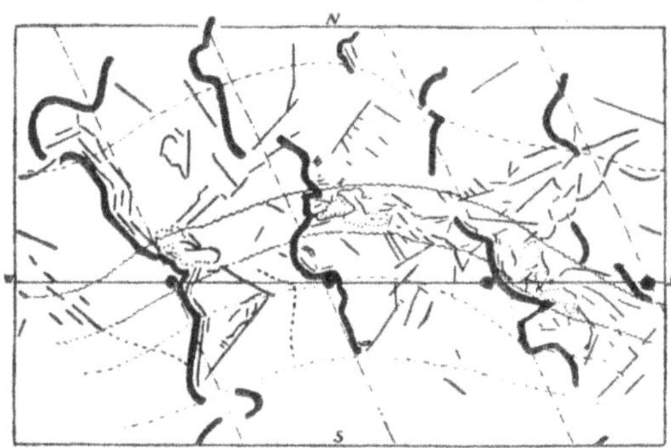

Fig. 184. The oblique course of the main geographical lines (After Prinz).

According to Hobbs "the ultimate cause of the common type of deformation is presumably the continued secular cooling of the planet" [2]).

Modern theories

In later years a hypothesis was advanced by Sonder, a Swiss geologist. He joins in our opinion that the lineaments did not form at an earlier date than the consolidation of a world-encircling elastic crust. Special emphasis is laid by him on the anisotropic constitution of the earth's crust. According to his opinion the elastic compressibility in the continental parts is about 50 percent higher than in those parts of the crust which are built up of simatic material. The extremes of maximum and minimum stresses in the crust would vary as much as 5000 atmospheres. Hence the existing anisotropies of the crust would cause torsional effects and these, in turn, would give rise to the formation of joints and dislocations. Accordingly their orientation is considered as largely dependent on the original distribution of continents and oceanic depressions. And gradually, in the course of time, they became more and more apparent or rejuvenated during the growth and decay of the structural units of the globe. Sonder thinks that each continent is characterized by a set of lineaments of its own, radiating from the continental nuclei towards its periphery. So he speaks of African, Atlantic and Asiatic lineament-systems, the first nearly NE–SW, the second slightly turned clockwise, the other one slightly anti-clockwise. At the same time he finds these three systems closely associated in one and the same district, for example in the surroundings of the Rhine-graben. My impression is that he

[1]) Lapworth 1892, p. 697.
[2]) Hobbs, 1911, p. 164.

has somewhat overrated the significance of these directions, the more so because he himself mentions the occurrence of intermediate directions between his three sets. Such cases are met with everywhere. One may speak of the principal strike of a lineament-set, but allowance should always be made for some deviation on both sides of the main trend(cf. fig. 188).

Apart from the NE and NW system Sonder recognises a NS and EW system but he pays little attention to it in his further disquisitions.

Remarkably enough most authors emphasize the opinion that the diagonal system prevails over the "meridional and east-west" system. However, some of the most conspicuous fault-systems of the earth belong to the latter. Among them are, e.g., the East African rift-zone and the Rhine-graben although in detail parts of them are diagonal. Some authors have tried to correlate their formation with the neighbouring orogenic belts, arguing that the orientation of these fault zones is nearly perpendicular to the folded chains. This hypothesis, however, does not hold good. The trend of the Limagne-graben in France, for example, is quite conspicuously parallel to the Alpine belt. Moreover, large rift-zones like those of East Africa and the Rhine probably began to originate as early as the Paleozoic. Cloos pointed out that they are situated in zones of elevation and he further holds that the direction of these structures is controlled by pre-existing anisotropies of the crystalline basement. According to his opinion the directions are inherited from early Pre-Cambrian structures. Time and again these structural lines revived as rejuvenated phenomena. It is therefore not surprising to notice that traces of great rifts have seldom, if ever, been found from older periods, except in the regions of the present graben systems. The meridional direction is regarded by Cloos as a planetary feature of great antiquity. The same opinion was held by Suess.

The meridional trend characterises several other well-known ridge-shaped lineaments. To mention only two: the marginal ridge which separates the continental basins of Africa from the submarine basins of the southern Atlantic, and the Mid-Atlantic Rise between 0° and 55° S. Lat. and between 20° and 10° N. Lat. Both lineament systems were found characterizing the arrangement of the volcanoes in the East Indies.

It may be noticed also on fig. 194, that the lineaments marked a_3 and a_5 are ridge-shaped boundaries of the North Sea basin. Here too a respectable antiquity for these structures is very probable. Similar axes striking NS are known also from the deeper geology of the United States. According to Lugn their Pre-Cambrian origin has been proved and since that remote past they have repeatedly played an active "posthumous" role as rejuvenated structural elements.

The most recent theory is that advanced by Vening Meinesz in 1943. It contains an attempt at correlating the origin of planetary systems of lineaments to a large displacement of the poles in the early infancy of the earth's history. Thus for the second time we come across a revolutionary theory of polar displacements — the first time being in Chapter IX, on the origin of ice-ages. Hence it seems worth while to consider the problem of the poles at some length.

Wandering of the poles

The earth's axis of rotation has not a fixed position in space. It exhibits the movements known as nutation and precession. But apart from these well-defined movements the equator is thought to have made — since primeval times — an angle of some 22° to 24¹/₂° with respect to the ecliptic. Nevertheless astronomers think the position of the earth's axis has not been subjected to large changes. And we may leave to them the question how and when the remarkable inclination of the earth's axis to the plane of its orbit round the sun originated. For the moment it is sufficient to know that no astronomer would accept a hypothesis involving large deviations from this state of affairs during geological times. As already mentioned, (p. 279) the rotating earth is like a spinning top or gyroscope which has a powerful tendency to keep the position of its rotation-axis constant. Any drastic alteration of its direction would cause a world-wide catastrophe. That nothing of the kind has ever happened is known from historical geology.

Another question is whether the outer elastic and rigid shell of the earth may possibly have moved as a whole over the deeper plastic layers of sima. This is the kind of polar displacement advocated by some geologists. Loading and unloading of ice-caps, displacements of sediment by erosion, etc. are factors that may have had some influence of this kind though it is questionable whether such processes can have had far-reaching consequences. At any rate some authors eagerly introduced large displacements of the crust, as they were struck by the contrast between the climatic evidence of older strata and that of the conditions now prevailing. The resulting hypothetical displacements of the poles are widely known. Many textbooks on geology reproduce a figure showing an earth-globe with something like the track of a snake on it, a curved line running from the area of the present South-pole (which would have been the site of the present north-pole in Pre-Cambrian times) via the Pacific, Alaska, South and East Greenland, towards its present position. It represents the hypothesis of polar displacement of Kreichgauer. Another hypothesis mentioned in nearly every textbook is that of Simroth. There are a few others and, of course, every one knows the paleoclimatic maps of Köppen and Wegener. The reasons why their theory, too, has to be abandoned were given in Chapter IX.

It was also mentioned there that a mathematical treatise by Milankovitch pretended to furnish strong support to the theory of Köppen and Wegener. According to our opinion that would be only the worse for the mathematical treatise. The question had to be left at this point in the first edition of *the Pulse of the Earth*. However, we may now add that it is much worse than we could have thought at that time. This appeared when, in 1943, H. Kuiper wrote a thesis which contains a critical re-calculation of the theory treated by Milankovitch.

It will hardly be necessary to recall to memory that according to the theory of Köppen and Wegener, the North-pole shifted from a spot near the Hawaiian Islands towards its present position, since Paleozoic times. It was this suggestion that seemed to be confirmed in such a remarkable way by Milankovitch. Now, to begin with, this result was not so striking as many thought it to be, for Milankovitch based his calculations on certain premisses which themselves involved the theory of Köppen and Wegener. However, Kuiper showed [1]) that a very remarkable error has crept into one of Milankovitch's equations. When the equation is corrected it transpires that the North-pole followed its path along the Hawaiian route in a direction diametrically opposed to the direction mentioned by Milankovitch and postulated by Köppen!

Since Kuiper's thesis will not be available to everyone who is interested in it, and moreover as it is published entirely in the Dutch language I will outline here the crucial point in some detail.

In each part of the crust $c=2R^3\rho Dz_0$, R representing the earth's radius taken as constant; D the thickness of the floating crust (fig. 185) and ρ its specific gravity. Hence the product ρD represents the mass of a column of diameter of such material z_0 represents the height of the unit centre of gravity of this floating part of the crust above the centre of gravity of the displaced column of the substratum (fig. 185). In comparing equations of c_1 for the continents and c_2 for the oceanic sectors the question arises in which of these two cases the product ρDz_0 is the larger one. Now doubtless this product is larger for the continents than for the oceans since the surface of a continent floats at a higher level than the bottom of the ocean (fig. 186 A). Assuming the specific gravity of the floating column to be the same in both cases, it follows that the mass of the continental column surpasses that of the oceanic column. So we get: $c_1 > c_2$.

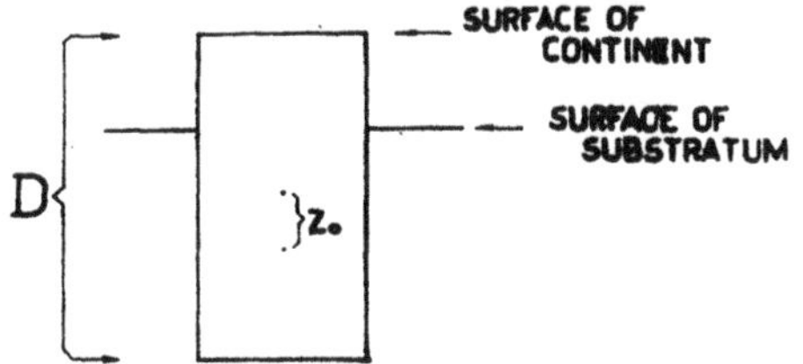

Fig. 185. Schematic representation of a column of sial floating in a simatic substratum. Further explanation in the text.

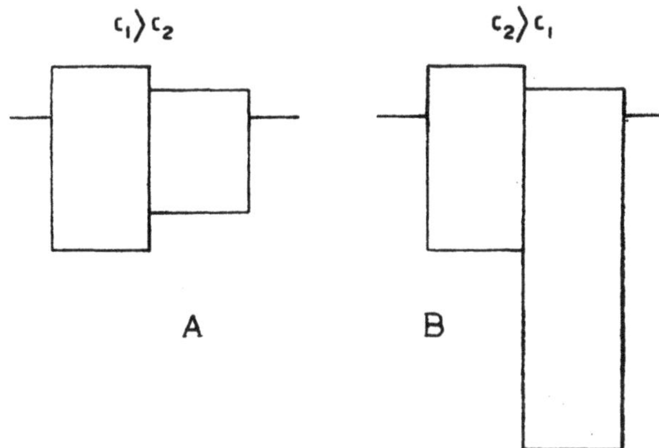

Fig. 186. Schematic representation of continental and oceanic columns of sial. A according to the usual conception; B according to the assumption of Milankovitch.

Milankovitch, however, writes $c_2 > c_1$. In order to introduce this formulation — which apparently comes into conflict with common sense — Milankovitch assumes that the specific gravity of the oceanic crust differs only very little from the specific gravity of the substratum but very much from that of a continent. He chooses these values in such a way that the height of

[1]) Kuiper, 1943, p. 54, 64, passim.

the crustal column in the oceanic sector becomes very large. Hence the mass of such a column and therefore the product $\rho D z_0$ will surpass the continental value (fig. 186 B). So he writes: $c_2 > c_1$.

Now the high degree of improbability of this assumption is obvious since one would have to admit a crustal (sial) thickness of the ocean floor of some 80 kilometers, etc. But Milankovitch's manipulation has an interesting background, because of the fact that the value of a so-called "Polfluchtkraft" depends on the values of c. If $c_2 > c_1$. the pole would travel in accordance with the theory of Köppen and Wegener, i.e. from Hawaiï towards its present position. If, however, $c_1 > c_2$ the pole — admitting the other premisses for the moment — would also travel along the Hawaiïan track but in the opposite direction!

In Kuiper's thesis the problem of shifting poles seemed to have reached a safe resting-point. Indeed for one moment geological science seemed to be freed from the nightmare of hypotheses based on wandering poles. However, only a few months later, the idea of large polar displacements was formulated once more in a new hypothesis by Vening Meinesz.

Two things will be clear from what has been said about hypotheses of shifting poles in Chapter IX; at any rate I hope so. Firstly that the older theories of large polar displacements are inadequate to explain paleoclimatic conditions in the earth's history; that they are even contradictory to the facts known at present about past climates. Secondly that the former climatic changes of the earth may be understood by quite different processes. And hence that in respect to paleoclimatology hypotheses of polar displacements — anyhow of large displacement of the sort advanced by Kreichgauer, Simroth, Köppen, Wegener and Milankovitch — are superfluous constructions of no value whatever to science.

It will be seen that in this respect Vening Meinesz' hypothesis of polar displacement is dissociated from this severe judgment on account of the fact that the processes advocated by him had already come to an end at a time when the earth had not yet begun to record climatic conditions — and problems.

Vening Meinesz investigated the stresses brought about by a change in position of the earth's rigid crust with regard to its rotation-axis. He assumes the crust to be everywhere of equal thickness and to behave as an elastic plate. The flattening at the poles, when displaced over the plastic substratum, must have led to the formation of shear-planes in the crust. The resulting curves of shear in the earth's crust as determined by Vening Meinesz' equations are shown by fig. 187. The net, corresponding to a displacement of the pole — from near Calcutta over 70° along the meridian of 90° E.L. in early Pre-Cambrian times — shows some remarkable correlations with many of the topographic configurations and structural lines of the earth's surface. His theory asserts that probably the crust indeed shifted with regard to the poles in the manner just mentioned and that the crust underwent a corresponding block-shearing. The direction and the amount of the polar displacement were chosen by Vening Meinesz in a special way. A short time before he had studied the lineaments of the Azores and the sea-

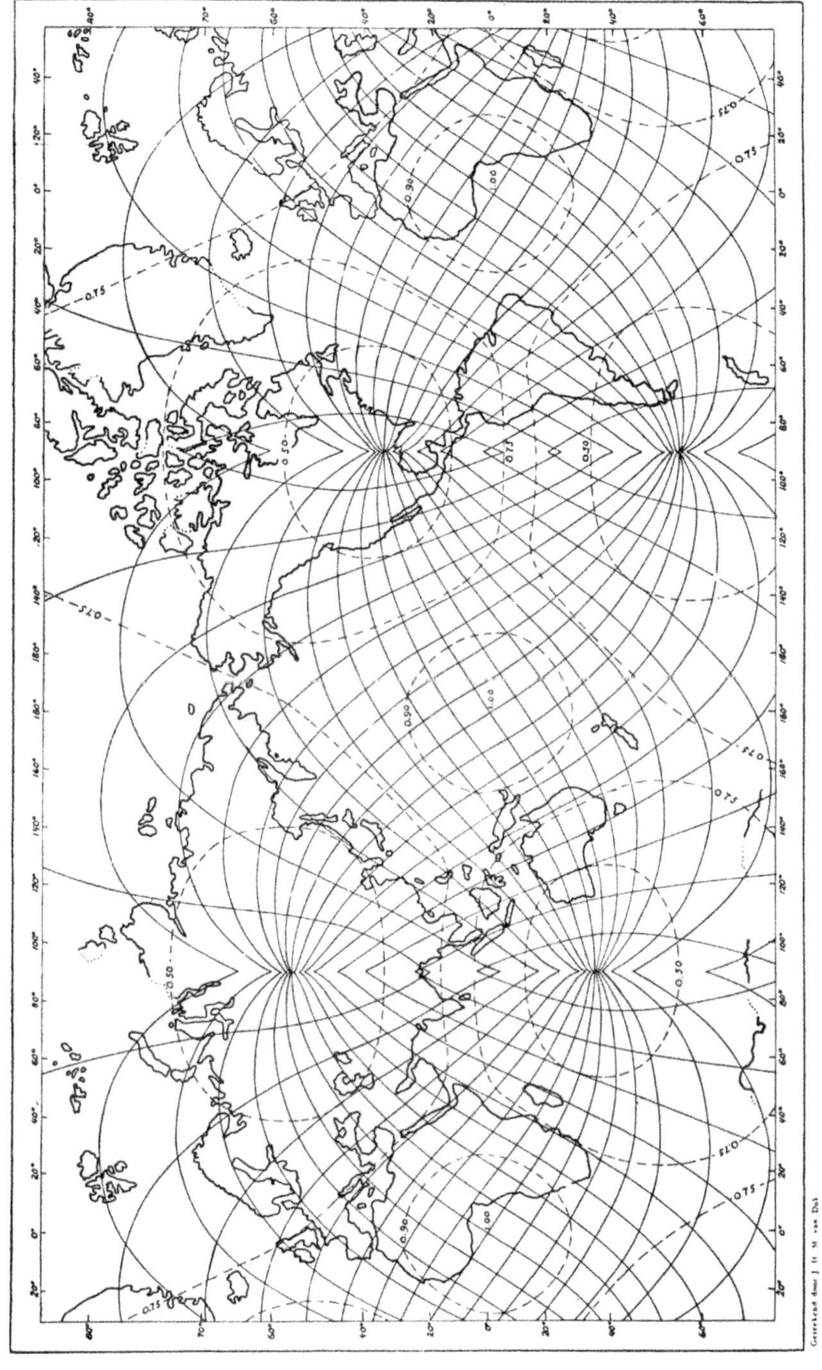

Fig. 187. Shearing net based on the assumption of a displacement of the North Pole from near Calcutta over 70° along the meridian of 90° E. L. (From Vening Meinesz).

floor in the vicinity of these islands. He therefore chose the theoretical displacement of the pole in such a manner that the shear-net would coincide with the prevailing lineaments of the Azores. Once the shear-net was completed he was struck by the fact that many other lineaments fit rather well in the net of shear-lines. A number of striking coincidences could be enumerated.

The question at once arises whether we are dealing with a case of fortuity or with a correlation of far-reaching consequences regarding the structural history of the earth. A statistical investigation of the morphological lineaments in the topography of the ocean-floors — taking deviations up to 12° on either side of the trend of the net as positive correlation — induces Vening Meinesz to believe that it is not a question of fortuity. I agree with him that the congruence is not fortuitous. But I am very doubtful in respect of the idea that this would be in favour of the theory of a large polar displacement, and that the earth's lineaments could only be explained by such a happening. For suppose we construct another net consisting of two bisecting series of lines running more or less diagonally in respect of the meridians; whatever the motives of its construction, it will show positive correlation to the earth's diagonal lineament system. But this does not imply that the motives which lead to its construction have to be regarded as representing the true cause of the origin of the lineament system.

Moreover the net of fig. 187 has some disadvantageous properties. To mention only a few: The areas bounded by the 0.50 line are regions where the stress has only half its maximum value. According to the theory this means that no serious stress-phenomena have to be expected within these boundaries. Nevertheless the great San-Andreas fault, the enormous Bartlett deep and several other large faults are situated within this area. And there is some congruence along the east and south coast of North America where the lineament along which Appalachia subsided coincides with a non-useful line of the net. Similar difficulties arise when Central Asia is examined. If these regions were structurally different from other parts of the globe — e.g. if they revealed much fainter lineaments or no lineaments at all — that would fit well into the theory. Now, however, these regions stand as evidence that opposes the suggested explanation of polar displacement. For, if the usual tectonic phenomena and lineament-sets could originate in these regions without the action of a well-pronounced shear-net due to polar shift, why should we need such a hypothesis to explain similar phenomena in regions outside the 0.50 line?

Therefore I suggest that the congruences found are not a merit of this special shear-net but that the congruence is simply inherent to the fact that the main trends of the net are diagonal. And further: that each net of diagonal lines will shown a certain degree of positive correlation simply because one of the earth's lineament-sets runs diagonally!

This in fact, brings us to a further difficulty. Granted for a moment that the diagonal system of lineaments could be explained by a large polar shift as suggested by Vening Meinesz. Then, what can we say about the system of meridional and east-west lines? It finds no explanation in the theory and therefore would imply a second planetary effect which caused — among others — some of the most powerful rift-zones of our globe, including some conspicuous lineaments like part of the Mid-Atlantic Rise etc. (cf. p. 301). East-west lineaments may be seen even on the Altair-chart of the Azores which formed the starting point of Vening Meinesz' considerations. As a further illustration, a concrete example is reproduced in fig. 188. It

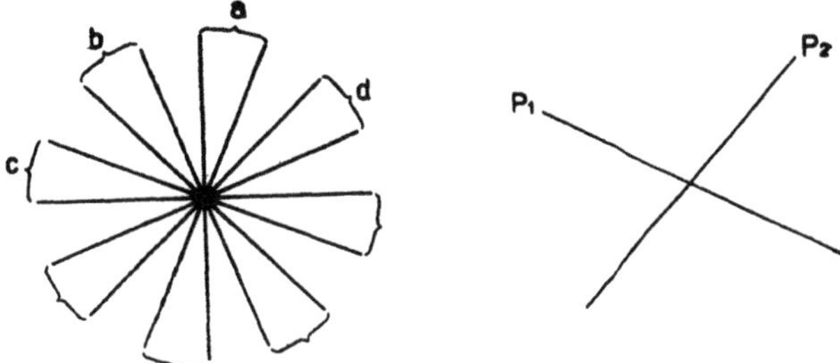

Fig. 188. To the left principal directions of the four lineament groups of N.W. Europe. a, b, c and corresponding to the same notations in fig. 194. To the right the two directions of the shear-net for the same area shown by fig. 187.

shows the four principal directions of the major lineaments of fig. 194. Two of these four sets — viz. those marked *c* and *d* — might be correlated to the directions p_1 and p_2 of fig. 188 which represents the trend of Vening Meinesz' shear-net for this area. But the directions *a* and *b* remain unexplained. They belong, however, to very important lines in the structural history of Western Europe at least since the Paleozoic, and probably since much older, Pre-Cambrian, times.

Another difficulty of a quite different nature arises if we would accept the hypothesis of polar displacement. The present tetrahedral arrangement of continents and oceanic depressions clearly demonstrates a correlation with the position of the earth's axis of rotation. Now there are two possibilities. One is that this situation has come into being after the pole migrated to its present position. If this view be accepted it becomes incomprehensible why the shear-net should be correlated with the arrangement of the continents and ocean-floors.

The other possibility is that the tetrahedral configuration already existed before the beginning of polar displacement. In that case, however, it would be incomprehensible why the earth should show exactly that coincidence of its rotation-axis with the line that connects the centre of the Antarctic continent and the Arctin basin; in other words why the earth's axis should be so clearly correlated with the tetrahedral pattern. I suppose not many would like to ascribe this remarkable coincidence to mere fortuity!

These are the principal reasons why I cannot adhere to Vening Meinesz' interesting hypothesis of a large shift of the poles in the early Pre-Cambrian. In one respect this means that — according to my conviction — the pole surely did not wander, not even from Calcutta to its present position in Pre-Cambrian times. On the other hand it means that the origin of both the planetary lineament-sets remains an unsolved problem. For in my opinion all that can be said with any certainty is that they originated in the earth's early infancy. Probably we might add that in some way their origin was connected with the formation of a world-encircling solid, elastic and rigid crust (cf. p. 248 and p. 300). And, finally, that the old planetary pattern was repeatedly rejuvenated during the later history of the earth's crust. This last point is worthy of a closer inspection.

Periodical Rejuvenation

In order to elucidate the periodical manifestations of the rectilinear elements of the earth I should like to consider firstly the phenomenon of dome-shaped elevations and rift-valleys. A limited region will then be treated in more detail. Finally some problems of a more general nature will come up for discussion.

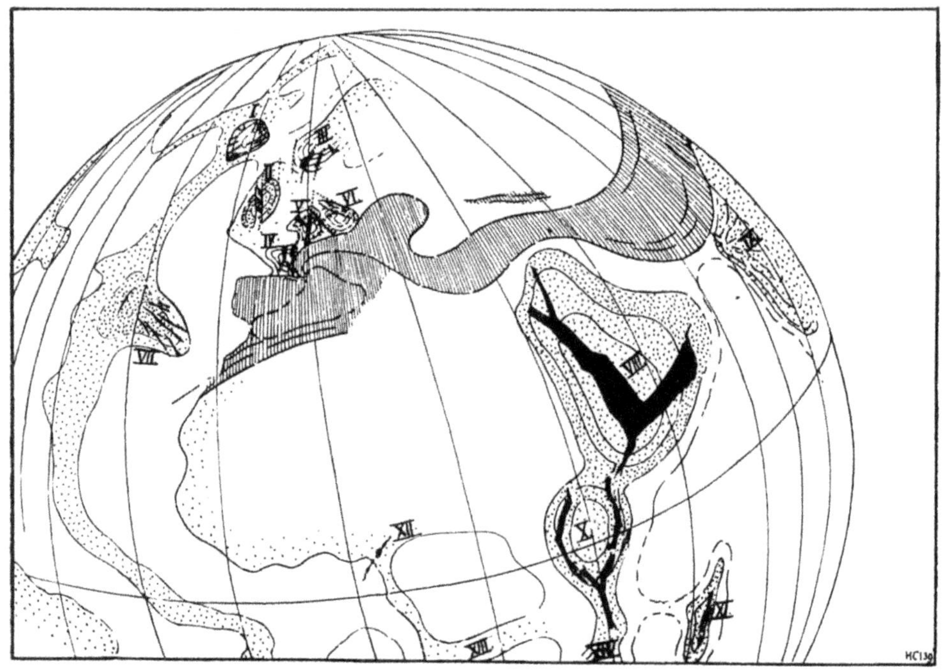

Fig. 189. Dome-shaped elevations, with rift-valleys and volcanism (After H. Cloos). I Iceland, II Great Britain and Ireland, III Graben and faults of Oslo etc., IV Plateau Central of France with the Limagne-Graben, V Rhine-shield and graben, VI Bohemia-Saxonia, VII Azores, VIII Arabia — Nubia, IX India, X East Africa, XI Madagascar, XII Camaroon.

Dome-shaped elevations and rift-zones

Cloos published an extensive study on the dome shaped-elevations, and attempted to show that fault systems and rift-valleys accompanied by volcanism are as a matter of fact the direct result of (and locally dependent upon) slowly updoming nuclei. Examples of such nuclei are the Vosges and the Black Mountains with the Rhine graben, the Plateau Central with the Limagne graben, the African-Arabian Shield with the rift-zone of East Africa, the Red Sea, etc. (fig. 189). Cloos is of the opinion [1]) that the first dome shaped elevation of the Rhine Shield probably originated in the Upper Paleozoic, and that it was chronologically related to Variscian mountain-building, whereas the present phenomena would be due to a much more intensive process of elevation in the Tertiary, and this would have been

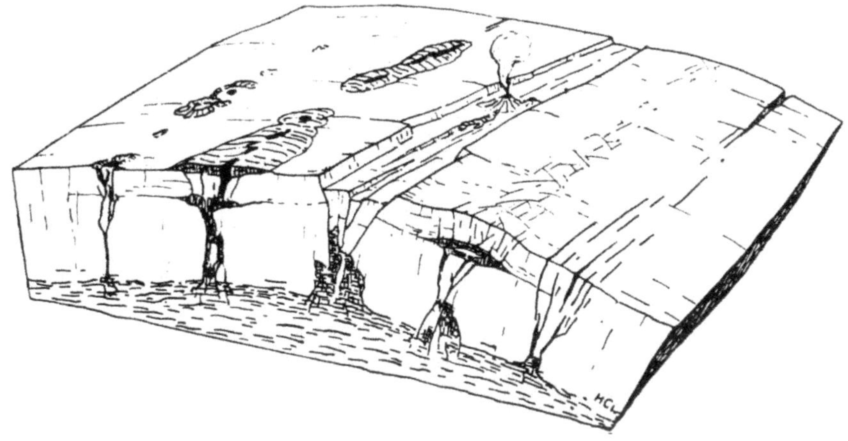

Fig. 190. Block-diagram of a dome-shaped elevation with faults, graben and volcanism (From H. Cloos).

chronologically related to Alpine epochs of movement. Deep tension-faults and rifts affected the upper layers of the vault as a result of this updoming (fig. 190). An older updoming tendency had probably also preceded the formation of the Arabian-African Shield and their rift-systems [2]).

Many authors, including Taber, who wrote a clear review of the subject, were of the opinion that a graben and its elevated margins originated as the necessary result of the formation of two faults dissecting the earth's crust obliquely and converging towards the simatic substratum (fig. 191). In the theory of Cloos, however, the dome-shaped elevation precedes the formation of the faults and graben-structure. In a later stage the effect of isostatic adjustment of the trough manifests itself [3]). According to Cloos the Rhine Shield (fig. 194) already existed during Mesozoic times in the shape of a dome-shaped elevation, whereas the Rhine graben in its present configuration originated in the Oligocene. This opinion receives confirmation from the paleogeographic map of Von Bubnoff. These reconstructions

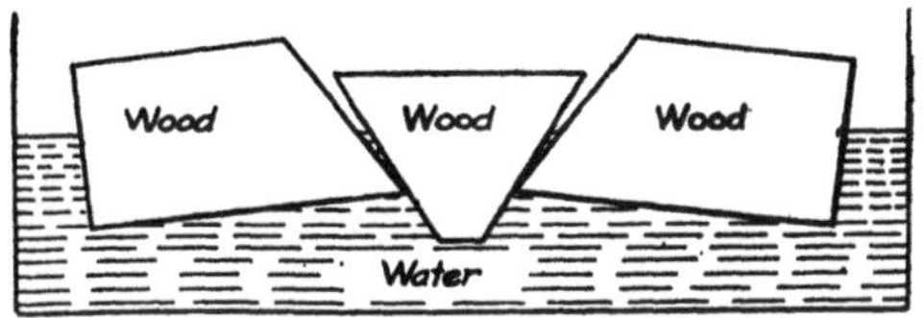

Fig. 191. The formation of rifts according to the theory of Taber (From S. Taber).

[1]) Cloos, 1939, p. 462.
[2]) Cloos, 1939, p. 442.
[3]) Cloos calls that effect: antithetic.

show the area of the Rhine Shield as land amidst the Mesozoic transgressions (fig. 192). Similar data are available for the Plateau Central of France (fig. 193).

This induced Cloos to take an entirely different view of the subject than that advocated by Van Waterschoot van der Gracht in 1938. The latter held

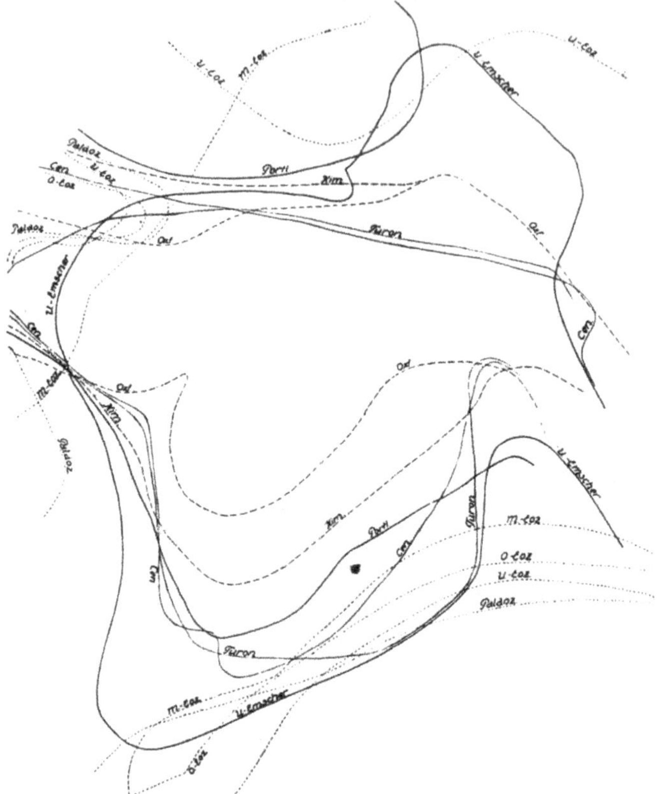

Fig. 192. Paleogeographic reconstruction of the Rhineshield in Mesozoic times (From Cloos, after Von Bubnoff).

that the rift-systems of Europe and Africa represent the first stage of the disruption of blocks, which — as Wegener postulated — are beginning to drift towards the Atlantic or Indian Oceans.

My own opinion, however, is that the updoming is brought about by a phenomenon similar to the depression of the basins, viz. by a displacement of sima in the substratum. We have even fewer data to go by, here, than in other cases, but the dome-shaped elevations, too, would appear to give indications of their own periodical occurrence and chronological relation to the tectonic and magmatic cycles discussed previously. A great deal more will have to be known of the earth's history, and especially of its Pre-Cambrian history, before an explanation can be given why the crust arches up in one area, while basin-shaped depressions are formed in another.

We shall now proceed to elucidate the periodical rejuvenation of linear

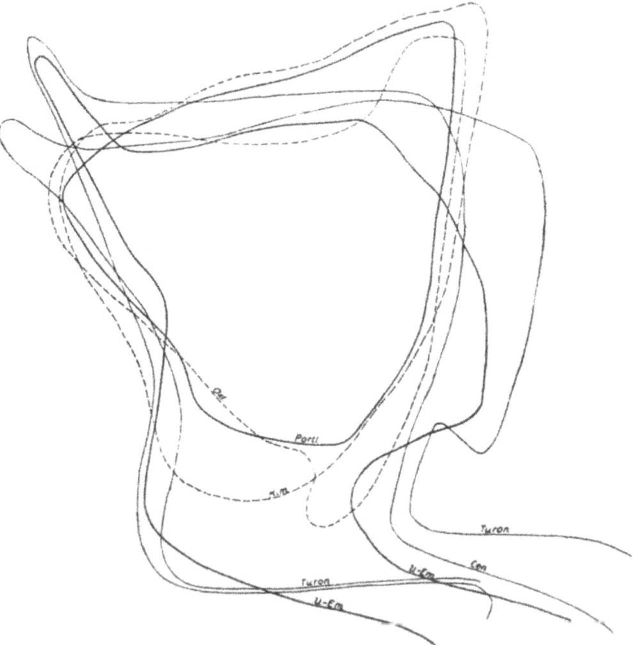

Fig. 193. Paleogeographic reconstruction of the Plateau Central, France, during Mesozoic times (From Cloos, after Von Bubnoff).

structures by making a close inspection of one well-defined area, the North Sea Basin and its surroundings.

The North Sea Basin

The North Sea Basin is enclosed on its western and eastern sides by two axes of elevation, viz. the Pennines (a_5 on fig. 194) and the axis of Erkelenz (a_3). The fault-zone of the Limburgian coal district broadens towards the northwest into the graben of the central Netherlands. In the opposite direction it is connected with the Rhine graben. The southern border of the North Sea Basin is formed by the Brabant Massif. It dates from at least pre-Carboniferous times, but its influence as a geanticlinal ridge of elevation was manifest in many Mesozoic and Cenozoic epochs. In a voluminous memoir Stevens has pointed out that the present morphology of Belgium — e.g. the pattern of the rivers — still reveals the updoming influence of this important element.

It is worth while to consider three different aspects of the geological history of the basin. For it will be seen that three groups of data converge towards the same conclusion.

In the first place the rate of sedimentation in the basin will be considered, and it will be demonstrated that the notion of certain periods of accelerated and retarded bottom-movement follows simply from a comparison of the sedimentation during various time-intervals.

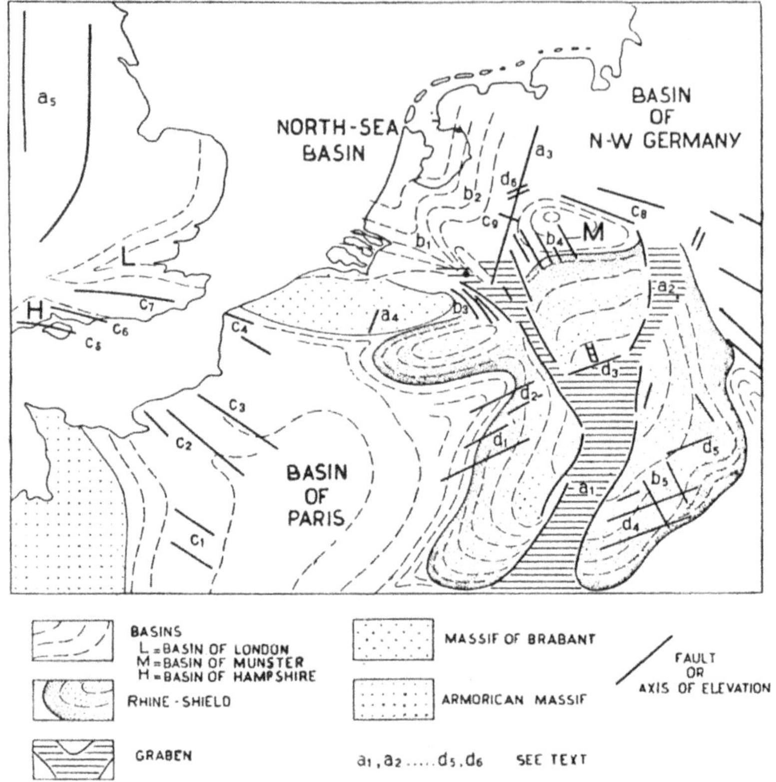

Fig. 194. The North Sea basin and its surroundings.

Secondly, this result will be confirmed by a further consideration of the paleogeographic development of the basin.

Finally it will be shown that the principal phases in the structural history of the region are synchronous with tectonic epochs of world-wide importance.

(1) Rate of sedimentation.

The basin of the North Sea [1]) contains a sedimentary sequence ranging from Upper Permian deposits to Pleistocene. In the opinion of Van Waterschoot van der Gracht the post-Carboniferous sediments total about 7,500 meters, or possibly even 9,000 meters. These figures correspond to an average rate of sedimentation of 0,4 centimeters per century, as shown by Table IX.

---

[1]) No. 11 on plate 6.

TABLE IX

Rate of Sedimentation in the North Sea Basin

Formation	Duration in million years	Maximum thickness of sediment in metres	Average rate of sedmentation in cm. per century
Cenozoic	62·5	Largely exceeding 1,000	Largely exceeding 0·16
Cretaceous	40·6	> 3,500	> 0·86
Jurassic.	43·7	appr. 1,500	appr. 0·34
Triassic.	37·5	appr. 1,300	appr. 0·35
Permian	37·5	250–600	0·07–0·16
Post-Carboniferous formations	222	7,500–9,000	0·4
Minimum rate of sedimentation			0·07
Maximum rate of sedimentation.			> 0·86

Now the older formations are known only from deep borings along the edge of the basin. The data for the Mesozoic have been interpolated from what is known of the basin of north-western Germany. On the other hand the figures for the Cenozoic pertain to the North Sea basin proper, where sedimentation in Tertiary and Pleistocene times [2]) largely surpassed the sedimentation in the basin of north-western Germany. So the figure 0.4 of Table IX probably is on the high side [3]).

Table X reviews a number of data enabling a comparison with the rate of sedimentation in some other basins and troughs. Some examples have been gathered from regions which generally are thought to show large amounts of sediments deposited in a comparatively short time.

The average rate of subsidence and sedimentation in the North Sea Basin is only half of that of the Gulf Coast Basin, but surpasses by twice the movement of the Paris basin in Jurassic times. The movement of the East Indian geosynclinal oil-troughs largely surpassed that of the North Sea Basin.

Table XI offers a further means of orientation and comparison. It summarizes some data concerning the rate of sedimentation in some of the present oceanic areas.

---

[1]) No. 12 on plate 6.

[2]) The subsidence of the basin in the Tertiary and Pleistocene is illustrated by fig. 195–198. See also the coloured map accompanying Van Waterschoot van der Gracht's paper of 1938.

[3]) In the southwestern part of the Netherlands the thickness of post-Paleozoic formations amounts to only 1,200–1,500 meters, which would correspond with an average of 0.05–0.07 centimeter per century.

TABLE X

Rate of Sedimentation in a few "fossil" Basins and Troughs

Idiogeosynclines of the East Indies	Duration in million years (approx.)	Maxim. thickness in metres	Average sedimentation in cm. per century
Neogene of Indragiri, Sumatra .	30	2,000	0·66
,, Djambi, Sumatra. .		4,000	1·3
,, Palembang, Sumatra		6,000	2·0
,, Atcheen, Sumatra. .		6,800	2·2
,, Koetei, Borneo. . .		7,000	2·3
Tertiary of S. Celebes	60	> 4,000	> 0·66
,, Tidoengsche landen, Borneo		7,000	1·1
,, Atcheen, Sumatra. .		9,500	1·6
,, Pasir, Koetei, Borneo		15,000	2·5
Minimum rate of sedimentation			0·66
Maximum rate of sedimentation			2·5
Jurassic in basin of Paris. . . .	50	1,200	0·24
Low-Cretaceous to Pleistocene of Gulf Coast Basin	100	8,000	0·8

TABLE XI

Rate of sedimentation in the deep-sea in centimeters per century. From data of the Lord Kelvin, Meteor, and John Murray Expeditions

Type of deep-sea sediment	Since close of the Pleistocene 20,000 years ago		During last glaciation	During the Pleistocene
	In tropical part of Indian Ocean	In tropical, equatorial part of Atlantic Ocean		In Northern part of Atlantic Ocean
Blue mud	—	0.178	0.33	—
Globigerina-ooze	0.059	0.12	0.21	0.03–0.06
Red deep-sea clay.	—	0.086	—	—
Diatom-ooze.	0.054	—	—	—
min. rate of sedimentation	0.025	0.05–0.09	0.21	0.03
max. rate of sedimentation	0.07–0.1	0.13–0.33	0.33	0.06

It is generally known that sedimentation in the deep-sea is very slow. Now the figure 0.33 which was found for the equatorial regions of the Atlantic differs only very slightly from the average figure 0.4 of the North Sea Basin. But the first-named figure bears on recent marine sediments, which hardly suffered the influence of compaction as is the case with the sediments of the North Sea Basin and with other examples of older strata.

Much higher figures, however, result if only the more recent history of the North Sea Basin be considered. Again some data have been summarized in Table XII.

TABLE XII

Rate of subsidence and sedimentation in the North Sea Basin during the Pleistocene and Holocene

Average subsidence of the bottom in the N.W. of the Netherlands	Duration in years	Positive shift of the coastline in meters	Average subsidence in cm per century
Pleistocene plus Holocene.	600,000	300	5
From base Riss-glacial to recent.	200,000	80	4
Holocene.	20,000	22	11
According to data from tide gauges.	appr. 50	0.15	30

These data point to an abnormally high rate of sedimentation if compared with the figures of Tables IX-XI [1]). It seems obvious to conclude that the high figures of Table XII are the expression of the abnormal sequence of events in the Pleistocene. Three factors may have been of special influence viz. (1) subsidence of the peripheral belt around the last Scandinavian ice-cap, (2) compaction of the sediments, (3) rise of the sea-level as a result of the melting of the continental ice-caps. At any rate these figures clearly show that certain epochs are characterized by a much accelerated movement of the bottom as contrasted to the average 0.4 centimeter per century. This result is well in accordance with the paleogeographic data known from the same region.

(2) Paleogeographic development

The maps for Lower-, Mid-, and Upper Pliocene reproduced in fig. 195-197 show a gradual retreat of the coast line. This phenomenon has to be ascribed to a rising floor, which movement proceeded from the south (Rhine Shield) in a northern direction. But in the Pleistocene ,—fig. 198 — when the growing ice-caps caused a world-wide lowering of the sea-level, the coastline moved in a southern direction instead of continuing its retreat. Hence, we may conclude that the rising floor of the Pliocene was followed by a subsidence of the floor in the Pleistocene. The influence of strong differential movements along the faults bordering the graben (b_1, b_2 and b_3 on fig. 194) is well illustrated in the maps for the Pleistocene (fig. 198). A similar action along the major faults is known also from the Pliocene.

[1]) As a matter of fact Kuenen and Neeb found an average rate of sedimentation of as much as 5 (even 6.5) centimeter per century for the deep-sea basins of the East-Indies. Again, however, this concerns recent marine sediments which are not influenced by compaction.

The map for Günz-glacial and Günz-Mindel interglacial show a retreat of the sea beyond the present North Sea coast of Holland (fig. 199 and 200).

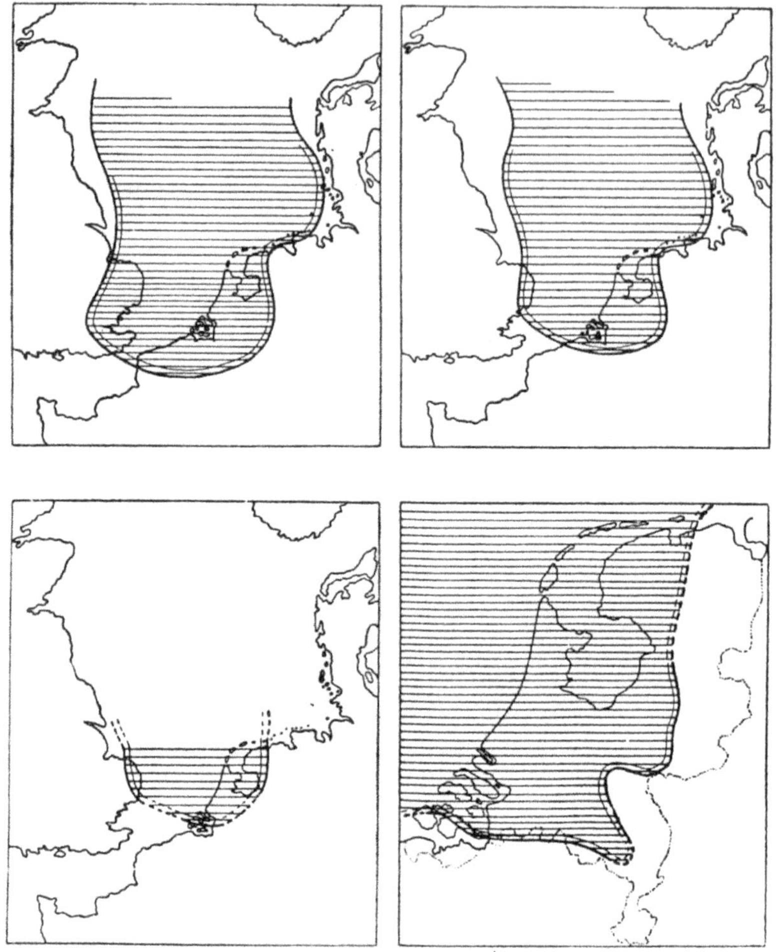

Fig. 195–197. Paleogeographic maps of the Netherlands for Lower-, Mid-, and Upper Pliocene, showing a gradual retreat of the coastline. Fig. 198 Pleistocene paleogeography, demonstrating a rising movement of the bottom when compared to fig. 195–197 (After P. Tesch).

It follows from these data that the rate of the subsiding bottom-movement at that time was at least retarded.

In the boundary region of the North Sea Basin a Pleistocene tilting may be deduced from the arrangement of fluviatile deposits. The rivers entrenched themselves in the deposits of a previous period. Thus the well-known sequence of fluviatile terraces was formed along the incised channels of the streams, the age of the terraces diminishing from Lower Pleistocene for the highest to subrecent for the lowest. In the subsiding basin, however, the sediments have been deposited in the opposite order, the more recent

Fig. 199–200. Paleogeographic maps of the Netherlands for Gunz glacial times (fig. 199) and the Gunz-Mindel interglacial. To be compared to fig. 195–198 (After P. Tesch).

sediments on top of the older ones. The result of these circumstances is the remarkable crossing of the terraces which is illustrated by the schematic block-diagram of fig. 201 [1]).

From the available data a similar tilting of the bottom may be deduced

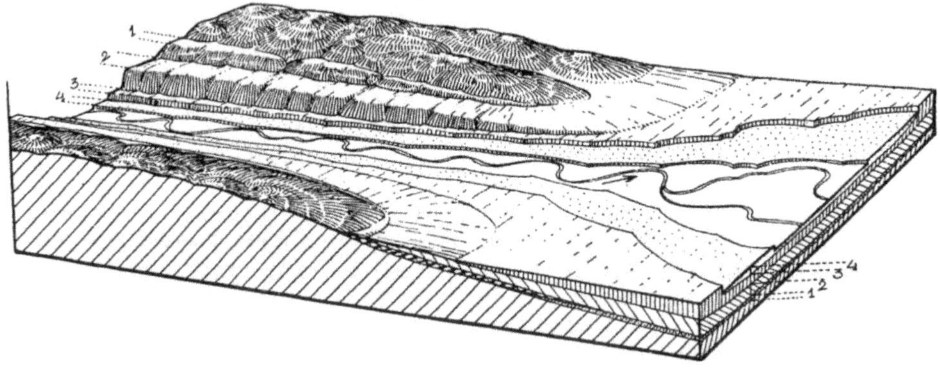

Fig. 201. Schematic block diagram of crossing river-terraces caused by a tilting movement (From K. Oestreich).

for two other epochs, viz. (1) post-Lias and pre-Senonian, (2) Oligocene. The pre-Senonian warping was correlated by de Sitter with the Upper Cimmerian epoch of compression; the Oligocene movement was synchronous

[1]) From K. Oestreich 1938 p. 555 fig. 1. See also R. A. Daly Changing world of the Ice Age, 1934, p. 189 and fig. 100 on p. 190.

with one of the principal phases of folding of the Alps. Moreover both the epochs are marked by major phases of movement along the faults of the southern Netherlands and by the formation of the present Rhine graben, etc.

(3) Structural history

In short, the most important phases of movement of the North Sea Basin, the Rhine-shield and the intervening fault-zone appear to be inter-related, while these movements can in turn be correlated with movements in the earth's crust, which are known to be of world-wide importance. Again and again we see that special epochs, marked by compression and folding in geosynclinal belts, are characterized by important processes outside the geosynclinal zones as well.

Finally a few examples concerning the structural history of the same district may be added. The time of origin of the North Sea Basin as such, especially of its eastern and western boundary, coincides with the Lower Cimmerian phases (a^3 and a^5 on fig. 194), a preliminary indication of their formation coincides with the Saalian phase. The formation of the Paris basin dates from the Lower Cimmerian phase. The Munster Basin originated contemporaneously with the Upper Cimmerian origin of numerous faults and so-called anticlinal ridges in the basin of Paris, the Boulonnais, the Weald and the Egge and Osning (c^1–c^8). The same epoch is marked by major tectonic movements along the Limburgian faults (b^3). The first appearance of the London Basin, as a south-western prolongation of the North Sea Basin coincides with the Laramide phase, etc.

General conclusions

From these facts we may deduce that all these phenomena are inter-related and depend on a common cause which has its source in the deeper substratum. The chronological relation between epochs of mountain uplift — i.e. epochs of decreasing compression in the earth's crust — and the origin of basins was already mentioned as a striking result of our examination in Chapter IV (p. 63). Now the dome-shaped elevations with their faults and rift-systems are incorporated into the series of periodic events. Once a basin or a zone of elevation has come into being it is rejuvenated during one or more later epochs. And these epochs are marked generally by a tendency for folded mountain-chains to rise, i.e. the epochs immediately succeed a time of compression. On the other hand — it is clear for evident reasons — that the origin of faults and the rejuvenating movement along existing faults may commence contemporaneously with a certain epoch of compression. The relation between these two phenomena is necessarily a very close one. In order to get a clear idea of their mutual connection fig. 47 and fig. 49 should be looked once more. When considering the processes associated with a buckling crust emphasis was laid on the well-founded supposition that the earth's crust forms one world-encircling rigid and elastic whole. One phenomenon associated with an epoch of compression of the crust, is the folding of geosynclinal belts and the formation of a sialic root below these folded zones. Accordingly, epochs of compression are at the same time epochs of crustal shortening. One possibility is that the process is accompanied by a contemporaneous shortening of the earth's radius as well as by a simultaneous shrinking of the total volume of the earth's interior by a certain well defined and coordinated amount. Another possibility is that the process of crustal shortening may give rise to the formation of transcurrent faults, fissures and rift-valleys. And as the planetary system of dislocations has probably already existed in the earth's crust since early Pre-Cambrian times, the faults would be rejuvenated periodically during epochs of compression. It is only one step further to the supposition that the host of the earth's lineaments — some of them bearing a chain of impressive volcanoes like those of Hawaii, others appearing as seismo-tectonic structures like the Bartlett trough — owe their present aspect to one of the last periodical rejuvenations of certain very old features of the earth's crust. A similar opinion was expressed by Betz and Hess in their paper on the floor of the North Pacific Ocean.

Summary

Three major problems are associated with repetative linear patterns of the earth's surface.

Firstly the question arises as to the time of origin of the two existing lineament-sets of world-wide importance — one being NS and EW, the other one diagonally, viz. mainly NE–SW and NW–SE. In the second place the cause of the special directions of the lineaments was examined. And, finally, the repeated rejuvenation of the linear structures was considered.

As to the first point there seems little doubt that the planetary systems date from primeval times. Probably their formation was associated with the consolidation of a continuous and solid crust around the earth.

The cause of the special directions of the two lineament-sets still remains an unsolved problem. Several theories, including those of large polar displacements, were discussed but found to be unsatisfactory.

Clearly, however, the later history of the lineaments — especially their repeated rejuvenation — is ultimately connected with various periodical events in the earth's crust. Probably the same holds good for the formation of dome-shaped elevations with their faults and rift-valleys.

References

ANDERSON, E. H. *The dynamics of faulting and dyke formation with applications to Britain* (Oliver and Boyd, Edingburgh, 1942).

BETZ, F. and HESS, H. H. *The floor of the North Pacific Ocean* (The Geogr. Review, 31, 1942).

BUSK, H. G. *On the normal faulting of Rift Valley structures* (Geolog. Magazine, 82, 1945).

CLOOS, H. *Zur Mechanik grosser Brüche und Graben* (Centr. f. Mineral Abt. B. 1932).

CLOOS, H. *Hebung, Spaltung, Vulkanismus* (Geol. Rundschau, 30, 3A 1939).

DAUBRÉE, A. *Etudes synthétiques de Géologie Experimentale* (Dunot, Paris, 1879).

GREGORY, J. W. *The plan of the Earth and its causes* (The Geogr. Journal, 13–1899).

HOBBS, W. H. *Repeating patterns in the relief and in the structure of the land* (Bull. Geolog. Soc. America, 22, 1911).

JEFFREYS, H. *On the mechanics of faulting* (Geolog. Magazine, 79, 1942).

KUENEN, PH. H., *Volcanic fissures with examples from the East Indies* (Geolog. en Mijnbouw, 7, 1945).

KUIPER, H. *Poolbewegingen tengevolge van Poolvluchtkracht* (Dissert. Utrecht 1943).
LAPWORTH, C. *The heights and hollows of the Earth's surface* (Proceed. R. Geogr. Soc. 14, 1892).
LONGWELL, CH. R. *Classification of faults* (Bull. Americ. Assoc. Petrol. Geolog. 27, 1943).
LUGN, A. J. *Pre-Pensylvanian Stratigraphy of Nebraska* (Bull. Americ. Assoc. Petrol. Geol. 28–1934).
MILANKOVITCH, M. *Säkulare Polverlagerungen* (Handb. d. Geophysik I. 1936).
MILANKOVITCH, M. *Astronomische Mittel zur Erforschung der Erdgeschichtlichen Klimate* (Hand. d. Geophysik, IX – 1938).
OESTREICH, K. *La genèse du Paysage naturel* (Tijdschr. Kon. Nederl. Aardrijksk. Genootschap 55–1938).
PHILIPP, H. *Das O.N.O.-System in Deutschland* (Abh. Heidelberger Akad. d. Wissensch. 1931).
PICKERING, W. H. *The place of origin of the Moon* (Journ. of Geology–15–1907).
RAISZ, E. *The Olympic-Wallone lineament* (Americ. Journ. Sci. 243A, 1945).
SITTER, L. U. DE, *The alpine geological history of the S. Limburg coal district* (Jaarversl. 1940–41 Geolog. Bur. Mijngebied 1942).
SONDER, R. A. *Die Lineament-tektonik und ihre Probleme* (Eclogae Geolog. Helvetiae 31–1938).
STEVENS, CH. *Le Relief de la Belgique* (Mém. de l'Institut Geolog. de l'Université de Louvain, 12-1938).
TABER, S. *Fault troughs* (Journ. of Geology, 35, 1927).
TESCH, P. *Het voetstuk van Nederland* (Tijdschr. Kon. Nederl. Aardrijksk. Genootsch. 54, 1937).
TESCH, P. *De Noordzee* (Nederl. Rijks Geol. Dienst A. 9, 1942).
UMBGROVE, J. H. F. *Periodical events in the North-Sea Basin* (Geolog. Magazine, 82, 1945).
UMBGROVE, J. H. F. *Recent theories on polar displacement* (Americ. Journ. Sci., 244, 1946).
VENING MEINESZ, F. A. *Topography and gravity in the North Atlantic Ocean* (Proceed. Nederl. Akad. Wet. 45, 1942).
VENING MEINESZ, F. A. *Spanningen in de aardkorst tengevolge van poolverschuivingen* (Verslag Nederl. Akad. Wet. 52, 1943).
WATERSCHOOT VAN DER GRACHT, W. A. J. H. VAN, *The paleozoic geography and environment in N.W. Europe* (C. R. Congres Stratigr. Carbon Heerlen 1938).
WATERSCHOOT VAN DER GRACHT, W. A. J. H. VAN, *Lateral movements of the Alpine foreland of northwestern Europe* (Proceed. Kon. Akad. Wetenschap. Amsterdam 41, 1938).
WILSON, J. T. *Structural features in the north west territories* (Americ. Journ. Sci. 239, 1941).

CHAPTER XII

THE PULSE OF THE EARTH

"And the secret of it all is in the heart of the earth, forever invisible to human eyes". (R. A. DALY)

Introduction

The preceding chapters showed time and again, that the different groups of phenomena are genetically related to one another, and that they have a common, deeper cause in a figurative as well as a literal sense, to which we referred repeatedly by means of such neutral designations as "processes in the substratum", "changing conditions in the interior of the earth", "deep-seated forces", "pulsating rhythm of subcrustal processes", etc. These vague descriptions designate the paramount source of all subcrustal energy, which manifests itself with a periodicity observed in a whole series of phenomena in the earth's crust and on its surface, viz. the alternating decrease and increase of compression of the earth's crust and the closely related epochs of folding, the process of mountain-building and submersion of borderlands, the periodic formation of basins and dome-shaped elevations, the magmatic cycles and rhythmic cadence of world-wide transgressions and regressions, the pulsation of the climate, and — lastly — the pulse of Life.

Chronological relation of periodic events

The object of this last chapter is to recapitulate the above phenomena, and to give a condensed synopsis of their correlation. A diagrammatic view of the preceding events will be found in Table II. The latter finds its explanation in fig. 202 which contains a schematic survey of the chronological relation of the periodical processes of one cycle. An arbitrary lapse of time has here been subdivided into five stages, the division being based on a certain chronological order represented by A, B, C, D, and E, and this is followed by a new cycle (A^1 etc.).

A. This first stage is marked by a world-wide regression of the epicontinental seas, resulting either from the elevation of the continents, or the subsidence of the ocean-floors, or even — and this seems the most probable course of events — from the simultaneous but opposed movements of both the continents and the ocean-floors. A period of intensive erosion ensues on the continents, and the climatological zones begin to show a greater differentiation than they did previously. This period is also characterized by an increased compression of the crust, which consequently buckles in geosynclinal areas. One of the consequences of this event is that the contents of the geosynclines become folded, and this is in turn accompanied —and later succeeded— by the intrusion of acid batholiths. A simultaneous phase of movement generally occurs in those basins which were already in existence at that time. We are at present concerned with the surface of the earth in its consolidated state. The hypothetical processes responsible for the origin and arrangement of continental bucklers and oceanic depressions during the early Pre-Cambrian will not be considered here (cf. Chapter VIII).

B. The first stage is followed by a period of decreasing crustal compression. By this time the processes of folding and regression have already attained their full effect. The folded belts now emerge as mountain-chains. Basins and new geosynclines begin to form. Dome-shaped elevations originate, and the sea-level begins to move in an opposite direction (positive movement). The interrupted line in fig. 202 represents a period of only slight regression, and the dot-dash line and full-drawn graph in the same column depict periods of extensive and exceptionally extensive regression respectively.

C. The mountain-belts have grown to their full height, and the relief of the continents is very accentuated during this stage. The formation of dome-shaped elevations, rift valleys and contemporaneous volcanism are

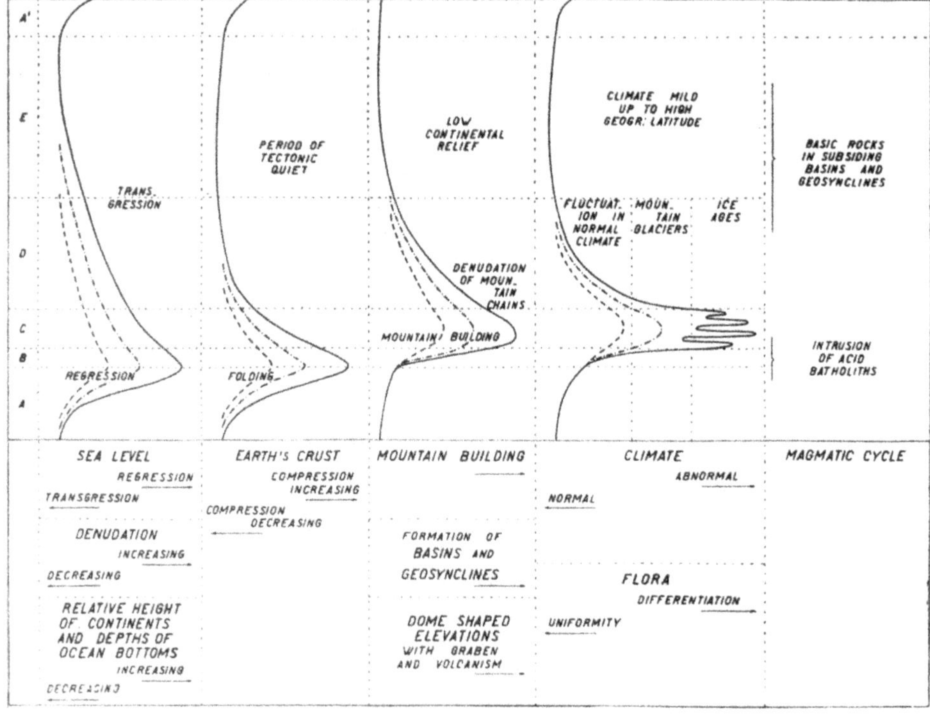

Fig. 202. Graphs showing the chronological relation of the various periodic phenomena.

now completed. The newly formed basins and geosynclines have already subsided to a considerable depth.

The three curves representing the more or less pronounced deviations of the climate from normal conditions correspond with the greater or lesser intensity of the processes indicated by the three series of curves on the left. An extreme regressional stage and period of very marked and widespread mountain-building will thus correspond with a considerable differentiation of climatic zones and the occurrence of glaciation. An additional cooling power of the ice-sheets causes them to extend very rapidly. This is illustrated by the steepness of the curve on the extreme right. Cosmic factors are now responsible for the division of the glaciation into a series of ice-ages. Naturally, this curve only represents the glacial and interglacial stages very schematically.

These very abnormal climates have been observed twice since the Cambrian, i.e. during the Upper Paleozoic and Pleistocene, and it was during these two periods that the differentiation of the flora was so much more marked than during the intervening time.

The same major periodicity of approximately 250 million years is reflected in the tremendous Variscian and Alpine mountain-building and the formation of basins and troughs. The Upper Paleozoic phase of mountain-building lasted for about 50 million years, and the Alpine belts began to rise 200 million years later [1]). During the intervening time periods of minor activity occurred, with an average cadence of 50 million years, and were accompanied by lesser manifestations of an abnormal climate as shown in Table II [2]). Our galaxy rotates in about 200 million years. Other figures can be found in astronomical literature, viz. 3C0 and 250 million years. This cosmic cycle may thus be said to correspond on the whole with the major periods of terrestrial activity. It is difficult to decide for the moment whether this coincidence is a mere question of chance, but it is obvious that this point will have to be taken into consideration in the further development of science.

D. The above period of abnormal climatological conditions ends even more rapidly than it began. The mountains have been eroded to a considerable degree during the intervening time, and the sea-level has risen slowly but surely.

E. At this stage we find a low relief on the continents, together with vast epicontinental seas and a mild and equable climate extending to high

[1]) These figures agree fairly well with those cited by Kuenen in his article of 1941 on major geological cycles.

A similar attempt was made by Wahl (1943). His results are rather well in accordance with those of Kuenen. The duration of a period of folding and mountain-building was estimated at 50–80 m.y. by Kuenen and 60–80 m.y. by Wahl. Their estimates regarding the culmination points of the major geological cycles are as follows, — the first figures being from Kuenen's paper, those between brackets from Wahl. Alpine 50 (59), Variscian 250 (286), Caledonian 370 (381), East-African-Samian 600 (612), Karelian-Huronian 775 (801), Gothian-Algomanian 880 (941). Svekofennian-Laurentian 1050 (1122).

Remarkably enough, neither Kuenen nor Wahl have found any evidence of the Mesozoic revolutions. A few older cycles than Laurentian are plotted in Kuenen's graphs.

[2]) It need perhaps hardly be emphasized that the graphs in Table II should not be considered to be mathematical constructions. The positions of the tops of the curves have only been indicated roughly. Their amplitudes are only of value from a diagrammatic point of view and are unsuited for measurement.

latitudes. A maximum transgression means a minimum difference in the height between the continents and the ocean-floors. Moreover, this stage is pre-eminently a period of tectonic quiet. In the meantime, however, certain belts and basins continue very gradually to subside. Abyssolithic injections of basic melts break through the crust and adjust themselves into the steadily accumulating strata of the geosynclines and basins as eruptive and extrusive products.

The chronological correlation of A, B, C, D and E as represented in fig. 202 is wholly diagrammatic. A, B and C combined cover a much shorter lapse of time, however, than the longer stages under D and E.

Our own cycle has already advanced to some extent into stage C.

A fundamental problem

Behind the results arrived at so far there still lurks one essential question, viz. to what deeper impulses must we attribute these phenomena? In what manner might these problematic subcrustal processes produce a periodic recurrence of phenomena such as those represented schematically in Table II?

We might compare our quest to that of the detective in criminal fiction, who attempts to discover the essential qualities of the mysterious culprit whose traces all lead in one direction. We are no longer content with the classical image of Hades, the old and crippled god of the subterranean realm, for his image is now replaced by a description as found in Table II and explained by fig. 202. We know the culprit's haunts as well as some of his more salient features. Very little has been ascertained as regards his restless infancy, but it is possible to show the main outlines of his cardiogram for the last 500 million years as derived from the movements of the terrestrial crust and the sea-level. His finger-prints are supplied by the magmatic phenomena. What creature is this that breathes so heavily once in every 250 million years, and why does its pulse beat approximately four times during the intervening period? Why are the wrinkles in its face arranged according to the intricate pattern in Plates 1–5? Its skin is not only lined by mountain-belts, but is also pock-marked by numerous basins. Yet nothing is known of the events that brought about these marks during certain specified periods. And there are several hypotheses to explain what was responsible for one of its most striking features — the oceanic depressions.

To return from this geopoetical paraphrase to geological prose: can the subcrustal periodicity be said to be of a physico-chemical nature? Might we attribute it to a periodic system of convection currents, or might it be the result of a combination of these diverse activities? Other processes, too, of which nothing is known at present, may be active in the interior of the earth.

Everyone will have to admit that all views on this subject must necessarily retain a speculative character.

Most readers will be acquainted with the interesting suggestion which Joly and Holmes put forwards some years ago, when they attempted to derive periodic phenomena from radioactive processes, and will probably also have read the opinions of such authors as Schwinner, Escher, Holmes, Pekeris and Vening Meinesz on convection currents. Some may have come across Grigg's publication in which this author tries to show that convection currents may in fact occur intermittently, and will have admired the ingenious model by which he sought to demonstrate the activity of cyclic convection currents. In short, geologists and geophysicists are already striving for a solution and the hypotheses that have appeared from time to time may perhaps already contain a germ of truth. Nevertheless, the mysterious interior of the earth involves more than just a single periodic process. It is far more complicated than the tentative theories have supposed so far. I hope that the preceding chapters and the accompanying graphs will have made this clear. Geology will have to proceed hand in hand with Geophysics in a combined effort to unravel this most absorbing problem. A solution will only be found when it will be possible to indicate that a logical and necessary correlation of cause and effect exists between certain events in the earth's interior and those diverse phenomena to which I referred briefly in the title as *the Pulse of the Earth.*

References

BARRELL, J. ***Rhythms and the measurement of geologic time*** (Bull. Geol. Soc. of America, 28, 1917).
BUCHER, W. H. ***Deformation of the Earth's Crust*** (Bull. Geol. Soc. of America 50, 1939).
DALY, R. A. ***Our mobile earth*** (1929).
GRIGGS, D. A. ***A theory of mountain building*** (Americ. Journ. Sci. 237, 1939).
HOLMES, A. ***The thermal history of the Earth*** (Washington Acad. of Sci., 23, 1933).
JOLY, J. ***The theory of thermal cycles*** (Gerlands Beitr. z. Geophysik 19, 1928).
KUENEN, PH. H. ***Major geological cycles*** (Proc. Nederl. Akad. v. Wetenschappen 44, 1941).
LONGWELL, CH. R. ***The mechanics of Orogeny*** (Americ. Journ. Sci., 243A, 1945).
UMBGROVE, J. H. F. ***Periodicity in terrestrial processes*** (Americ. Journ. Sci. 238, 1940).
VENING MEINESZ, F. A. ***Evenwichtsverstoringen in de aarde*** ("De Ingenieur" 1940).
WAHL, W. ***Altersvergleich der Orogenesen und Versuch einer Korrelation des Grundgebirges in verschiedenen Teilen der Erde*** (Geolog. Rundschau 34, 1943).

APPENDIX

DESCRIPTION OF PLATES 1-8 AND TABLES I AND II.

The following pages will discuss and enumerate the data on which the maps have been based. This part should only be consulted by those geologists who wish to examine questions of detail. The citations of literature refer to the bibliographies at the end of Chapters II and III.

Plate 1. Caledonian epochs of compression

We will begin with Spitsbergen. The epochs of compression in this part of the world are probably Ardennian, according to Frebold, and possibly also Taconic and Erian. The eastern limit of the geosyncline is situated in North-eastland, where the Caledonian strata become less and less disturbed. The northern continuation of the geosyncline is a problematical one. It may possibly run via the submarine ridge uniting Spitsbergen and the north of Greenland with the so-called Smith-Sund geosyncline of northern Greenland. There are no data, however, to prove the exact age of the folding in this area. It might therefore be a Caledonian geosyncline, but it might also be a zone of later folding (Teichert 1939, p. 146, 147). A feebly folded Caledonian area exists south of Spitsbergen, on Bear-Island (Devonian resting unconformably on Hekla-Hoek formation). A later epoch (Erian) than that observed in the western sector (e.g. the Trondheim area), where the principal phase is the Ardennian, was reported in Scandinavia, around Oslo. Bailey asserts (1938, p. 39, 73) that the Ardennian also constitutes the principal epoch of the Highland border-zone of Scotland. Other epochs too, seem to have been observed in England (von Bubnoff 1930, p. 131, 132), the Taconic being particularly evident in South Wales. A section of the Pentland Hills near Edinburgh shows a strong Erian and weak Mid-Devonian phase of movement (fig. 22). The complicated history of the Caledonian zone of England and Wales is clearly demonstrated by Owen Th. Jones in his address of the year 1938.

Though the Caledonides in southern England now lie buried beneath the sediments of a Variscian geosyncline, there are still many indications in Europe that Caledonian movements extended over a large portion of the European Continent. The presence of a Caledonian unconformity was revealed in many parts of the so-called Saxo-Thuringian zone, forming part of the later, Variscian sector of Europe. A Taconic unconformity, extending as far as the central mountains of Poland, is clearly visible in the Rheno-Hercynian zone.

Schwinner claimed as a result of microseismic observations that the Scottish Caledonides should not be connected with the Scandinavian as the former appear to bend around towards Iceland. The Caledonian folds of Scandinavia seem to turn in a southerly and south-easterly direction.

It is still doubtful whether a branch of the Caledonian geosyncline of Scandinavia and Scotland extended towards western Europe, as Stille assumed (1924, p. 72). (The massifs of Brabant, Rocroy, Stavelot and Condroz, and the late Caledonian folding in the Boulonnais). This was denied by Van Waterschoot van der Gracht (1938, p. 1393), who regarded the Brabant Massif, to mention but one example, as a Pre-Cambrian area, which was folded in a NS. direction, and the enveloping Early Paleozoic sediments as strata which were down-folded into the massif during the Variscian epoch. Many others, however, believe that the intensely folded Early Paleozoic sediments are in fact unconformably covered by Upper Paleozoic deposits, and that the plane of unconformity itself was again folded with the later sediments. An example of this is found in the French Ardennes, near Fépin (fig. 21). Nevertheless, the absence of traces of a Caledonian unconformity in the Moldanubian zone appears to be indisputable (see under the Variscides). A Caledonian unconformity can be found in the south of Spain, and a Taconic in Portugal (Born, p. 667). The above shows that the situation is far from being a simple one, and scarcely more comprehensible. For Caledonian movements have at any rate been observed north as well as south of the Moldanubian zone in Europe.

While Caledonian movements, and partly

folding, are thus known to occur on the eastern side of the Atlantic and to cover a large area, all that is found at the other end is a comparatively narrow zone, directed approximately NE.-SW. The latter can be followed from Newfoundland to New York, and the Caledonian zone of folding around the coast appears to have been folded at an earlier date, i.e. during the Taconic epoch (cf. Van Waterschoot van der Gracht). Another important point is that the principal epoch of folding in North America is the Taconic. Later Caledonian phases are not observed in these parts, though the Scottish and Scandinavian Caledonides have often been taken for a continuation of the American belt. It may be that the Newfoundland Caledonides continue in the direction of the eastern coast of Greenland. A certain similarity in the marine faunae would seem to prove this. It is as yet impossible, however, to give an accurate description of the Caledonian epochs of folding in this last area (Teichert 1939, p. 150). The Caledonian age attributed by Koch and others to a zone of folding in the north of Greenland does not appear to be sufficiently well-founded. The period of folding might be Variscian, or it might even be a more recent one (Teichert, 1939, p. 146, 147). However, even if it were true that a connection existed between the American Caledonides and those in Greenland, this would not preclude the existence of a branch extending towards Europe. Still as explained above, the area of Caledonian folding is a far more extensive one in Europe than in America. And, secondly, the principal epochs of folding of the Newfoundland and the British-Scandinavian Caledonides are different ones. Hence the extremities could not at any rate have been directly connected.

Very little is known of a possible westward continuation of the Caledonides of the Atlantic coastal area of North America. Taconic movements in large areas in the Rocky Mountains, and locally in the mountains along the coast, are indicated by the absence and feeble development of the Silurian. These indications are lacking in the Canadian Rockies.

However, Caledonian unconformities are recorded in many parts of the more northerly ranges of Alaska. Their distribution as shown in Plate 1 is based on Stille's summary (1940, p.p. 44—48, fig. 12). The ages of the movements is late Caledonian (Ardennic, Erian). Taconic movements are not recognised with certainty in this area as yet.

As we move south, evidence of Caledonian movements is only found in a few places in the South American Andes. Gerth (1932, p. 104) mentioned a Taconic unconformity in the Argentine Province of San Juan, and also reported the presence of a Taconic and Late Caledonian unconformity in Peru. Another fragment of Caledonian folding may possibly crop out in the Venezuelan Sierra de Merida. Further evidence of such folding may — apart from that observed in the Cordillera — possibly be found in the southern part of the Sierra of Buenos Aires (Gerth 1932, p. 187–190). The Brasilides and Pampine Sierras, which were once regarded as an area of Caledonian folding, are clearly of Pre-Cambrian age (Gerth, 1932, p. 80). The remarkable appearance of folded Cambrian quartzite in the north of the Argentine should furthermore be noted. This quartzite is covered unconformably by Ordovician deposits, beginning with a basal conglomerate. The phase of folding is in this case probably Stille's Sardic epoch, which was also observed in Sardinia, the Himalayas, China, Central Asia and New Brunswick (Stille 1939, p. 771).

As very few data are available at present, nothing can be said for the moments of the trend of the Early Paleozoic mountains, nor can anything be said of a possible link between these mountains and others situated on other continents. It may be assumed, however, that the Caledonides probably extended (roughly) in the form of an arc, like the Variscides.

Three areas in Africa are known to contain evidence of probable Caledonian folding. Hennig (p. 59, 60) reported the presence of Caledonian folding in the massif of Ahaggar, but Krenkel has pointed out since then that its Caledonian age has been refuted (p. 1442). Hennig also showed that terrestrial deposits, corresponding approximately in age with similar deposits in the Cape System, occur in folded chains in the Congo, and that analogous and intensely folded Paleozoic beds have apparently been covered by Permian in the Katanga. He regarded the latter as Variscian chains but Krenkel allows for the possible occurrence of Caledonian instead of Variscian folding in the "Katangides". In the latter's opinion the "Congolides", as well, consequently, as the Lower Silurian folding ("intensive Rahmenfaltung") of the "Griquaides" (p. 684, 912), are undoubtedly Caledónian. These belts have not been indicated on Pl. 1.

Leuchs observes that part of a Caledonian zone of folding can be followed around the Pre-Cambrian nucleus of Angaraland (1, p. 176). This zone shows later Variscian movements along its eastern margin. Another Caledonian area is likewise known to extend along the eastern and north-eastern margin of the Pre-Cambrian massif part of which subsided into the Kara Sea. A small portion of such a zone — which is presumed to have existed around the massif of Tsuchktschen — may apparently still be found in the New Siberian Islands. Upper Silurian was folded with the rest in the Caledonian zone around the massif of the Kara Sea. The oldest Caledonian epochs observed along the south-western edge of the massif of Angara appear to be of Cambrian age (Obrutchew, p. 134–137, and Leuchs, 1, p. 153) and to date from the close of the Silurian in the Salair (Leuchs 1, p. 155).

Caledonian folding also played an important part in Manchuria, the Mongolian Altai and the Russian Altai (Leuchs, 1 p. 160, 163, 188). The Caledonides of South and South-East Asia probably occupied a considerable expanse, for Caledonian unconformities are observed in many mountains in which evidence is found of Variscian or younger folding e.g. in the Kirgizes (Leuchs 2, p. 29), the Kwenlun (1, p. 33 and 2, p. 189), the Tianchan (2, p. 117), the Arkatagh (2, p. 231) the Nanchan (2, p. 227) and the south-eastern part of the Asiatic continent. Faint Taconic movements were reported by Lee (p. 108) in the whole area north of the Tchingling mountain-chains. Early Caledonian movements, on the other hand, have been observed in the south of China and in Indo-China (Lee, p. 115), but these epochs cannot be determined accurately. No Caledonian belt has been observed in the East-Indian Archipelago, though such a mountain-chain is known to exist in Australia. Indications of Caledonian folding in Borneo were found by Albrecht (Acad. Thesis Utrecht 1946).

A geosyncline extending along the eastern edge of Pre-Cambrian western and central Australia, and apparently coming to an end in this last area is known to have existed in Australia in former times. Very intensive Taconic and post-Silurian movements (?Erian) occurred, according to Andrews, and a Mid-Devonian unconformity was also reported in the eastern sector of Victoria. Generally speaking the movements appear to grow more recent as we move from the west to the east, and the structure of the Caledonian belt seems to become more and more intricate, consisting of consecutively folded geosynclinal zones. Folding in Flinder's Range, Adelaide, apparently occurred as early as the Mid-Cambrian, and the geosyncline subsequently receded towards Victoria. It remains to be seen whether the same or whether different Cambrian epochs are involved in Asia. On this last continent movements also succeeded Lower Ordovician time. The youngest Caledonian movement in the eastern part of Victoria occurred towards the close of the Upper Devonian (Bretonic epoch). These influenced the earlier folded zones. Andrews still classified phases as Caledonian epochs (p. 124, 173), since they represent the final epoch of these ranges.

Plate 2. Variscian epochs of compression.

The analyses of the Variscides of Western Europe by Kossmat and other German geologists clearly illustrate the complicated zonal structure of this area. When considering the huge expanse occupied by Variscian folding in Asia, it should be born in mind that these areas are at least quite as complicated from a structural point of view as those of western Europe, and that they should not be regarded as merely one, vast geosyncline of tremendous width; moreover the different zones were folded in different epochs. In the European Variscides, too, the geosyncline proper is confined to a narrow zone — the so-called Rheno-Hercynian zone (1 on Plate 2). Yet even here three distinct units can be observed. In the southern sector (e.g. the Schiefergebirge of the Rhineland and Brittany) the folding is Bretonic and north of this sector — in Normandy, Rocroy, Hohe Venn and Siegerland — probably Sudetic. More recent folding is found in the most northern coal-bearing zone, where the principal epoch is probably the Asturian. Kossmat mentions yet another zone: the Saxo-Thuringian, situated south of the Rheno-Hercynian area (2 on Plate 2). Folding in this zone is of Sudetic age. The Moldanubian zone (the Plateau Central, the southern regions of the Vosges, the Black Forest and Bohemia) is found south of the Saxo-Thuringian zone (3 on Plate 2); it consists largely of Pre-Cambrian rocks. No Caledonian unconformity has been observed here, nor is it possible to say whether Variscian folding occurred in this sector. There is no doubt, however, that it was influenced to a considerable degree by Paleozoic movements. Intensive faulting occurred, and was probably accompanied by overthrusting of entire blocks over folded areas.

The Sudetic is the principal epoch of movement in the central region of the Variscides. This phase did not affect the outer zone, which was formed at a later date. Folding in this area was probably Asturian. The central part supplied the outer zone with sediments, and the latter rose as a mountain-chain towards the close of the Carboniferous (the Saalian epoch is scarcely noticeable).

How are we to account for the continuation of these belts? Stille and Born were both of the opinion that they bent around within the European shelf and emerged again in the Spanish Meseta. So little is known of the detailed structure of this last area that it is impossible to say whether it contains any evidence of Central European zones. One thing should certainly be noted, however, viz. the structure of the Meseta runs NW.-SE., and the Asturian Mountains actually do bend around, forming the so-called Asturian Knee. Sudetic, Asturian and Saalian epochs are found in Asturia. I am of the opinion that if the Variscian Mountains of Western Europe curve back towards the south of that continent this would tend to show that the Asturian Carboniferous should not be connected with the northern marginal zone of the Rheno-Hercynian belt, but that it should be regarded as a formation similar to the northern paralic zone (but situated south of the Variscides).

The presence of Variscian folding was partly revealed and partly assumed in the Western and Eastern Alps, the Carpathians, the Iberian-Islands and — finally — Corsica, where the most influential epoch appears to have been the Sudetic. The most important phase in Asturia and the northern marginal deep of the Variscides was the Asturian; and in the Pyrenees, the Donetz basin and the Ural, the Saalian.

The Variscian area of North Africa, which was later influenced by Alpine folding, was connected with Europe. This applies to a belt south of this area, extending as far as the Touareq Massif. Folding in these sectors is Bretonic in the south, and Sudetic further north (Born, p. 794). The folds run, generally speaking, in a NS. direction, bending eastwards at their northern extremity (Hennig, p. 61).

Though it may be correct to assume that the European Variscides form an independent system, this does not necessarily mean that some of the folds did not branch off towards the west continuing on the other side of the ocean. As an example we cite a connection which may have existed between northern Europe and the Appalachians, and to this might perhaps be added a second link between the southern part of North Africa and the northern part of South America.

Before dealing with America, the possible existence of a connection between the European and Asiatic Variscides should be considered. Very little can be said on this subject. Variscian folding seems to have been observed in the Dobrutcha, and is presumed to exist in the basement of the Carpathian Mountains and that of the Pannonian Basin (Cornelius, p. 358). Ancient massifs are found in the Balkans (e.g. the massifs of Rhodos, Pelagon and the Cyclades), but nothing is known as regards the age of the folding. It is probably Variscian in the massif of Pelagon and the massifs of the eastern Alps, where Sudetic and Saalian epochs occur according to Stille. Folding was Saalian in Euboea (Stille, p. 105). Variscian, and especially Saalian unconformities were observed in many parts of Asia (Philippson and Stille, 1924, p. 118). The same epoch also characterizes the southern Variscides of Asia, and can for instance be observed in the Himalayas and Burma (Stille, p. 119).

Thus only vague indications can be found in the ancient massifs and exposures of the basement-complex of the Alpine Mountains of a possible previous connection between the European and Asiatic Variscides. The Ural, which is situated on the border of these continents, will be discussed subsequently, together with Asia. It should be noted now, however, that the principal epoch in the Ural, as in the Pyrenees, was the Saalian.

A probably Variscian zone, stretching from the Caspian Sea into the Donetz Range can be traced via a group of smaller exposures which von Bubnoff christened the "Ammodetic Mountains". Folding in this zone is not intensive and can be compared to that in the Jura Mountains. The Donetz basin was formed between the Podolian Mass and the Mass of Voronez. This took place in the Upper Devonian. The first epoch occurred towards the end of the Carboniferous and the beginning of Permian time, the principal phase between the Permian and Upper Triassic (von Bubnoff 1, p. 213). It has apparently not yet been possible to determine the exact age of these epochs. Stille (1934, p. 105, 207, 141) speaks of Saalian and Upper Cimmerian folding. The principal epochs of compression in Europe would hence have been the Sudetic and Asturian, except in the regions which have just been mentioned. The Sudetic was particularly influential. It will subsequently be seen that this phase also constituted the major epochs in widespread areas of Asia. Opinions vary as regards the principal phase of folding in the Timan Ranges (von Bubnoff 1, p. 196).

Two geosynclines can be observed in the North American Variscides, i.e. the geosyncline of the Wichita and Arbuckle Mountains, with a W.N.W.-E.S.E. strike, and the great Appalachian geosyncline. The first dates from the Cambrian. Intensive folding in the Wichita mountain-chains was Sudetic (Wichita epoch), and subsequently Asturian (Arbuckle epoch). The intervening Ardmore-Anadarko basin was only influenced by the last epoch. This Lower Paleozoic geosyncline probably continued in a S.E. direction, and was overthrust by the mountain-chains of an Upper Paleozoic belt, running from the Marathon Mountains to Newfoundland via the curve of the Ouachitas and the Appalachians. The epoch of folding in the Marathon Mountains is the Asturian (Arbuckle epoch). A fainter and later stage of folding probably occurred during the Saalian epoch (Appalachian). The Ouachitas, on the other hand, probably overthrust great stretches of the foreland, as well as the chains of the Wichita-Arbuckle geosyncline (a considerable part of which had already been worn down by erosion) during this same Appalachian revolution. The Acadian (Bretonic) phase, too, was of exceptional importance in the Appalachian chains proper. For detritus (molasse) from the chains which were then just beginning to develop was later gathered into one of the foreland basins. The basin, into which all the products of erosion were thus assembled, was folded during the so-called Appalachian revolution (Saalian epoch) and simultaneously overthrust by the older folds of the Appalachian System. The Appalachian epoch may therefore be said to have also influenced most of the earlier Appalachians.

Available data on the structure of the North American and European Variscides clearly show that both parts cannot be joined without assuming that some fragment has disappeared. Meanwhile, the American Variscides indicate in a most decisive manner that epochs of folding may vary within a relatively small superficies. The former existence of a connection between both formations should therefore not be denied *a priori*, though the epochs of folding of the European and American Variscides cannot be said to coincide entirely.

Mention should here be made of Stille's hypothesis that the American Variscides bent around in a southerly direction like those in Europe. The North American Variscides, however, extend arc-wise from the Marathon Mountains in a westerly and northerly direction, spreading along the boundary of the Canadian Shield, in which last area it becomes difficult to disentangle them from the more recent structure of the Rocky Mountains. An unconformity has been observed in several places between the Paleozoic and Triassic (north and east California, Oregon, Ochoco Mountains), while many of the batholithic intrusions appear to date from the Paleozoic (in the Klamath Mountains e.g.). The Variscian geosyncline may consequently have continued around the Canadian Shield and joined up with the Variscian girdle of northern Asia. Weak indications of Variscian unconformities have been found in a few places in British Columbia and Alaska (Stille, 1940, pp. 90–100). Another question is whether a Variscian branch can be said to have extended southwards. This question is even more difficult to answer than the preceding one. Waters and Hedberg (p. 13) wrote the following in this connection: "In the Caribbean region, because of the absence of Early Mesozoic sediments, exact dating of the diastrophism which deformed all Paleozoic sediments before Cretaceous time is not possible. In northern Central America folding and faulting of the Permian sediments accompanied by intrusion occurred before deposition of the overlying Cretaceous. In Jamaica, Hispanolia and northern South America, Cretaceous sediments rest unconformably on older rocks, which may be either of Early Mesozoic or of Paleozoic age. The extensive deformation in the Caribbean region is commonly correlated with the Appalachian revolution. Whether it coincides with this or occurred somewhat later is questionable". It is therefore not improbable that a communication existed between the Variscides of North and South America, running through the area which now constitutes the archipelago of Central America.

Later investigations will also have to solve many problems in South America, but our knowledge of the Variscides in this continent is such, that the general outlines of Variscian chains can be traced from Venezuela to the Argentine along the western edge of the Pre-Cambrian mass of Guyana and Brasil, where the ranges turn towards the Gondwanides of Buenos Aires and meet the Atlantic. The epochs of folding have not yet been identified in the greater part of the territory occupied by the South American Variscides. They appear to date mostly from the Upper Paleozoic. In the ranges of Buenos Aires, an Upper Carboniferous (Asturian?) epoch and a final phase which Gerth assumed to be a Permian epoch in 1932 (p. 190), but which he described later (1935, p. 249) as a Trisassic (or so-called Gondwanides) epoch, is found side by side with probably Caledonian folding. This final epoch, though younger than the phases usually observed in Variscian Mountains, thus coincides with the most recent epoch of folding found so far in the South African Cape Mountains. According to Stille, however, the Saalian epoch was the most recent epoch of folding in the Gondwanides of Buenos Aires (Stille 1940, p. 133).

In this case the epochs of compression would consequently correspond on both sides of the ocean. However, Krenkel and Gerth (p. 190) declared the structures of the Sierra of Buenos Aires and the Cape Mountains to be fundamentally different ones, though, as seen above, an Upper Carboniferous and a Triassic epoch are evident in the last area (Krenkel, p. 593). An area of Variscian folding is also found in East-Island (Falkland Islands). The Permian layers are still folded steeply in this region and the latter also appears to contain indications of pre-Permian (post-Devonian) folding. Baker's latest observations show that an extremely faint movement can be followed in those parts of West Island where a continuation of the folds of East Island might be expected. Hence the Sierra of Buenos Aires should not at any rate be imagined to be connected with the Western part of the Falkland group by means of sharp curves, but should be considered to extend arc-wise into its eastern sector. Another branch would then connect them with Africa. Besides, these connections do not even appear to be necessary. For not only do the Cape Ranges expire near the Atlantic coast (Krenkel p. 613), but one branch turns around in a northerly direction in the Cedar Mountains, running approximately parallel to the West Coast of Africa. Moreover, the actual Cape Mountains would seem to form a mere northern fragment, of a far larger Variscian Mountain System which — together with its connection with another fragment: the Natal Mountains — is believed to have foundered in the ocean south of Africa.

Which epoch of compression should be considered to have constituted the most recent phase of folding in the Cape Mountains and the Gondwanides? Gerth and Du Toit are somewhat vague in this respect. In "Our Wandering Continents" (1937, p. 81), the latter wrote: "several pulses which attained the maximum during the Early Triassic and which appear to have corresponded approximately with the Asturian, Saalian and Pfalzian phases in Laurasia", and on page 310: "The Gondwanide foldings about the closing Permian and early Triassic". To this he added in 1939 (p. 505): "Before the close of the Triassic the bulk of the Cape foldings and much of their subsequent erosion took place". Gerth seems to imply that the Lower Triassic was the youngest epoch in the Sierra of Buenos Aires. I consequently referred to this epoch in 1939 as a separate "Gondwanide phase" of the Triassic (Table 1, Geol. Mag. Vol. 76). Henning speaks of a Rheto-Liassic epoch (1938, p. 128, 134). In 1924 Stille referred to it as the Lower Cimmerian phase, since the most recent epoch of folding in South Africa occurred subsequently to the deposition of the Beaufort beds (Permian and Triassic), but prior to that of the Stormberg beds (Rhatic or Liassic). Stille was probably right, and in this case, too, (cf. the Dobrutcha), the most recent epoch of the Variscian sequence probably occurred towards the close of the Triassic. Baker, too, was of the opinion (1937, p. 27) that the youngest epoch observed in the Falkland Islands was "possibly about the close of Triassic time although it may have been pre-Rhatic". It is still questionable whether a southern continuation of the Variscides existed in Chile (cf. Stille 1940, p. 137).

Leuchs assumed that Caledonian and Variscian zones had formed around the three largest Pre-Cambrian massifs in the same way as in Europe, where the Baltic-Russian Shield was first partly surrounded by the Caledonides and subsequently by the Variscides. The mountains of Nova Zembla and the Ural constitute the most western Variscides, and as a low area, containing Mesozoic and Tertiary sediments, extends between the Ural and the Angara Shield (the basin of Western Siberia), Leuchs concluded that its basement had been folded during the Variscian epoch. The arcuate structure of the Variscides in the Kirgize steppes east of Lake Aral would indeed seem to bear this out (Leuchs, 2, fig. 10 and 14).

As the Saalian represents the most recent epoch of folding in the Ural, folding and mountain-building may consequently be expected to have occurred at an earlier date further east as we approach the Angara Shield (the latter is surrounded by the Caledonides). Sudetic folding has in fact been observed in the Russian Altai in the immediate neighbourhood of the Caledonides in the south-west corner of Angara (Leuchs 1, p. 157, 160). Similar folding is found in the Tarbagatai (Leuchs 2, p. 33), and (south of these in the Dsungarei (2, p. 41). The Tianchan Range, forming the northern boundary of the Tarim basin, was folded during a Late Variscian epoch. The same folding is observed between the Nanchan mountain-chains and the massifs of Tarim and Ordos. The principal epoch of folding in the Kwenlun appears to have been the Bretonic. Earlier phases (Sudetic? Appalachian?) seem to have had but little influence (2, p. 190). In the area of the eastern confines of the massif of Angara, folding apparently occurred during a Lower and a Mid-Carboniferous epoch.

Since the massif of the Kara Sea is known to have first been surrounded by a Caledonian and subsequently by a Variscian zone, this last belt may reasonably be expected to belong to that massif (and not to that of Angara). In that case the Variscian zone would be more recent than the Variscides that formed around the nucleus of Angara. The youngest epoch in Taimyr appears indeed to be the Saalian (Leuchs, p. 110). Similar folding seems to occur in the Ural. Even overthrust-sheets have been observed in these mountains. The Pfalzian phases in these sectors do not appear to be very pronounced.

In the south, Variscian folding was even observed in such Alpine chains as the Himalaya. The Variscides bend around the ancient massifs of Ferghana, Tarim and Ordos. Younger Cimmerian movements greatly obscure the

Variscian phases along the southern boundary of the shield of Angara (Transbaikalia).

In south-eastern China the Variscides continue around the massifs of eastern China and Indo-China. Lee summarized all areas in which Sudetic, Asturian and Appalachian unconformities are known to occur (p. 125, 133, 139 and 149).

The Appalachian phase is believed to have been the closing and principal epoch in the Tsingling (south of the massif of Ordos). There is evidence that the Variscian zone probably extended beyond the actual continent, reaching as far as the Riu-Kiu Islands, Japan and Sachali, but these indications are very vague. A post-Paleozoic and pre-Eocene unconformity is mentioned by Hanzawa in the Rui-Kiu Islands, and a post-Carboniferous though pre-Turonian one has been reported in the eastern part of Sachalin (Leuchs 1, p. 189). The folded strata of the Japanese Variscides include Upper Carboniferous sediments (Stille, 1924, p. 122).

Little can be said of the former connection between the Variscides of Asia and Europe, for all that is left of this link is now buried beneath the Alpine zone of folding of Iran and Asia Minor. This matter was referred to when we were dealing with Europe.

Another question at this point is whether the Variscides of south-western Asia had at one time been connected with those in Australia. Nothing definite can be said on this subject, as a supposedly Variscian area of folding has so far only been observed in a few regions in the Indian Archipelago, e.g. in Malaya, Borneo and a small number of other islands scattered in between these areas. This question has a direct bearing on the age of the Donau-formation and similar rocks. They were originally defined as Jurassic formations by Molengraaff. I considered them to be older than the Upper Triassic, and believed them to have been folded during the Early Cimmerian. Zeylmans van Emmichoven, however, has pointed out since then (in "De Ingenieur in Ned. Indië 1938") that they had probably been folded during a Late Variscian epoch, and subsequently refolded during the Early Cimmerian phase. A Variscian unconformity has also been observed in Sumatra, near Lake Toba. It is not thought unlikely that Late Paleozoic movements affected a number of other islands.

The Australian Variscides extend along the coast of Queensland and New South-Wales. On the whole the epochs become more and more recent as the area extends further east. A post Lower Carboniferous phase appears to have been met with on the coast of Queensland, between Cape York and Townsville, and an Upper Paleozoic epoch (which Andrews claimed to be the Appalachian) was observed in the area south of Townsville. David (p. 61) was of the opinion that the first phase corresponded with the Asturian in Europe.

Paleozoic movements have likewise been reported in areas outside the continent. Piroutet and Wilckens mention Permian lying unconformably on Paleozoic formations devoid of fossils in New Caledonia. A Variscian movement of an indefinite age is found in New Zealand (Stille 1924, p. 123). It had little influence as will be seen later when dealing with Mesozoic movements.

It may therefore be concluded that large areas of the Variscian Mountains were already in existence during the Upper Carboniferous, and that a later Saalian phase confines itself to the marginal areas in all continents, lying on the inside in America and on the outside in Europe, Asia and Australia. The Ural, the Taimyr Mountains and the Pyrenees are examples of Saalian mountain-chains. The Pfalzian had very little influence on mountain-building and is scarcely noticeable in the belts in which it occurred. The still more recent Early Cimmerian epoch appears to have constituted the last phase in the Cape Mountains and the Sierra of Buenos Aires.

Plate 3. Mesozoic epochs of compression.

Geosynclines which were folded during the Mesozoic characterize the marginal areas of the Pacific.

A typical example of a geosyncline which was folded towards the close of the Jurassic, and which is known to contain much basic volcanic material, is found along the West Coast of the United States, in the Sierra Nevada. The epoch of compression in this area is called the Nevadian by American geologists. It coincides with Stille's Late Cimmerian phase. The Early Cimmerian epoch probably did not affect the geosyncline to any important degree, though an unconformity may be observed locally between the Triassic and Jurassic (e.g. in Central Oregon). The rocks were intensely folded and metamorphosed during the period of transition between the Upper Jurassic and Lower Cretaceous, and many batholiths penetrated the mountains which were developing at that time. The geosyncline extended northwards from the Sierra Nevada, spreading over the area which the Cascade Mountains, the Coast Ranges and the Alaska Ranges occupy at present. From there it probably continued by way of the Aleutians, and probably joined the extensive area of Cimmerian folding in Asia. Austrian movements also affected the northern sector, particularly in Oregon, Washington and British Columbia, and

Laramide movements may be observed still further north (Stille 1936, p. 141 and 1940).

Little is known at present of the southward prolongation of the Nevadian geosyncline through Mexico and Baja California, but a Nevadian unconformity would appear to exist in these areas as far South as Sonora (see Stille 1936, p. 148 and 1940, p. 584). The same period of folding and intrusion is observed again in Cuba (cf. Waters and Hedberg, p. 16), and is also believed to occur in the northern part of South America.

Gerth does not consider it improbable that the area which he described as the "Zentralandine Sedimentationsraum" had been folded towards the close of the Jurassic, and is also of the opinion that a second phase probably occurred towards the beginning of the Upper Cretaceous (1939, p. 14). This view is not shared by Stille, who — though admitting that very few local Nevadian unconformities are found is this area — considers that an important part of the Andes was folded at a later date, i.e. during the so-called Andine or Subhercynian epoch in the Upper Cretaceous, but pre-Senonian. Stille asserts that the Anders are the only existing folded chains in which Subhercynian compression was manifested so strongly. He has given an extensive review of South American epochs of folding in his book of the year 1940. We will mention a few of the arguments which this author set forth. One of these was that the geosynclinal stage was accompanied by vulcanism (the tremendously thick basic porphyrite formation, also known as the diabase-melaphyr formation) up to Cretaceous time, and that plutonism occurred in the form of a multitude of granitic intrusions during and after the folding. It will be seen that the mechanism in this case is the same as that in the earlier folded Sierra Nevada of North America. On the other hand, Stille compared the eastern part of the Andes to the Rocky Mountains in North America. However, folding, volcanism and plutonism in this sector are of a considerably fainter type than that in the Sierra Nevada. Stille was of the opinion that there might still be evidence in Central America of the existence of a Subhercynian epoch, though no trace of this phase is found in the North American Cordillera. Conversely, neither an Austrian nor a Nevadian epoch of any importance has so far been observed in South America. Plate 3 illustrates Stille's view on this subject.

Burckhardt, including many others who followed his example, regarded the porphyrite-breccias which are distributed on such a large scale over the whole area of the South American Andes as products of erosion of a land mass situated in the eastern part of the Pacific. Gerth objected that such an assumption would be incompatible with all that is known at present on the subject of these rocks. He attempted to furnish another explanation of the origin of the formations (1935, p. 279–282). Born mentions an additional area — a "Pacific coastal mass" — which is presumed to have foundered during the Lower Tertiary (1932, p. 842). Part of the Cordillera, consisting of gneiss, granite and crystalline schists and extending along the Peruvian coast would then represent a fragment of this coastal mass.

Another investigator who dealt with these submerged borderlands was Schuchert (1935, p. 636). One of these masses is, for instance, "Choco", which is said to have constituted the western margin of the northern Cretaceous geosyncline of Columbia, supplying this trough of sedimentation with products of denudation.

In the opinion of Stille (1940, p. 616, 617) a "borderland" has to be considered as part of a geosyncline that was folded during some previous epoch of compression and subsequently acted as a geanticlinal ridge.

Folded chains which the coast intercepts are likewise found in Peru. The Peruvian coast turns north-west, and the ancient Cretaceous lines of the structure of the Andes curve around in a similar direction. The coast bends north at a certain point, but the structures continue to extend north-west and are consequently cut off by the coastline. Born outlines a short connection between the abruptly ending belts (1933, p. 399), but Steinmann surmises that this N.W. branch of the Andes might possibly even have extended far into the Pacific. The hypothetical constructions of Steinmann, von Ihering, Koto, Repelin and others were opposed by Schaffer.

The arc of the Southern Antilles, like the area of South America, had probably already been influenced by Subhercynian folding, but the Laramide phase appears to constitute the principal epoch in Patagonia and further south (Stille, p. 149).

Wilckens, too, who summarized the various geological data of this region (1933), asserted that the Andes continued towards Antarctica via the loop of the Southern Antilles, and based this conclusion on both the sedimentary and eruptive rocks. Nothing definite is known of the subsequent trend of the West Antarctic orogenetic zone. An attempt has been made to trace them via Antarctica to New Zealand with the aid of morphological data. This area covers no less than sixty degree latitude, of which no geological data are known! The reader is referred to the figures in the publications by Born (1932, p. 855, fig. 381) and Taylor (1940, 3, 2, fig. 1 and p. 5, fig. 2). The little that is known at present of the principal epoch of folding in New Zealand seems equally vague. Born regarded it as the fragment of a Late Cimmerian chain (1931, p. 763). Stille described it as an Austrian zone of folding (1924, p. 150). The various and conflicting arguments appearing in geological literature on the above matter caused some doubt to rise within me as regards the right age of the principal epoch, and I consequently turned to Dr. Marshall, the well-known authority on New Zealand geology, requesting him to enlighten me on the subject. Dr Marshall's honoured and extensive reply is reproduced below. It should first be noted, however, that Early Paleozoic strata, resting unconformably on crystalline schists and beginning with Ordovician sediments, occur in New Zealand. A second unconformity can be observed between the Matai series, which comprises Jurassic strata, "basal Cretaceous" (Benson 1924, 1, p. 128) and the later Oumara series. The Senonian fossils closely resemble an analogous fauna in New Caledonia, Grahamland, Antarctica and Chili (cf. Benson, p. 130, and Marshall). Faulting occurred during the Tertiary and Pleistocene, and was locally accompanied by very faint folding. Dr. Marshall wrote me as follows in 1939:

"I certainly think that the pre-Ordovician folding was less important than that of Late Mesozoic. This opinion is based on my view that the only folded Paleozoic rocks are in the northwest and in the extreme southwest of the South Island. I regard the schists of Marlborough (Blenheim) and of Otago as altered Triassic greywackes though some geologists think that they may be pre-Ordovician.

"As to the date of what I think is our critical period we have the following guides:

1. The youngest of the folded rocks are (*a*) Nugget point (*b*) Kawhia. At both of these place *Inoceramus* of a large deeply sulcated type occurs and in the latter ammonites which I have referred to the genera *Phylloceras* and *Aegoceras*.

2. The oldest of the rocks deposited after the critical period of folding are found (*a*) at Amuri Bluff on the east coast of the South Island, (*b*) at Kaipara and Whangoroa at the North Island (*c*) at Wangaloa, close to the mouth of the Clutha river (*d*) at Hampdee just north of Oemaru on the east coast of the South Island.

(*a*) The fossils found here have not been fully described. However, they definitely include saurians with nearly flat vertebrae, *Inoceramus*, *Belemnites* and a few ammonites; they seem to be late Cretaceous.

(*b*) The fossils at Kaipara include a number of ammonites which I have classified (Transact. New-Zealand Institute Vol. 56, p. 226). These ammonites seem to be close to the Trichinopoly fauna of Lower Senonian age in India.

(*c*) Wangaloa. The fossils from here I have described. (Transact. New Zealand Institute vol. 49, p. 450). A small degenerated *Belemnites* is the only Cephalopod. *Puguellus* and *Perissaloa* suggest a Senonian age.

(*d*) Hampden (Transact. N.Z. Inst. Vol. 51, p. 226). Heie theie are two Trigonias".

It cannot be decided for the moment whether the Upper Cimmerian, Subhercynian or Austrian epochs, or wether several of these phases are involved, for the most recently folded strata contain *Aegoceras*, an ammonite suggesting a Liassic age, and the oldest strata on top of the unconfoımity date from the Lower Senonian. A noteworthy feature in this area is that Upper Creaceous strata (to all appearances Lower Senonian) rest unconformably in several localities.

If it is correct to suppose, as Benson claimed, that Lower Cretaceous is still found in the Matai series, this would show that an Austrian or Subhercynian epoch is concerned. Dr. Marshall, however, observed that present data do not for the moment justify such a conclusion. The three phases which we mentioned above have thus been marked with question-marks on the map. Hence New Zealand may at any rate be said to constitute a fragment of a Mesozoic chain.

Along the eastern coast of Australia "immense pressures operated from east-north-east to west-south-west along the present Queensland coast from at least as far south as the mouth of the Brisbane River to beyond Broadsund" (David p. 86, 87). This process of folding (the youngest known so far in Australia) occurred in post-Cenomanian time. David was of the opinion that it might possibly be the Laramide epoch (p. 86), but its age has not yet been determined accurately, and it is therefore still possible that we are concerned in this case with the Subhercynian phase. Both epochs have consequently been indicated with question-marks (see Pl. 3 and Pl. 4). Mention should also be made here of a hiatus between the Triassic and Portlandian in New Caledonia. It may therefore be assumed that this region was influenced by either the Early or the Late Cimmerian epoch. The last intensive folding in this area, however, was of Tertiary age.

Cimmerian mountain-chains occupy large tracts in Asia. The following facts, which appear in a publication by Leuchs, to which have been added additional remarks from Lee, will illustrate this. The Triassic and Jurassic sediments are particularly thick between the nuclei of Angara and Tschuktschen and display folding of a very pronounced character. Powerfully overthrust-structures of the Alpine type are met with especially in the marginal areas. The Early

Cimmerian epoch, and a late Cimmerian phase, can be observed in the Werchojansk and Tscherski Mountains (Leuchs 1, p. 175, 177). This area continues around the south-eastern corner of the Angara shield and is then connected with China, where the principal epochs of compression in widespread areas have also been indentified as Cimmerian movements. The marginal chains of the nucleus of Ordos contain folding which is at least Variscian, but the final epoch of compression is the Late Cimmerian. Folding and overthrusting are even known to occur in the Upper Cretaceous (Leuchs 2, p. 95). Laramide and Austrian phases have been observed in the Nanling mountain-belts and all along the coast (Lee, p. 189, 190). Further south the Early Cimmerian epoch was particularly pronounced in Indo-China, and can be traced as far as Malaya. The same may probably be said to have constituted the final epoch in Banka Billiton and West-Borneo. The south of China is the area of the so-called mesocathaysian geosyncline (Lee, p. 212, fig. 61). Another geosynclinal area (this time in Eastern Siberia) has been mentioned previously. Cimmerian movements are even found in the massif of Angara, viz. in the Jennessei-Lena zone, where the sediments of the basin of Wilui were compressed in slightly undulating anticlines and synclines (Leuchs 1, pp. 166). Cimmerian unconformities are known to occur in several sectors of the Tethys zone, e.g. in Pamir (Leuchs 2, p. 155, 158), and Early and Late Cimmerian movements were observed in the Gulf of Karabugas, along the Caspian Sea (2, p. 9, 11). Late Cimmerian unconformities are found in a zone extending approximately parallel to the Ural (1, p. 200). Cimmerian folding has also been reported in Iran, Anatolia, Korea (Kobayashi) and Japan (Ozawa).

The Dobrutcha is the only part of Europe where no folding occurred after the existing Early and Late Cimmerian movements (Born p. 705). Early Cimmerian unconformities have been observed in other places (in the Carpathians and the Alps). A Late Cimmerian unconformity occurs in the Crimea, Caucasus and Apennines, and an extremely faint one in the Alps. Both epochs have been identified in the Balkans, and it is known at present that these unconformities do not only occur in the region Tethys, but that they are also found in France, Germany and even in other continents. To these areas should also be added the Gondwanide chains, which were dealt with above under the Variscides. The Cimmerian unconformities occurring in mountain-chains which were folded during the Tertiary will be found in Plate 4.

Plate 4. Cenozoic epochs of compression.

The North American Continent, besides furnishing a splendid example of mountain-chains that were folded during the Late Cimmerian phase (e.g. the Sierra Nevada), also provides a typical example of Laramide folding (e.g. the Rocky Mountains). Many overthrusts of blocks which were not folded to a very intense degree, all of which dip westwards, are met with north of the basin of Wyoming. Vertical movements along a set of normal faults dominate south of this area. The Laramide movements, however, are not confined exclusively to the Rocky Mountains. A great number of synclines and anticlines occur in the less mountainous region east of these belts. Some details are now known of the synclinal depression in, for instance, Alberta. This depression runs along the eastern part of the Rocky Mountains (Waters and Hedberg, p. 23, fig. 3). Laramide movements extending as far as and into the Coast Ranges, have also been observed in the western sector. Yet it is at times doubtful whether the movements involved are Laramide or post-Eocene (Pyrenean) epochs.

It is impossible to trace the exact southward course of the Laramide belt. Pronounced Laramide movements, accompanied by the formation of granodioritic batholiths, are known to have occurred in Mexico (Schuchert, p. 34, 35; see also Stille 1936, p. 148). The exact age of the folding and intrusions in many parts of the Caribbean area have not yet been determined. The age in Hispaniola, however, and in Jamaica, Cuba, Bonaire and Curaçao appears to be post-Upper Cretaceous and pre-Upper Eocene. Though sedimentation continues without a break from the Cretaceous to the Eocene in the geosynclines of Venezuela, the Andes, the Cordillera Oriental and the Sierra de Perija, a decidedly Laramide unconformity has been observed in adjacent areas such as northern Central Columbia (Waters and Hedberg, p. 28).

Stille mentions Cretaceous and Tertiary strata resting conformably (and folded locally) in the "Subandine", eastern part of the Andes of Bolivia and Ecuador. The same may be observed in the Cordillera of Venezuela, except in the coastal Caribbean Cordillera. Younger movements occurred in the northern area during Oligocene and Miocene time, and a series of smaller idiogeosynclinal basins (such as that of Maracaibo) were formed subsequently.

Two additional geological papers have appeared on the structure of the Antilles since Rutten's review (1935) and that of Sapper (1937; the papers referred to are those of Senn, 1940, and Weyl, of the same year). Senn —

unlike Suess, Rutten and Hess — assumes that a land or shallow water communication, extending from the Greater Antilles and the Curaçao Ridge, formerly connected the coasts on either side of the Atlantic Ocean. This link would have been shaped as a series of geanticlinal ridges in the Upper Cretaceous, and the ridges would have been influenced by Laramide folding. Senn argued that a connection, which assumed the form of "a submarine plutonic arc in the Lesser Antilles", was brought about between the Greater Antilles and the Curaçao Ridge during said period. The Pyrenean epoch of folding would have elevated this arc above sea-level, causing another volcanic arc to be created on its inner side. The previous connection between the Antillean and Mediterranean region would then have been intercepted. It should at any rate be noted that Weyl reports that a series of mountains turns around in San Domingo, coming to an abrupt end on the boundary of the present basin. It is hard to believe that the Greater Antilles would not have been connected with South America by means of a comparatively ancient arc-shaped structure. The post-Turonian (Subhercynian) epoch observed by Rutten and his collaborators in Cuba may possibly bear out this statement. In the meantime, it is still to be doubted whether another old connection had not existed between the Antilles and the Mediterranean. Senn and Gerth both drew attention to this question in 1940. The possibility of trans-Atlantic belts was dealt with in Chapter VIII. The data on epoch of folding in Plates 3 and 4 have been taken from publications by Rutten c.s., Sapper, Stille, Senn and Weyl.

Younger movements — particularly the Miocene and Pliocene — had more influence in the northern and central part of South America (e.g. cf. Gerth, 1939, p. 36, 52) than the Laramide revolution. Though the latter may possible have had some significance locally, Laramide folding certainly did not constitute the final epoch in these areas (see Gerth, e.g. 1939, p. 42, 53, 58, Stille 1940, pp. 587– 589). This phase, however, appears to have been the last one of importance in Patagonia, and also — probably — in the extension towards Antarctica via the Southern Antilles (Stille, p. 149, Gerth 1939, p. 28, Stille, 1940, p. 586).

All that is known of the younger epochs of folding in the Cordilleras of North and South America can be summed up briefly as follows.

The most western folded chains of North America are also the most recent ones. The zone which these belts occupy can be followed from the Californian coastal chains to the Klamath Mountains and the coastal chains of Washington Oregon, the Olympic Mountains, Vancouver Island and other islands along the coast of British Columbia, the Alexander Archipelago, the St. Elias Mountains (in which the strike may be seen to turn west, south of and parallel to the Aleutians and the Konyage Islands).

A series of Upper Mesozoic and Tertiary idiogeosynclinal troughs formed in between these ranges and the Sierra Nevada-Cascade-Alaska mountain-chains. Enormously thick deposits filled most of the troughs. In the Ventura basin the sediments attain a thickness of approximately 20,000 meters; the Pugget Trough contains Cretaceous and later deposits up to Oligocene, and the California Valley (which is still in a geosynclinal stage) 10,000 meters of thick deposits ranging from Miocene to recent strata.

The age of the unconformities and epochs of folding cannot be determined accurately as a whole. The Miocene and Pliocene epochs in California probably correspond with the Styrian, Attic and Rhodanic epochs in Europe and the last intensive phase probably occurred during the Pleistocene. Stille refers to this last epoch as the Pasadenian. A thorough and lucid description of the intricate structural history of California can be found in Reed and Hollister. Present data show that these recent areas of folding and these basins originated in regions that had already been subjected to folding during the Nevadian phase, and that they also reveal the influence of Laramide compression.

Movements were observed towards the close of the Eocene in many places in the vicinity of the Caribbean. Waters and Hedberg assert that the Cordillera Oriental, in the northern part of South America, and the Sierra de Perija had evolved during an Upper Eocene or pre-Upper Eocene epoch of folding and of contemporaneous elevation of that particular portion of the Andine geosyncline. The broad geosyncline of Venezuela and the Andes, extending as far as Peru, would have undergone many important changes during this period. "The broad Venezuelan-Andean geosyncline was broken up into the smaller, narrower Orinoco, Maracaibo-Falcon and Magdalena basins by the rise of the Cordillera Oriental and the Venezuelan Andes, and throughout much of northern South America Oligocene and Miocene sediments rest with angular unconformity on Eocene strata" (Waters and Hedberg, p. 39). One of Rutten's recent publications gives a clear picture of the extremely complicated history and structure of these areas (Proceed. Kon. Akad. v. Wet. Amsterdam, XLIII, 1940). A Lower Miocene movements known to have occurred during the subsequent history of the Orinoco geosyncline, but the most important Tertiary epoch in this area and the regions of the Maracaibo-Falcon and Magdalena basins dates from the Mid-

Pliocene (?Rhodanic epoch), "and they were largely raised above sea-level....". "The Bolivar geosyncline, in Western Columbia, was folded and uplifted into mountains" (Waters and Hedberg, p. 41). It was at this time, too, that frequent faulting affected the islands of the Caribbean, and this gave rise to the surmise that Barlett Trough, Anegada Passage and other exceedingly deep parts had begun to form during the above-mentioned period, which also witnessed the submersion of "Antillia".

Certain movements, consisting principally of a vertical elevation accompanied by a slight warping of the strata or folding of a widely undulating character are known to have occurred in Central and South America during the Pleistocene, and to have continued up to the present day. Elsewhere the process of subsidence and sedimentation is still in progress (e.g. in the basin of Maracaibo).

The areas of folding on the other side of the Pacific can be followed from Kamschatka to the East Indies via Sachalin, Japan, Formosa, the Riu-Kiu Islands and the Philippines, and a Tertiary area of folding may be traced eastward from the East Indies to the Fijis, and westwards to Europe and North Africa via Burma, the Himalayas, the Trans-Himalayas, the Karakorum, Pamir, Iran and Asia Minor. It would be impossible to describe the diverse epochs of folding of these areas exhaustively. We will consequently confine ourselves to a few remarks. In the Ka-akorum and Trans-Himalayan mountain-chains the final folding is the Laramide revolution. (Leuchs 2, p. 244, 247). Laramide folding (though not as a final phase) is also known to occur in many other Cenozoic belts (e.g. in part of the East Indies, New Caledonia? and the Fijis). A zonal migration of phases of folding was observed in several belts. In 1938 I attempted to indicate the position of these zones in the East Indies, and to trace their continuation to the Asiatic continent (Burma). An attempt to correlate the Tertiary stratigraphy of Asia (as well, consequently, as the Asiatic Tertiary epochs of deformation) with those in Europe generally gives rise to serious difficulties, and I have therefore refrained from using the names Stille gave to Tertiary unconformities.

However, in this connection the following should be mentioned. In an interesting paper by Reed (1937) some of the East Indian transgressions and epochs of folding were tentatively correlated to similar phenomena and sequences in California. Three distinct zones may be observed in the western part of the Indian Archipelago. They are (1) the areas of Malaya, the Riouw Archipelago, Banka and Billiton, all of which were subjected to folding during the pre-Tertiary (Early Cimmerian); (2) the idio-geosynclinal basins of Atjeh and Southern Sumatra, which were folded towards the close of the Tertiary (Wallachian?); (3) areas in which the most recent process of folding dates from the Miocene, viz. Western Sumatra and the group of islands west of this region (Styrian epoch?), where younger movements reveal very unpronounced folding, and much faulting along the surface. These three zones can be traced over the Asiatic Continent (in Burma, namely; for a further discussion of epochs of compression in the East Indies see my publication of 1938). One point, however, has to be discussed here since Brouwer and some of his students expressed some doubts regarding the influence of the Miocene epoch of folding on the island Timor.

More than one Tertiary epoch of compression — as defined by the present author — has been recognised in Timor by Brouwer and his collaborators. Remarkably enough this fact was not mentioned in their publications. On the contrary Brouwer only emphasizes that on Timor the period of severe overthrusting dates from pre-Miocene times. Why not? I would not have the slightest objection if it were demonstrated by convincing data. The more so since the only remarkable and puzzling point would be the exceptional position of Timor as compared to the most recent epoch of severe compression on other islands of the same belt. Therefore, we might expect to find the phenomenon amply demonstrated with the aid of accurate maps and convincing geological profiles. This, however is not the case.

In the first place Brouwer and his collaborators admit the activity of a more recent epoch of compression after the pre-Miocene folding. "The pre-Miocene structures", Brouwer wrote (Exped. Lesser Sunda Isl. IV, p. 376) "have been influenced by later movements, which have caused folding and faulting, later upthrusting or overthrusting has also been observed" (see also IV, p. 381). And one of Van West's illustrations even shows a "post-Oligocene overthrust structure" (Ibidem III, fig. 8) which he considers, however, "of local importance" (Ibidem III, p. 120). Tappenbeck, another of Brouwer's students, asserted that the Miocene was not influenced by strong folding, but he gives an illustration of steeply folded Miocene strata (Ibidem I, p. 56, fig. 12). Moreover on his map (p. 55 fig. 10) Miocene limestones show a dip varying between 25° and 50° and Miocene marls and limestones up to 70°, whereas the greatest dip of Eocene and Mesozoic strata on the same map amounts no more than 45°. Wanner found Miocene Foraminifera in the brecciated limestones of the zone of strong overthrust masses of the so-called Fatu-nappe.

Tappenbeck tried to explain away these facts (I, p. 51) but his argumentation is the reverse of what might be called convincing and he had to admit that some facts do not fit very well in his hypothesis (I, p. 102). At any rate it is clear, even from the few quotations just given that Brouwer and his students found several indications of a severe epoch of compression in the Miocene. It was mentioned in Chapter IV that several epochs of strong compression played a role in the history of the Banda geosyncline. The most recent epoch was in the Miocene and obviously this was stated again by Brouwer c.s. in the course of their local explorations on Timor.

In 1934 Chhibber published a geological review of Burma, in which all that is known of the structure of this area is set forth clearly.

Three separate physiographic zones occur in Burma. These zones (which also differ from a geological point of view) can be summed up as follows. (I) The area of the Shan Plateau, extending southwards to Tenasserim. (II) The so-called Central Belt of Burma. (III) The Arakan Yoma, bounded in the west by the basin of Assam.

(I) This first zone is composed of pre-Tertiary strongly folded strata, striking NS. The latest folded rocks date from the Cretaceous (p. 2 p. 210; Laramide folding?). This area was land during the whole of the Tertiary (p. 110, fig. 16, p. 211). The continuation of this Tertiary land is found in Malaya, the Riouw Archipelago Banka and Billiton, and as far as Borneo. Moreover, when discussing the pre-Tertiary history of the East Indies it was seen that folding occurred during an earlier period (the Early Cimmerian) in Malaya, etc., and not during a later (Laramide) period, as in the Shan Plateau. Burma's eastern zone is bounded further west (in the direction of the Central Belt), by a fault appearing in the landscape as a morphologically very fine fault-scarp.

(II) The Central Belt is sometimes also referred to as the "Basin of the Burmese Gulf" (p. 2). This Tertiary basin subsided to a considerable depth, and its sedimentation bore an intensive character. It constitutes a geosyncline of the same type as that of the geosynclinal basins in Sumatra, northern Java, etc. It should furthermore be noted that folding occurred simultaneously in both areas, i.e. towards the end of the Tertiary and the beginning of Pleistocene time (Wallachian?). Cotter is of the opinion that the present Gulf of Martaban constitutes a remnant of this Tertiary geosynclinal area (p. 212). In the same way, the East Indian basin of Atjeh and the basin of Eastern Java clearly continue below sea-level.

(III) The Arakan Yoma and the Naga and Manipur Hills represent an area of folding in which no Paleozoic sediments have been found, though Mesozoic formations (Triassic and Cretaceous) are known to occur in this region. This area was folded towards the end of the Cretaceous (Laramide epoch?) and thus already formed a barrier during the Eocene between the Burmese geosyncline and the Gulf of Assam (p. 3, 5, 210). It persisted as an ever-rising geanticline (p. 212), i.e. an uplift, during the Tertiary, in which three epochs of crustal movements can be observed, one occurring towards the close of the Mesozoic, a second during the Mid-Miocene, and a third during post-Pliocene time (p. 216). The question at present is: which is the corresponding zone in the Indian Archipelago? Chhibber writes in this connection: ".... continues southwards through the Andaman and Nicobar-Islands to Sumatra and Java" (p. 3, 5). The western part of Sumatra (the so-called Barisan geanticline), and the series of islands west of this region probably formed a single zone, which ought to be regarded as the extension of the Arakan Yoma. The formation of the two "geanticlines" (Barisan and the series of islands west of Sumatra) and their separation by a fairly deep deep-sea basin would then date from a very recent post-Pliocene period, and have accompanied the youngest movements of upheavel in the Arakan Yoma. This suggestion probably contains a great deal of truth, for, as stated above, the Tertiary epochs of folding concur in both Sumatra and the islands west of it. Chhibber describes five different zones of volcanic rocks in Burma. The volcanic strip in Central Burma, extending to Sabang by way of the island of Narcondam and Barren-Island and from there to the Lesser Sunda-Islands via Sumatra and Java is especially noteworthy.

These East Indian zones cannot be followed to western Asia and Europe beyond Burma, for there are too few data to guide us on this subject at present. Nevertheless, the little that is known of this matter gives us good reason to hope that it will become possible in the future. For several epochs of folding which are absent from the main chain, occurring in one special zone only, have already been observed in the Himalayas and Karakorum. This was made clear by de Terra (1936). This author states that after the intensive folding towards the end of the Cretaceous "the northern sector apparently continued to be land, the southern or Himalayan part resumed its geosynclinal evolution" (p. 865). Very pronounced Tertiary folding affected what remained of the geosyncline in post-Mid-Eocene time, but before the Lower

Miocene, and movements may also have occurred simultaneously with the Miocene folding in the East Indies, for de Terra goes on to say: "This orogeny may have continued to the end of the Burdigalian epoch (Lower Miocene), or at least it may have been locally revided at that time....". After this folding, which may be said to be "possibly subdivided into an Oligocene and a Lower Miocene sub-phase", the whole area of the Himalayas remained above sea-level. A marginal deep consequently formed in the Mid-Tertiary along the southern margin of the Himalayas, and the Siwalik-layers, which attained a thickness of many thousands of meters (these layers constitute the erosion products of the Himalayas and might be compared to the molasse of the Alps) were deposited within the trough. The layers were folded during the Pleistocene. The southern chains of the Himalayas were overthrust to the south, and "the Karakorum and adjoining regions suffered a broader uplift, during this process" (p. 867). The chronology of the various epochs, together with the fact that the latter only occur in particular areas, remind one of conditions in the East Indies. De Terra, on the other hand, drew attention to certain similarities between these regions and the Alps (p. 868).

A zonal succession of crustal movements is again observed in Iran, where the chains emerge from the narrow strip of strongly compressed mountains along the Pre-Cambrian massif of India, attaining a freer development in this region. The youngest movements in Iran have been reported in the outer or most southern chains. De Böckh speaks of a Wallachian phase (p. 155 in Gregory 1929), and also indicated the presence of Laramide, Pyrenean and Savian epochs (p. 156). The most recent formations in this sector consist of the Mesopotamian marginal deep and the Persian Gulf. The latter may be compared to the Siwalik trough south of the Himalayas. Former authors mention a Pre-Cambrian nucleus in Persia, which was supposedly surrounded by folded chains. De Böckh draws a parallel between this "nucleus" and the Pannonian Basin in Europe. Baiers' latest statements, however, show that there can be no question of the existence of such a massif. The idea of an Iranian "nucleus", or of the existence of the massifs of Gobi and Tibet, to which many authors formerly adhered, should therefore be abandoned, as the same is incompatible with present data. In 1939 Arni observed the following five structural units in the westward extention (Anatolia). From north to south we find: the marginal folds of Anatolia-Iran, the Iranides, Taurides, Anatolides and Pontides. The youngest epoch of compression i.e. the post-Miocene (Attic?) and post-Pliocene (Wallachian?) appear again in the southern area, while a hiatus characterizes all zones in the Oligocene (it is difficult to determine whether we are concerned with a Pyrenean epoch in this case, or a Savian). Other (Laramide and Late Cimmerian) phases, too, are known to occur in the northern Anatolides and the Pontides. The most recent paper on the structure of Iran is by Schroeder (1944). The massifs of Asia Minor probably affected the trend of the Tertiary chains. Leuchs (1938) assumed the existence of three separate nuclei, and referred to them as the Karic-Lydian, Lyakonian and Halys Massifs.

The older phases also had considerable influence on the formation of chains that branch off towards the Caspian Sea north of Persia. The Oligocene Savian epoch seems to have been responsible for the most intensive and also for the final process of folding in this area, while Early Cimmerian, Late Cimmerian, Austrian and Laramide unconformities, too, have been observed in this sector (Leuchs 2, p. 9, 11). Von Stahl asserts that the principal epochs of folding in the Caucasus are post Cretaceous (Laramide) and post-Miocene (Attic?). A Miocene (post-Burdigalian) unconformity has also been reported in this region.

The Caucasus ends on the shores of the Black Sea. The Dobrutcha is the only range that can be regarded as its continuation west of the Black Sea via the Crimea. However, Tertiary formations appear to lie quite undisturbed in this area (Born, p. 705). The most recent epoch of this region appears to be the Late Cimmerian.

I will deal briefly with the extension of the Tertiary chains of Asia Minor in the direction of Europe, for the Alpine chains in this last continent (especially the Balkan Mountains, the Carpathians and the Apennines) still raise many problems which can only be solved by an extensive study of the various details. Moreover a complete review of data on this subject can be found in one of Born's publications. The Karic-Lydian Massif may possible continue into the nucleus of the Cyclades in the Aegean sea. The pre-Mesozoic massifs of Pelagon and Rhodope are found in the Balkans. The Tertiary folded chains formed in the intervening area and around the massifs in the same way as the Carpathians settled around the Pannonian Massif. The latter has since subsided and now constitutes a basin filled with Late Tertiary deposits. Folding in this sector was affected by neighbouring Alpine movements and is not very noticeable. It seems highly probable that another massif, which lay above sea-level during the Pliocene and supplied the geosynclinal area of the Apennines with erosion products, foundered to form the Tyrrhenian Sea (Born, p. 714).

It is exceedingly difficult to reconstruct the original connection between fragments which are found to day in the surrounding areas of the Mediterranean. Their problematical character may perhaps best be illustrated by the many conflicting hypothetical constructions which are met with in geological litterature (see e.g. Born 1932, p. 722, fig. 284). All the epochs on the map have been adapted from Stille (1924).

According to Heybroek the Liban and Antiliban are horst mountains, whereas the Békaa is a graben. Both would belong to the great system of faults and rifts which form an extension of the African system along the Mediterranean coastal districts of Arabia. Folded chains, as supposed by Kober, Krenkel and Blanckenhorn, do not exist in these regions.

The Tertiary area of folding in Spitsbergen (Frebold) is not of a deep-rooted Alpine type.

Our next step will be to examine the northward and eastward continuation of the East Indian Alpine zones. It cannot be denied, however, that our inadequate knowledge of the Tertiary and pre-Tertiary history of the Philippines makes it very difficult to follow them north. Much more is known of Taiwan (Formosa). The backbone of this island consists of the so-called "Slate-formation", in which metamorphosed Lower Tertiary deposits were observed. These contain *Assilina*, *Discocyclina* and *Camerina*. It should be noted that a conformable sequence of Miocene and Pliocene is also apparent is this island. Yabe and Hanzawa reported the existence of foraminifera suggesting a Tertiary-f age in the lower strata of this series (*Nephrolepidina* was found together with *Miogypsina*, but no *Eulepidinae* were observed).

The exact correlation of the "Slate-formation" and the younger Tertiary series cannot be defined at present, but it is generally believed that they are separated by a hiatus. Yabe and Hanzawa expressed a similar opinion when they wrote that"....in the Neogene time (excluding the very early part), an island of Eocene rocks, intensely folded and variously metamorphosed, came to existence at the present site of the backbone range; the island was surrounded by the Neogene seas in which the sediments of the Kaisan-, Byôritsu- and Shokkôzan beds were deposited in upward succession".

This Tertiary folding must therefore have succeeded the Lower Eocene and preceded the "later part of the Miocene" (Tertiary-f). We find it impossible to indicate it more accurately. The history of the Riu-Kiu Islands corresponds in some respects with that of Taiwan. It is marked by the same period of post-Eocene regression and folding. Hanzawa assumed that a Stampian or Aquitanian epoch was involved in these islands, as in Taiwan (Hanzawa 1935), and that the same "Burdigalian" transgression, coinciding approximately with Mid Tertiary-f or Bebuluh transgression, affected both regions.

The Tertiary areas of compression clearly continue eastwards in New Guinea, though it is impossible to outline their exact course and extent, as there are too few data on this subject. There can be no doubt that the New Hebrides and the remaining fragment, situated towards the extreme east — the Fiji Islands — form an integral part of the continuation. The geological survey which Ladd published in 1934 concentrated particularly on the most southern of the two larger Fiji Islands, i.e. on Viti Levu. The geological history of this island contains interesting points of analogy when compared to that of the East Indian Archipelago, e.g. an Upper Tertiary-e (or Bebuluh) transgression, movements in the lower part of Tertiary-f plus g, which a former author — Brock — regarded as an important epoch of compression, but which Ladd preferred to classify as local disturbances (important movements in the East Indian Tertiary-f). Other examples are: (1) the thick deposits of marl of the Suva formation, containing many molluscs and foraminifera (dito in e.g. north Java and the eastern part of Sumatra) (2) the occurrence of important faulting towards the end of the Tertiary, which is generally supposed to have accompanied the foundering of large parts of Melanesia (the same phenomenon was observed in many islands in the Indian Archipelago, and is in this case regarded as a contemporary process of the formation of the deep-sea basins), and (3) intensive pre-Tertiary folding followed by a subsequent period of erosion, i.e. the presence of land which was first flooded by a shallow sea in Neogene time. This flooding occurred at an earlier date in other regions, e.g. in the area which is now occupied by Eua in the Tonga group (the oldest Tertiary formations in the East Indies rest unconformably over this whole area; they are of Eocene age in some sectors, but Neogene elsewhere, e.g. in the greater part of Sumatra).

All that is known at present of the New Hebrides has come down to us from earlier investigations. Mawson reported that "extensive submarine beds were accumulating above the folded Miocene series" (l.c., p. 471). Chapman's research has shown that these Miocene strata belong to the *Lepidocyclina*-bearing Tertiary deposits. According to Mawson the folds point in the direction of the New Caledonian "foreland". A series of tuffs and marls several thousand meters thick, which probably cover the youngest

Miocene and Pliocene rocks, appear to rest unconformably on the folded Miocene strata. Above this sequence we find coralline limestone which in some places has been elevated approx. 700 meters above sea-level.

Miocene folding in the New Hebrides may possible have occurred simultaneously with the movements in Viti Levu. Remarkable enough, there appears to be good reason to suppose that this same epoch of folding is also present in New Britain (= N. Pommeren). For "steeply dipping older Miocene sediments overlain unconformably by a gently-folded Pliocene series of foraminiferal sediments and tuffs" have been observed in this island, and older rocks were found in addition to these formations.

I still agree with Mawson that "if at any time the New Hebrides ridge formed continuous land connected in the north or elsewhere with other land-masses, these conditions are most likely to have prevailed in the early history" (Mawson, 1905).

It seems improbable that these Late Tertiary areas of folding would have continued towards New Caledonia and New Zealand, for the last intensive compression occurred at an earlier date in New Caledonia, i.e. towards the close of Eocene time (this area has since continued above sea-level, according to Piroutet), and the principal epoch of folding is also a far older one in New Zealand, as shown above. A second epoch of compression (the Cimmerian) succeeded the Triassic, preceding the Portlandian transgression. A third (Laramide) succeeded the Cretaceous and preceded the Lutetian transgression. A fourth and final phase, which had considerable influence, occurred after Eocene time (these facts are from Piroutet; see O. Wilckens 1925).

Plate 5. Chronological analysis of the continents.

The facts laid down in Plates 1–4 have been combined in Plate 5 so as to form a single picture. In those cases in which regions were subjected more than once to geosynclinal subsidence or folding, only the most recent epochs of compression have been indicated in each specific continental zone. Older movements and matters of detail will be found in Plates 1–4.

The general aspects of continental structure were outlined in Chapter II. Caledonian and later zones were dealt with in the descriptions of Plates 1–4. The vast Pre-Cambrian areas depicted on Plate 5 consequently remain to be discussed.

A division of the Pre-Cambrian foundation of the continents was recently made possibly by the determination of the absolute age of several rocks and the result is that we are now able to distinguish between at least some groups of folded belts. This subject was reviewed by A. Holmes in "The Age of the Earth" and by Wahl. Yet in spite of the promising nature of these first results it is not yet possible to write a more or less connected history of the Pre-Cambrian areas. In dealing with the structural history of the continents, we began, therefore, with the Paleozoic and for the moment merely wish to draw attention to the huge expanse which the Pre-Cambrian areas occupy. The latter are not only known to extend over the greater part of the continents (the Baltic-Russian Shield, Canadian Shield, Africa, Arabia and the largest parts of Australia, Antarctica and South America), but fragments are also observed in mountain-chains that were subjected to folding during more recent periods. Extensive areas are also known to lie beneath the sea.

The Pre-Cambrian basement occurs in the intervening area between the Cordilleras of West America (British Columbia, Colorado, etc.), and locally in the folded mountain-chains themselves, e.g. in the Bighorn Mountains and the Black Hills of the Rocky Mountains geosyncline. Pre-Cambrian rocks are also found in the more recent European belts, large fragments of which may for instance be observed in the Moldanubian zone of the Variscides. In the Scottish Highlands Pre-Cambrian rocks with a normal sequence of Cambrian strata may be seen plunging beneath the Caledonian overthrusts. Pre-Cambrian massifs presumably also exist in the North American Variscides.

In South America, Pre-Cambrian formations appear in the Pacific coastal areas of Chili and Peru, and are also met with between Trinidad and the most southern point of the continent in the Cordillera. The former expanse of Pre-Cambrian formations in South America was not only a far greater one than that of the present continent, but it also had a totally different shape (Gerth 1932, p. 80, 81). Pre-Cambrian rocks are also found protruding here and there from the mountain-chains and younger deposits in the basins of Australia and Africa.

A number of shields, situated between folded mountain-chains that formed around them subsequently, exist in Asia, as in Europe (two large shields and a few smaller nuclei are found on the European Continent, viz. the Baltic Russian Shield, stretching southwards to the massif of Ust-urt (Leuchs 2, p. 12), and the Barent Sea Massif with its south-eastern spur: the massif of Putkow Kamen). East of the Variscides of Nova Zembla lies the Kara Sea Massif. Its Pre-Cambrian basement may still be observed in Taimyr and a few islands. This massif, however, is covered for the most part by a shallow sea. The same applies to the massif of Tschuktchen (Leuchs 1, p. 180). Pre-Cambrian rocks from the shield of Angara (Asia's central nucleus) crop out in the Anabar, Baikal, Aldan and Jennessei horsts, while basin-shaped depressions characterize other parts of this shield. These depressions were filled with later sediments (e.g. the graben of Tundra and Lena, and the basins of Tungus and Wilui). The massifs in eastern and southern China, too, occupy a considerable expanse. It may probably be assumed that the basins of Tarim and Ferghana originally constituted nuclei, situated at a greater height and subsequently surrounded by the Variscian folded chains. These nuclei have since subsided and been covered by a thick layer of sediments. Mention should also be made of the Pre-Cambrian mass of India (and Ceylon), which unlike the areas previously cited, only played a passive role in the geological history of Asia (this had already been shown by Leuchs). Pre-Cambrian rocks are also found in the area between the younger formations — in the Paleozoic and later mountain-chains extending among these massifs (e.g. in the Kwenlun, Karakorum and Himalayas; see Leuchs 2, p. 189, 243 and 258).

The Barent Sea Massif is at present a shelf and lies at a depth of between 200 and 400 m. below sea-level. The course of a few rivers can still be traced in the shelf relief. This area (Frebold, Spitsbergen, p. 172, fig. 81), together with that of the Kara Sea Massif (Frebold, p. 171, see also Holtedahl's bathymetrical chart for the northern seas) lay above sea-level in comparatively recent times. In this last case, it is also possible, when east of Nova Zembla, to reconstruct the course of a few rivers in the submarine relief of the floor. The following indications lead Frebold to conclude that the Barent Sea Massif was probably part of a Pre-Cambrian area. In the eastern portion of North East Land the dislocation of the rocks of the Hekla-Hoek series gradually diminishes. The earlier movements and the Variscian and Tertiary epochs only occur in the western part of Spitsbergen. The eastern sector has not been influenced by these movements. And finally, it should be noted that Pre-Cambrian formations crop out along the coast of Putkow Kamen.

Plate 6 and Table I. Basins of the second group.

Opinions differ widely regarding the application of the names geosyncline, trough and basin, as stated in the introduction of Chapter III.

Orthogeosynclines in the sense of Stille are elongated trough-shaped zones of subsidence filled with sediments out of which mountain chains, in structural type comparable to the Alps, were born.

Many examples show that an orthogeosyncline may be divided into at least two zones. The central strip is characterized by the origin of basic to ultrabasic magmatic products in its initial stage of subsidence and by strong folding and overthrusting in its final stage. This zone is called eu-geosyncline. It often happens, however, that a narrow subsiding trough originates along one or both of its margins at a later stage. The contents of the newly formed trough — named marginal deep in Chapter III — are folded during a later epoch of compression.

This marginal geosyncline, the development of which forms a final stage in the evolution of an orthogeosyncline was named miogeosyncline by Stille, as contrasted to the central eugeosyncline [1]). Initial ultrabasic products are absent in the miogeosyncline.

If a subdivision into geanticlinal ridges and secondary troughs took place during the long enduring development of an orthogeosyncline the name polygeosyncline (Schuchert) may be applied to such a stage. An example given by Stille is the Variscian geosyncline of Europe. Another example is the Tethys-geosyncline of the Alps which Schuchert, however, classified as a mesogeosyncline. As shown by fig. 93 the Alpine strata formed in elongated basins that were separated from each other by geanticlinal ridges. Deep-sea sediments are found on some of the islands of the East Indies, and submarine troughs are now on the sites of former land areas. The present geanticlinal ridges, often showing thick accumulations of Tertiary and older strata, are now the source of clastic sediments which are transported into the submarine troughs. Therefore the present author tentatively classified the East Indian regions marked by notation I on Plate 8 and corresponding to the zone of strongly negative values of isostasy found by Vening Meinesz, as a polygeosyncline in the sense of Schuchert [2]). Temporarily, however, parts of such an area, if considered apart may show the characteristics of a monogeosyncline [3]).

The labile strips of the earth known as orthogeosynclinal zones were contrasted by Stille to the "kratons" i.e. the rigid areas that had a stable function during long eras. They are divided by him in two types (1) the highly situated continental areas and (2) the deeply situated ocean floors.

Troughs and basins originating within a kraton were named parageosyncline by Stille

[1]) Stille 1940, p. 15.

[2]) Umbgrove 1934, p. 145.

[3]) Umbgrove 1933, p. 41.

(1936). These are structures of more local importance, shorter duration and more limited and irregular shape in contrast to the orthogeosynclines. The Mesozoic and Cenozoic basins of Europe north of the Alps are classified as parageosynclines by Stille [1]). They were classified as basins of the second group in chapter III. The folding of the sedimentary contents of these basins and troughs is usually not of the Alpine type but is less intense and is called "germanotype".

However, it appears that after the folding of an orthogeosyncline, basins and troughs may also originate in this zone. Their sedimentary sequence may contain many thousands of meters including mostly neritic, hemipelagic and terrestrial sediments, and also volcanic tuffs and lavas. Their shape is elongated or irregular, their period of development comparatively short-lived and the folding not-intensive ("germanotype"). The limited extent, basin-shaped morphology, the site of the basins in earlier folded areas of a complicated polygeosyncline, the longest axis of the basins usually coinciding with the prevailing structural trend of basement and surroundings, the enormous thickness which the sediment-layers possess locally, and which they have obtained in a geologically relatively very short time and finally the type of folding induced the present author to introduce the name idiogeosyncline for intramontane basins of this type in the East Indies [2]). It appears that analogous intramontane geosynclines were called parageosyncline, too, by Stille [3]). This name, however, is a homonym. For among the basins classified as basins of the second group in Chapter III are examples of what Schuchert called parageosynclines.

A different classification of geosynclines introduced by Kay was mentioned in chapter III (p. 43 note [1]). It is clear, however, from the short discussion just given that the classification of basins and troughs of chapter III comprises many different types of subsided and subsequently folded areas that were classified as geosynclines by other authors. However, the eugeosynclines in the sense of Stille were excluded from our classification. These were discussed in chapter II.

Chapter III reviewed the different types of basins, but only a few examples of the second group were included in the discussion. These will now be dealt with in the order in which the basins have been numbered on Plate 6. The figures inserted between brackets refer to the corresponding numbers of the basins. Attention will also be paid to a few submarine basins towards the end of the Appendix.

It is not always easy to determine the chronological correlation of the formation of discordant basins and certain internal processes, for all kinds of difficulties arise when we attempt to do so. One is that few data are known of the stratigraphy and structures of the basins except rarely, and a second that their contents have in many instances only been folded very faintly and at times even to an almost imperceptible degree. We are consequently faced with the question whether the oldest sediments cropping out along the edge of such basins ought in fact to be regarded as the oldest depositary strata, as it might be that older deposits, which settled during the first subsiding movement, have since been covered unconformably and therefore hidden from sight by more recent layers. As subsidence is known to have been repeated in some basins after a pause in the first downward movement, the possibility of its occurrence in other less well-known basins should not be discarded altogether. The result is that the beginning of subsidence cannot be determined accurately in all cases. Still, it is apparently possible to indicate the large outlines of the beginning of subsidence in spite of these difficulties. Some may feel that the notations of a few basins on Plate 6 need altering. To mention one example — the Australian North and Desert Basins may have begun to subside during the Devonian instead of the Carboniferous and the beginning of the subsiding movement would then have to correspond with a late Caledonian instead of an Early Varician epoch. The history of the basins of western Germany and the North Sea probably began in Triassic (Rhetic) time, and some might perhaps wish to connect it as such especially with the Early Cimmerian phase. Yet marine sedimentation and that of the movement of the adjacent positive elements commenced as early as the Permian, and this may explain why I chose to connect their origin with a Late Variscian (Saalian) epoch. A variety of other examples might be given, but to my mind the existing facts — though very incomplete and presenting many difficulties — nevertheless corroborate the conclusions arrived at in Chapter III.

We will deal with North America first. The Canadian Shield contains a number of more or less extensive areas covered by Paleozoic strata. These should be considered to represent remnants of a veneer which was originally still much more extensive. That such remnants should be found in those areas in which they

[1]) Stille 1940, p. 8.

[2]) Umbgrove 1933, p. 36 and 1934, p. 158.

[3]) See e.g. Stille 1940, p. 12.

actually occur may be due either to a basin-shaped depression or to later down-folding or foundering along faults. Morley Wilson (in Ruedemann, 1939, p. 234) supposes that either the first or the last possibility, or a combination of both, affected the neighborhood of Hudson Bay and Southampton Island during the Mid-Ordovician, Upper Ordovician, Silurian and Devonian, and expresses the view that all three may have jointly been responsible for the preservation of the Paleozoic sediments in Ottowa Valley. The central basin of Hudson Bay (no. 1) reminds one very strongly of an original depression.

The foreland of the Appalachians and the Ouachitas consists of the "interior lowlands". The basement is composed of the prolongation of the Pre-Cambrian Canadian Shield, and is buried under Paleozoic and younger strata. The Pre-Cambrian foundation has a very irregular surface and is marked by many basins and domes or "uplifts". In the neighbouring area of the Variscides the most prominent examples of the positive elements are the Adirondack and the Cincinnati, Nashville and Ozark domes (see Plate 5). For further details the reader is referred to the tectonic maps of King, Longwell and Van Waterschoot van der Gracht. These include a clear survey of the intervening negative elements such as the basin of Pennsylvania (no. 5), which is directly opposite the Appalachians, containing a maximum of 6,000 feet of Carboniferous and Permian strata. Other basins are those of Michigan (no. 4), Illinois (no. 3), the large mid-continental basins (no. 2), etc. The anisochronous frame of the Permian Texas basin (no. 6) was mentioned in Chapter III.

As mentioned above, the Marathon-Ouachita Mountains probably form the frontal chains of a large Variscian belt which was buried under recent sediments of the Gulf Coast. These deposits assumed a semi-circular basin-shaped structure (no. 7) and include a sequence which originated in the Lower Cretaceous after the Nevadian epoch of folding. It continued up to the Pleistocene with many a hiatus, the most important being the Laramide unconformity. The thickness of these strata, which continue under the sea, as far as the edge of the shelf varies between 500 and 8,000 m (Stephenson 1939, p. 530, Pl. 1). While sediments were thus elevated above the sea-level in the Gulf Coast area and faintly folded and intersected by numerous faults, the central part subsided in the recent geological past, ultimately forming the Gulf of Mexico. The Gulf Coast basin and Gulf of Mexico probably originated on a Variscian basement and should therefore be classified as formations of Type IV. The Gulf Coast basin was formed after the Nevadian phase of folding, and its subsiding tendency was resumed fairly recently (Gulf of Mexico).

Two large basins occur in the Pre-Cambrian block of South America, viz. the Amazon and Parana basins (no. 8 and 9). The first originated after the process of Taconic folding (Gerth, 1, p. 96), and includes unfolded Gothlandian up to Carboniferous. The basin of Parana, however (see Gerth, 1932, 1, p. 144), was formed after a late Caledonian epoch. It was filled with Devonian strata; Carboniferous deposits are absent. Its subsidence was resumed towards the close of the Paleozoic, resulting in the deposition of Upper Permian and Triassic strata, and this was succeeded by large outflows of lava. The sediments have not been affected by folding and total approximately 2,000 m.

The most northern European basin, i.e. that of the Barent Sea (no. 10), has an anisochronous frame, as shown in Chapter III.

The area of subsidence in the north-western part of Europe extends E.W. towards England from the Russian Shield, and NS. to Belgium from Denmark. This area extends along the edge of the massif of Brabant in Belgium. Three separate basins are observed here, each one having a more or less distinct history. Ridge-shaped elevations of the Paleozoic basement with a WNW-ESE. trend emerge in the intervening areas. N.-S. and NNE-SWS. trends too, are found in the later Mesozoic and younger history as faintly undulating structures, faults and graben. The horst and rift-forming movements of blocks in the foundation were also accompanied by simultaneous folding of the sedimentary layers with a general WNW.-ESE. trend (the so-called Saxonian folding in the Jurassic, Cretaceous and Tertiary). Typical examples are the axes of the Teutoburger Forest and of Egge-Osning, the anticline of Bray, and that of the Artois-Boulonnais in the basin of Paris. The prolongation of the last anticline is found in that of the Weald (S. England).

In north-western Europe, subsidence is known to have begun in the Upper Permian (Zechstein), but the division into separate basins dates from the Upper Triassic. The basin of north-east Germany has since then been separated from that of north-west Germany by the so-called axis of the Elbe, or Pompeckji axis. The geanticline of Erkelenz is in turn situated between the basin of north-west Germany and the most western Tertiary basin of the North Sea. Its most western boundary is formed by the Pennine axis in England.

The basin of north-west Germany (no. 12), which is described by Stille as the "Niedersächsische", originated in Rhatic time and contains an almost complete sequence of marine

Jurassic and Cretaceous. In northern Hannover the sediments amount to some 6,000 m excluding Permian and Triassic strata (the Triassic total at least 1,300 m, while the Upper Cretaceous in the basin of Munster — which was formed during the Late Cimmerian epoch — attains a maximum thickness of more than 1,400 m.) The southern margin along the edge of the "Schiefergebirge" of the Rhineland and Harz and "Flechtinger Höhenzug" consists of a series of faults.

The sediments are considerably thinner in the basin of eastern Germany (this is not indicated separately on Plate 6). Moreover, marine Jurassic and cretaceous strata are absent. This lead Van Waterschoot van der Gracht to doubt whether it were possible to speak of a basin in this case (1938, p. 1379).

The basin of the North Sea (no. 11) was filled by a sedimentary sequence ranging from Zechstein deposits to Pleistocene. The older formations are only known from deep borings along the edge. A fine illustration of the subsidence during the Tertiary and Pleistocene is found in a map of Van Waterschoot van der Gracht of 1938.

The south-western corner constitutes the basin of London, but the latter only originated as such from post-Upper Cretaceous time (Laramide epoch) onward. It contains a comparatively thin veneer of sediments (approx. 500 m).

In the opinion of Van Waterschoot van der Gracht the post-Carboniferous sediments of the north-western sector total about 7,500 m or possibly even 9,000 m.

The phases of movement observed in the above basins are chronologically related to those appearing in the folded mountain-chains. The subdivision of the large north-western European area into three basins, each separated from the other by SSW.-NNE. geanticlines, was due to the Early Cimmerian epoch. The Late Cimmerian phase (which Stille subdivides into three further phases: the Dilster, Osterwald and Hils epochs) was responsible for the formation of the WNW. anticlines of Germany and England and the basin of Paris. The Paleozoic structure of the basement may be said to be reflected in these "folds"; see fig. 35 (in an extensive publication, Stevens showed that the present geomorphology of Belgium is still influenced by the structure of the foundation). The horst-shaped elevations of Harz, Osning, etc. where formed during the later Subhercynian epoch. (Stille divided the latter into the Ilseder and Wernigerode phase). It was this Subhercynian epoch that raised the salt-domes in the basin of north western Germany. The Laramide epoch then caused many new movements to occur along the existing fault-lines and also produced many new faults (the same holds good for the basin of Paris, and the district of Mons, in South Belgium).

It should finally be noted that Tertiary movements are responsible for various unconformities and for the large rift-valleys of the Rhine, the Rhone, the Limagne and the Central Graben of the Netherlands. These originated largely in the Oligocene (Pyrenean, Savian? epochs). Important movements occurred along these fault-system and the newly formed faults at the end of the Upper Miocene (Attic epoch?) and at the end of the Pliocene (Wallachian epoch?). In an area like that of southern Limburg, where movements of a group of horsts and graben occurred along N.W. faults as early as the Mesozoic, Tertiary movements generally took place in the opposite direction.

Another group of large rift-valleys is found on the other side of the Alpine chains, viz. the huge fault- and rift-systems of Africa, the Red Sea and Western Arabia (fig. 140). Both are accompanied by vulcanism. Reference was made to these graben in Chapter XI.

The above broad outlines of the general principles of the formation and deformation of the North-European basins will now be followed by a brief summary of the remaining basins.

No. 13′ — the basin of Paris — was discussed as an example of a discordant basin in Chapter III. No. 19, on the other hand — the Pannonian basin — was discussed as a type of nuclear basin.

No. 14 is one very similar to the basin of Paris and known as the basin of Aquitania. This area also shows the influence of numerous phases of movement corresponding with epochs of compression in the Pyrenees and the Alps.

The basins of the Iberian Peninsula (the Ebro-, no. 15; Duro-, no. 16 and Tajo basins no. 17) are considerably younger (Born p. 662, fig. 249). The Ebro Basin originated during the Laramide phase and was filled with Tertiary sediments, beginning with Eocene strata, while the Castillian basins of the Variscian Meseta were filled chiefly with Oligocene and Miocene deposits.

Five basins may also be observed in the area of the Baltic-Russian Shield. The Pre-Cambrian basement crops out in Fennoscandia and the massifs of Podolia and Woronesh. This basement is covered by a relatively thin layer of sediments in the massif of the White Sea and the East-Baltic Ufa and Ust-urt Massifs. The Polish (no. 18), Moscow (no. 20), East-Russian (no. 21) and South-Russian basins (no. 22), together with the area of subsidence of the Caspian Sea (no. 23), originated in between these positive elements. The first three are the oldest. They were filled with Paleozoic sediments; these amount to a maximum of 850 m in the basin of Moscow (Devonian up to Permian strata) and a maximum of 1,000 m in the East-Russian basin (Carboniferous up to Permian). The Paleozoic sediments are covered unconformably by Mesozoic strata (v. Bubnoff, Europe, 1, p. 194 and 203). The contents of these basins were affected by feeble folding, resulting in weakly undulating anticlines with an approximate NS. trend coinciding in origin with the epochs of folding in the Ural. These same anticlines, however, were also influenced by more recent movements.

The Polish basin (no. 18) formed in the Caledonian period according to Von Bubnoff (Europa 1, p. 197, 198, fig. 47 and 48), but the subsiding tendency migrated in a SSW. direction during the post-Jurassic. The Upper Cretaceous sediments rest unconformably on the older sequence and the younger sedimentation consists of Oligocene and Miocene deposits.

The depression of the Caspian Sea (no. 23) forms a southward prolongation of the East-Russian basin (no. 21), as it were, and contains sediments ranging from Permian to Upper Tertiary strata totalling some 2,400 m (v. Bubnoff Europa, 1, p. 194). Faint folding occurred in the Upper Triassic, Lower Turonian and Lower Senonian. Its subsidence must have been particularly intensive during the Maastrichtian, and the bottom of the depression, as shown by the deposits of cocolith-ooze, must have lain more than 1000 m below sea-level (fig. 40). Where the depression intersects the Tertiary folded chains, there are clear indications that the movement was continued until the Late Cenozoic. The present shape of the Caspian Sea might be compared to that of a double basin, of which at least the northern half may be said to date only from the Pleistocene (Leuchs 2, p. 16, 23).

On one side the Caucasian Mountains come to an abrupt end along the Caspian Sea, and these chains are intersected on the other side by the coast of the Black Sea (no. 24). I have found very little as regards the history of this basin in geological literature, but there is no doubt that it should be classified as a basin of Type IV, for it intersects the chains of the Crimea on the northern side and the mountains of the Dobrutcha at the western extremity.

Von Bubnoff is of the opinion that the South-Russian basin (no. 22) north of the Donetz Ranges originated in the Lower Liassic (Europa 1, p. 213). Nevertheless, its subsidence is known to have been of an intensive nature in the Senonian, Upper Eocene and Oligocene. Several Mesozoic phases are observed in its history, one of the strongest being the Laramide (ibidem, p. 214). This same investigator writes (p. 208) that "Man hat den Eindruck als hätte sich nach Auffaltung des Donetzgebirges die Südrussische Senke gleichsam als Fortsetzung und Kompensation für den unterbrochenen Senkungsvorgang ausgebildet". This basin might indeed have to be regarded as a marginal deep of the Donetz Ranges, but its shape and extension over the north-western part of the Donetz Ranges, together with the absence of such a fore-deep along the Ammodetic chains, would render this classification a somewhat doubtful one.

The formation of the basins of Europe shows no direct correlation with the areas of folding in which they are situated. The basin of Paris extends inside the Variscian zone. Outside the latter lies the North Sea Basin, the foundation of which consists of Caledonian and older zones. Both have sunk transversally into existing folded structures of different periods. They are not confined to any given zone of folding, and the time of the initial formation differs but slightly in each case.

Attention will now be paid to the basins in other continents, particularly to their position and the time of their origin. We will first take a few examples in Africa. Krenkel (p. 560) writes that the site of the Karroo basin was already occupied by a depression before the beginning of sedimentation in the Karroo-beds. This depression might have formed "als Begleiterscheinung der ältesten Phase der Kapidenfaltung". The phase to which Krenkel refers here (p. 593) is the Upper Carboniferous. The Southern margin of the basin was later folded by the Triassic epoch of the Cape Mountains. The latter had probably been much larger during Carboniferous time than the fragment of the chain is at present, as part of these ancient belts have foundered in the ocean since then.

Sedimentation in the Karroo Basin consists of the so-called Karroo system, i.e. a sequence of strata totalling as much as 6,700 m (Krenkel, p. 805). On this rests as much as 1,400 m of volcanic rocks. The sediments range from the Carboniferous Dwyka series to the Triassic Stormberg-beds. An analogous series is found in the Congo basin (no. 27) but the sediments appear to be far thinner in this last area. This should be attributed to the different ways in which the basins originated. Veatch (who published a memoir on the Congo Basin in 1933) writes that "the present Congo geological basin is not an original basin of deposition, but a basin of subsequent formation". During the deposition of the Middle and Upper Lukuga beds, representing the Permian Ecca and Lower Beaufort-beds of South Africa, the Congo region had a topography of no great relief, sloping in general from NW to SE.

"Following the deposition of the Lower Beaufort (Upper Lukuga) Permian beds, this region was subjected to the disturbances which marked the end of the Permian and continued through the Early Triassic, and which on the one hand produced a great east-west mountain range in what is now the ocean south of the Cape and on the other the uplift, with associated faulting and folding, which resulted in the general absence of the Middle and Upper Beaufort (Lower and Middle Triassic) in Africa north of about the 26th Parallel, except near the present east coast. During this Early Triassic erosion period all the Lower Beaufort, Ecca, and Dwyka beds (that is, the Upper, Middle, and Lower Lukuga of the Congo) were removed by erosion, except where they had been infaulted or infolded.

"Incidents of this period of folding, faulting, and erosion were the production in the Stanleyville region of the Congo and the Cassange region of Angola, in the eastern and southwestern part of the present Congo Basin respectively, of local fault basins, which may appropriately be called early rift-basins".

The region was subjected to peneplanation which produced successively the Mid-Cretaceous peneplain, the Miocene peneplain, and the end-Tertiary peneplain. "The present Congo hydrographic basin is entirely post-Miocene and owes its origin to the uplift and warping of the Miocene peneplain".

The present shape of the Congo basin (like that of the basin of Paris, see Chapter III) differs entirely from its original tectonic design.

The formation of the plate-shaped depression (no. 28) of the Kalahari seems to date from post-Upper Cretaceous time and to contain strata from Tertiary up to recent (Krenkel, p. 677). A still younger formation is the basin of the Tshad (no. 25). This contains Upper Tertiary (Miocene? or Pliocene, see Krenkel p. 1379), Pleistocene and recent deposits. A boring in the Late Tertiary or so-called Tshad-beds revealed that the thickness of the latter amounts to 100 m (Krenkel, p. 1384). Another very young formation is the basin of Ghasal (no. 26); see Krenkel p. 16, 132.

The most northern basins of Asia — the Kara Sea Basin (no. 30), and that of Tschuktchen (no. 31) — were discussed in Chapter III.

The basins of Tungus and Lena-Wilui are situated in the Angara shield. Cambrian and Silurian were deposited in the basin of Tungus (no. 33), but no Devonian or Lower Carboniferous strata have been observed here so far, though Upper Carboniferous and Permian deposits are known to occur in the area. Very faint folding marked the close of the Paleozoic, but the beds on the western side were folded during the Caledonian and Variscian revolutions. It would consequently seem that a subsiding tendency occurred on two occasions, i.e. one in the Cambrian and a second in the Upper Carboniferous. Two subsiding movements are known to have affected the basin of Wilui (no. 34). The first began in the Cambrian and continued up to the Permian, with an epoch of folding towards the Upper Cambrian. The second occurred in the Jurassic (Leuchs, 1, p. 68, fig. 6 p. 87 and 167, fig. 54).

One of the most important basins of the Asiatic Caledonides is that of Gobi (no. 40). Its foundation was subjected to intensive folding and forms the continuation of the surrounding chains (Leuchs 2, p. 72). This basin began to subside in the Lower Cretaceous, after the late Cimmerian epoch. The series of non-marine deposits (totalling approximately 4,000 m, see Leuchs 2, p. 78) continued up to recent times and shows a faint Laramide unconformity.

A much older basin is that of Minussinsk (no. 35). It was depressed into the arcuate structure of the Caledonian East-Sajan and the Kusnezki-Alatau (Leuchs 1, p. 151, 152). Marine Devonian sediments were deposited in the basin after the Late Caledonian movements, and were accompanied by continental Old Red, Carboniferous and coal-bearing Upper Paleozoic strata. These sediments reveal the effects of only faint Variscian folding. The Carboniferous strata amount to 1,000 m (see Obrutchew p. 442 for the preceding details).

The basin of Urjanchai (no. 36) originated in the Devonian contemporaneously with the Minussinsk basin. Deposition of marine sediments continued up to the Lower Carboniferous. The later deposits, however, (Permian and Jurassic), were terrestrial. "Die Faltungen und Brüche wurden durch periodische Senkung und Druck der Umrahmung geschaffen" (Obrutchew, p. 445).

The basins of Umrutschi (Leuchs 1, p. 123, fig. 42) and Chikuching subsided transversally over the folded chains of the Variscian Tianchan. Lower Permian up to Jurassic sediments of neritic up to terrestrial facies were deposited on their basements (these two basins do not appear on Plate 6, as they are too small for insertion).

Type IV must be considered to include the huge basin-shaped depression of Western Siberia (no. 32). This extends parallel to the Ural and in the south joins the northern part of the Kysylkum and of the Lake Aral district (Leuchs 2, p. 19). The foundation of the West-Siberian Basin, which was probably folded during the Variscian epoch, is not exposed. Marine sedimentation is known to have occurred from the Upper Jurassic up to the Tertiary (Leuchs 2, p. 32).

Too little is known of the history of the basin of Dsungarei (no. 37) to be able to determine with any amount of certainty whether this formation should be regarded as a nuclear basin (as that of Tarim) or a type of basin analogous to that of Gobi. The last alternative seems the most probable one, since remnants of the prolongation of the Sudetic Tarbagatai presumably still protrude here and there. Steep overthrusts bound the basin of Dunsgarei towards the south (the Variscian Tianschan) and the north (the Caledonian Altai). This basin presumably originated as such after the process of Variscian folding, but this is still uncertain. No mention will be made here of smaller basins such as those of Krasnojarsk, Tschulym and Ubsa Nor, for there are as yet very few data on these formations.

Deposition in the basin of Tarim (no. 39) began in the Permian or Upper Carboniferous and continues up to this day. The neritic but chiefly terrestrial deposits are presumably extremely thick and show several unconformities and "germanotype" folding.

The subsidence and sedimentation of the massif of Ordos (no. 41) was particularly important in the south-east (basin of Shensi), where a series of terrestrial strata ranging from Permian to Cretaceous accumulated to a thickness of 4,400 m (the subsidence subsequently ceased). Movements occurred after the Permian. The later deposits lie undisturbed and were consequently uninfluenced by the Cimmerian phases affecting the Variscian chains around the massif and resulting in marginal thrusts in the direction of the basin. Analogous marginal fault-planes also frame the basin of Ferghana (no. 38). Its contents of Cenomanian up to Neogene strata about 2,000 m (Leuchs 2, p. 136), but it is not yet known when subsidence began, nor is anything known of the real thickness of the sediments. This last also applies to the deposits in the Kara Sea (no. 30) and Tschuktchen Massifs (no. 31), for both are largely covered by a shallow sea.

The "red basin of Szechuan" (no. 42) probably ought to be included in this group, for it is framed by folded chains which were overthrust during the Cimmerian epoch, while marine Cambrian and Ordovician strata — though no Silurian, Devonian or Carboniferous sediments — have been shown to be present within its foundation. The Permian deposits are not very thick (400 m), but the presence of sediments ranging from Lower Triassic to Cretaceous and attaining a thickness of 5,000 m has been reported. The contents of the basin (including pre-Triassic basalt) were folded either towards the close of the Cretaceous (Laramide folding) or at an earlier date, i.e. towards the close of the Jurassic. A fainter and older Cimmerian phase was observed in addition to the above movement (Lee p. 174, 235).

Two of the five larger basins in Australia — the Desert (no. 44) and North (no. 43) basins — have been filled with several thousand meters of Carboniferous and Permian strata (approx. 3,000 m in the Desert basin, cf. Clarcke p. 33). These receptacles are quite far removed from the Paleozoic belts, but the great Artesian basin (no. 45) occupies the marginal areas of the Variscides as well as part of the marginal zones of the Caledonides, all of which have been peneplained to a considerable degree since their formation. This Artesian basin contains on an average 1,000 m of Jurassic and Cretaceous sediments, but the latter sometimes total 2,300 m (Andrews, p. 122). The Murray River and Eucla basins (no. 47 and 46), with their Miocene and Pliocene deposits, are even younger than those already mentioned (David, p. 121).

Teichert recently announced that the North Basin (no. 43) not only contained Carboniferous and Permian deposits, as was formerly supposed but that successive Mesozoic transgressions had also left deposits behind them in this sector. A continuous sequence of sediments, ranging from Upper Cretaceous to Upper Tertiary strata, was deposited subsequently. These were only folded very slightly towards the close of the Tertiary, and Teichert consequently regards the North and even the Desert basin (no. 44) as part of a geosyncline representing a branch of the pre-Tertiary Banda geosyncline which I reconstructed in the eastern sector of the East Indian Archipelago. I would like to point out, however, that so far as the Desert basin is concerned (1) the Upper Paleozoic sediments observed in this region consist mostly of estuarine and fluvio-glacial deposits, and that only one part of these sediments is composed of marine deposits (Clarke, p. 33); (2) that the sediments were deposited basin-wise, and that the geosynclinal shape of a trough is therefore lacking in this case; and (3) that the sediments have not been affected by folding, and that I consequently find it impossible to follow Teichert when the latter attempts to connect the Desert Basin with the Banda geosyncline in the East Indies. Although the marine fauna proves that an open sea-connection existed between the two areas, this does not mean that they were connected by a geosyncline.

Though sedimentation in the North Basin continued during the Tertiary, I find it equally impossible to determine its geosynclinal character in this case. For (1) a Paleozoic basin of the same type as that of the Desert basin appears to have existed in this area; (2) rare evidence of apparent transgressions has been observed in the Bajocian and Upper Jurassic, and these do not in themselves constitute an argument in favor of geosynclinal conditions; (3) Neocomian strata appear to be absent, and transgression of Albian, Cenomanian and Santonian seem to be present; a complete series ranging from Upper Cretaceous to Pliocene

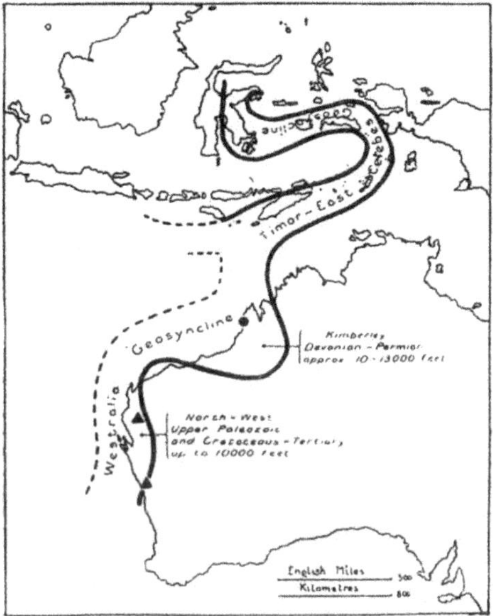

Fig. 203. The "Westralian geosyncline" of Teichert and its supposed connection with the Banda geosyncline in the East Indies (After C. Teichert).

deposits devoid of any influence of Laramide and Miocene compression, was deposited. Hence this area cannot be connected with the above East Indian belt. Still, Teichert's publication is important in so far that it mentions the occurrence of Cretaceous up to Tertiary sedimentation in said area, a sedimentation which was subjected to faint folding towards the close of the Tertiary. This "slightly folded" (p. 86) sequence of sediments may possibly explain the conspicuous and relatively intensive negative anomaly of the isostasy observed by Vening Meinesz in this region.

The above enumeration will now be followed by a brief examination of a few deep-sea basins. Born, Leuchs, Lawson, and others are known to have not only regarded the East Indian basins as young depressions, but to have also held the same view as regards the other basins along the eastern margin of Asia. Leuchs (1, p. 209) and Born (1932, p. 747) both even assume that the arc of the Aleutians and Tertiary strata of Alaska and Kamschatka surround a foundered pre-Tertiary "massif", part of which is at present buried beneath the sea. For a further discussion of these regions see Chapter VII, p. 203.

The Mediterranean consists of a series of basins, the largest being the basin of the Baleares (situated between Spain, the Baleares, North-Africa, Sardinia, Corsica and the southern coast of France), the Tyrrhenian basin, the Adriatic and the eastern part of the Mediterranean: the Ionic Sea, which is known to contain several deeper portions (see v. Seidlitz, p. 17, fig. 3). All these areas foundered in comparatively recent times and I cannot therefore endorse the view that part of the Mediterranean represents a remnant of Tethys (cf. v. Seidlitz, p. 57, 58). The recent origin of the basin of the Baleares can be deduced from the abrupt expiration of the folded chains of the Baleares, Pyrenees and Maritime Alps. The missing connections between these Alpine chains seem to have foundered. The Baleares basin must have formed after these Alpine chains had been folded, and must consequently be at least of post-Oligocene age. Fallot (v. Seidlitz p. 35) is of the opinion that the basin was formed in Miocene time. The Tyrrhenean area still lay above sea-level in the Pliocene (Born 1932, p. 711-714) and would thus constitute a basin of very recent date. This was also concluded by Suess (2, p. 349) and by Coster (1945). The trend of the folded chains in the Apennines and Corsica seem to indicate that the Tyrrhenian basin should be classified under Type IV. The comparatively shallow Adriatic may possibly have had the same origin and may perhaps be of the same age (cf. Born p. 713). It cannot be said for the moment whether it is connected with the basin of Po or whether the latter should be regarded as a marginal deep of the Alps. It is equally difficult to estimate the age of the eastern sector of the Mediterranean. Von Seidlitz holds (p. 365, 366) that the Aegean Sea began to form in the Miocene. The subsiding movements however, occurring along a series of faults, would have been particularly pronounced in the Pliocene and Pleistocene, and the eastern part of the Mediterranean, too, was very probably transformed into a deep-sea area in Miocene time (cf. also Suess 2, p. 353).

The Caribbean Basin and such deep graben as the Bartlett and Anageda troughs probably originated in the Mid-Pliocene contemporaneously with the occurrence of many faults on neighbouring islands and folding of several idiogeosynclines in northern South America (Orinoco, Maracaibo, Falcon, Magdalena and Bolivar).

With the aid of the character of the sediments and the trend of the folded chains intersecting the present coasts of the respective islands, Rutten showed that many deep-sea regions in the Antilles only formed after Cretaceous time and that some parts even originated quite recently (1934, p. 556–558). Still, the Caribbean can be subdivided into at least five separate basins and troughs (with an additional submarine graben: Barlett Trough), and marginal deeps extend along the external side of the Antilles. Several preliminary remarks on the origin of these formations and their correlation to geological and gravimetrical data appeared in papers by Hess in 1937 and 1939, but the final publication has not yet gone to press. We will therefore refrain from entering into a detailed examination of this highly interesting region, which is in many respects so similar to the area of the East-Indian Archipelago.

Born (1932, p. 836 and fig. 366 on p. 830) wished to regard the Caribbean as a foundered block known as Antillia, around which the folded chains would have settled later. The facts put forward by Rutten, however, make it doubtful whether this area may be assumed to represent a structurally homogenous unit.

An analogous and recent area of subsidence is also found inside the so-called arc of the Southern Antilles. The South Georgia and South Orkney-Islands, both of which were well-known to Holtedahl, were described thus by this author: "Die grossen Mengen von klastischem zum Teil grobklastischem terrigenen Material, das in diesen ganz schmalen Gebirgsketen vorkommt (das vulkanische Tuffmaterial natürlich nicht mitgerechnet, weil es ja aus der unmittelbaren Nachbarschaft stammen kann) setzen ein Muttergebiet im jetzigen Tiefseegebiet voraus" (1930, p. 56).

I will confine myself to the above examples since I have no intention of attempting to give an exhaustive survey of deep-sea troughs, basins and submarine furrows. Of course, other examples might very easily be added — for instance: the marginal deeps of Eastern Asia, Western India and the Southern Antilles, the relief of the foundered part of Melanesia (indicated schematically on Pl. 6), etc.

All deep-sea basins discussed so far are either situated inside or immediately alongside continental folded chains and island-arcs. It should be remembered, however, that basin shaped structures are also found in the floors of the large oceanic sectors proper. These were dealt with in Chapter VIII.

Plate 7. Magmatic clans.

Plate 7 shows the distribution of magmatic clans. The data are simplified and redrawn after a map published by Rittmann in his book "Vulkane und ihre Tätigkeit". The distribution of magmatic clans is treated on pp. 70–75.

Plate 8. Geology and Gravity in the East Indies.

A coloured map showing the relation between gravity field and structural history of the East Indies was published by Vening Meinesz and Umbgrove in 1934. A similar map was published by Umbgrove in 1946. The gravity field of Plate 8 is based on the revised data of Vening Meinesz' map of the year 1940. In a few places, however, the construction of the isogammes is different. This concerns more especially the region of Sumba, Obi, Ceram and Atcheen (N. Sumatra). Compare p. 181 note [2]).

An explanation of the map is given in pp. 180–200 and it should be compared to fig. 117, fig. 118 (distribution of earthquake foci), fig. 122 (Pre-Tertiary history), fig. 132 (Tertiary paleogeography), fig. 39 and 122 (deep-sea relief), fig. 121 (Tertiary stratigraphy) and fig. 123 and fig. 130 (schematic sections and block-diagram of a double island festoon).

The distribution of volcanoes is reproduced according to the official list of volcanoes (Bull. of the Netherl. Ind. Volcanol. Survey no. 75, 1936). However, solfatara fields, volcanoes in a fumarolic stage and volcanoes which showed eruptive action since the year 1600 are all indicated with one and the same notation.

Table I, see under Explanation of Plate 6 p. 342.

Table II

The graphs of Table II have been explained in the previous chapters. It hardly needs mentioning that their value greatly depends on the length of the geological periods i.e. on the geological time scale. The remarkable fluctuations of age determinations during the last ten years are summarized in the accompanying fig. 204.

The determinations of absolute age as listed in the third and fifth column from the left of fig. 204 under the headings helium and lead time-scale are taken from publications of Holmes (1937) and Urry (1936). These age estimates show such a remarkable consistency in their appropriate positions that Holmes regarded it "as the final proof that the ages calculated from the lead and helium ratios are at least of the right order and that no serious error is anywhere involved". However, two years later Evans c.s. came to the conclusion that Holmes' opinion was too optimistic. Many of the ratios had arisen.

HELIUM AGES OF MAGNETITES IN MILLIONS OF YEARS AFTER EVANS, GOODMAN AND KEEVIL Aº 1942-1944	REVISED HELIUM TIME SCALE IN MILLIONS OF YEARS AFTER URRY AND HOLMES Aº 1941	HELIUM TIME SCALE AFTER HOLMES Aº 1937	STRATIGRAPHIC COLUMN OF TABLE II	LEAD TIME SCALE		
				AFTER HOLMES Aº 1937	AFTER EVANS AND GOODMAN Aº 1944	AFTER HOLMES Aº 1946
MIOCENE {26, 37 EOCENE {59, 60	PLIOCENE 8 MIOCENE 10 OLIGOCENE 20 CRETACEOUS – JURASSIC 60	[illegible] 32	PLIOCENE, MIOCENE, OLIGOCENE, EOCENE } TERTIARY	35 70	34 58	59
EARLY CRETACEOUS {81, 88	TRIASSIC {93, 100, 108	83 96 107	CRETACEOUS			
LATE TRIASSIC – EARLY JURASSIC 126-128 JURASSIC 130-131 LATE TRIASSIC – EARLY JURASSIC {133-135, 145	(POST TORRIDONIAN) {128, 130 UPPER CARBONIFEROUS 135 LOWER CARBONIFEROUS 140		JURASSIC	123	123	
	DEVONIAN 180	[illegible] 165 175	TRIASSIC			
		224	PERMIAN	220		220
		234 254	CARBONIFEROUS	232 268 278	[illegible] 240 268 276	
CARBONIFEROUS 300 DEVONIAN 318		282	DEVONIAN			280
		345	SILURIAN		333	
DEVONIAN ? {386, 396		365	ORDOVICIAN	348 [illegible] 371	348 341 – 381 371 [illegible]	380
		427 453	CAMBRIAN	[illegible] 408		440
	KEWEENAWAN {486, 520-525		PRECAMBRIAN		531	

CUMULATIVE MAXIMUM THICKNESSES OF STRATA PLOTTED AGAINST AGES FROM LEAD RATIOS (HOLMES Aº 1946)			
STRATIGRAPHIC COLUMN	AGES	MAX. THICKNESS IN THOUSANDS OF FEET	CUMULATIVE MAXIMUM THICKNESSES
PLIOCENE		4	4
	16	18	22
MIOCENE			
	32	21	43
OLIGOCENE			
	46	15	58
EOCENE PALEOCENE			
	69	23	81
CRETACEOUS			
	142	64	145
JURASSIC			
	166	20	165
TRIASSIC			
	196	25	190
PERMIAN			
	220	18	208
CARBONIFEROUS			
	274	40	240
DEVONIAN			
	322	37	285
SILURIAN			
	350	20	305
ORDOVICIAN			
	426	40	345
CAMBRIAN			
	506	40	385

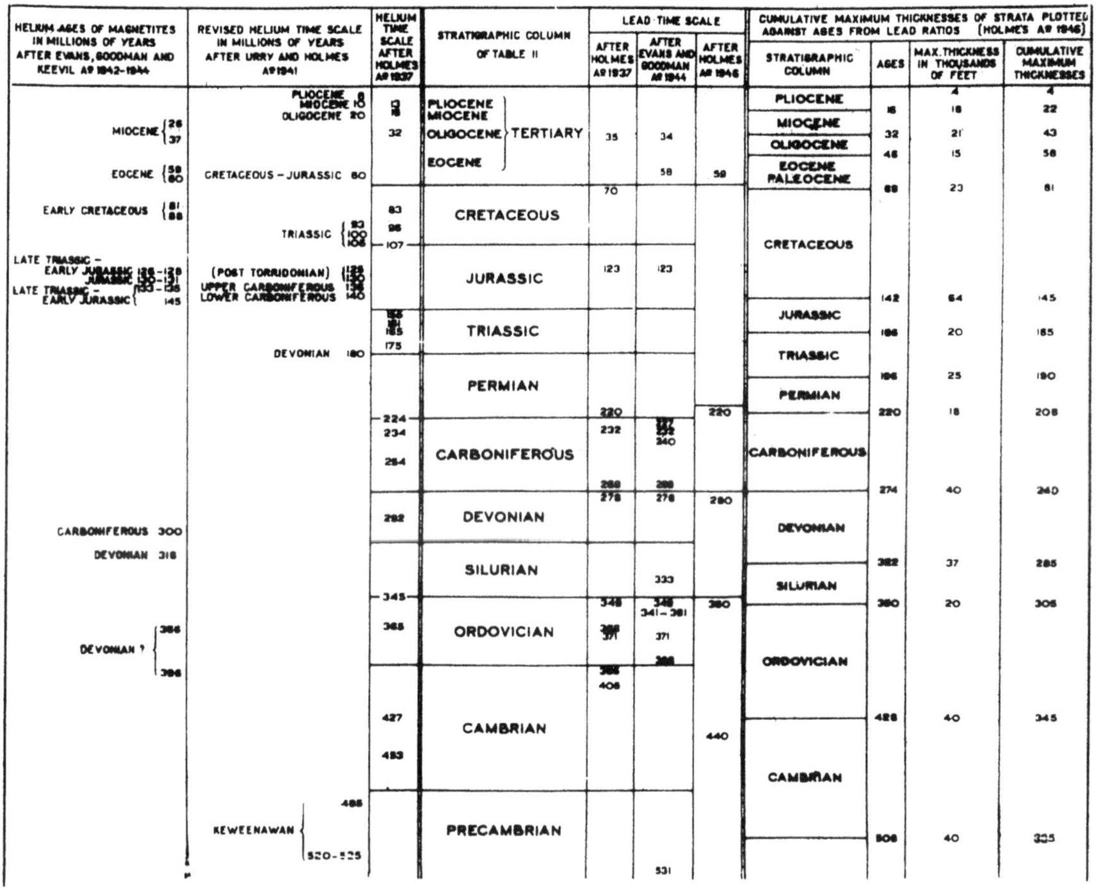

Fig. 204. Determinations of geological ages.

helium ratios were found to be too high because of previously unsuspected errors in radium standards and possibly also the use of very dilute standard solutions. According to their revised and more accurate methods of investigation the helium ratios have to be lowered. A new time-scale of helium ratios was published by Urry and Holmes in the year 1941. These data, with additions of a few Keweenawan data mentioned in a previous paper [1]) are represented in the second column from the left of fig. 204. Most of these geochronological data differ considerably from the "old" helium ages. Moreover an embarrassing discrepancy between the helium time-scale and that based on lead-ratios had arisen.

[1]) Evans c.s. 1939, p. 34 Table XI

Determinations of absolute age by the helium method are likely to be in error owing to loss of a large fraction of radiogenic helium. Furthermore, Hurley and Goodman found that mineral species differ considerably in their specific retentivity of helium. Pyroxene retains more of its helium than feldspar and probably magnetite retains most, if not all, of its helium. Accordingly, magnetite has been adopted for helium age determinations. And a new helium time scale based on measurements of magnetites was published by Evans, Goodman and Keevel in 1942[1]). It is represented by the left-hand column of fig. 204[2]). The helium ages of magnetite approach and some of them even exceed "the lead ages obtained on radioactive minerals from corresponding geologic horizons. Individual helium ages cover the same range of value as lead ages, i.e. from 15 million years for Tertiary rocks to 1650 million years for eaily pre-Cambrian". Hence, the magnetite results have restored the position!

Regarding the lead-ratios Evans and his coworkers wrote[3]) "Among the large number of lead ages on radioactive minerals, only four have been definitely established which are (1) free from alteration, (2) of well established geological horizon, and (3) based on isotopic abundance or atomic weight determination of the lead". Although the span of possible ages indicated by the $\pm$ figures of these four minerals is rather great, it was impossible to bring their appropriate positions into agreement with the helium scale of 1941. On the other hand no disagreement exists between the new lead ages and the "old" lead time-scale nor with the "magnetite time-scale".

In a more recent paper by Evans and Goodman (1944) fourteen lead-ages are mentioned. These are plotted in the sixth column from the left of fig. 204.

The periodicity of 250 m.y. and the cadence of about 50 m.y. that have been mentioned in previous chapters (pp. 283 and 323) and the approximate delimination of the geological formations as shown by fig. 10 and Table II are based on the revised lead scale. Still more recent lead ages were published by Holmes in 1946 (Seventh column from the left). Fig. 204 shows these data and the length of the geological periods as deduced graphically by plotting cumulative maximum thicknesses of strata against ages from lead ratios[4]).

In doing so a slightly different representation of the stratigraphic column is obtained. It may be noticed from fig. 204 that the boundaries of the Cretaceous, Jurassic, Triassic and some other formations have shifted downwards along the time-scale. Obviously the figures for the thicknesses of strata and the average rate of sedimentation depend on many rather uncertain factors. Further age measurements on well-authenticated minerals will enable an ever increasing accuracy in drawing the boundaries in the stratigraphic column.

The publications mentioned above are cited in the bibliography at the end of Chapter I.

---

1) Evans and Goodman 1944, p. 227.

2) A longer list of data was published by P. M. Hurley and Clark Goodman in their "Preliminary Magnetite Index" (Bull. Geol. Soc. America 54, 1943). Their index contains 70 determinations, the age ranging as follows: 15–25 for Miocene, 43–45 for Late Eocene or Oligocene, 50–75 Laramide, 63 Mid-Cretaceous (?), 100–132 Nevadan, 140–174 Late Triassic, 200–245 Upper Carboniferous, 340–365 Devonian, 390 Late Silurian, 500–1650 Pre-Cambrian.

3) Evans, ibidem, 1939, p. 945.

4) Professor Holmes kindly gave me these data in advance of publication, in 1945. Afterwards, however, he received still more recent data and had to revise the figures he gave me. Accordingly, the figures in the third column from the right of fig. 204 should be changed as follows: 12, 26, 38, 58, 127, 152, 182, 203, 255, 313, 350, 430, 510.

INDEX

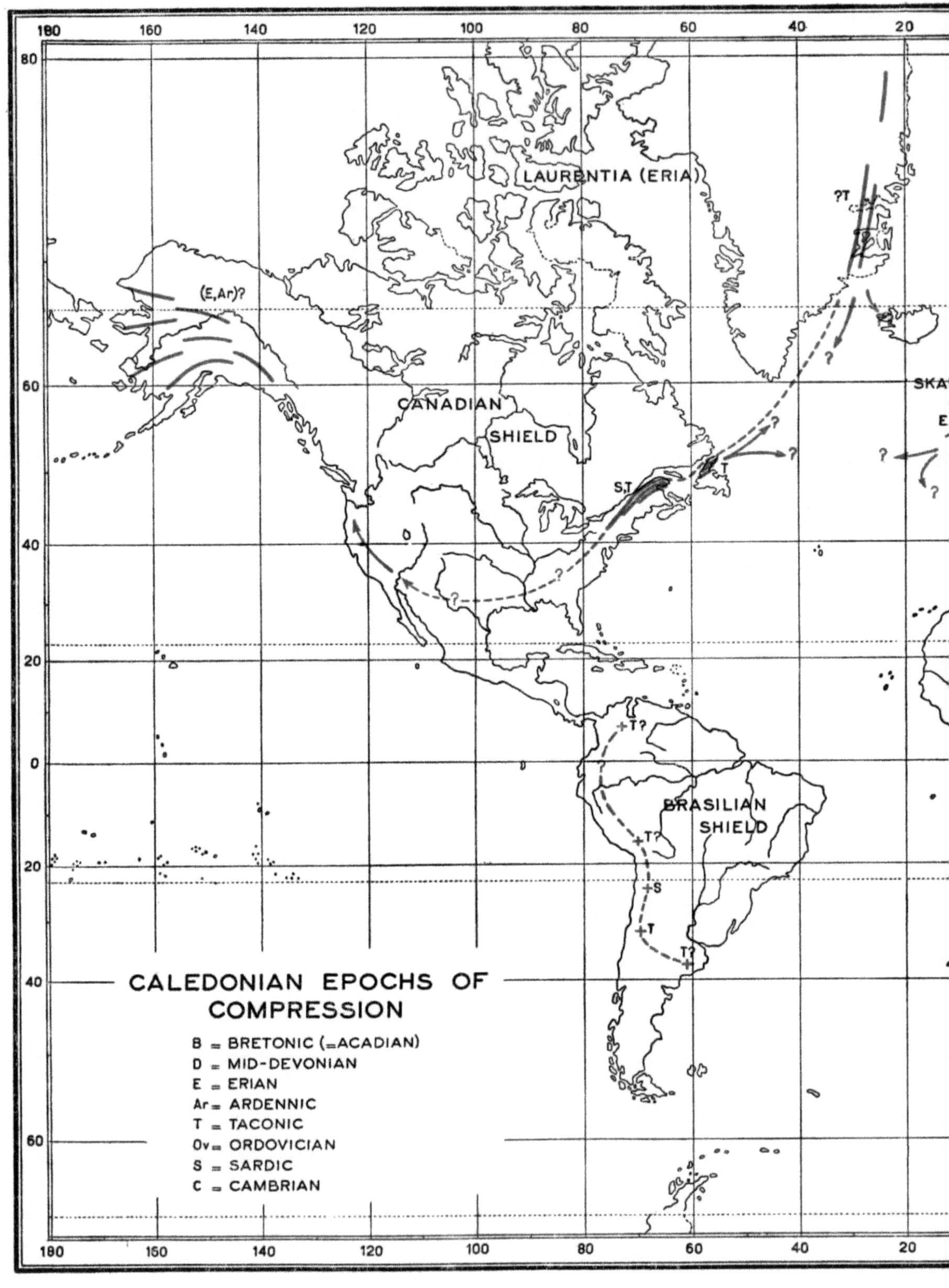

LAURENTIA (ERIA)
CANADIAN
SHIELD
BRASILIAN
SHIELD
(E,Ar)?
S,T
T
?T
T?
S
T
T?
SKA
E
CALEDONIAN EPOCHS OF COMPRESSION
B = BRETONIC (=ACADIAN)
D = MID-DEVONIAN
E = ERIAN
Ar = ARDENNIC
T = TACONIC
Ov = ORDOVICIAN
S = SARDIC
C = CAMBRIAN

NUCLEUS OF BARENTS SEA
NUCLEUS OF KARA SEA
NUCLEUS OF TSCHUKTSCHEN
ANGARA-SHIELD
BALTIC RUSSIAN SHIELD
AN
ELD
AUSTRALIAN SHIELD
S
T?
S
E?
C,Ov
B
T,E,D
T
T,E

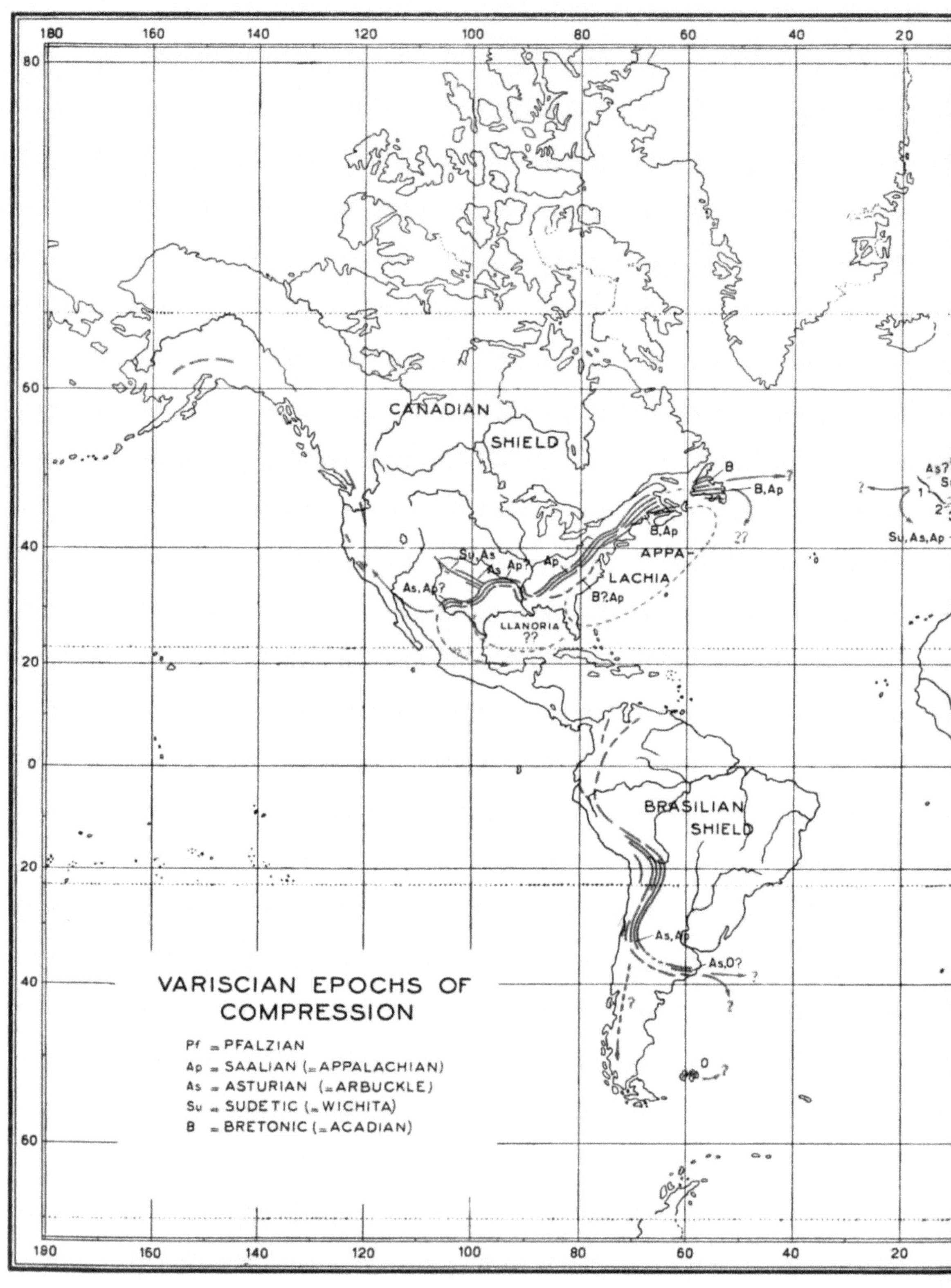

VARISCIAN EPOCHS OF COMPRESSION
Pf = PFALZIAN
Ap = SAALIAN (= APPALACHIAN)
As = ASTURIAN (= ARBUCKLE)
Su = SUDETIC (= WICHITA)
B = BRETONIC (= ACADIAN)
CANADIAN SHIELD
APPALACHIA
LLANORIA
BRASILIAN SHIELD
B, Ap
B?,Ap
Su, As
As, Ap?
As, Ap
As, O?
Su, As, Ap

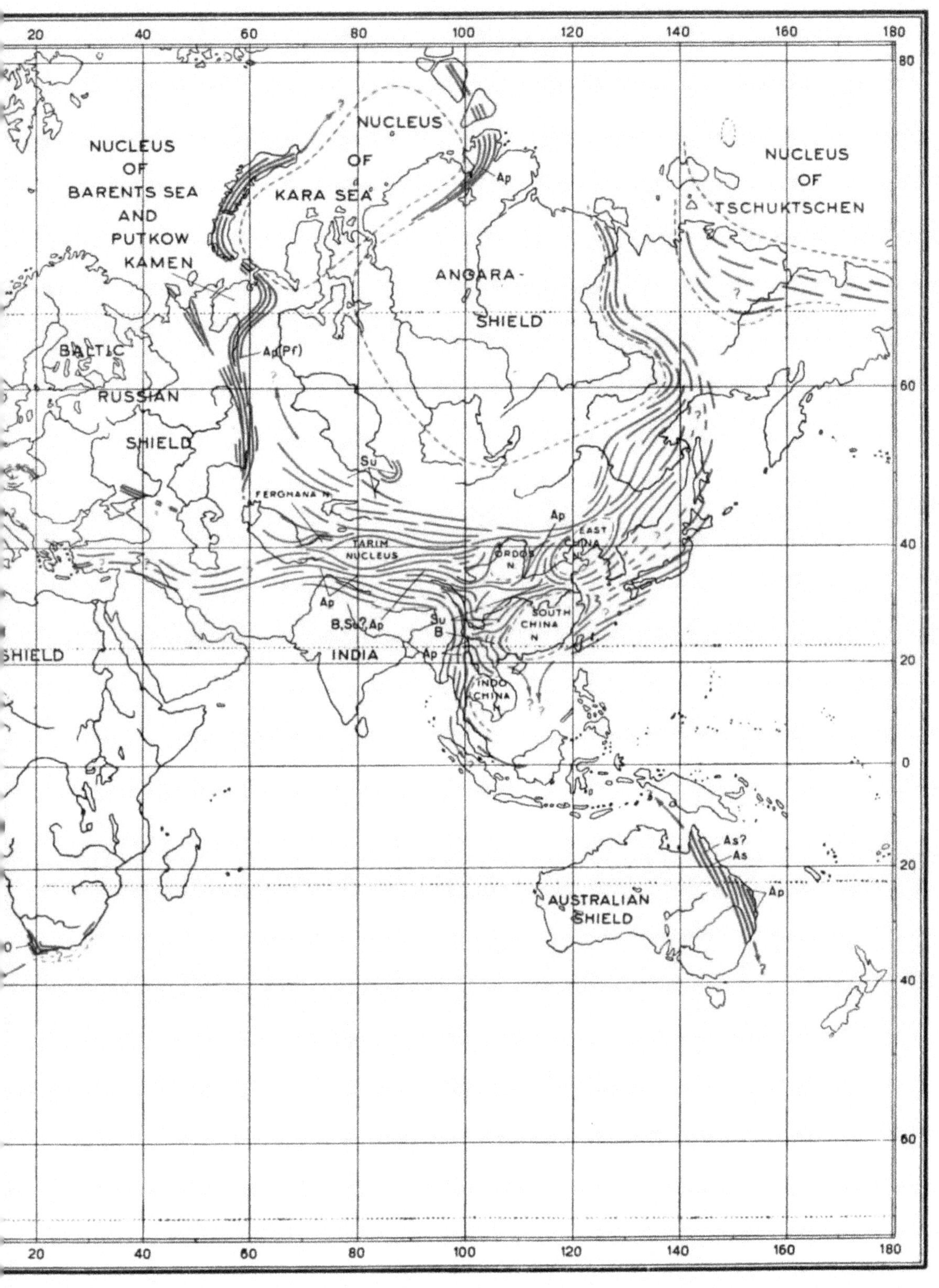

NUCLEUS OF BARENTS SEA AND PUTKOW KAMEN
NUCLEUS OF KARA SEA
NUCLEUS OF TSCHUKTSCHEN
ANGARA-SHIELD
BALTIC RUSSIAN SHIELD
FERGHANA N
TARIM NUCLEUS
ORDOS N
EAST CHINA N
SOUTH CHINA N
INDIA
INDO CHINA N
AUSTRALIAN SHIELD
SHIELD
Ap
Ap(Pf)
Su
B,Su?,Ap
B
As?
As

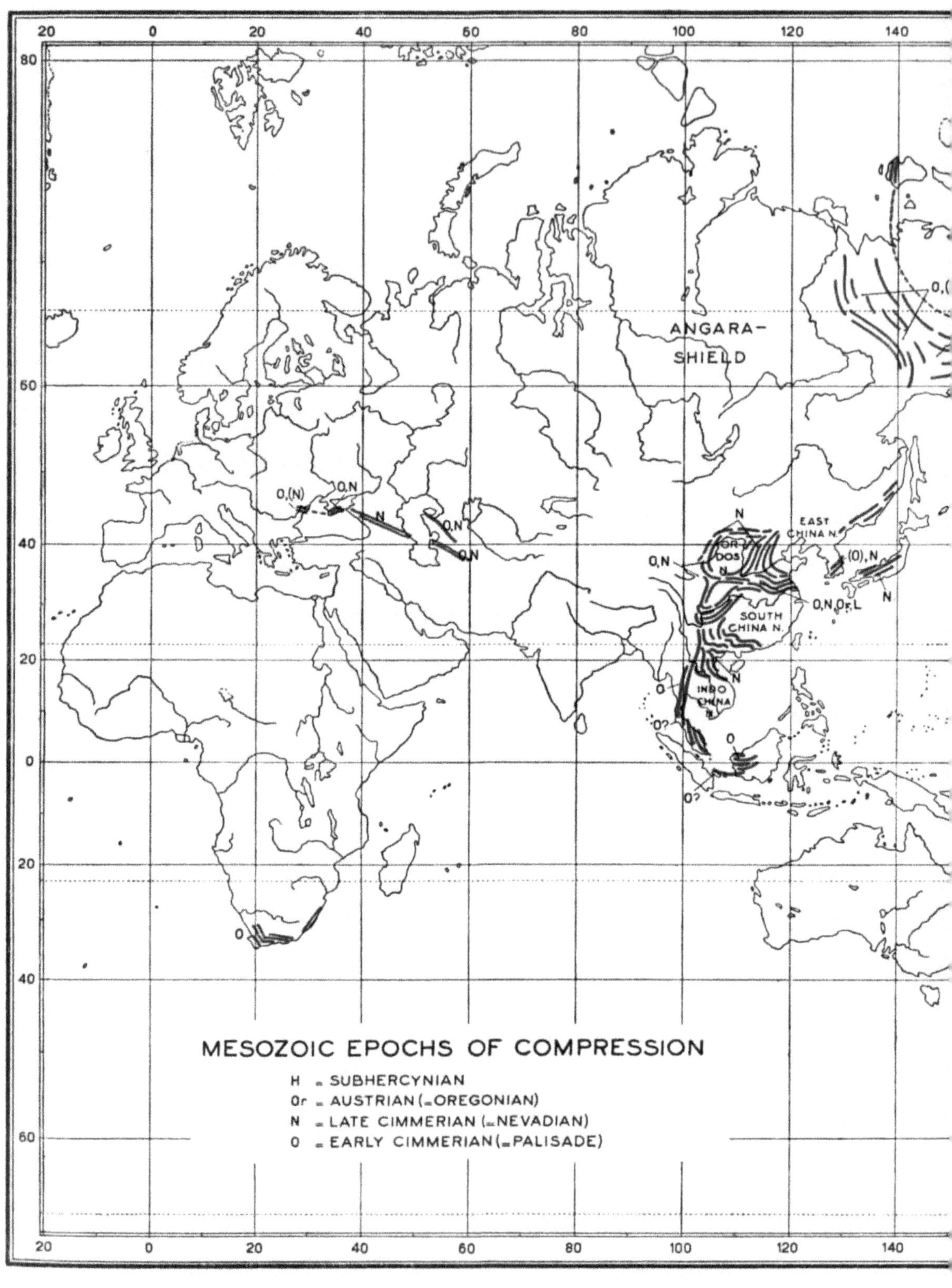

ANGARA-
SHIELD
EAST
CHINA N.
OR-
DOS
N
SOUTH
CHINA N.
INDO
CHINA
N
0,(N)
0,N
0,N
0,N
(0),N
0,N,0r,L
MESOZOIC EPOCHS OF COMPRESSION
H = SUBHERCYNIAN
Or = AUSTRIAN (=OREGONIAN)
N = LATE CIMMERIAN (=NEVADIAN)
0 = EARLY CIMMERIAN (=PALISADE)

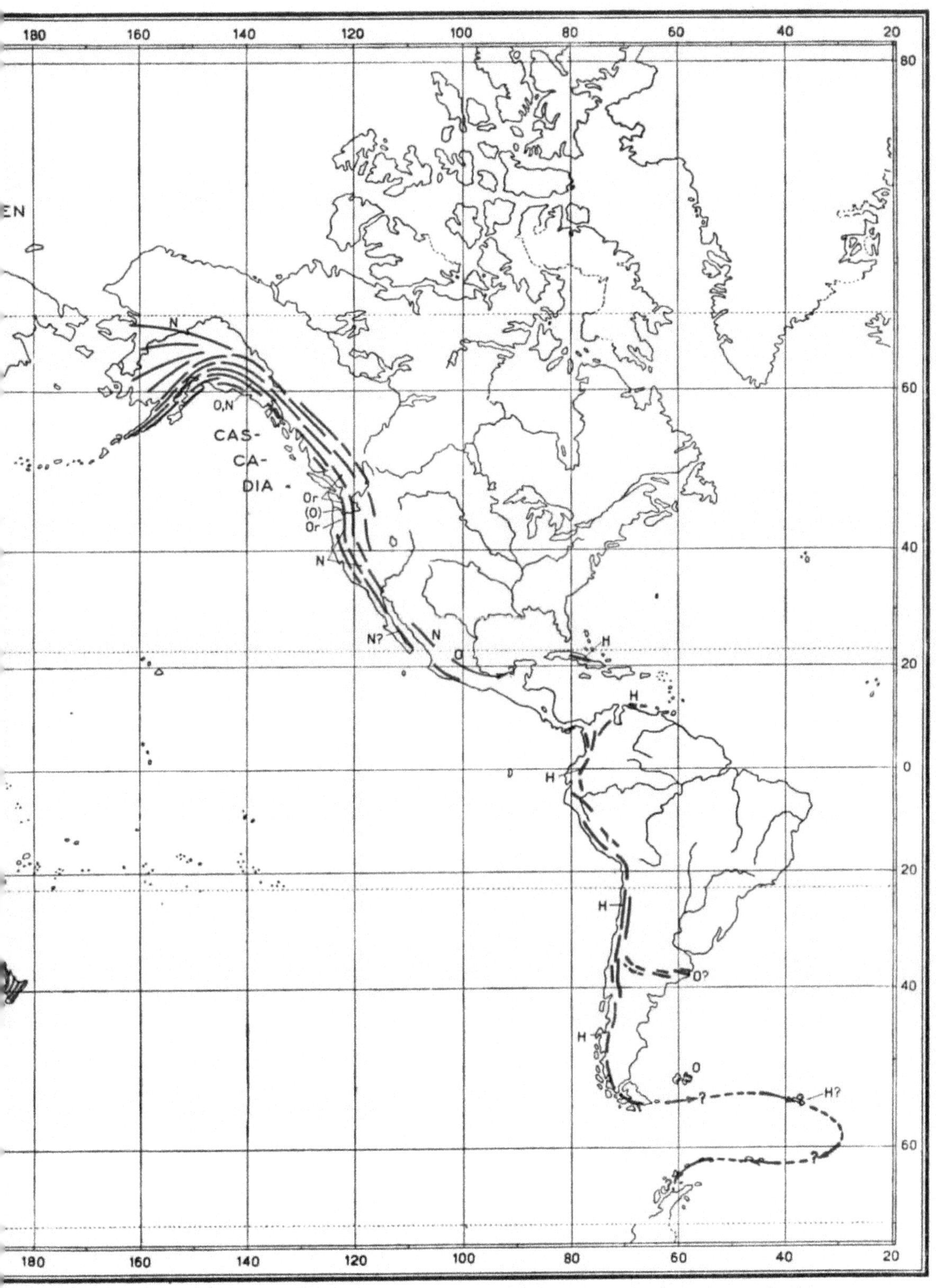
CAS-
CA-
DIA
EN
N
O,N
Or
(O)
Or
N
N?
N
O
H
H
H
H
H
O?
O
H?

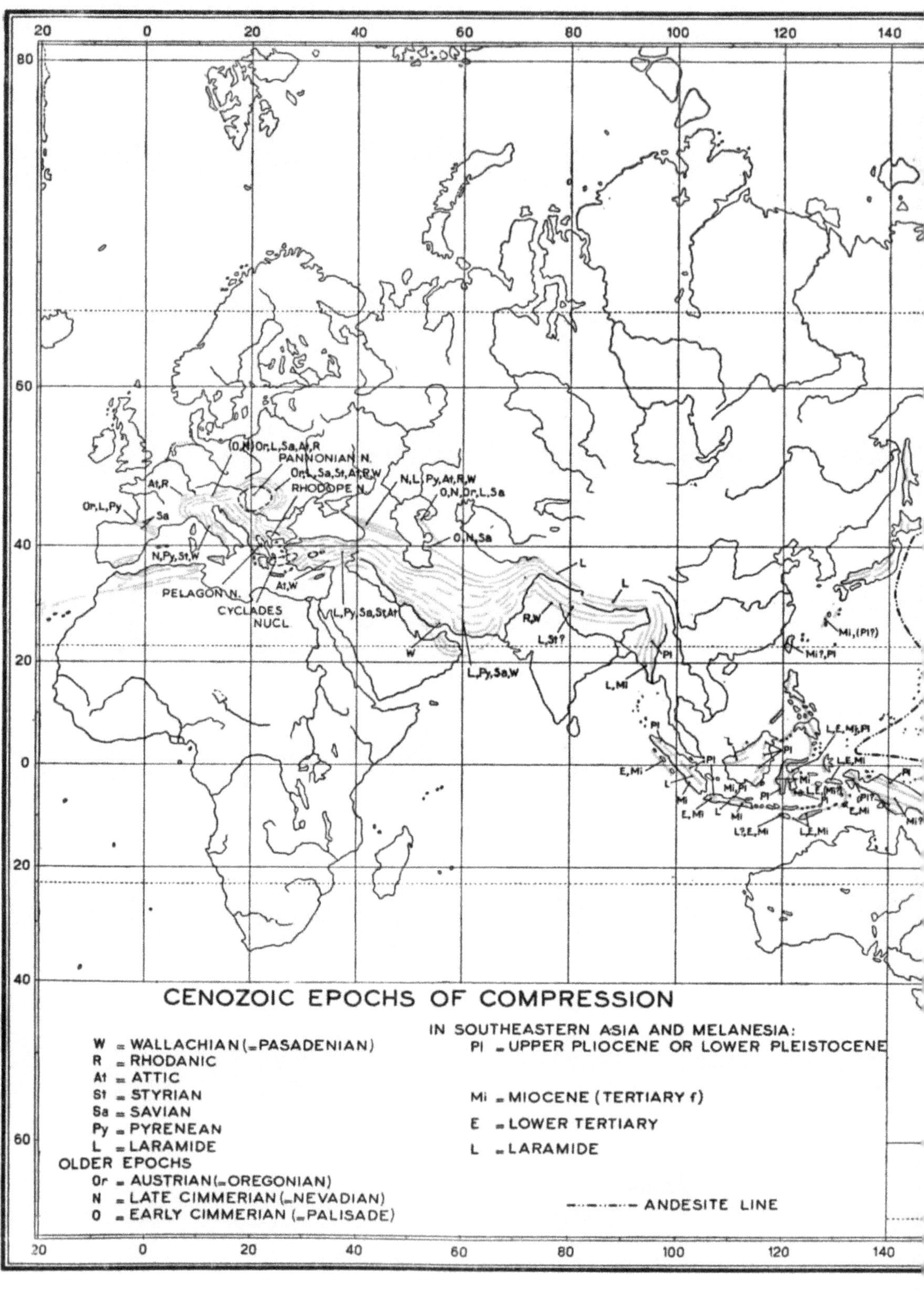
CENOZOIC EPOCHS OF COMPRESSION
W = WALLACHIAN (= PASADENIAN)
R = RHODANIC
At = ATTIC
St = STYRIAN
Sa = SAVIAN
Py = PYRENEAN
L = LARAMIDE
OLDER EPOCHS
Or = AUSTRIAN (= OREGONIAN)
N = LATE CIMMERIAN (= NEVADIAN)
O = EARLY CIMMERIAN (= PALISADE)
IN SOUTHEASTERN ASIA AND MELANESIA:
Pl = UPPER PLIOCENE OR LOWER PLEISTOCENE
Mi = MIOCENE (TERTIARY f)
E = LOWER TERTIARY
L = LARAMIDE
ANDESITE LINE
PANNONIAN N.
RHODOPE N.
PELAGON N.
CYCLADES NUCL.
(O,N) Or,L,Sa,At,R
Or,L,Sa,St,At,R,W
N,L,Py,At,R,W
O,N,Or,L,Sa
O,N,Sa
At,R
Or,L,Py
Sa
N,Py,St,W
At,W
L,Py,Sa,St,At
W
L,Py,Sa,W
R,W
L,St?
L,Mi
Mi,(Pl?)
Mi?,Pl
L,E,Mi,Pl
L,E,Mi
E,Mi
L?,E,Mi
L,E,Mi

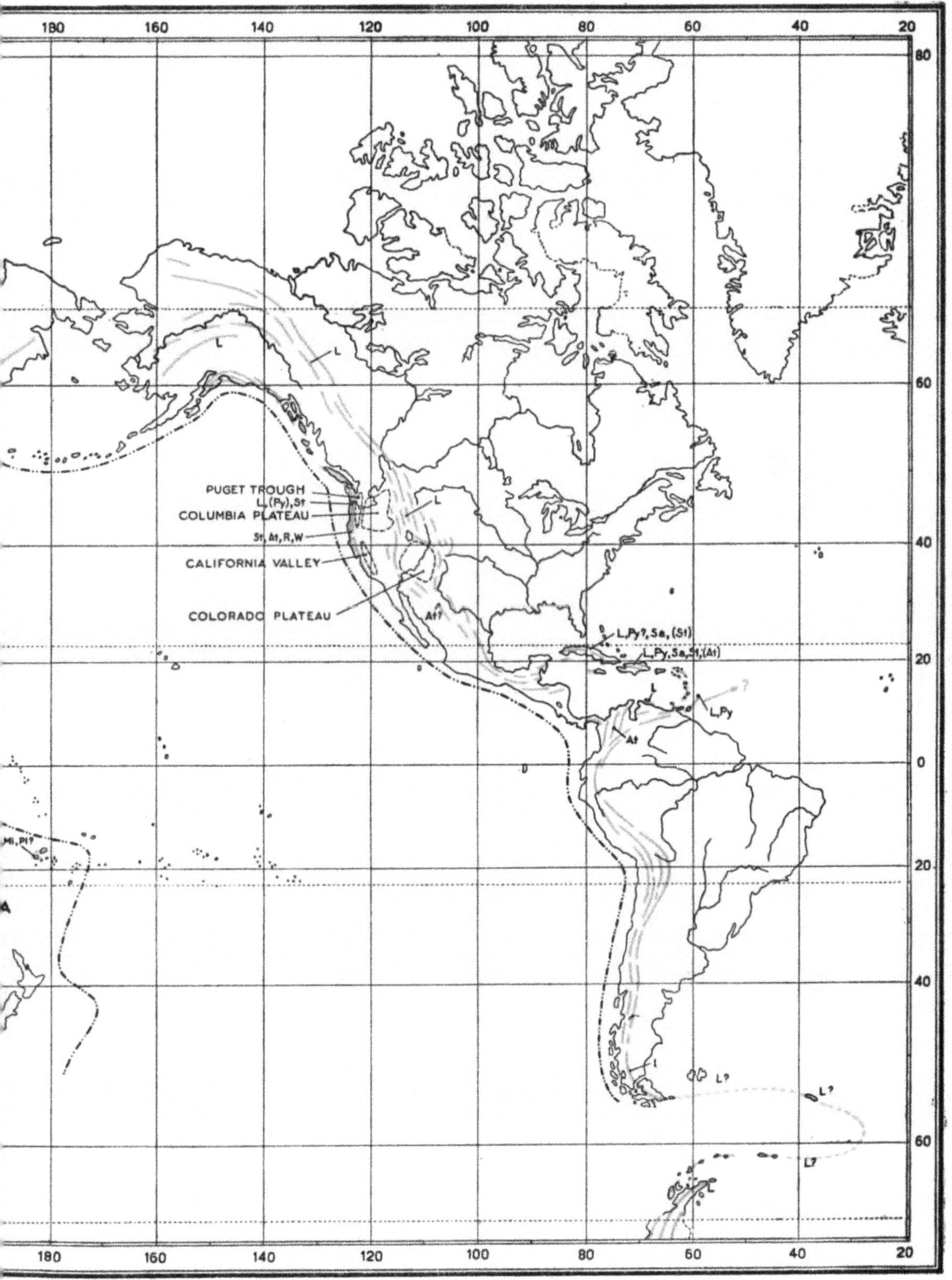
180
160
140
120
100
80
60
40
20
PUGET TROUGH
L,(Py),St
COLUMBIA PLATEAU
St,At,R,W
CALIFORNIA VALLEY
COLORADO PLATEAU
At?
L,Py?,Sa,(St)
L,Py,Sa,St,(At)
L
L,Py
At
L?
Mi,Pl?

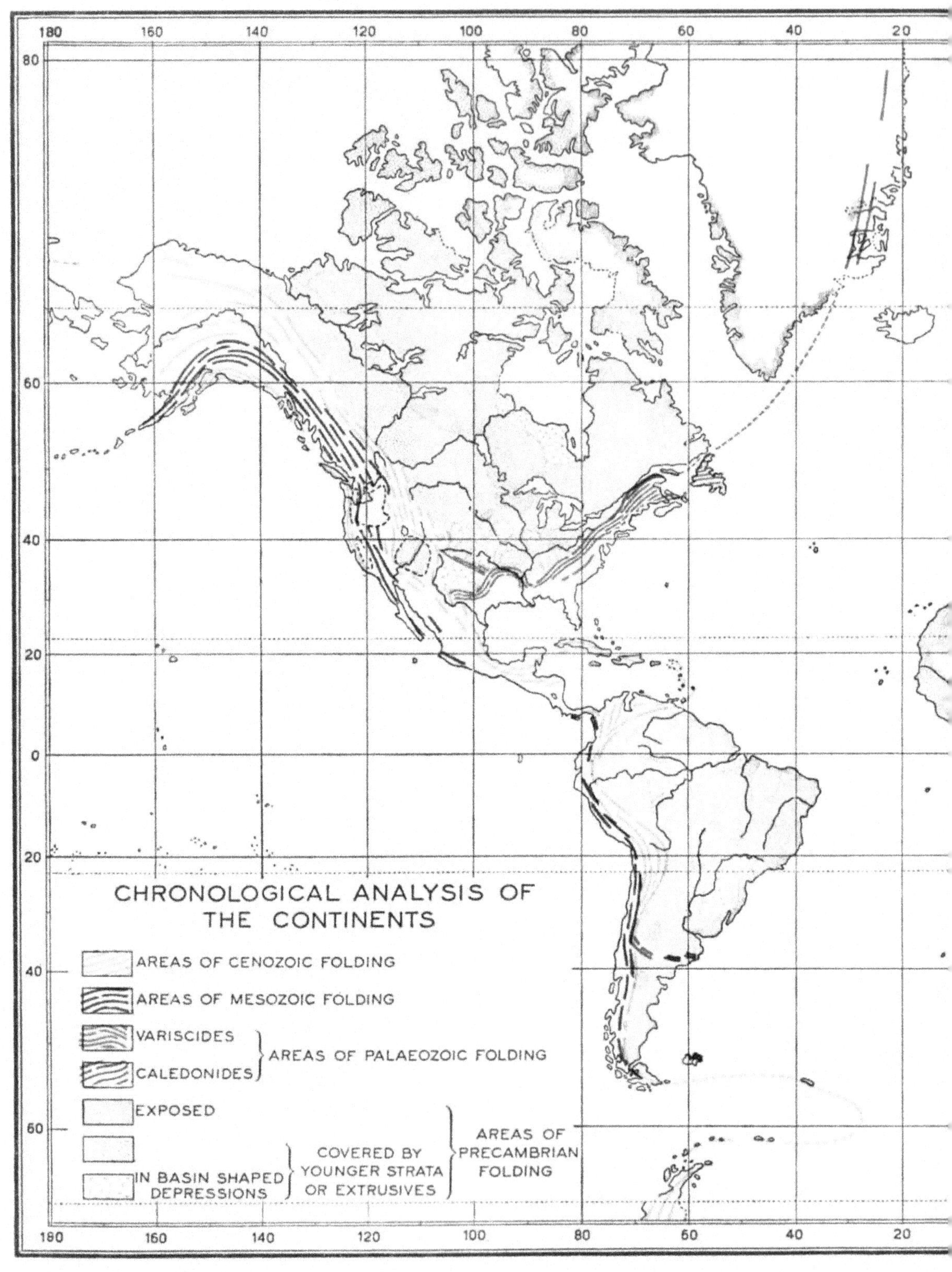

CHRONOLOGICAL ANALYSIS OF THE CONTINENTS
AREAS OF CENOZOIC FOLDING
AREAS OF MESOZOIC FOLDING
VARISCIDES
CALEDONIDES
AREAS OF PALAEOZOIC FOLDING
EXPOSED
IN BASIN SHAPED DEPRESSIONS
COVERED BY YOUNGER STRATA OR EXTRUSIVES
AREAS OF PRECAMBRIAN FOLDING
180
160
140
120
100
80
60
40
20
0

PLATE 5

20 40 60 80 100 120 140 160 180
80
60
40
20
0
20
40
60
20 40 60 80 100 120 140 160 180

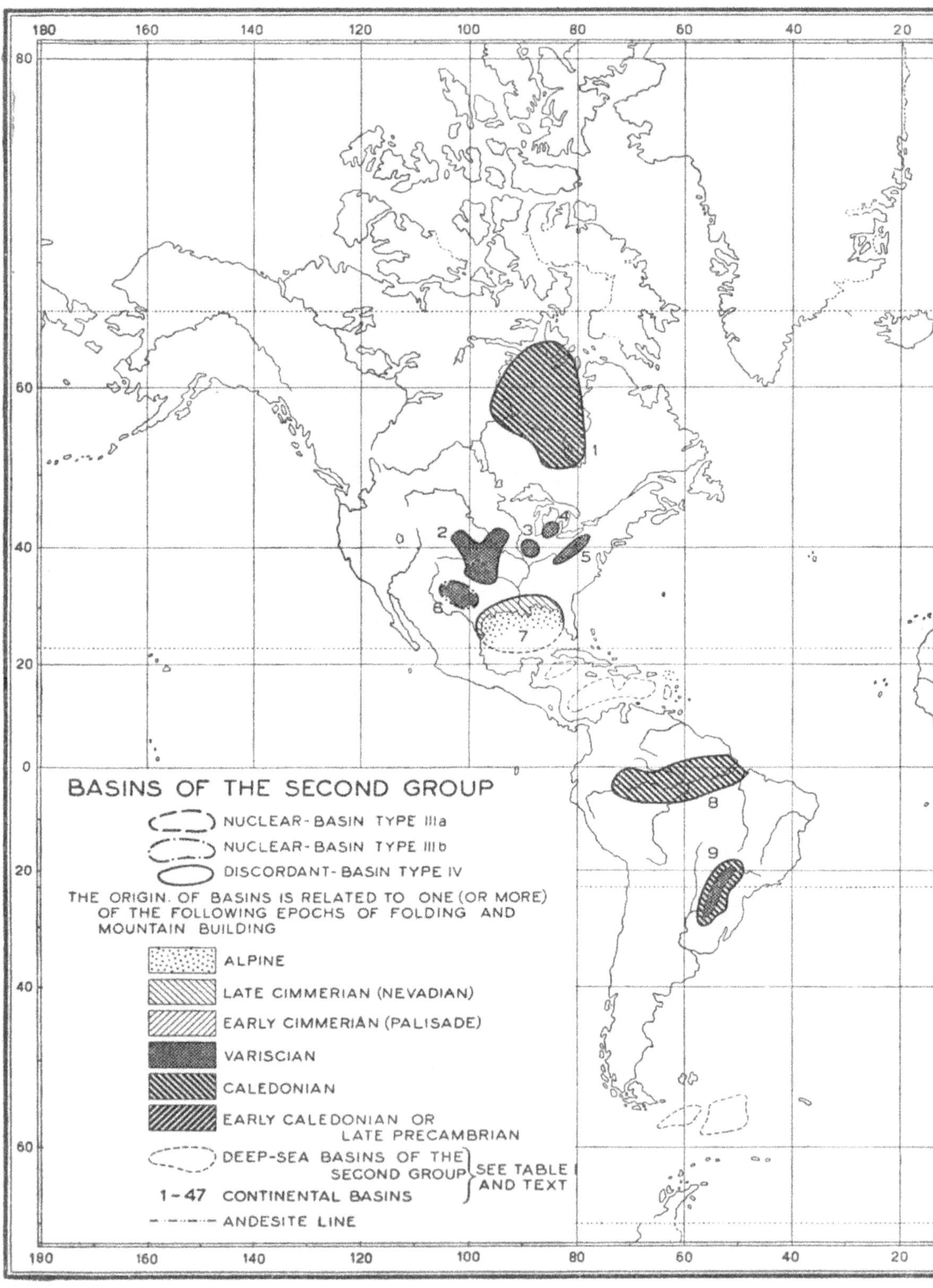

BASINS OF THE SECOND GROUP
NUCLEAR-BASIN TYPE IIIa
NUCLEAR-BASIN TYPE IIIb
DISCORDANT-BASIN TYPE IV
THE ORIGIN. OF BASINS IS RELATED TO ONE (OR MORE) OF THE FOLLOWING EPOCHS OF FOLDING AND MOUNTAIN BUILDING
ALPINE
LATE CIMMERIAN (NEVADIAN)
EARLY CIMMERIAN (PALISADE)
VARISCIAN
CALEDONIAN
EARLY CALEDONIAN OR LATE PRECAMBRIAN
DEEP-SEA BASINS OF THE SECOND GROUP
1 - 47 CONTINENTAL BASINS
SEE TABLE I AND TEXT
ANDESITE LINE
1
2
3
4
5
6
7
8
9
180
160
140
120
100
80
60
40
20
0

20
40
60
80
100
120
140
160
180
80
60
40
20
0
20
40
60
10
30
31
33
34
20
21
32
35
36
12
18
22
19
24
23
37
38
39
40
41
42
25
26
27
28
29
44
43
45
46
47

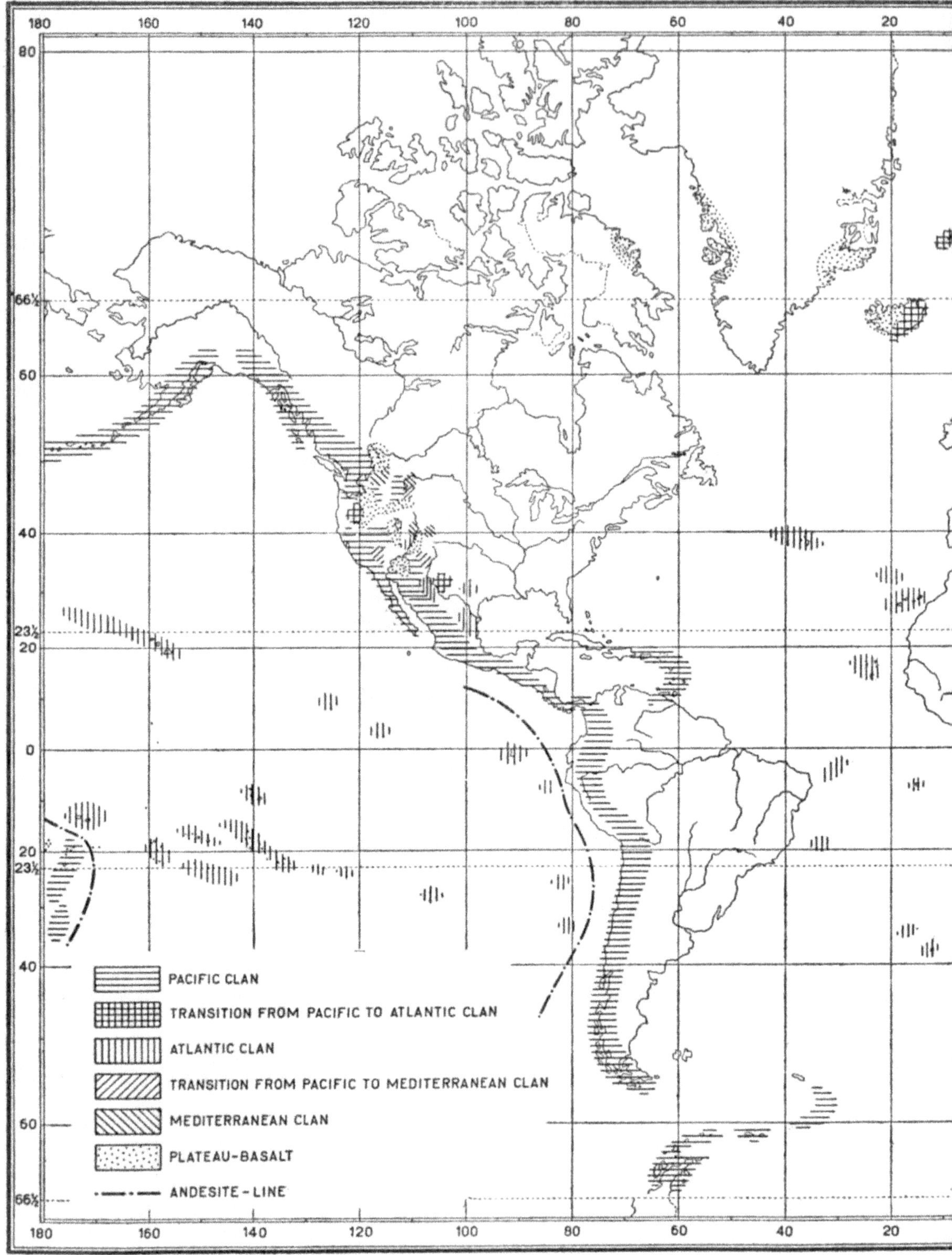

180
160
140
120
100
80
60
40
20
80
66½
60
40
23½
20
0
20
23½
40
50
66½
PACIFIC CLAN
TRANSITION FROM PACIFIC TO ATLANTIC CLAN
ATLANTIC CLAN
TRANSITION FROM PACIFIC TO MEDITERRANEAN CLAN
MEDITERRANEAN CLAN
PLATEAU-BASALT
ANDESITE-LINE

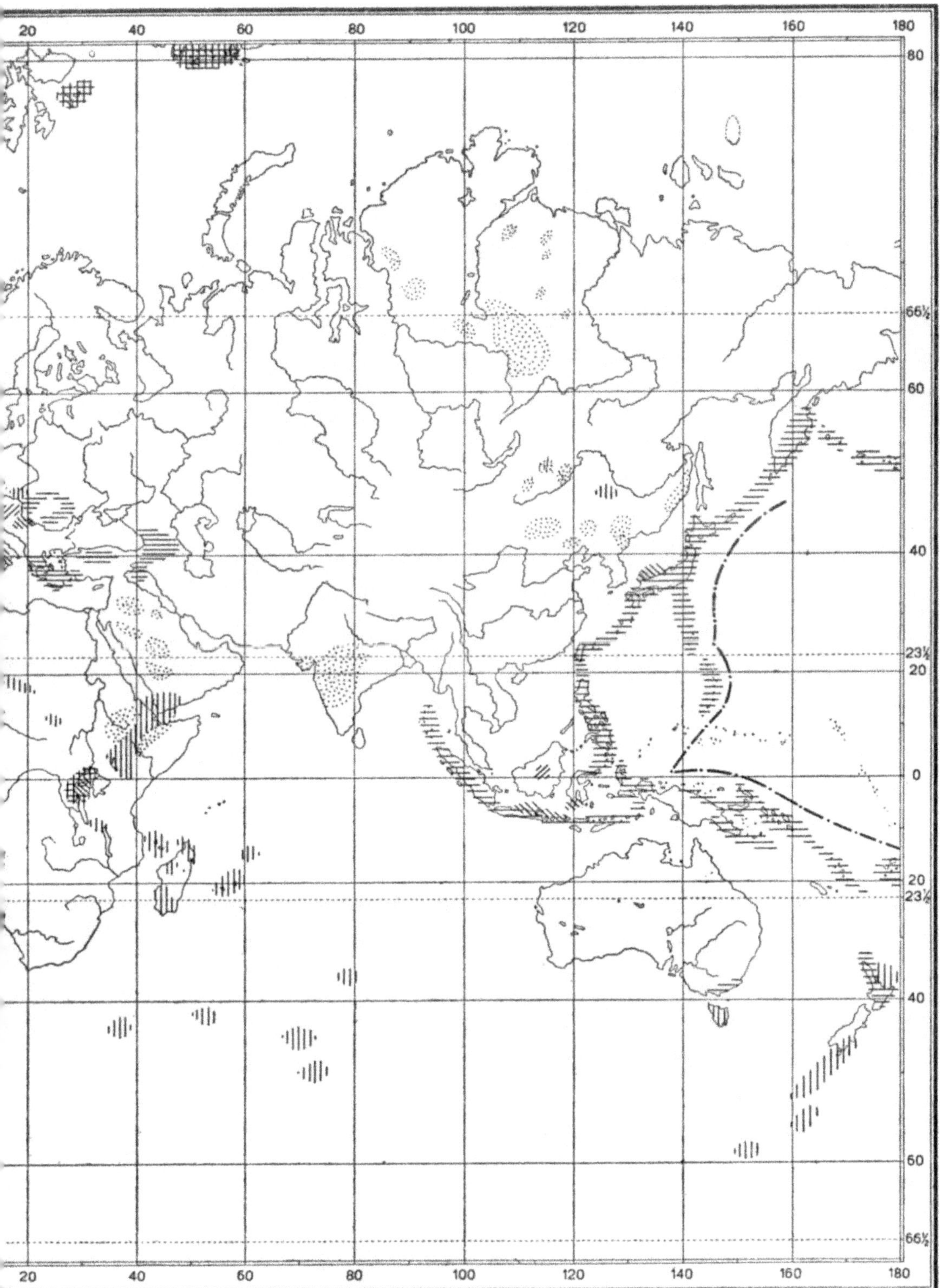
20
40
60
80
100
120
140
160
180
80
66½
60
40
23½
20
0
20
23½
40
60
66½

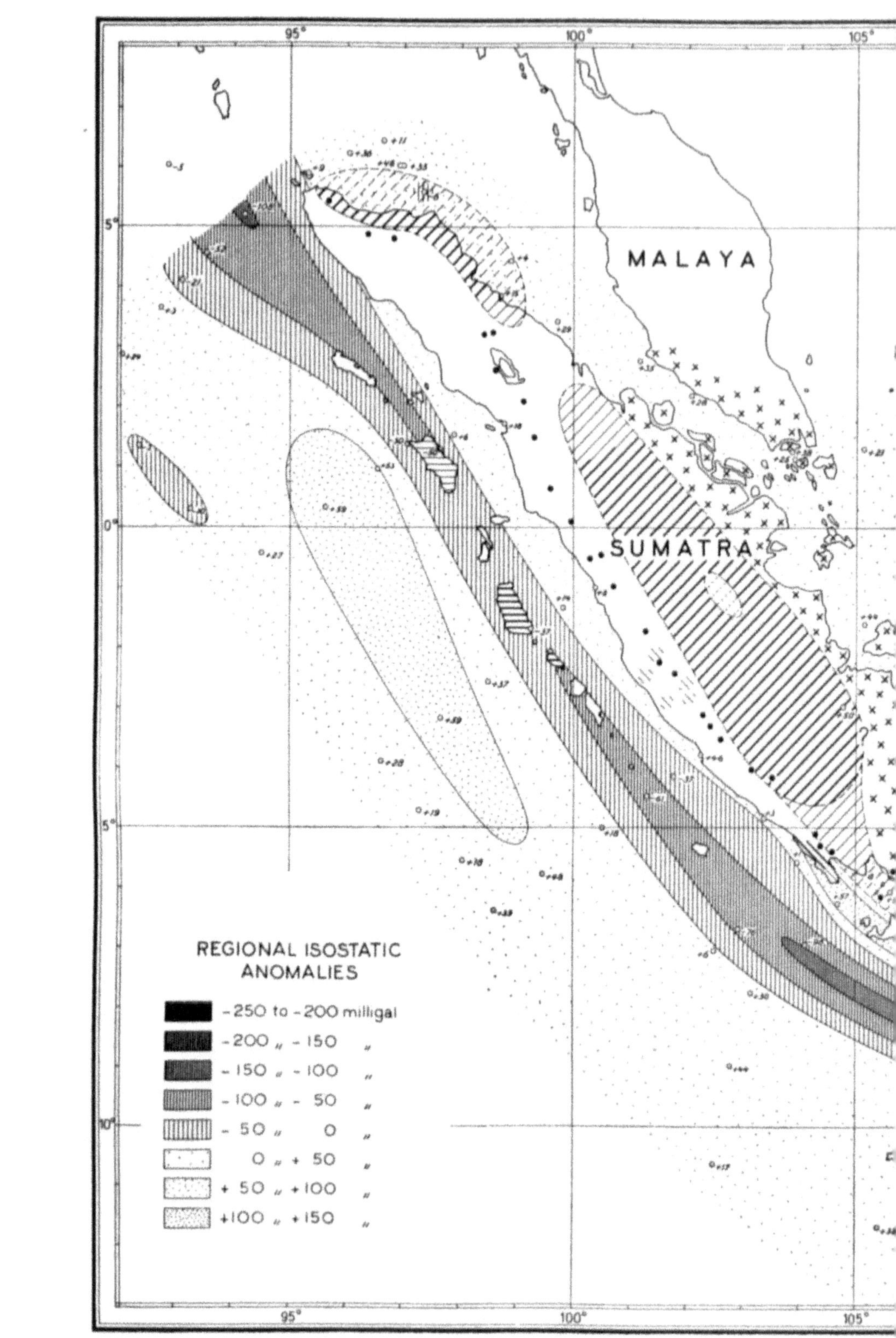

MALAYA
SUMATRA
REGIONAL ISOSTATIC ANOMALIES
-250 to -200 milligal
-200 " -150 "
-150 " -100 "
-100 " - 50 "
- 50 " 0 "
0 " + 50 "
+ 50 " +100 "
+100 " +150 "
95°
100°
105°
5°
0°
5°
10°

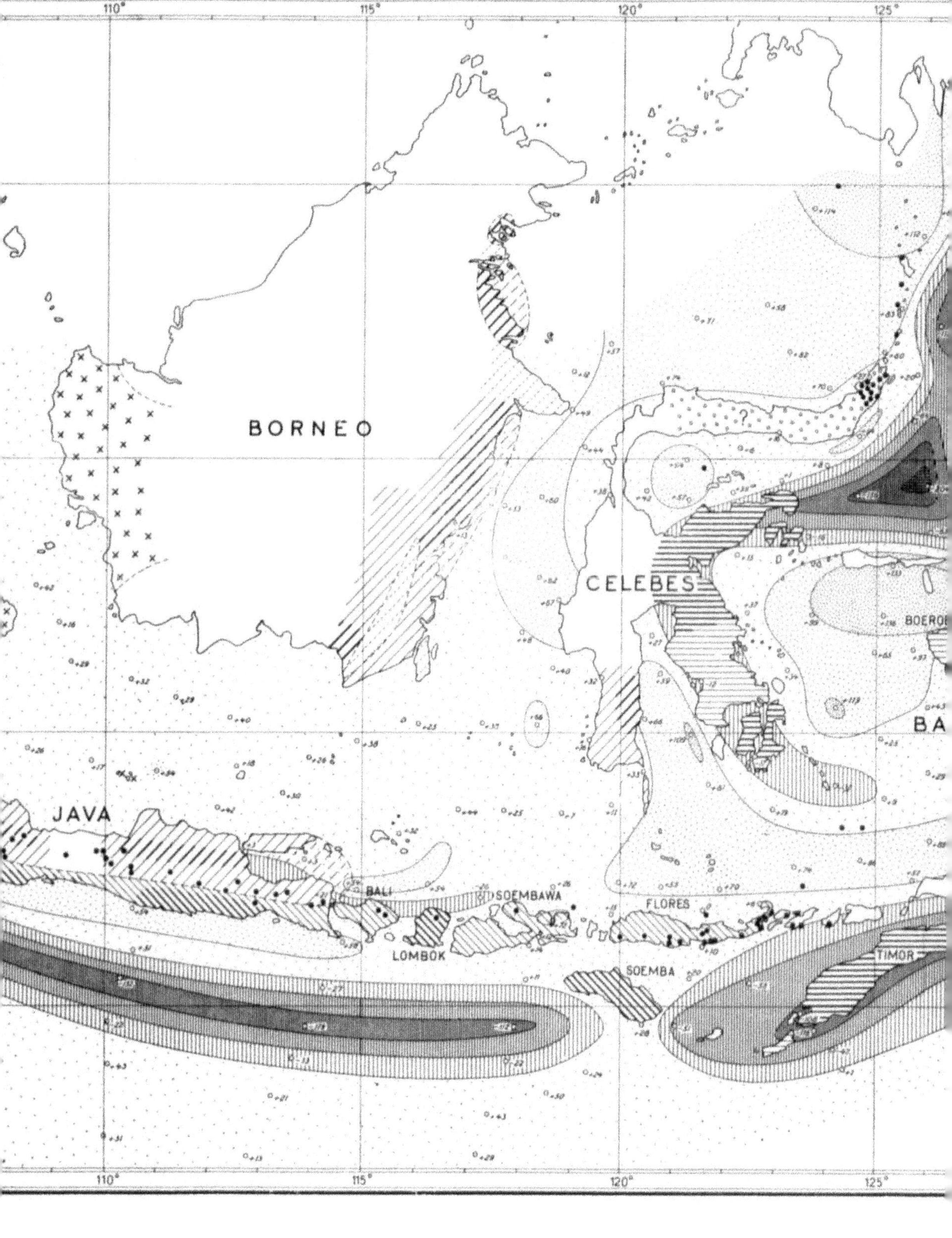

110°
115°
120°
125°
BORNEO
CELEBES
BOEROE
BA
JAVA
BALI
SOEMBAWA
LOMBOK
FLORES
SOEMBA
TIMOR

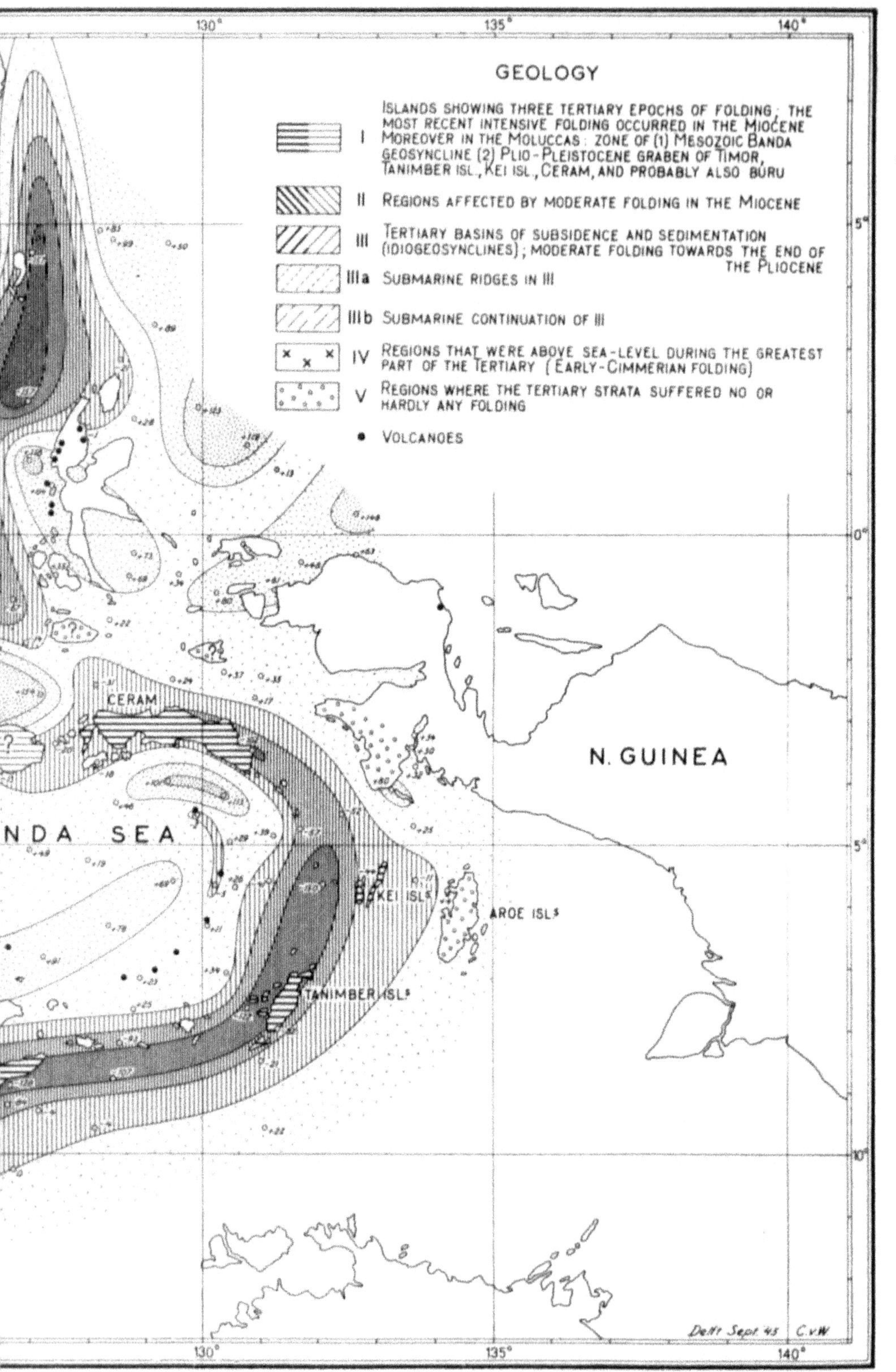
GEOLOGY
I Islands showing three tertiary epochs of folding; the most recent intensive folding occurred in the Miocene Moreover in the Moluccas: zone of (1) Mesozoic Banda geosyncline (2) Plio-Pleistocene graben of Timor, Tanimber isl., Kei isl., Ceram, and probably also Buru
II Regions affected by moderate folding in the Miocene
III Tertiary basins of subsidence and sedimentation (idiogeosynclines); moderate folding towards the end of the Pliocene
IIIa Submarine ridges in III
IIIb Submarine continuation of III
IV Regions that were above sea-level during the greatest part of the Tertiary (Early-Cimmerian folding)
V Regions where the tertiary strata suffered no or hardly any folding
Volcanoes
130°
135°
140°
5°
0°
10°
CERAM
NDA SEA
N. GUINEA
KEI ISLS
AROE ISLS
TANIMBER ISLS
Delft Sept. '45 C.v.W.

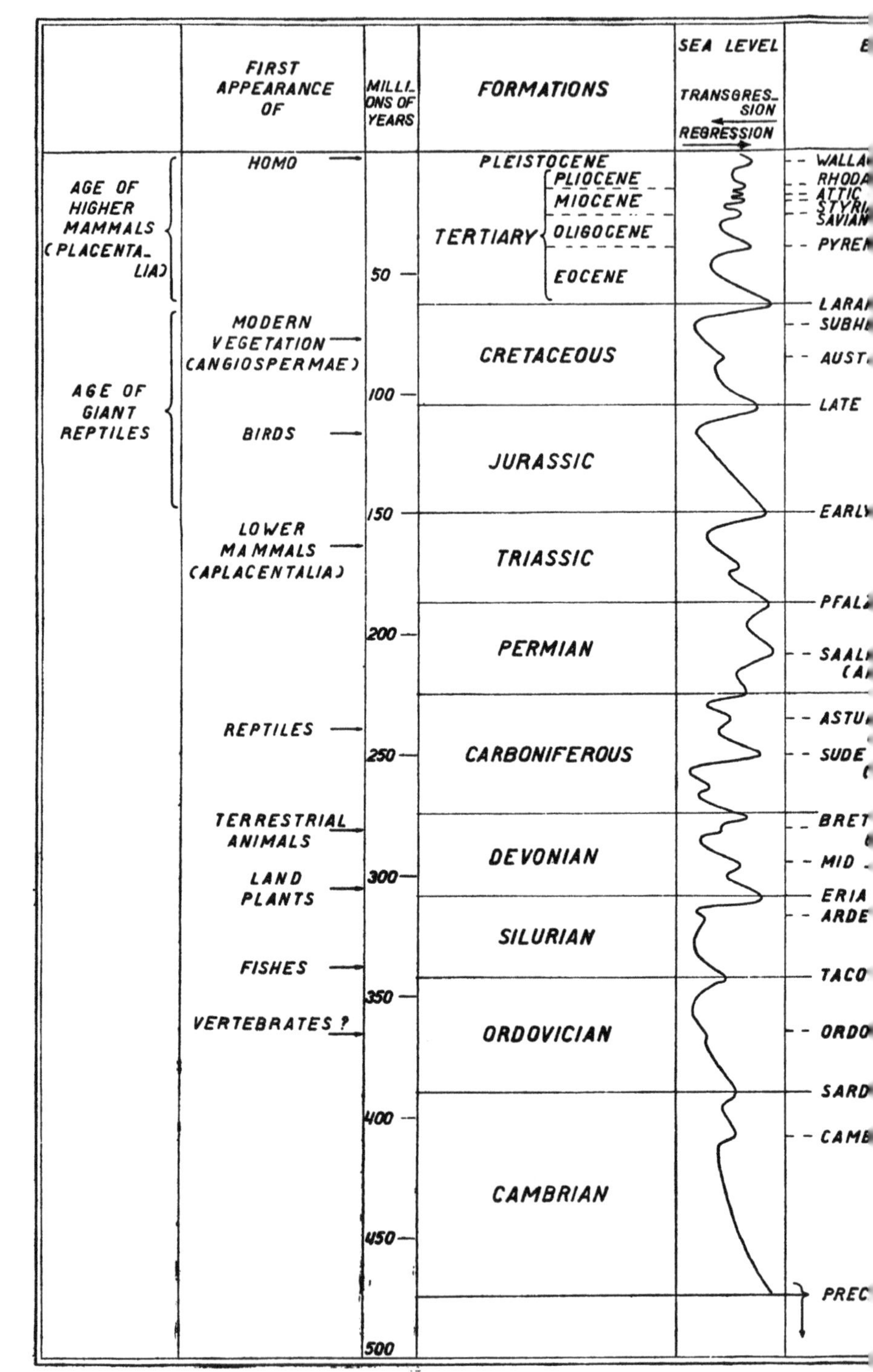

Diagrammatic synopsis

TABLE II

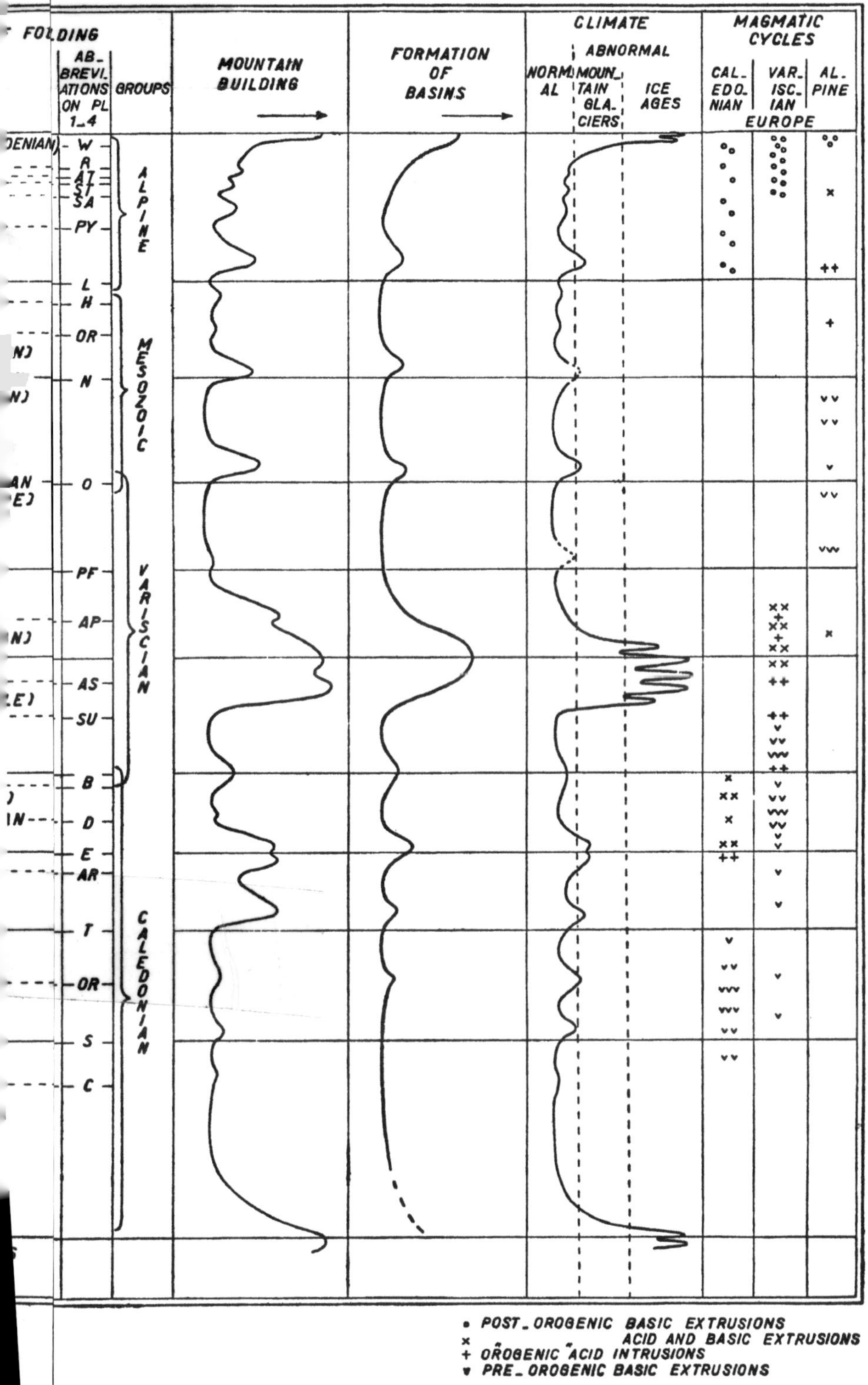

… FOLDING	AB-BREVI-ATIONS ON PL 1-4	GROUPS	MOUNTAIN BUILDING →	FORMATION OF BASINS →	CLIMATE: NORMAL	CLIMATE: ABNORMAL – MOUNTAIN GLACIERS	CLIMATE: ABNORMAL – ICE AGES	MAGMATIC CYCLES, EUROPE: CALEDONIAN	VARISCIAN	ALPINE
…DENIAN)	W	ALPINE								
	R									
	AT									
	ST									
	SA									
	PY									
	L									
	H	MESOZOIC								
	OR									
…N)	N									
…N)										
…AN …E)	O	VARISCIAN								
	PF									
	AP									
…N)										
…E)	AS									
	SU									
	B	CALEDONIAN								
…)										
…IN	D									
	E									
	AR									
	T									
	OR									
	S									
	C									

• POST-OROGENIC BASIC EXTRUSIONS
× " ACID AND BASIC EXTRUSIONS
+ OROGENIC ACID INTRUSIONS
v PRE-OROGENIC BASIC EXTRUSIONS

…ulse of the earth"

TABLE I. REVIEW OI

Nomenclature and division	Expressions used by other authors	E
		submarine
FIRST GROUP Type I *Marginal deep*	Monogeosyncline (Schuchert), Geosynclinal (Haug, Pruvost), Geosyncline (Grabau), Miogeosyncline (Stille), Saumtiefe (Stille), Kuenen's basins of the fourth group.	Persian gulf, Deep-sea trough west of Nia and the Mentawei-Islands, Java trough, Flores trough(?), Timor-Ceram trough, Mindanao trough, Aleutian trough, Kuriles-Japan-Bonin throu Mariana trough, Kermadec-Tonga trough, Brownson trough, South Antilles trough.
Type II *Intramontane trough*	Intramontane geosynclinale (Born), Stenogeosynclinal (Scupin), Becken (Stille), Kuenen's basins of the third group, Parageosyncline (Stille).	Gulf of Martaban, Madura straits, Mentawei-Java trough, Sawu deep, Wetar deep, Weber deep.
SECOND GROUP Type III *Nuclear basin* a. with isochronous frame	Innensenke, Basin, Parageosyncline (Schuchert), Kuenen's basins of the first group.	Kara-Sea b. (no. 30), Tschuktschen b. (no. 31) Banda-Sea basins, Celebes Sea b., Sulu-Sea b.,
b. with anisochronous frame		Barents-Sea (no. 10)

SIIS AND TROUGHS

ples continental	Number on Plate 6	The formation of basins is related to one (or more) of the following epochs of folding and mountain building: Early Caled: or older	Caledo-nian	Varis-cian	Early Cimme-rian	Late Cimme-rian	Alpin
Bain of Kusnezk (?),			+	+			
Paalic Upper Carboniferous tough in W. Europe,				+			
Applachian monogeosyncline,				+			
Sivalik trough,							+
Mcasse trough,							+
Gudalquivir trough,							+
Kaakum trough,							+
Bain of Mesopotamia							+
Caedonian basins of Scotlnd,			+				
Bain of Gorlowsk (Siberia),				+			
,,Rtliegende'' basins of Gernany,				+			
Bain of Kusnezk,					+		
Caifornia Valley,							+
Ova's of Asia-Minor,							+
Idigeosynclines in the East-Indies (pro parte),							+
Idigeosynclines in S. America							+
Ferghana basin,	38			+			
Tarim basin,	39			+			
Ordos (Shensi) basin,	41			+			
Szechuan basin (?),	42			+			
Pannonian basin,	19						+
Texas basin	6			+			

Type IV *Discordant basin*	Geosynclinal (Pruvost), Labile shelf (Von Bubnoff), Parageosyncline (Schuchert), Kuenen's basins of the second group.	Gulf of Mexico, Caribbean basins (pro par Mediterrane basins, Bone basin, Makassar basin, Tomini basin (?), Black-Sea b. (no. 24).

gus basin,	33	+	...	+			
ui basin,	34	+	...	...	+		
lson basin,	1	...	+				
azone basin,	8	...	+				
ana basin,	9	...	+	+			
in of Pologne,	18	...	+	...	...	+	
in of Moscou,	20	...	+				
in of Minussinsk,	35	...	+				
in of Urjanchai,	36	...	+	+	...	+	
-Continent basin,	2	...	...	+			
nois basin,	3	...	...	+			
higan basin,	4	...	...	+			
in of Pennsylvania,	5	...	...	+			
th Sea b. (+ London),	11	...	...	+	...	...	+
in of N. Germany (+ Munster),	12	...	...	+	...	+?	
t Russian basin,	21	...	...	+			
pian basin,	23	...	...	+			+
go basin,	27	...	...	+	+?	+?	+
rroo basin,	29	...	...	+			
in of Dsungarei,	37	...	...	+			
in of Urumtschi,	...	...	...	+		...	
in of Chikuching,	...	...	...	+			
rth basin,	43	...	...	+?			
sert basin,	44	...	...	+?			
in of Paris,	13	...	...	...	+		
uitania basin,	14	...	...	...	+		
th Russian basin,	22	...	...	...	+		
eat Artesian basin,	45	...	...	...	+		
lf Coast basin,	7	...	...	...	...	+?	+
in of W. Siberia,	32	...	...	...	...	+?	
bi basin,	40	...	...	...	...	+?	
ro basin,	15	...	...	...	...	...	+
ro basin,	16	...	...	...	...	...	+
io basin,	17	...	...	...	...	...	+
had basin,	25	...	...	...	...	...	+
asal basin,	26	...	...	...	...	...	+
lahari basin,	28	...	...	...	...	...	+
cla basin,	46	...	...	...	...	...	+
rrary river basin.	47	...	...	...	...	...	+

GPSR Compliance
The European Union's (EU) General Product Safety Regulation (GPSR) is a set of rules that requires consumer products to be safe and our obligations to ensure this.

If you have any concerns about our products, you can contact us on

ProductSafety@springernature.com

In case Publisher is established outside the EU, the EU authorized representative is:

Springer Nature Customer Service Center GmbH
Europaplatz 3
69115 Heidelberg, Germany

www.ingramcontent.com/pod-product-compliance
Ingram Content Group UK Ltd.
Pitfield, Milton Keynes, MK11 3LW, UK
UKHW051325070726
13610UKWH00014B/82

* 9 7 8 9 4 0 1 0 3 0 1 8 2 *